Design of
Concrete Structures

Design of Concrete Structures

Christian Meyer
Columbia University

PRENTICE HALL INTERNATIONAL SERIES
IN CIVIL ENGINEERING AND ENGINEERING MECHANICS

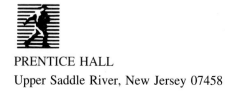

PRENTICE HALL
Upper Saddle River, New Jersey 07458

Library of Congress Cataloging-in-Publication Data

Christian Meyer
    Design of concrete structures / Christian Meyer.
        p. cm.
    Includes bibliographical references and index.
    ISBN 0-13-203654-1
    1. Reinforced concrete construction.   I. Title.
TA683.M587   1996
624.1′8341—dc20   95-24569
                                          CIP

Acquisitions editor: Bill Stenquist
Editorial/production supervision: ETP/Harrison
Project manager: Irwin Zucker
Buyers: Donna Sullivan and Julie Meehan
Copy editor: Corleigh Stixrud
Cover director: Jayne Conte
Editorial assistant: Meg Weist

The author and publisher of this book have used their best efforts in preparing this book. These efforts include the development, research, and testing of the theories and programs to determine their effectiveness. The author and publisher make no warranty of any kind, expressed or implied, with regard to these programs or the documentation contained in this book. The author and publisher shall not be liable in any event for incidental or consequential damages in connection with, or arising out of, the furnishing, performance, or use of these programs.

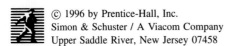 © 1996 by Prentice-Hall, Inc.
Simon & Schuster / A Viacom Company
Upper Saddle River, New Jersey 07458

Printed in the United States of America

10 9 8 7 6 5 4 3 2 1

ISBN 0-13-203654-1

Prentice-Hall International (UK) Limited, *London*
Prentice-Hall of Australia Pty. Limited, *Sydney*
Prentice-Hall Canada Inc., *Toronto*
Prentice-Hall Hispanoamericana, S. A., *Mexico*
Prentice-Hall of India Private Limited, *New Delhi*
Prentice-Hall of Japan, Inc., *Tokyo*
Simon & Schuster Asia Pte Ltd., *Singapore*
Editora Prentice-Hall do Brasil, Ltda., *Rio de Janeiro*

To Hwa Soon

Contents

Preface

Design of Concrete Structures is the result of the continuous expansion and improvement of lecture notes for a course on reinforced concrete design, which I have been teaching to seniors at Columbia University for 15 years. Its primary purpose is to fulfill the conventional role of a textbook, namely, to assist students in absorbing the material presented in class. This means it should free students from the traditional chore of dutifully copying each word and number from the blackboard, so that they can concentrate on understanding the difficult material instead. The conscientious student will attend class well prepared by having read in advance the material to be covered.

In addition, this book includes a few topics that we never were able to cover within a 14-week, 3-point semester course. These are, for example, torsion, slab design, footings, and retaining walls. But, with a solid knowledge of the basic principles of reinforced concrete design, there is good reason to believe that the average student can master these subjects through self-study.

This book does not claim to revolutionize the way in which design is taught, but it does advance a few new ideas and it does have several noteworthy features:

1. Appendix A contains a Concrete ABC, a comprehensive glossary of terms common in concrete engineering and the construction industry, which should be of value to students new to the field. The Cement and Concrete Terminology Report of ACI Committee 116 was a valuable source.

2. Much effort went into the design of problem sets, which draw on homework assignments and exams given over the years. Rather than simply asking students to

apply formulas or procedures by rote (as is all too common), many of the problems are designed to help the students think critically and appreciate the various physical aspects of theoretical concepts.

3. The incorporation of the computer into design courses is a controversial topic among educators. The approach reflected in this book is to proceed at this point without such incorporation as much as possible. Once students have mastered the basic design principles, they should be encouraged to automate some of the procedures as a fail-safe test to see whether they have in fact understood them. The consistent step-by-step layout of the various design procedures paves the way for such programming assignments. Alternatively, students familiar with spreadsheet programs should be encouraged to use them to solve some of the problem sets. This is prudent incorporation of the computer into the classroom. I am strictly opposed to the use of ready-made design software at the early stages of an engineer's career. Students who have mastered the basic principles of design will quickly and easily learn to intelligently use such CAD systems in an engineering office—after graduation.

Chapter 1 gives a comprehensive overview of modern design philosophies and discusses the role of the structural engineer in the construction industry. An amply illustrated tour of the world of concrete is meant to expose students to the large range of possibilities and to whet their appetite for entering a career with fascinating design experiences. Even though most students are likely to have already had a first design course, typically in structural steel, they will appreciate another opportunity to be exposed to the larger picture before being consumed by a sea of technical details.

Chapter 2 is more than the traditional description of materials. First, many schools no longer offer a basic materials course with adequate coverage of concrete. Next, the field of concrete technology has made impressive advances in recent years. Much of the up-to-date technological information is rather shortlived and therefore not particularly suitable material for a design course. Yet, its importance is well-recognized and therefore some of it is included in this chapter, primarily as a source of reference information and for self-study. A third point is the fact that most students, after graduation, will find themselves occupied with the repair, rehabilitation, and strengthening of existing structures rather than with the design of new structures. Many of the problems related to our country's crumbling infrastructure are really problems of materials and their durability (or lack thereof). Therefore, a thorough knowledge of material properties is a prerequisite for solving these problems.

In Chapter 3, a foundation is laid for a basic understanding of reinforced concrete in bending. Building on this foundation, Chapter 4 presents logical design procedures for reinforced concrete flexural members.

Chapter 5 covers the three diverse topics of shear, bond, and torsion design. Emphasis is again placed on the basic concepts, on which current design practice is based, rather than on all the details. Although these details are very important in practice, they are likely to change with the next edition of the ACI Code, whereas the basic concepts will not.

The interrelationship between design and analysis is a source of confusion for students confronted with statically indeterminate structures. Even though structural analysis is taught as a separate course, it has been my experience that students generally have a hard time absorbing and applying its rather difficult concepts. Repeated exposure is often a great help. For this reason, Chapter 6 places heavy emphasis on the review of some structural analysis concepts for continuous beams and frames, especially of those methods that designers will find useful for checking purposes. This applies primarily to moment distribution, which in my view is no longer receiving the attention in structural analysis courses that it deserves. An effort is made to explain the interplay between analysis and design and to pave the way for the rewarding experience of young engineers suddenly finding themselves "in the driver's seat," by controlling the behavior of structures through design decisions.

The design of columns is covered in Chapter 7. Starting with a review of the basic theory underlying interaction diagrams and classical column buckling, the ACI Code design procedures for short and slender columns are derived and illustrated by example.

Chapter 8 deals with the design of slabs and plates. Here, I felt it useful to include sections on basic plate theory and yield line theory as supplements to the common design procedures of the ACI Code. This material should appeal to more ambitious students as preparation for a more advanced concrete design course.

A good number of additional chapters could be included in this book—chapters covering the design of buildings, pavements, dams, bridges, etc. But this book is meant as an introductory text, dealing primarily with basic principles of concrete mechanics and design. Buildings, bridges, etc. are very specific examples of *applications*. Students who have mastered the basic principles should have no difficulty applying these to specific applications. Chapter 9 was added as just one such example, namely, foundations and retaining walls. This topic is in the truest sense of the word so *fundamental* that it is a fitting subject for the final chapter. No new design principles are introduced, except for some concepts of soil mechanics. Obviously, the material presented is far from exhaustive, as entire books are devoted to this topic alone.

I have tried to reference the ACI Code as sparingly as possible to prolong the "shelf-life" of this text. But, given today's design practice, this is a difficult task. In particular, the Code requirements for shear design are still (unfortunately) largely empirical. However, there is hope that future editions will build on new insights into shear behavior of concrete members (gained, for example, through advances in fracture mechanics) that will form a more rational foundation for shear design requirements. The most recent Code changes proposed for ACI 318–95 pertain primarily to the minimum flexural steel requirements, development length specifications, and the design of slender compression members. They have not yet been adopted by ACI at the time of this writing, but have been incorporated into the book. Students should be encouraged to have ready access to the latest edition of the Code. The commentary contains valuable explanatory material, which was omitted here to avoid unnecessary repetition.

I am greatly indebted to my wife, Hwa Soon, and to my sons, Peter and Guenter, for encouraging me over the years to continue the ambitious undertaking of writing this book and actually completing it. I also would like to thank my great teacher, Alex

Scordelis, Professor Emeritus of the University of California at Berkeley. He not only advised me during my doctoral studies, but also taught me how to think as an engineer and how to transmit this knowledge to students. Finally, I would like to express my gratitude to various staff members of Prentice-Hall and ETP/Harrison: Doug Humphrey, for active support during the early stages of the project; Bill Stenquist, for seeing the book through its production; Dena Kaufman; and especially Corleigh Stixrud, for his painstaking care and patience in editing the entire work.

New York City   *Christian Meyer*

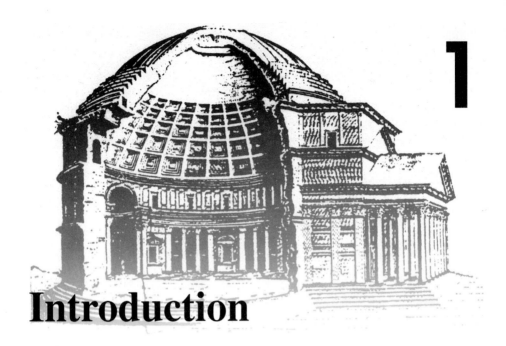

Introduction

1.1 HISTORICAL REMARKS

Most of us assume we can identify concrete when we see it. Yet, if we were to search for its origin or determine when it was "invented," it would soon become apparent that we would need to define concrete before we could identify it. Deferring a more elaborate definition until the next chapter, let us start with a common frame of reference, namely, the definition in Webster's New World Dictionary:

> **concrete:** a hard, compact substance made of sand, gravel, cement, and water, used in the construction of bridges, dams, buildings, etc.

Taking the gravel out of this mixture results in *mortar*, and if lime serves as the cementing or binding agent, we get lime mortar, which was most likely invented in prehistoric times. It was certainly known to the earliest civilizations, such as the Babylonians, who used it in stone masonry. The Egyptians also added gravel to mortar, that is, they knew how to make a material that by the above definition is concrete. Recently, a fascinating theory has been proposed by the French chemist, Joseph Davidovits, that the great pyramids are actually made of geopolymeric limestone concrete blocks rather than natural, quarried limestone [1, 2]. Although this theory has not been universally accepted by the scientific community, the various arguments brought forth in the current discussion lead to the certain conclusion that only very detailed scientific examinations can establish the difference between natural and synthetic stone.

1

Advancing a few millenia brings us to ancient Roman civilization. Even though the cement the Romans used was not the portland cement used today, there is no question that, by the above definition, they used concrete in their various constructions. The most detailed account of Roman architecture and construction in general and their building materials in particular, is found in *De architectura libri decem* (Ten Books of Architecture) by Vitruvius. On the kind of cement the Romans used for their concrete, we can read in book 2:

> There is also a kind of powder which by nature produces wonderful results. It is found in the neighborhood of Baiea and in the lands of the municipalities around Mount Vesuvius. This being mixed with lime and rubble, not only furnishes strength to other buildings, but also when piers are built in the sea, they set under water. [3]

The Roman mortar consisted of quicklime and burnt clay or natural materials, predominantly of volcanic origin, which contained reactive silica. The best known deposit of these materials was found near the village of Pozzuoli, near Naples. This is why materials that are capable of reacting with hydrated lime (due to their reactive silica content) are generally called *pozzolana*.

One of the most impressive Roman concrete structures is the Pantheon (Fig. 1.1). Completed in 126 A.D., its dome span of 43.5 m was not surpassed until the nineteenth century. Parts of the port of Ostia, built during Trajan's reign (52–117 A.D.), withstood centuries of pounding surf until the coastline shifted, and survive to this day [4].

Modern portland cement was invented in 1824 by Joseph Aspdin, a builder from Leeds in England. He patented his process of grinding limestone, mixing it with finely divided clay, burning the mixture in a kiln oven, and finely grinding the resulting clinker. He gave his invention the name portland cement because of its resemblance to the natural building stone quarried near Portland, England. The commercial success of Aspdin's cement led to the rapid spread of hydraulic cements throughout Europe. This was the beginning of the modern cement industry, which in 1990 produced 85 million tons of cement in the United States alone.

The relatively low tensile strength of concrete and the resulting susceptibility to cracking prompted builders from early on to look for proper means of reinforcement. The most successful of these pioneers, however, was not a builder but a gardener. Joseph Monier of Paris was interested in reinforcing his concrete tree planters and in 1850 patented a system of using steel wire reinforcement. Although other patents had been filed earlier, Monier's patents are generally credited with having had the most significant impact on the development of what is now called *reinforced concrete*. Some of this credit is shared by G.A. Wayss and J. Bauschinger of Germany, who purchased the German and American rights to Monier's patents and established the basic principles for the application of reinforced concrete. By the turn of the century, the behavior of reinforced concrete was relatively well understood, and together with advances in materials and structural engineering, the use of reinforced concrete spread rapidly. Another fundamental advance was made with the introduction of *prestressed concrete*, generally credited to Eugene Freyssinet of France. By precompressing the concrete, usually by means of embedded high-strength steel tendons, it is possible to subsequently subject it

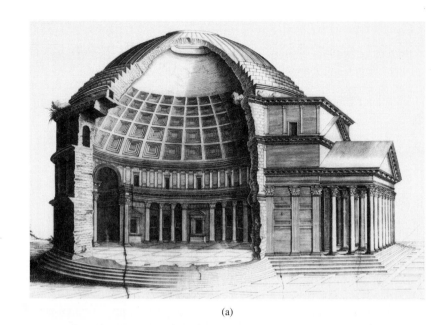

(a)

(b)

Figure 1.1 The Pantheon, Rome: (a) exterior (courtesy of Photo Bibliotheca Hertziana, Rome); (b) interior (photo: Luciano Pedicini).

to tension without causing a net tensile stress or cracks. Principles of prestressed concrete design are generally covered in more advanced concrete design courses and will not be discussed in this book. Today, for better or for worse, we find ourselves literally surrounded by concrete, from road pavements and buildings to bridges and pipelines.

1.2 ADVANTAGES AND DISADVANTAGES

Concrete is produced by mixing cement, aggregate, and water, resulting in a viscous fluid. Because of its liquidity, freshly produced concrete can be formed into virtually any shape. At the same time, it needs to be contained by *formwork* to maintain the desired shape until it hardens and can carry load. In any case, its formability makes concrete adaptable to a range of applications limited only by one's imagination.

In residential building construction, a smooth ceiling finish is an advantage. This can be achieved easily with appropriate formwork. Steel decks, in contrast, may offer other advantages, but often need to be covered up by hung ceilings. Another advantage of concrete in residential construction is the good sound insulation property associated with heavy mass. Light construction is more difficult to soundproof over a wide frequency range.

An important advantage of concrete is its good fire resistance. Compared with the serious threat that fire poses to timber and steel construction, concern in concrete construction is mostly reduced to the necessary protection of the embedded steel reinforcement.

Concrete offers not only limited protection of the steel against high temperature effects, but also against corrosion. The alkalinity of the cement creates an ideal chemical environment for corrosion resistance. However, this is impaired by the intrusion of chlorides, caused, for example, by the use of deicing salts on highway bridges and roadway surfaces.

Concrete structures are often very massive and rigid, so the deformations remain relatively small. Even in steel buildings, concrete walls are frequently used to reduce lateral deformations and vibrations caused by horizontal loads, such as wind.

Compared with most other building materials, concrete is relatively inexpensive. The raw materials are generally available locally, and the need for skilled labor is quite limited. Also, the relatively low maintenance cost improves the long-term economy of a structure. However, comparative economic or cost evaluations should always take into consideration a large number of other factors besides pure material costs.

Although the strengths of commercially available concretes have increased markedly in recent years, the weight-to-strength ratio of concrete is still considerably higher than that of structural steel. Thus, steel is likely to remain the material of choice for structures with very long spans. The heavy weight of concrete may be advantageous at times, but more often it is not. For example, at sites with poor soil conditions, larger and more expensive foundations are needed.

Another disadvantage of concrete is its low tensile strength and the resulting tendency to crack. As we will discuss in detail later, concrete must crack in order to activate the strength of the steel reinforcement, but the designer can minimize the adverse effects of cracking by distributing the tensile strains over a large number of small cracks rather than a few large ones. Large cracks not only look awful, they also reduce the protection offered by the concrete cover against steel corrosion.

A further source of potential problems is the viscoplastic nature of concrete, which causes it to undergo creep deformations under long-term loads. These creep deformations can be a multiple of the instantaneous elastic deformations and must be taken into account in the design of most structures.

Because cement requires water to hydrate, it is important that the water not freeze before the concrete has gained sufficient strength. For this reason, good quality concrete cannot be obtained in subfreezing temperatures unless special precautions are taken, a factor that adds to the cost of construction. At the other extreme, concreting in hot and arid climates requires measures to assure that none of the water required for cement hydration is lost to the environment.

Last but definitely not least is the fact that it is much more difficult to control the quality of concrete produced in the field compared, for example, with rolled-steel shapes produced in a mill under tightly controlled conditions. Even with good quality control, concrete properties exhibit a considerable amount of statistical scatter—a situation that we just have to live with.

1.3 THE WORLD OF CONCRETE

As in each special discipline, the field of concrete engineering has developed its own vocabulary over the years. To the layperson, much of it is either unknown altogether or only vaguely familiar, if not misunderstood. For this reason, Appendix A contains the "Concrete ABC," a glossary that the reader can consult whenever in doubt about the meaning of a specific term or definition. Much of this glossary is taken from [5].

Before we immerse ourselves in the technical details of designing reinforced concrete structures, it is appropriate to take a brief tour of the world of concrete and have a look at some of the marvels of modern engineering. After all, it is the inspiring examples set by the master builders that entice us to follow in their footsteps. As a result, some of today's novices will become tomorrow's masters.

Buildings.   Concrete buildings can be found all over the world. The decision to use reinforced concrete instead of other building materials depends on too many factors to enumerate here. For example, in some parts of the United States, a frequently heard rule holds that concrete is more suitable for multistory residential construction, whereas office buildings are more likely to be built with steel. An equally general rule holds that nobody can predict a priori the building material most suitable for a particular project, as each case needs to be evaluated on its own merits. The cost of formwork is often a decisive factor. Figure 1.2 shows typical sets of plywood forms supported by timber shoring. Both forms and shoring need to remain in place until the concrete has been placed and has gained sufficient strength to carry its own weight as well as the loads

(a)

(b)

Figure 1.2 Formwork for concrete: (a) (courtesy of Patrick Brunner); (b) (courtesy of the Portland Cement Association).

imposed during construction. The construction loads can be significantly higher than any other loads a building is subjected to during its remaining service life. Therefore, the construction phase can often be considered a full-scale load test, on the premise that if the building survived the heavy (and often unauthorized) overloads during construction, it is likely to serve safely during its remaining life. In fact, most failures of structural members or entire buildings occur during construction.

Because the formwork contributes so much to the total cost of a project, it requires special attention [6]. It is generally custom-built for each project, and the more efficiently it can be reused, for example, by adopting repetitive floor plan layouts, the greater the reduction in construction cost. The Cityspire Building in New York City is not exactly the most typical residential building, but Fig. 1.3 clearly illustrates how the floor layouts permit such efficient reuse of one set of formwork.

Trump Tower in New York City (Fig. 1.4) is an even more unusual building. With 60 stories, it is one of the tallest concrete buildings in Manhattan. Its layout in the lower stories is extremely irregular—there is only one column in the entire building, which runs straight from the roof into the foundation without any setbacks, interruptions, or transfers. The 757-ft-tall Carnegie Tower in New York (Fig. 1.5) is distinguished by its record height-to-width ratio of 12.5 to 1. Buildings with half this ratio are considered slender. Because of sky-high land values in midtown Manhattan, developers try to fit as much floor area as the zoning ordinance allows into building lots that were previously

Figure 1.3 Cityspire Building, New York City (courtesy of Rosenwasser/Grossman Consulting Engineers).

Figure 1.4 Trump Tower, New York City (courtesy of Swanke, Hayden, Connell Architects and The Office of Irwin G. Cantor, Structural Engineers).

Figure 1.5 Carnegie Tower, New York City (photo: Stephen L. Senigo/Architectural Photography).

considered unsuitable for the erection of high-rise structures. This poses a challenge
for engineers to develop new and more efficient structural schemes. The 311 South
Wacker Drive Tower in Chicago (Fig. 1.6) with 71 stories and a height of 969 ft, was
the world's tallest concrete building for merely three years. This record was broken
in 1993 by the Central Plaza Building in Hong Kong. With 78 stories and a height
of 1228 ft, it is the world's fifth tallest building, surpassed only by Chicago's Sears
Building, the twin towers of New York's World Trade Center, and the Empire State
Building [7].

    Prior to the 1970s, concrete could not compete seriously with steel for high-rise
buildings in urban areas, primarily because of major differences in the speed of construction. Especially in times of high interest on construction loans, the time needed
for construction is a decisive economic factor, because the developer does not receive
a return on the investment until the building is ready for occupancy. If we can erect
ten floors of a steel frame building in the same time needed to build the formwork and
falsework for a single concrete floor, place the reinforcing steel, place the concrete, and
wait until the concrete has hardened to permit the building cycle to proceed, then there
is obviously no contest. However, advances in the last twenty years, both in construction methods and in the development of concretes with high-early strengths, have led
to construction cycle times of as little as two or three days per floor. This compares
directly with the erection times possible in steel-framed construction. A more novel
type of construction attempts to combine the advantages of both steel and concrete.

Figure 1.6   311 South Wacker Drive
Tower, Chicago (courtesy of the Portland
Cement Association).

This so-called *composite construction* appears to be particularly economical for very tall buildings.

In a completely different method of concrete construction, building components are manufactured in stationary factories and then shipped to the site for assembly. Such *precast concrete* construction has the advantage of both cost savings and improved quality control. Parking garages are a particularly popular application of precast construction (Fig. 1.7). Mass-production techniques can increase the cost-effectiveness of both residential and commercial buildings (Fig. 1.8). These techniques are widely used in countries with acute housing shortages. Taking one step further toward industrialized manufacturing techniques leads to the prefabrication of entire building modules, such as bathroom and kitchen units with all plumbing fixtures and piping already factory-installed, tiles and windows in place, and everything else but the pictures on the walls (Fig. 1.9).

Shells.   A beam carries loads primarily by bending. An arch, by virtue of its curvature, transmits the same load to the supports through direct action, that is, in axial compression, with little or no bending if the load is uniformly distributed. In two dimensions, the equivalent of the beam is the plate, which can carry loads only by out-of-plane bending. A shell structure, in contrast, supports the loads primarily through in-plane or membrane action. This is much more efficient, and as a result, thin shells

Figure 1.7   Precast parking garage (courtesy of Walker Parking Consultants/Engineers).

Figure 1.8   Precast panel building construction (courtesy of FABCON, Inc).

Figure 1.9   Precast building module (courtesy of the Portland Cement Association).

Figure 1.10 Experimental shell by Dischinger, 1939 (courtesy of Dywidag Systems International).

can carry heavy loads over very large open spaces. The first modern thin shell concrete structures were built in the 1920s for the Zeiss Company of Jena in Germany, based on a theory developed largely by Franz Dischinger and Ulrich Finsterwalder (Fig. 1.10) [8]. Less than thirty years separate this relatively modest beginning from Nicolas Esquillan's Exposition Palace in Paris (Fig. 1.11). With a span of 790 ft and a covered floor area of over 80,000 ft^2, the Exposition Palace is still one of the world's largest shell structures [9]. Thin shell construction has opened up completely new dimensions and possibilities for engineers and architects. The Palazetto built by Luigi Nervi for the Rome Olympics in 1960 demonstrates for all to see, both inside and outside, how it carries loads into the foundations (Fig. 1.12) [10].

Another example is the St. Louis Priori Church (Fig. 1.13), designed by the architectural firm of Hellmuth, Obata & Kassabaum and engineered by Mario Salvadori. It illustrates the considerable amount of skilled carpenter labor needed to construct custom-built formwork. Because of the high cost of labor, such shell structures cannot be built economically in the United States unless the designer or contractor devises some innovative scheme to reduce construction cost. Dante Bini's solution is shown in Fig. 1.14: he places a canvas on the ground, assembles the reinforcing steel, places the concrete, covers it with a second canvas, and then inflates the entire structure, raising it to its predetermined shape. After the concrete has hardened, the can-

Figure 1.11 Paris Exhibition Hall.

(a)

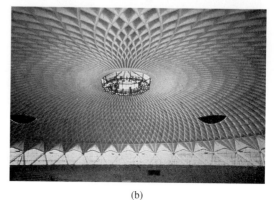

(b)

Figure 1.12 Palazzetto, Rome: (a) exterior; (b) interior.

vas membranes are removed and openings cut out for entrances, windows, etc. [11]. We can also take a hint or two from Mother Nature: a chain, suspended off two points, assumes an "ideal" shape, known as a *catenary* (from the Latin word *catena* for chain), because the tension forces in the chain are in perfect equilibrium with the gravity loads, without any bending. Turning the chain upside down results in an arch of equally optimum shape, except that the forces are now compressive. This simple

(a)   (b)

Figure 1.13   Priori, St. Louis: (a) finished; (b) under construction.

(a)   (b)

Figure 1.14   Bini shells (courtesy of Dante Bini).

fact underlies the design of Eero Saarinen's St. Louis Arch. By similar reasoning, Heinz Isler suspended a sheet of fabric off a few selected points and sprayed it with water. After the cold winter temperatures froze the sheet stiff, he could turn it upside down to obtain a shell that was ideal for the given plan and support conditions (Fig. 1.15) [12]. Determining such a shape by analysis would have been rather difficult. To actually build large shells Isler is dependent on skilled carpenters, but a client willing to pay a premium for a beautiful structure can always be accommodated (Fig. 1.16).

(a)   (b)

Figure 1.15 Icesheets by Isler: (a) single icesheet suspended; (b) icesheets assembled. (Courtesy of Heinz Isler.)

(a)   (b)

Figure 1.16 Isler shells (courtesy of Heinz Isler).

Bridges. Concrete bridges have evolved over the last hundred years and constitute an important element of today's road network in the United States and elsewhere. As is true for all other concrete construction, the expense of formwork and the supporting falsework poses a problem that constantly calls for improved and innovative construction techniques. This is especially so when a major waterway, deep valley, or busy transportation artery is to be bridged. Here, rapid progress was made possible by the evolution of prestressed concrete, which permits the erection of long-span bridges without the need for falsework. One such construction scheme, called *cantilever construction* for obvious reasons, advances simultaneously in two opposing directions (Fig. 1.17). This method of construction was pioneered by Ulrich Finsterwalder in the 1950s. The placement of reinforcing steel, the placement of the concrete, and the application of prestress all take place in *travelers*. These are movable segments of durable formwork, which are adjustable to accommodate varying cross-sectional dimensions and can be enclosed and heated for

(a)   (b)

Figure 1.17 Cantilever construction: (a) Wand River, South Carolina; (b) Genesee River, New York. (Courtesy of the Portland Cement Association.)

Figure 1.18 Traveler for cantilever construction (courtesy of Dywidag Systems International).

year-round construction (Fig. 1.18). Alternatively, in *segmental bridge construction*, the bridge is assembled from precast segments, which are manufactured either in a stationary factory or in a temporary casting yard in the vicinity of the bridge site (Fig. 1.19). A noteworthy example is the Linn Cove Viaduct (Fig. 1.20). It passes through very

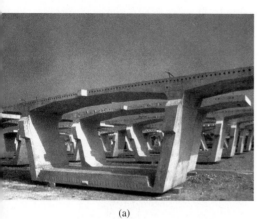

(a)   (b)

Figure 1.19 Segmental bridge construction, Kishwaukee Bridge, Illinois: (a) bridge segments (courtesy of the Portland Cement Association); (b) bridge construction (courtesy of DRC Consultants).

environmentally sensitive terrain, so that only an absolute minimum of disturbance was acceptable during construction [13]. The solution was *top-down construction*. Beginning at an abutment, the precast segments were brought across the completed part of the deck and post-tensioned in place. The piers were constructed by lowering precast segments into place from the cantilevered deck, thereby leaving the surrounding vegetation undisturbed.

In another method, called *incremental launching*, the bridge is constructed one segment at a time in a fixed casting bed located at one abutment. When a segment is completed, it is pushed out by jacks in the direction of the span, so that the casting bed is available for construction of the next segment (Fig. 1.21). Before the tip of the cantilever reaches the next pier, the negative bending moment is considerably larger than any moments in the completed bridge. To reduce this very temporary negative bending moment, a lightweight "nose" is attached to the bridge front. Incremental launching has been used to construct bridges over six and more spans, even on a horizontal curve. If a downhill grade is involved, precautions must be taken to prevent a runaway situation.

There exists a large amount of literature on modern bridge engineering and on the great bridge designers [14–17]. Figure 1.22 shows the notable Salginatobel Bridge, designed by Robert Maillart. Well known both for the innovative use of the three-hinged arch and for its beauty, this bridge is enhanced by its dramatic setting in the Swiss landscape.

(a)   (b)

Figure 1.20 Linn Cove Viaduct: (a) construction (courtesy of DRC Consultants); (b) finished bridge (courtesy of the Portland Cement Association).

Other structures.   The versatility and durability of reinforced concrete make it the material of choice for marine structures, infrastructure and industrial facilities, and even sculptures and artwork. In fact, the applications are so diverse and numerous that only a few examples shall be given here by way of illustration.

A traveler on American highways may get the impression that a water tank, being a utilitarian industrial structure, must by necessity be ugly. Figure 1.23 shows that this need not be the case. Communities all over the world place great emphasis on the esthetic appearance of even utilitarian structures. Why not add an observation deck or a cafe to the needed water tank or TV tower and let the general public enjoy the view of the landscape!

Figure 1.24 shows a cooling tower for a power plant. Such towers have been built to heights exceeding 650 ft. They are typically of hyperbolic shape to facilitate a natural upward draft of the airflow, which causes the steam from the generator turbines to cool down as it rises and condense to water. Such shells are built using *slip-forming*

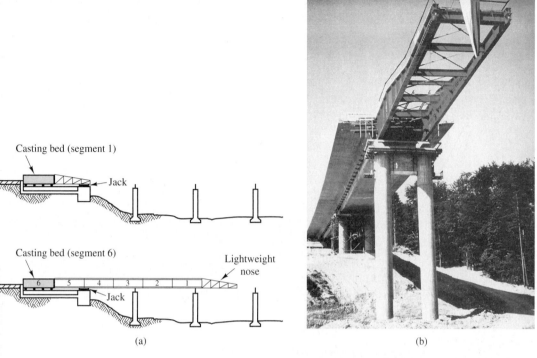

Casting bed (segment 1)

Jack

Casting bed (segment 6)

Lightweight
nose

6 5 4 3 2 1

Jack

(a)   (b)

Figure 1.21 Incremental launching: (a) scheme; (b) actual bridge (from *Welt des Betons*, courtesy
of Deutscher Beton-Verein).

techniques, where the formwork is advanced upwards at a rate that allows the young
concrete to gain sufficient strength to support the construction loads.

Nuclear power plants have received a mixed reception in various parts of the world.
They hold promise as an economical solution to a country's energy needs, but it has been
questioned whether the benefits are worth the risks associated with any kind of failure
or malfunction, not to mention the problems associated with nuclear waste disposal. In
order to prevent the release of fission products, such as was experienced in Chernobyl
in 1986, nuclear reactors in the Western world are housed in protective containment
buildings. In the event of a loss-of-coolant accident, for example, a major pipe break
that interrupts the flow of coolant into the reactor core, a considerable build-up of pressure
and temperature can be expected. A containment building must therefore be designed
as a pressure vessel capable of withstanding the enormous internal pressures (as high as
50 psi and more) and still remain leak-tight to prevent the release of radioactive fission
products to the environment. Almost all containment buildings are concrete structures,
with exterior wall thicknesses of typically 5 or 6 ft (Fig. 1.25). The reactor itself is
usually a steel vessel, although some power plants, for example in Europe, have been
built with prestressed concrete reactor vessels. Figure 1.26 illustrates a design for a
plant that was never built. It consists of a cylindrical vessel that is 91 ft high, has an

Figure 1.22   Salginatobel Bridge (courtesy of David Billington).

Figure 1.23   Water tank, Finland (courtesy of DRC Consultants).

Figure 1.24   Cooling tower (from *Welt des Betons*, courtesy of Deutscher Beton-Verein).

(a)

(b)

Figure 1.25   Nuclear containment building: (a) foundation mat; (b) building under construction. (Courtesy of Stone and Webster Engineering Corporation.)

outer diameter of 101 ft, a wall thickness of about 15 ft, and weighs over 100 million lbs [18].

Some of the world's greatest engineering achievements are concrete structures. Major dams typically serve several purposes—water storage for irrigation, flood control, power generation, and the creation of new recreation areas. Hoover Dam, for example, with a height of 726 ft and 4.3 million yd^3 of concrete, impounds Lake Mead with a

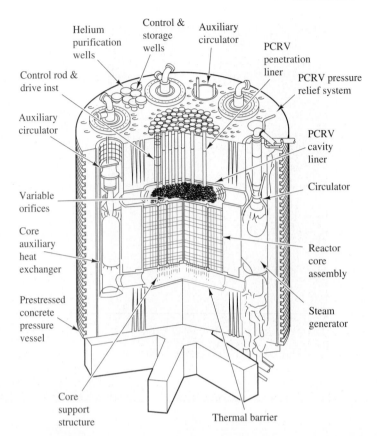

Helium purification wells

Control & storage wells

Auxiliary circulator

PCRV penetration liner

Control rod & drive inst

PCRV pressure relief system

Auxiliary circulator

PCRV cavity liner

Circulator

Variable orifices

Core auxiliary heat exchanger

Reactor core assembly

Prestressed concrete pressure vessel

Steam generator

Core support structure

Thermal barrier

Figure 1.26 Prestressed concrete reactor vessel. (From C. Meyer, "Design Life Simulation of a PCRV," *ASCE, J. of the Struct. Div.*, vol 108, no. ST5, 1982. Reprinted with permission of the American Society of Civil Engineers.)

capacity of over 6 trillion gal. (Fig. 1.27). Its construction during the height of the Great Depression provided jobs for thousands of workers.

Figure 1.28 illustrates another example of the world of concrete in hydraulic engineering. The Dutch people have been battling the North Sea for centuries. At present some 40% of their small, densely populated country lies below sea level and needs to be protected by dikes and dams. The southernmost part of the country contains the deltas of the Rhine, Meuse, and Scheldt rivers, with hundreds of miles of vulnerable riverbanks and coastline. Following the last great flood of 1953, the government undertook the ambitious Delta Project to drastically shorten the length of the coastline. Some riverarms were closed permanently by dams, others controlled by flood gates. The most difficult project was the closure of the five-mile wide Eastern Scheldt. Environmentalists and local fishermen opposed the construction of a permanent dam. A storm surge barrier was built instead, which would be closed only in the event of a storm flood. The final solution incorporates an artificial island, which

Figure 1.27   Hoover Dam (courtesy of the U.S. Bureau of Reclamation).

(a)   (b)

Figure 1.28   Eastern Scheldt Storm Surge Barrier: (a) overview (courtesy of Rijkswaaterstaat, The Netherlands); (b) assembly line construction of piers (photo by author).

was also used to construct the project's concrete piers. These piers are spaced every 45 m and support a total of 62 movable steel gates and a roadway. The prestressed concrete piers, each weighing 18,000 tons, were constructed in the manner of an assembly line in special compartments of the artificial island. After completion of all the piers in one such compartment, it was flooded and the piers floated into their final positions in the river. This giant project was dedicated by the Queen in 1986.

Among the largest structures ever built are off-shore oil drilling platforms. Following the discovery of oil in the North Sea in the 1960s, several concrete platforms have been built (typically in one of the deep fjords of Norway) and then towed into place. Figure 1.29 shows a sketch of the largest of them, the Troll Platform, which is scheduled to be completed in 1995. With a total height of 430 m (1410 ft), it ranks as

Figure 1.29   Troll Platform (courtesy of Norwegian Contractors).

one of the tallest structures ever built. Weighing more than one million metric tons, it contains enough reinforcing steel to build 15 Eiffel Towers.

Figures 1.30 and 1.31 illustrate a problem especially familiar to those of us who live in urban areas. Lest anybody think otherwise: Concrete structures need to be maintained on a regular basis. Lack of maintenance will allow deterioration due to natural causes and normal service effects. Concrete highway bridge decks in climates with harsh winters are particularly vulnerable, as municipalities tend to rely on salt for snow and ice removal. The salt can penetrate the cracks in the concrete and corrode the reinforcing steel. There are many other examples of the high price to be paid when we neglect to properly maintain our infrastructure facilities.

Figure 1.32 points to a problem faced by designers of structures in seismically active regions. The brittle nature of concrete makes it very vulnerable to the loads associated with strong earthquake ground motions. In fact, until the 1950s, it was standard practice not to build tall concrete buildings in seismic regions. Figure 1.32 shows a multistory building that collapsed during the Mexico City earthquake in 1985 and caused

Figure 1.30 Deteriorated bridge abutment (courtesy of the American Society of Civil Engineers).

Figure 1.31 Deteriorated bridge deck (courtesy of the American Society of Civil Engineers).

Figure 1.32 Mexico City earthquake, 1985 (courtesy of John Kariotis).

the tragic death of hundreds of occupants. It was poorly designed by today's standards, but even structures that were designed according to the code standards of the time have often fared poorly when subjected to seismic disturbances. However, it is now possible to make concrete ductile by properly reinforcing it with steel [19]. The result is a proliferation of tall concrete buildings in seismically active regions, such as California. Some of these buildings have as many as 32 stories, supported by ductile moment-resisting frames.

The versatility and formability of concrete lends itself to artistic expression. By proper choice of materials (aggregate, cement, color additives), formwork, and surface treatment an unlimited number of different surface textures and special effects can be achieved (Fig. 1.33). This poses a challenge for architects and artists [20]. Concrete that is modified specifically for architectural or artistic appearance is called *architectural concrete*. The direct cooperation between engineer and artist can lead to surprising results. For example, Fig. 1.34 shows the Thomas Road Bridge in Phoenix, Arizona. The city paired the bridge design firm of Cannon & Associates with sculptor Marilyn Zwak for this project. The bridge piers, as a result, serve not only structural purposes—they are also works of art, which pay homage to the Hohokam Indians of the area [21].

(a)   (b)

Figure 1.33   Surface textures (from *Welt des Betons*, courtesy of Deutscher Beton-Verein).

(a)   (b)

Figure 1.34   Thomas Road Bridge, Phoenix, Arizona (courtesy of Cannon & Associates, Inc.).

1.4 WHAT IS DESIGN?

By the time civil engineering students enroll in a concrete design course, they have usually already finished some other design course, typically in steel structures. Yet it seldom hurts to be exposed to some basic concepts more than once; this way it is easier to understand and remember them. Therefore, let us start by addressing one of the fundamental questions: What is structural design all about? The following definition, although unpolished, will suit our present purpose:

> Structural design is the creative and rational activity that results in specifications for structures, which satisfy certain safety, serviceability, and other requirements throughout their service life and can be built under economical constraints.

If this definition reads somewhat awkwardly, it is because of the deliberate accumulation of key words. These key words require further comment.

A *creative* activity is simply one that results in something where nothing existed before. Within the context of structural engineering, this "something" must fulfill a number of functional requirements. The grandest design of all is powerfully described in the Book of Genesis. The ambitions of engineers are more modest, yet their creations, which improve people's living conditions, rank unquestionably among the noblest and highest achievements of humankind.

A *rational* activity is best identified by juxtaposing it to its opposite, namely, an *empirical* activity. (For the purposes of our discussion this definition is much more restrictive than its use in common English, in which an empirical activity can also very well be a rational one.) In our context, the rational approach implies the use of principles of mechanics to predict the behavior of a design under load and other external effects and to adjust the design such that it satisfies a set of prescribed requirements. In contrast, the empirical approach is more of the trial-and-error kind. In this sense, the engineers of ancient Rome followed mostly the empirical approach. Even though they built marvelous structures, some of which serve their purposes to this day, the science of mechanics was unknown to them. They could not rationally predict or "compute" the behavior of their structures under load beforehand. They had to rely on experience, knowing what worked and what did not.

One of the first instances in which the rational approach was applied to a major practical problem was the strengthening of St. Peter's Basilica in Rome in 1743 [4]. Shortly after its completion, Michelangelo's masterpiece developed cracks, which grew worse in time, until Pope Benedikt XIV was concerned enough to call for corrective action. However, instead of following the tried and true methods of the past, such as asking a builder to use personal experience and judgment to propose a strengthening scheme, the Pope pursued an approach that was rather unconventional, if not radical, in the eyes of his contemporaries: he commissioned three mathematicians, the Jesuit Ruggero Giuseppe Boscovich and two Franciscan monks Thomas le Seur and Francois Jacquier, to study the problem and propose a strengthening scheme. These scholars were familiar with the latest advances in the theory of mechanics and employed the Principle of Virtual Displacements, established in its general form not long before by Johann Bernoulli, to compute the forces in the dome's ring beam at the springline. They found that the two iron tension rings provided when the dome was erected were not strong enough to resist the horizontal thrust. They recommended additional iron rings, considering a safety factor of two—another hallmark of modern engineering thinking. In other words, they employed scientific methods to *predict* the behavior of the dome under load with and without brace, rather than relying on some builder's practical experience. The size of the dome was so much larger than anything built before that the Pope had good reason to suspect that the collective know-how of the builders of the day was not sufficient to assure a reasonably safe strengthening scheme. After all, this collective experience had proved to be inadequate in the first place, although fortunately without catastrophic consequences.

Returning to our definition of structural design, the next keyword to require elaboration is *safety*. This is probably the most important concept in all of structural engineering. It implies the prevention of harm to the users and occupants of the structures we design. It contains the ultimate dictum that we simply cannot afford to make mistakes. It is said that doctors bury their mistakes, architects grow ivy over theirs, but engineers must live with their mistakes for the rest of their lives. Because of this absolute requirement, we cannot afford the trial-and-error approach of pre-engineering days. This is not to say that earlier designers could make fatal mistakes with impunity. It is said that in some countries a bridge designer was expected to show his confidence in the design by standing below the bridge during the load test. A section of Hammurabi's Code (from about 2100 B.C.) contains rather drastic measures:

> If a builder build a house for a man and do not make its construction firm, and the house which he has built collapse and cause the death of the owner of the house, that builder shall be put to death. If it cause the death of the son of the owner of the house, they shall put to death a son of that builder. [22]

Our society is not so extreme in its retribution but it is litigious enough to require structural designers to never forget the immense responsibility that rests on their shoulders. A "small" mistake, such as a misplaced decimal point, may cause a catastrophe. Therefore it is extremely important that students learn, from the very beginning, to work carefully and to conscientiously avoid all those "little" mistakes, which in the real world may turn out to be not so little after all.

Safety is commonly provided for by the use of *safety factors*. The strength requirements, with which we shall be concerned in great detail throughout this book, can be written in the general form

$$\text{Strength} \geq \text{Safety Factor} \times \text{Load Effect} \qquad (1.1)$$

The safety factor provides our margin of safety against the structure's strength being exceeded by the effect of the applied load. Frankly speaking, the factor of safety is more or less a factor of ignorance, because there are considerable uncertainties associated with both the structure's strength and applied load effects. Strength is a function of material properties, which exhibit much statistical scatter. There are also dimensional tolerances, craftsmanship, and quality control—all of special concern in concrete construction. The uncertainties associated with the loads are even greater. Some loads can be determined rather well, such as the structure's own weight (known as dead weight). However, externally applied loads due to traffic, wind, or earthquakes vary widely, not to mention the possibility of accidental loads. All of these uncertainties are lumped into factors of safety in a pragmatic if not simplistic manner. In most design situations, it is not the responsibility of the individual engineer to determine reasonable values for such safety factors—minimum values are prescribed in the various codes and reflect the collective judgment of the profession.

Turning to the next keyword, *serviceability*, it is not difficult to identify with a client who spends millions of dollars to build a structure to expect that it serves its intended purposes and serves them well. If the roof leaks due to a faulty design detail, or if the tenants in a tall building feel nauseous on each windy day because of the swaying motion of the building, the owner is not likely to be happy. With regard to safety, the public has a legal basis to be concerned. This is why we have building officials charged to enforce building codes. The public is somewhat less concerned with questions of serviceability. It is the owner who is required to ensure that all performance criteria are satisfied by giving the design team a set of detailed *specifications*.

Economy is likewise a concept of little public interest, except, of course, when the public is also the client. The owner expects the engineer to *do it with less*, because this is one of the fundamental attributes of an engineer. However, it is important to distinguish between a minimum cost design and an optimum design. Designing a cheap building, in the truest sense of the word, is likely to cause trouble for the owner or tenants, often starting on day one or earlier. This may not serve the client well unless the plan was intentional, for example, if an unscrupulous client intends to sell the building immediately, speculating on a fast profit. Since owners pay the engineer's bills, it is their prerogative to demand a minimum-cost building, subject only to the Building Code's requirements (which are minimum requirements). However, in an open real estate market, word will spread sooner or later on the quality of such speculative offerings. A responsible professional engineer will advise the client of the true costs of certain design decisions, because these costs include many different components. Just as the wise buyer of a new car will base the choice not on retail price alone, but also on fuel economy, maintenance track record, resale value, etc., the owner of a building or bridge should be aware of the cost of maintenance and upkeep, repair, modernization, expansion, and, ultimately, demolition. The real concern should be the *life-cycle cost*, which may very well favor larger up-front investments to be recovered over the life of the structure. Owners may also want to place a value on future flexibility. Suppose the construction of a ten-story building is planned, and the engineer is instructed to provide a minimum-cost structure. A few years later, the owner wishes to add another two floors. Because of the initial shortsightedness, this option is now prohibitively expensive. The experienced engineer will always design for larger forces and moments than actually called for, "just in case." For buildings this is generally justified because the cost of the actual structure constitutes only a small fraction of the total cost of a building, sometimes as low as 20%. In the case of a luxurious office building in Manhattan, the granite cladding alone was reportedly more expensive than the entire structural steel frame. Small additional investments for future flexibility have therefore relatively little impact on the total cost of a building.

Another keyword implicitly contained in our definition of design is *constructability*. This term is self-explanatory, but the concept is often neglected in design offices. In years past, when engineering students were required to spend months (if not years) of practical training on construction jobs, they learned very fast about the constraints of real-world construction practice. Certain designs simply cannot be built, others only with heavy penalties in time or money. Young engineers are therefore well-advised to familiarize themselves with construction practices by consulting with contractors and construction managers about the constructability of their designs. The construction industry is highly competitive, and therefore such considerations are essential.

A keyword quietly hidden among the "other requirements" in our definition of design, and which is not even mentioned in most engineering texts, is *esthetics*. In the old days, there were only "builders." The emergence of modern engineering sciences has led to the unfortunate separation of builders into engineers and architects. In spite of this specialization, it is nowhere written that architects shall not concern themselves with structural engineering, or that engineers ought to be oblivious to the visual impact of their creations. In the case of most buildings, the client will hire an architect, who in turn retains an engineer as consultant. However, for the design of bridges, some buildings, and most industrial and utilitarian structures, the client hires the engineer directly, without an architect, and the results can be far from complimentary of the designer's esthetic sense. Our present engineering curricula do not seem to allow for courses in art appreciation or other ways of training in esthetics. The scientific rigor of undergraduate engineering education appears to leave little room for such "frills." The unhappy result is that many engineers neither develop a sense for beauty, nor appreciate its importance, and then show surprise when local citizen groups emerge to oppose the construction of an eyesore that would desecrate their environment for generations to come. Such insensitivity is often implicitly condoned by the client on the mistaken assumption that a beautiful structure is more expensive than an ugly or "utilitarian" one. This notion is simply not true. Great designers such as Maillart, Nervi, Candela, Torroya, and Leonhardt were also successful as low-cost bidders. A beautiful structure may indeed be economical. An ugly structure is more likely the result of careless design, or of unwise time or budget constraints placed on the engineer, which do not permit a search for more esthetically pleasing (and perhaps more economical) design alternatives.

Social and political considerations may be mentioned in passing as well, among other requirements. Engineers are often perceived as narrow-minded specialists who cannot look beyond their own fields of expertise and are insensitive to the effects of their creations in a larger societal context. As a result, major projects are often debated in public without much input from engineers, speaking either as technical experts or as concerned citizens.

1.5 DESIGN PHILOSOPHIES

A *design philosophy* implies a specific fundamental approach toward meeting the two separate requirements of safety and serviceability of a design problem. In modern structural engineering, there are two such basic design philosophies: *allowable stress design* and *strength design.* Other philosophies are basically a variation or combination of these two. To illustrate their fundamental differences, let us consider the simple example of Fig. 1.35a. A very stiff platform carrying a load $P = 60$ kips is supported by four cables, two of which are made of some material A, the other two of which are made of some material B. The stress-strain curves for these two materials are shown in Fig. 1.35b. Assuming that all four cables have the same cross-sectional area, determine the cable size required to safely carry the load.

First we shall use the allowable stress method. By allowable stress we mean that there exists a limit beyond which the material shall not be stressed. Let us assume that material A, which yields at a 40 ksi stress, shall not be stressed beyond 20 ksi. This effectively assures a "safety factor" of two against yielding. By the same criterion, we shall not stress material B, which yields at 30 ksi, beyond a limit of 15 ksi. Because the platform is assumed to be very stiff, all four cables must elongate by exactly the same amount, that is, the strain ϵ should be the same, regardless of whether a cable is made of material A or B. According to Fig. 1.35b, material A develops twice as much stress than material B for a given strain ϵ, because its Young's modulus is twice as high. Thus, when material A is stressed to the allowable limit of $\sigma_A = 20$ ksi, material B is stressed only to $\sigma_B = 10$ ksi, well below the allowable limit of 15 ksi. The equation of vertical equilibrium requires that $P = 2A_c\sigma_A + 2A_c\sigma_B$, where A_c is the area of a single cable. Knowing the cable stresses ($\sigma_A = 20$ ksi, $\sigma_B = 10$ ksi), and the design load ($P = 60$ kips), the equilibrium equation can be solved for the required cable size,

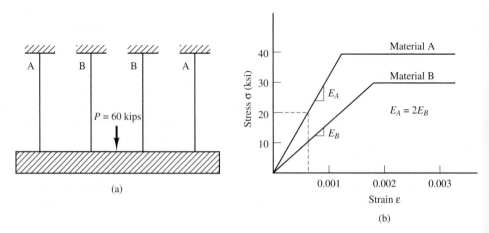

Figure 1.35 Platform suspended by four cables with two different materials: (a) structure; (b) stress-strain diagrams for materials A and B.

$A_c = \frac{1}{2}60/(20 + 10) = 1$ in^2. Thus, by selecting four cables of 1 in^2 cross-sectional area each, we assure with a safety factor of two that the cables of material A will not be stressed to their yield point.

Suppose now that we do not care so much whether the material-A cables yield or not, but rather that there is a safety factor of two against collapse of the platform. Such a collapse can occur only when all four cables are stressed to their limit. A glance at Fig. 1.35b reveals that the material-B cables cannot reach their yield limit unless the material-A cables reach theirs first and are strained well into the plastic range. When the material-B cables eventually yield, the equation of equilibrium at failure reads $2P = 2A_c(40) + 2A_c(30)$, where P is multiplied with the safety factor two. The required cross-sectional area for a cable now becomes $A_c = 60/(40 + 30) = .857$ in^2.

Thus, the strength design calls for slightly smaller cables than the allowable stress design, even though the same safety factor of two was used in both cases. How is this possible? The explanation is that in the first case we applied the safety factor only against the material-A cables reaching their yield point. This is not the same as requiring a safety factor of two against collapse (as in the second scenario above). In fact, the allowable stress design results in a hidden strength reserve, because the material-B cables are not stressed to their full allowable capacity.

Today, concrete structures are designed almost exclusively on the basis of strength to underscore the all-important safety against failure. Later in the book, we shall see in great detail, how this is done in specific design situations.

It is possible to integrate the various design goals in a unified approach, called the *limit states design*. Here the structure is designed to satisfy both service and ultimate limit states. A service limit state exists, for example, when a stress reaches an allowable value, a deflection exceeds a given limit, or the frequency and amplitude of vibration reach a combination that annoys building occupants—all under service loads, that is, those loads that a structure can actually experience during its service life. The loadings themselves are specified in the appropriate codes. For example, we can stipulate that a highway bridge, carrying a realistic truck, shall not exceed specified stresses in its various structural elements, that the deflections shall remain within specified limits, etc. It is the significance of a service limit state that typically nothing noticeable happens when it is exceeded. For example, if the allowable stress is exceeded by a few percent, the built-in safety factor is expected to prevent any distress. This is similar to exceeding the speed limit on a highway by a small amount. Ultimate limit states, on the other hand, are associated with physical events, such as the yielding of reinforcement, the sudden shear failure of a member, or the buckling of a compression member. When such limit states are reached, something noticeable happens, with potentially dramatic consequences. Limit states design assures, in a systematic manner, that the structure fulfills all design requirements, that is, that it can serve its intended purposes in an acceptable manner by having sufficient margins of safety against all limit states being exceeded, both in the service load range and under ultimate loads. By ultimate loads, we mean actual service loads multiplied with some overload factors, which are actually safety factors against ultimate limit states being exceeded. In mathematical terms, such requirements can be expressed as

$$\phi R_k \geq \gamma_o \left(\sum_i \gamma_i Q_i \right)_j \qquad k = 1, 2, \ldots \tag{1.2}$$

The symbol R represents a strength quantity, such as the flexural capacity of a beam or the axial load-carrying capacity of a column. The factor ϕ is a strength reduction factor that accounts for uncertainties with which the strength can be determined, for example, because of the statistical scatter of material properties. The symbol Q_i stands for the load effect that corresponds to the strength quality R, such as a maximum bending moment in a beam or the axial force in a column. The subscript i refers to the load source (e.g., dead load, traffic load, or wind). The γ_i factors are overload factors, that is, the factors of safety against the limit state being exceeded (e.g., failure). Since actual loads can be defined only with varying degrees of accuracy (e.g., comparing dead load and wind load), the γ_i factors must reflect the different uncertainties. The index j designates the specific load combination under consideration, such as dead plus live load, dead plus live plus wind load, etc. Therefore, the various γ_i factors will in general be different for different j-combinations. The factor γ_o accounts for the limitations and uncertainties associated with our analytical and modeling capabilities. The index k reflects the fact that we need to satisfy Eq. 1.2 for all kinds of different limit states. For example, the equation

$$0.9 M_u \geq (1.4 w_D + 1.7 w_L) \frac{L^2}{8}$$

may be interpreted in words as follows: Given a strength reduction factor of $\phi = 0.9$, the bending strength of a simply supported and uniformly loaded beam shall not be less than the combined maximum bending moments due to dead load w_D (with overload factor $\gamma_1 = 1.4$) and live load w_L (with overload factor $\gamma_2 = 1.7$), and $\gamma_o = 1.0$. For a serviceability limit state, a similar equation might read as follows: Assure that the combined deflection due to dead and live load, using overload factors $\gamma_1 = \gamma_2 = 1.0$, will not exceed the limit set by the play in the gasket of the window below the beam.

Such design methodology applied to ultimate limit states is now often referred to as *Load and Resistance Factor Design*, abbreviated LRFD, because factors for both loads (the γ-factors) and resistance (the ϕ-factor) are specified explicitly. In concrete design, this methodology has traditionally been called *ultimate strength design* (USD). Since the early 1980s, when this design philosophy was proposed for steel design, it has also been referred to as LRFD. The only difference between LRFD and USD is the way in which the various factors are determined. In the USD of the ACI Code, the factors were derived more or less in an empirical fashion, whereas in the LRFD specifications for steel structures, past experience was used only as a guide to set target reliabilities, and modern reliability theory was then used to compute the various load and resistance factors [23].

Aside from the selection of a basic approach toward design (i.e., strength or allowable stresses), there are other fundamental design principles that should be mentioned in this introductory overview.

The first is the concept of *redundancy*. This requires that a structure have alternative load paths in case of need. Physically, this is possible only if structural elements are arranged in parallel rather than in series. Consider two (elasto-plastic) springs, one with yield capacity $S_1 = 10$ kips, and one with yield capacity $S_2 = 15$ kips, arranged either in series (Fig. 1.36a) or parallel (Fig. 1.36b). A comparison between the two arrangements permits two conclusions:

1. While the capacity of the series arrangement is clearly equal to the smaller of the two springs, that is, $P_a = 10$ kips, the capacities of the parallel springs are additive as long as the springs have sufficient ductility to delay failure until both springs are stressed to full capacity.
2. Should one of the two springs fail for whatever reason, the capacity of the series arrangement is reduced to zero, that is, total failure is the consequence. In the parallel arrangement, the capacity is reduced to that of the remaining spring, that is, *the structure has an alternative load path.* Applied to a roof structure, a series arrangement will lead to catastrophic collapse, whereas a parallel scheme may retain sufficient reserve strength to permit evacuation and thereby reduce or avoid loss of life.

Current design practice with its stress on life-safety clearly favors redundant schemes. Some of the most recent design codes actually reward redundant structural systems with lower factors of safety. From a structural analysis standpoint, redundancy is equivalent to statical indeterminacy. In Chapter 6, we shall explore in some detail the mechanisms by which indeterminate concrete structures benefit from redundancy.

Another fundamental design principle concerns the *mode of failure*. A structural failure is such a serious event that the designer should take all possible precautions to prevent it. By following the strength design approach, it is possible to consider each potential failure mode separately and to assure that in each case the structure's strength exceeds the combined effect of all loads multiplied with their respective overload factors (Eq. 1.2). This requirement implies that the strength needs to be determined separately for each potential failure mode. Consider, for example, a beam that can fail either in

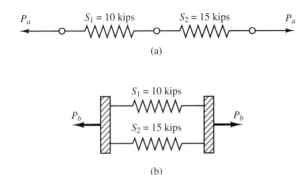

(a)

(b)

Figure 1.36 Series and parallel springs: (a) springs in series; (b) parallel springs.

bending, say at load P_1, or in shear, at load P_2. Obviously, the beam will actually fail at the smaller of the two load values, that is, the shear and flexural strengths should be thought of as springs in series. Whereas the shear failure of a concrete beam can be sudden and brittle, the beam can be designed to fail in flexure in a rather slow or ductile manner. Since a ductile failure mode is more "desirable" than a brittle one, the structural designer has the option to design the beam's shear strength more conservatively, so that in the case of overload, the beam can fail only in the preferable ductile flexural mode.

This point leads directly to the last fundamental principle to be mentioned here, namely *ductility*, which was referred to earlier in conjunction with seismic design. Ductility is the opposite of brittleness, and signifies the capability of a material or structural member to undergo large inelastic deformations without failure. While concrete is basically a brittle material, proper reinforcement with steel can make it very ductile and suitable for design. In fact, the significance of ductility cannot be overemphasized, and we shall return to this subject repeatedly throughout this book.

1.6 THE DESIGN PROCESS

The total structural design process can be thought of as consisting of three general phases: conceptual design, preliminary design, and final design [24]. In the conceptual design phase, the structural system for a particular structure is selected. Often more than one concept needs to be worked out in order to compare them on the basis of cost, functionality, or other important criteria. Decisions made in this phase have far-reaching repercussions on the subsequent phases. Therefore this is the most important phase and the most demanding on the designer's creative ability and experience. In the case of most buildings, the architect is usually the decision maker; but he or she is well advised to consult with the structural engineer from the very beginning of the conceptual design phase. The more unusual the building is in terms of height, floor spans, or irregular layout, the more important this consultation becomes. In the case of thin-shell structures, it is essential.

In the preliminary design phase, most of the specific structural design parameters (such as floor heights, column spacings, and beam and column cross sections) are selected if they have not been determined in the conceptual design phase. It is in this second phase that we encounter a fundamental dilemma of structural design: In order to design a structure, we need to know the forces and moments for which to design it. However, some of these forces and moments (e.g., those due to dead load) are not known until the structure has been designed. This dilemma is worsened in the case of statically indeterminate structures, because the analysis cannot be performed unless the individual member stiffnesses are known beforehand, in addition to the loads. This dilemma is resolved by the step-by-step procedure outlined schematically in Fig. 1.37. The key step is the preliminary design, because it entails most of the detail design decisions. To help the novice designer get started with this formidable task, it is worth remembering the golden rule of structural design:

What you don't know, assume.

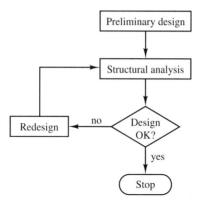

Figure 1.37   The design-analysis cycle.

    This dictum applies to just about all conceivable design variables—whether these are floor spans, beam depths, or slab thicknesses. Experienced designers can make such design decisions with either no analysis at all or with very approximate methods on simplified analytical models by hand without the aid of a computer. Young engineers are encouraged to sharpen their skills by making assumptions. Such assumptions are not only permissible, they are necessary. There is nothing wrong with making daring, even outrageous assumptions, because later on in the design process we check whether our assumptions are proper or not. Inexperienced designers will gradually overcome the initial difficulties as they learn which assumptions are good and which ones are not. By and by, they will become familiar with typical column spacings, floor slab thicknesses, beam dimensions, etc. Experienced designers have the added advantage of having access to a number of previous designs. Called upon to design a specific bridge, for example, they may have in their drawer the design of a bridge they have worked on previously, so they do not have to start from zero.

    Structural analysis, signified by the second block of Fig. 1.37, determines the response of the (preliminary) design to the prescribed loads. Structural analysis can be considered an engineering science, almost a branch of applied mathematics, with a claim to exactness within certain limitations. In the overall context of structural design, however, we are dealing with so many uncertainties with regard to external loads and material properties, dimensional tolerances, etc., that accuracy requirements can be relaxed, and often rather approximate analyses are acceptable, especially in the preliminary design stage. In current design practice, structural analyses are typically done with the help of computers, since powerful computers are now standard equipment in design offices, and user-friendly, efficient, and versatile structural analysis programs are widely available.

    The next block in the flow diagram of Fig. 1.37, the decision diamond, denotes the step in which all of our earlier assumptions are put to the test. The design is subject to the various previously enumerated requirements, many of them spelled out in building codes with the force of law. Such requirements refer to either *ultimate limit states*, which affect the safety and well-being of the public, or to *serviceability limit states*, which refer to the structure's ability to function properly and are therefore not related to life safety. Some serviceability criteria are spelled out explicitly in the building codes, others are not, and the concerned owner should include all of them in a detailed set of specifications. But the decision diamond of Fig. 1.37 covers other items as well. For example, a design may satisfy all strength and serviceability criteria but is so overdesigned as to be too expensive for the owner's budget, or the design cannot be built, as it fails the constructability test. Finally, let us not forget the esthetic requirements. These may be just as important as the others. For example, if the designer's pride is not sufficient reason, there could be neighbors so strongly opposed to an unsightly structure in their backyard that they might succeed in preventing it from being built.

The last block in Fig. 1.37 needs little comment. For each criterion that the design fails, be it strength, serviceability, economy, constructability, or esthetics, the design must be modified appropriately. If we have made unreasonable assumptions in the preliminary design phase, one of the tests signified by the decision diamond will tell us so. A young designer who has never designed anything before is at a definite disadvantage. It may take a discouraging number of cycles to arrive at an acceptable design. A seasoned designer, on the other hand, knows from experience what assumptions to make and, as a result, requires only few corrections, with seldom more than one additional iteration. This gradual accumulation of experience is the essence of practical training and the learning process. A college course is only the beginning.

1.7 ATTRIBUTES OF A STRUCTURAL DESIGNER

There exists no recipe for the making of a good structural designer. It is possible, however, to identify a number of prerequisites and capabilities that characterize one. The most important prerequisite is a thorough understanding of the principles of structural mechanics, often called "the basics." There is simply no substitute for being able to draw bending moment diagrams to verify static equilibrium, or to identify a torsionally stiff cross section. In the course of this book we shall review many of these concepts, even at the cost of repetition.

The knowledge required of an engineer can be divided into two general categories: (1) the basics, and (2) the technological details. One may use the half-life of an item of knowledge to determine to which category it belongs. Basic knowledge hardly ever changes. Newton's laws have been around for hundreds of years, and they will outlast all of us, Einstein's theory of relativity notwithstanding. On the other hand, a technological detail such as the required splice (overlap) length of reinforcing bars has a short half-life. It may change with next year's ACI Code. Concrete technology in general is a field that changes rapidly. Four years of engineering school are not enough to cover everything needed to become a professional engineer. As a result, many schools consider it their primary mission to transmit fundamental or basic knowledge. Familiarity with technological details will be gained during practical training, typically in an engineering office environment. This division of labor in education that has evolved over the years makes efficient use of the scarce and valuable time in engineering school. For example, detailed ACI Code provisions are deemphasized, as they can change soon after or even before graduation. Why should we concern ourselves in school with all the details of a particular prestressing system when our future employer may use a different system altogether! In this book, we shall try to adhere to the basics and refer to the ACI Code only to the minimum extent necessary. After graduation, students will have plenty of opportunity to familiarize themselves with all the details that the satisfactory performance of their job assignments requires. This allows us to concentrate on the primary task of undergraduate education, namely, the full understanding of basic engineering principles.

Another characteristic that is useful for the development of a designer is a natural curiosity, which is really a hallmark of engineers. Wherever you are, look at the various structures around you. The very fact that these have been designed and built is an indication that something about them is most likely "right." Conversely, if something does not look "right," it probably is not. If you look at existing bridges with a critical eye and try to estimate the depth-to-span ratios, you will soon notice that these ratios do not vary very much. With a little practice, you can readily recognize whether a beam is too shallow, too deep, or "just right" (Fig. 1.38). This statement does not preclude the occasion that requires an updating of the accepted norm or the notion of what constitutes "just right." Such an occasion may result from advances in technology. For example, the extreme slenderness of the Alm Bridge in Austria (Fig. 1.39), was made possible by an innovative combination of tensile and compressive prestress [25]. Thus, the fledgling student who is faced with a design problem and needs to assume a beam depth is encouraged to choose any value that appears to "look right." If it turns out to be too large, it will manifest itself in the finding that the beam is understressed, that is, material is wasted. If the assumed beam depth is too small, we shall also find out quickly, because the deflections under load will probably turn out to be unacceptably large.

Just by looking at structures, we can develop a sense for what looks right and what does not. Consider the two cantilever beams of Fig. 1.40. We do not have to be engineers in order to feel intuitively that the example in Fig. 1.40a looks right and that the one in Fig. 1.40b does not. It does not require the knowledge that the bending moment increases linearly with distance from the tip load to prefer the first solution, because nature is teaching us with each tree limb how to grow cantilever beams. Our

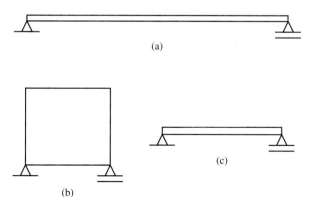

(a)

(c)

(b)

Figure 1.38 Beam span-to-depth ratios: (a) too high; (b) too low; (c) "just right."

Figure 1.39   Alm Bridge, Austria (courtesy of Hans Reiffenstuhl).

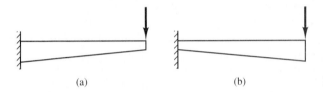

   (a)   (b)   **Figure 1.40**   Cantilever beams.

engineering knowledge supplies only a rational explanation of what everybody knows intuitively from nature's lessons. The same lesson applies to a vertical cantilever. As engineers we know that wind is one of the major horizontal loads acting on a building. And again, every tree trunk shows very convincingly how the increasing bending moment is resisted economically by an increasing cross-sectional area. For this reason we feel an immediate, almost esthetic, satisfaction of the building in Fig. 1.41a because it reminds us of a tree. On the other hand, the building in Fig. 1.41b does not transmit this feeling of satisfaction. Even if the designer took great pains to satisfy the building authorities that the design was safe, this building surely does not look right.

Thus, the lesson is to keep our eyes open and look at structures, to think about the reasons why one looks good and another one does not. What are typical column spacings in an office buildings? If this question is raised in class, "I don't know" is an unacceptable answer from aspiring engineers, because they have been in hundreds of buildings and must surely have had a chance to look at column spacings. Are they 2 feet apart, 20 feet, or 200 feet? Make assumptions, use common sense, and check your assumptions! It also helps to look at the creations of master builders. If we cannot afford to travel across the country to study their creations, we surely can afford to spend a few hours in the engineering or architecture library and look at some of the world's great bridges, buildings, dams, and towers.

It is also beneficial to the development of a designer to practice the art of sketching. Next to the spoken and written word, drawings and sketches form the third medium in

(a)   (b)

Figure 1.41 Buildings that (a) look right (9 West 57th Street, New York City); or (b) don't (courtesy of Guenter Meyer).

which we communicate our thoughts. Our colleagues, the architects, place much more emphasis on this skill than we do; but the ability to express structural concepts in sketches, to draw deflected shapes of structures under load, is an essential tool for the development of structural concepts.

Finally, one perennial complaint we educators hear from practicing engineers, the future employers of our students, is that civil engineering graduates may be very bright and hard-working and technically well prepared for their jobs, but their communication skills, both written and oral, are often lacking. Although this observation transcends the specific scope of this book, it might as well be mentioned, as we are enumerating the prerequisites for a developing structural engineer. The student is well advised to take this charge seriously—to become successful, designers must interact with clients and be capable of clearly communicating their thoughts, both orally and in writing. The reader may smile and wonder what this has to do in a book on reinforced concrete design. The answer is: a lot, unless you are prepared to be consigned to narrowly defined design tasks. Only if you can clearly express your ideas to an audience of nonexperts can you hope to become a truly successful designer.

REFERENCES

1. Davidovits, J., and Morris, M., *The Pyramids: An Enigma Solved*, Hippocrene Books, Inc., New York, 1988.

2. ACI, "The Great Pyramid Debate," *Concrete International*, Aug. 1991, pp. 28–44. (Discussion continued in *Concrete International*, Feb. 1992, p. 17.)

3. Vitruvius, *On Architecture*, Harvard University Press, Cambridge, MA, 1931, Book 2: Chapter 6.

4. Straub, H., *Geschichte der Bauingenieurkunst* (History of Civil Engineering), 2d ed., Birkhäuser Verlag, Basel, Switzerland, 1964.

5. ACI Committee 116, *Cement and Concrete Terminology*, American Concrete Institute, Detroit, MI, Report 116R-90, 1990.

6. *Formwork for Concrete*, American Concrete Institute, 5th ed., Detroit, MI, Special Publication SP-4, 1989.

7. Kelsey, J., "World's Tallest RC Building Completed in Record Time," *Concrete International*, Dec. 1993, pp. 46–48.

8. Dischinger, F., and Finsterwalder, U., "Eisenbeton-Schalendächer System Zeiss-Dywidag" (Reinforced Concrete Shell Roofs System Zeiss-Dywidag), *Der Bauingenieur*, Vol. 9, 1928.

9. Esquillan, N., "The Shell Vault of the Exposition Palace Paris," *ASCE, J. of the Struct. Div.*, Jan. 1960, pp. 41–70.

10. Nervi, P. L., *Aesthetics and Technology in Building*, Harvard University Press, Cambridge, MA, 1965.

11. Roessler, S. R. and Bini, D., "Thin Shell Concrete Domes," *Concrete International*, Jan. 1986, pp. 49–53.

12. Ramm, E., and Schunck, E., *Heinz Isler Schalen*, Catalog of Exhibit, Karl Krämer Verlag, Stuttgart, Germany, 1986.

13. Muller, J. M. and McAllister L. F., "Esthetics and Concrete Segmental Bridges in the United States," *Concrete International*, May 1988, pp. 25–33.

14. Leonhardt, F., *Bridges, Aesthetics and Design*, The MIT Press, Cambridge, MA, 1984.

15. Billington, D. P., *Robert Maillart and the Art of Reinforced Concrete*, Verlag für Architektur Artemis, Zurich, Switzerland, and Munich, Germany, 1990.

16. Plowden, D., "The Spans of North America," Norton, New York, 1984.

17. DeLony, E., "Landmark American Bridges," *ASCE*, New York, 1993.

18. Meyer, C., "Design Life Simulation of a PCRV," *ASCE, J. of the Struct. Div.*, vol. 108, no. ST5, May 1982.

19. Paulay, T., and Priestley, M. J. N., *Seismic Design of Reinforced Concrete and Masonry Buildings*, John Wiley and Sons, New York, 1992.

20. ACI, "Concrete in Art & Architecture," *Concrete International*, Sept. 1988.

21. Cannon, J. A., "A Bridge Between Engineering and Art," *Concrete International*, Sept. 1992, pp. 26–28.

22. Handcock, P., *The Code of Hammurabi*, The Macmillan Company, New York, 1920.

23. Ravindra, M. K. and Galambos, T. V., "Load and Resistance Factor Design for Steel," *ASCE, J. of the Struct. Div.*, Sept., 1978.

24. Scordelis, A. C., "Analysis of Structural Concrete Systems," IABSE Colloquium on Structural Concrete, Stuttgart, Germany, 1991.

25. Reiffenstuhl, H., "Druckspannbewehrung, ein neues Konstruktionselement zur entscheiden-den Steigerung der Tragfähigkeit von Stahlbeton- und Spannbetonquerschnitten" (Compressive prestress, a new structural concept to decisively increase the strength of reinforced and prestressed concrete sections), FIP Congress, New York, 1974.

Materials

Modern concrete technology is a highly specialized and rapidly changing discipline. For this reason, designers of concrete structures cannot be expected to be completely knowledgeable and current in this field—but they cannot afford to ignore it either. There are a number of reasons for this. First, as a matter of principle, designers should have a basic understanding of the important aspects and properties of the materials they are working with. These have profound influence on their designs, both short- and long-term. Also, current developments may impact the economics aspects of a design. For example, if a designer is not aware of a new admixture that greatly accelerates the strength development of concrete, the design may go out for bids, based on an unnecessarily long construction time, and as a result might lose out against a steel alternative.

For these reasons we have no choice but to familiarize ourselves with the major aspects of concrete materials and steel reinforcement. That is the purpose of this chapter. For a more detailed treatment of the subject, readers are referred to the specialized literature [1–3].

2.1 CEMENT

Starting with cement, it may be noted that laypersons and those who should know better often confuse the terms *concrete* and *cement*: a "cement truck" is usually a "ready-mix concrete truck," and "pouring cement" really means "pouring concrete." As

Pores	Solids			
Air	Hydrated cement	Aggregate		Chemical
Free water	Bound water	Fine	Coarse	and mineral
	Unhydrated cement			admixtures

Figure 2.1   Ingredients of concrete.

Fig. 2.1 illustrates, cement is just one ingredient of a mixture that contains coarse and fine aggregate, both hydrated and unhydrated cement, water, air, and usually some chemical or mineral admixtures, not counting the steel with which it is usually reinforced. Cement is basically the glue that binds the various ingredients together.

We shall restrict ourselves to portland cement, which is by far the most important. Its main constituents are four different minerals: tricalcium silicate C_3S, dicalcium silicate C_2S, tricalcium aluminate C_3A, and tetracalcium aluminoferrite, C_4AF.* These do not appear in nature in the required combination, but their basic ingredients are contained in limestone and clay, raw materials found in most parts of the world. In a cement plant, both the limestone and the clay are separately ground and combined with water to form slurries. These are mixed and blended to form the correct mixture and then burned in a kiln oven at about 1450°C. The result is a hard stone referred to as *clinker.* After cooling, the clinker is very finely ground, with particle sizes of the order of 75 μm (1 μm = .001 mm) and interground with a small amount of gypsum.

Since natural raw materials vary considerably, the actual manufacturing process must be adjusted accordingly. Even then, the chemical composition and proportions of the final product cannot be controlled to be exactly the same from batch to batch. Some variations are inevitable.

The production of cement is very energy intensive, and therefore the economy of a specific plant depends very much on its energy efficiency. The average fuel consumption for cement production in the United States in 1988 was 4.4×10^6 BTU/ton, compared to 3×10^6 BTU/ton in Japan [4].

When cement is brought into contact with water, it undergoes a chemical reaction, known as *hydration.* Because cement is primarily made up of the four chemical minerals discussed above, each with its own hydration characteristics, the hydration of portland cement is a combination of these individual characteristics. By changing the chemical composition, one can design a cement with certain properties, such as high-early strength or low-heat development.

The ASTM classification of common portland cements [1, 5] is reproduced in Table 2.1. The most common and standard cement is referred to as Type I. Type II cement has better sulfate resistance, that is, it is more suitable for structures that are in contact with seawater. Type III cement is characterized by development of high-early strength. Since C_3S hydrates much faster than C_2S, high-early strength can be

---

*In cement chemistry, the following abbreviations are commonly used: $C = CaO$ (calcium oxide or lime), $S = SiO_2$ (silicate), $A = Al_2O_3$ (alumina), $F = Fe_2O_3$ (ferric oxide), $H = H_2O$ (water), $\bar{S} = SO_3$ (sulfur trioxide).

TABLE 2.1   ASTM CLASSIFICATION OF CEMENTS [1, 5]

Cement types	I	II	III	IV	V
C_3S (%)	50	45	60	25	40
C_2S (%)	25	30	15	50	40
C_3A (%)	12	7	10	5	4
C_4AF (%)	8	12	8	12	10
$C\bar{S}H_2$ (%)	5	5	5	4	4
Fineness (m^2/kg)	350	350	450	300	350
1-Day Compressive Strength (Mpa)	7	6	14	3	6
Heat of Hydration (7 days; J/g)	330	250	500	210	250

obtained by replacing some of the C_2S by C_3S. Increasing the fineness of the end product has a similar effect because of the increased surface area. The high-early strength development is accompanied by an increased heat of hydration. Type IV cement generates less heat, but is also slower in strength development. This is achieved by reducing the amounts of C_3S and C_3A. Type V cement has been developed for improved sulfate resistance by lowering the C_3A content and replacing some of it by C_4AF. Only Type I and III cements are generally available; the others need to be specially ordered. The recent widespread use of mineral admixtures, such as fly ash and silica fume, with their important modification of cement hydration and other characteristics, has further eroded the significance of all but Type I and Type III cements.

High-strength concretes produced with high-early cement may reach peak temperatures of 100°F or more during hydration. But why should we be concerned about this heat of hydration? Consider the two blocks of Fig. 2.2, both made of the same concrete material and generating the same heat of hydration per unit volume. While the small block has 6 ft^2 of surface area to dissipate the heat generated by 1 ft^3 of concrete, the large block has a surface area of 600 ft^2 for a 1000 ft^3 volume. As a

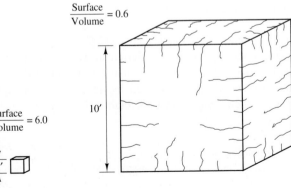

Figure 2.2   Temperature cracking.

result, the small block, with a surface-to-volume ratio of 6, cools down very fast with a relatively uniform temperature profile. The large block, with a surface-to-volume ratio of only 0.6, cools down much slower, and temperature gradients are much steeper. In this case, the surface concrete cools down and contracts, while the bulk concrete inside does not. The resulting tensile stresses in the surface concrete can lead to extensive cracking, which is particularly serious in massive structures like concrete gravity dams. Elaborate cooling systems need to be installed in such cases to remove the heat of hydration.

In addition to the specifications of Table 2.1, ASTM imposes other requirements on cements, such as limits on chemical impurities to assure acceptable performance. Also, the fineness requirement is important, because fineness increases the rate of hydration and therefore strength gain and allows the cement to hydrate more completely. It also reduces the durability of concrete, such as its resistance to freeze-thaw cycles.

The hydration of tricalcium silicate can be described by the following (simplified) chemical reaction:

$$2C_3S + 11H \longrightarrow C_3S_2H_8 + 3CH + \text{Heat} \tag{2.1}$$

or in long hand,

$$2(3CaO \cdot SiO_2) + 11H_2O \longrightarrow 3CaO \cdot 2SiO_2 \cdot 8H_2O + 3Ca(OH)_2 + \text{Heat}$$

In words: Tricalcium silicate + water turns into *C-S-H gel* + calcium hydroxide and heat. The C-S-H gel is a colloid with particles of the order 1 nm in size and widely varying composition. It is the mineral glue that gives the cement paste much of its strength.

The chemical reaction of dicalcium silicate C_2S during hydration is

$$2C_2S + 4H \longrightarrow C_3S_2H_3 + CH + \text{Heat} \tag{2.2}$$

Since the rate of hydration is proportional to heat development, which can be measured, a time history curve such as the one shown in Fig. 2.3 [1] provides useful insight into the strength development phases of cement (and concrete). An initial rapid

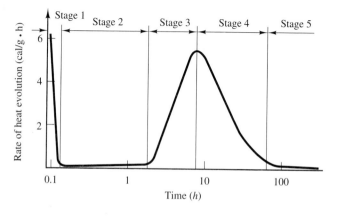

Figure 2.3   Rate of heat evolution. (From S. Mindess and J. F. Young, *Concrete*, 1981. Reprinted with permission of Prentice-Hall, Inc., Englewood Cliffs, NJ.)

evolution of heat, lasting up to about 15 minutes, is followed by a dormant period, during which the concrete remains plastic. Initial set starts in the next phase, two to four hours later. Final set is reached at about the time of peak hydration rate, and early hardening begins, followed by a steady-state period.

Some physical consequences of hydration can be visualized with the help of Fig. 2.4, which shows a cement particle surrounded by water. Initial hydration creates a thin C-S-H shell. Further hydration can occur only if water diffuses through this outside layer. As hydration continues, this layer gets thicker, slowing down the diffusion and subsequent hydration. Thus, hydration tends to approach 100% completion asymptotically. Finely ground cement with a large surface-to-volume ratio hydrates fast, while coarse grains may never hydrate completely. This process can also be used to explain the phenomenon of *healing*. If hardened concrete cracks, the cracks are likely to pass through cement particles, thereby exposing unhydrated cement to any existing unbound water. Thus, the hydration process can restart, thereby "cementing" the crack; that is, concrete can repair itself if free water is available and the cracks are not too wide.

The hydration of tricalcium aluminate is much more complicated than that of the calcium silicates. The first reaction is as follows:

$$C_3A + 3C\bar{S}H_2 + 26H \longrightarrow C_6A\bar{S}_3H_{32} \tag{2.3}$$

or in words: Tricalcium aluminate + gypsum + water turns into ettringite. After consumption of the gypsum, a second reaction takes place:

$$2C_3A + C_6A\bar{S}_3H_{32} + 4H \longrightarrow 3C_4A\bar{S}H_{12} \tag{2.4}$$

the end product being monosulfoaluminate. The addition of gypsum prevents flash set, the very rapid reaction, $2C_3A + 21H \longrightarrow C_4AH_{13} + C_2AH_8 \longrightarrow 2C_3AH_6 + 9H$, which generates a considerable amount of heat and an end product that is not very strong.

Pure C_3S and C_2S compounds do not exist in industrial cements. Small amounts of impurities are always present and have considerable effects on the hydration of the cement and the properties of the end product.

The various hydration products, collectively referred to as *cement gel*, have quite diverse properties, and the behavior of each compound contributes to the overall behavior of the cement paste. C-S-H does not have a well-defined chemical composition, and its particles are so small that they are barely identifiable under a scanning electron microscope (SEM). The gel gradually grows out from the cement grains, fills the capillary

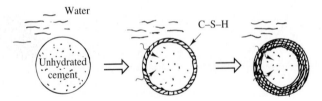

Figure 2.4   Asymptotic rate of hydration.

spaces occupied by water, and provides bond between cement grains. The gel water occupies about 25% of the gel volume and is an integral part of the gel. Even though it can be driven out by drying, it is incapable of hydrating fresh cement grains.

Calcium hydroxide, CH, in contrast to C-S-H, is a well-crystallized material. The crystals are sometimes large enough to be seen by the naked eye and have a distinctive hexagonal prism morphology. It is the most soluble phase and therefore can compromise the concrete durability. Figure 2.5 shows a micrograph of hydrated portland cement.

The ultimate strength of hardened cement paste depends strongly on the quantity of mixing water used, whereas the rate of cement hardening is affected by the chemical composition, fineness of cement, and the moisture and temperature conditions during the hardening process. The strength of cement paste drops as a semi-logarithmic function of the water-cement ratio (w/c ratio), assuming the mix is properly compacted. If incompletely compacted, air-filled pores lower its strength. In this case, we can predict strength by substituting "water + air" for water in the w/c ratio. The porosity of cement paste determines its strength; whether the pores are filled with air or water is almost irrelevant.

Portland cement chemically binds water about 25% its weight. By chemical combination (compression), the water volume is decreased by about $\frac{1}{4}$. Another 15% of its weight is held as gel water. For example, with $w/c = 0.4$ by weight, we combine

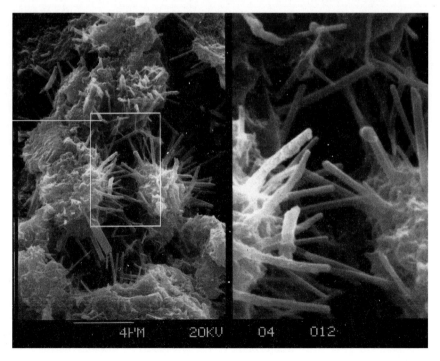

Figure 2.5   SEM micrograph of hydrated portland cement.

32 cm³ cement (or 100 g @ $\gamma = 3.2$ g/cm³) with 40 cm³ water (i.e., 40 g) to get 32 cm³ cement + 25 cm³ bound water (25% of 100 g cement) − 6 cm³ water (due to compression) + 15 cm³ gel water = 66 cm³ cement gel. The original cement volume more than doubles, but the total initial volume reduces by 6 cm³ because of "intrinsic shrinkage" or "contraction." As a result, air-filled pores are created, unless the cement is cured in water.

The theoretical minimum w/c ratio to hydrate all cement is thus 0.25 by weight. However, to assure that water reaches all cement particles, a ratio of 0.42 is needed for complete hydration [1].

Unhydrated cement is not wasted. On the contrary, lowering the w/c ratio below the minimum can lead to higher strength. Unhydrated cement reduces the porosity and is much more strongly bonded to the cement gel than any kind of aggregate.

The strength of the cement gel derives mostly from the C_3S and C_2S compounds. According to Table 2.1, the combined amounts of C_3S and C_2S are about the same for all five cement types. Therefore, the expected ultimate strengths are comparable, except that they are achieved at different rates. With greater fineness, more surface area is available for fast hydration, leading to high-early strength; but the cement gel itself slows down the hydration rate, which depends on diffusion through the gel. Thus, extra fine grinding has an effect only on early, not ultimate strength.

The term *curing* generally means the measures taken to keep the concrete from drying out during the early stages of hydration. Drying out terminates the hydraulic hardening process—no more water, no more hydration. Cement gel can form in water-filled capillaries only. Once the capillaries are dry, hydration stops but will restart after a resupply of water.

As a rule, higher temperatures accelerate the hardening rate, lower temperatures reduce it. *Maturity* is a concept that combines the effects of time and temperature on strength development [6]. It was originally introduced as a simple product of time t and temperature T (relative to some reference temperature T_0), that is, $M = t(T - T_0)$, and reflects the observation that concrete of the same maturity has approximately the same strength, whatever combination of temperature and time makes up that maturity (Saul's maturity rule). A given temperature-curing time history may then be converted to an equivalent age of curing at some reference temperature. For example, the equivalent age, t_e, may be expressed by the equation

$$t_e = \sum_0^t e^{-\frac{E}{R}\left[\frac{1}{273+T} - \frac{1}{273+T_0}\right]} \Delta t \tag{2.5}$$

where T_0 = reference temperature (°C); T = average temperature of concrete during time interval Δt; R = universal gas constant (8.3144 J/mol K); and E = activation energy = 33,500 J/mol (for $T \geq 20°C$) and $33,500+1470(20-T)$ J/mol (for $T < 20°C$). For example, if the reference temperature is $T_0 = 20°C$, the equivalent age at curing temperature $T = 30°C$ would be $t_e = 1.57t$, where t is the total curing time. At 0°C, the equivalent age would be $t_e = 0.15t$.

2.2 AGGREGATE

The bulk of hardened concrete is aggregate. It occupies 75% or more of the total volume and therefore contributes a large share to the strength of the finished product. Therefore, aggregate should have good strength and durability characteristics. It should be clean: free from silt, organic matter, oils, and sugar. Otherwise, it should be washed prior to use, because any of these impurities may prevent the cement from hydrating or reduce the cement paste-aggregate interface bond.

Before using aggregate it is necessary to determine its moisture content. If the aggregate is wet, that is, all its pores are filled with water and there is a film of water on the surface, then some of this water should be included when determining the w/c ratio, because less mixing water needs to be added than if the aggregate were dry. A completely oven-dry aggregate will absorb some of the mixing water, due to its natural porosity, so that its absorption capacity should be subtracted when determining the w/c ratio. Ideally, the aggregate should be saturated but surface dry, in which case the mixing water will neither be increased nor decreased during mixing. Sand is generally wet, so the extra moisture should be counted lest it adversely affect the w/c ratio and thereby the strength and durability.

For concrete to be sound and durable it is natural that the aggregate, its main constituent, also have these characteristics. Soundness generally designates a material's ability to maintain its mechanical properties when it is subjected to environmental influences that cause volume changes, such as freezing and thawing cycles and wetting and drying. This is often referred to as *weathering*. Aggregates must meet a number of requirements (including some of a chemical nature) for the concrete to be durable, that is, not to deteriorate during its intended service life. These will be discussed in more detail in Sec. 2.6, since questions of durability involve not only the aggregate but other concrete ingredients as well.

The more densely packed the fine (sand) and coarse (gravel) aggregate, the stronger the concrete will be. Thus, aggregate is subject not only to mechanical and chemical requirements—the size gradation is also important, a factor determined by sieve analysis. For this purpose, a sample of the aggregate is passed through a stack of standardized sieves with progressively smaller mesh sizes. The percentages of the aggregate passing through or retained by each sieve are recorded in a table or plotted graphically. The result is a grading curve (Fig. 2.6). With the sieve numbers spaced equidistantly on the abscissa and the sizes of subsequent mesh openings varying by a factor of two, the resulting plot is semilogarithmic. Coarse sieves ($\frac{3}{8}$ in and larger) are identified by the size of the sieve opening; the identification numbers of the fine sieves indicate the number of openings per inch. For instance, the No. 16 sieve has 16 openings per inch, each with a nominal size of 0.0469 in. The No. 4 sieve is used rather arbitrarily to separate the fine and coarse aggregate. Figure 2.6 includes the upper and lower bounds for the grading curve specified by ASTM [7] and the theoretical Fuller curve.

The problem of optimal packing of aggregate particles, if idealized as spheres, is a simple problem of geometry. Assuming that one set of spheres of radius R_i is ideally stacked, then the next set of smaller spheres needs to be of radius $R_{i+1} = 0.225R_i$ to

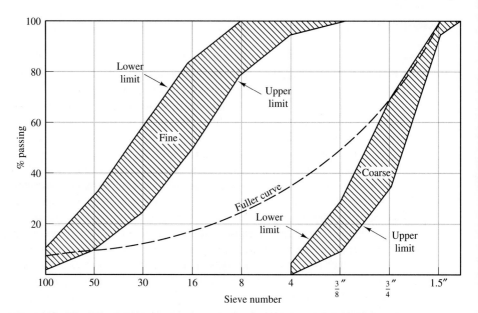

Figure 2.6 Grading curves with ASTM limits. (Copyright ASTM. Reprinted with permission.)

exactly fill the voids, and so on. In practice, aggregate particles are neither spherical nor of equal size. Besides, it is desirable to minimize the amount of cement paste needed to fill the remaining voids and coat the aggregate particles for bonding. For this reason, the limits of Fig. 2.6 basically serve as brackets for practical grading curves.

To avoid air pockets, known as *honeycombs*, upper limits on the aggregate size are dictated by the minimum dimensions of the structural components and the spacing of the reinforcing bars. Large aggregate particles reduce the amount of cement paste needed. Since rocks are cheaper than cement, their use reduces the cost. However, large aggregate particles impair the bond properties and therefore may affect the strength of the concrete. They provide more restraint to volume changes of the cement paste, thereby inducing additional stresses and cracks on the aggregate-paste interface, which weakens the concrete. On the other hand, large aggregate particles reduce the water content necessary to achieve adequate workability. As a result, the net effect of maximum aggregate size on strength is only slight, especially since aggregate is typically much stronger than the cement paste of normal-strength concrete.

At the other extreme, a large amount of very fine aggregate requires more cement paste for good workability because of the increased surface area. This is why ASTM prescribes a 5% limit for the fraction of sand that is allowed to pass the No. 200 sieve. In addition, the *fineness modulus* is a factor used in mix design. This is defined as the sum of the cumulative fractions retained on standard sieves. For example, if sieve Nos. 4, 8, 16, 30, 50, and 100 retain, respectively, 2, 5, 20, 30, 25, and 12% (the remaining 6% pass the No. 100 sieve), the cumulative fractions are .02, .07, .27, .57, .82, .94, and their sum is 2.69, which is the fineness modulus. This should lie somewhere between 2.3 and 3.1. For a finer grading, the fineness modulus is smaller. A good amount of fine sand may be needed for smooth surface finishes.

The omission of one or more sizes in the grading leads to *gap grading*, which can reduce the sand and cement requirements for a given workability.

The shape and texture of aggregate particles are of concern because they determine the amount of cement paste needed and often affect the concrete strength. The ideal particle is spherical in shape, with a smooth surface similar to that of natural gravel. Crushed stone aggregate is much more angular in shape, with larger surface-to-volume ratios and rough surface texture, which is advantageous for some applications.

Concrete made with natural stone aggregate typically weighs about 145 lb/ft^3. Thus, allowing an additional 5 lb for steel reinforcement, it is common to assume that reinforced concrete weighs 150 pcf. Basically, both weight and strength decrease in inverse proportion to the porosity of the aggregate and the finished concrete. Structural lightweight concrete, that is, concrete required to carry load, has a weight range of about 65 to 115 pcf. Concrete lighter than this is of insufficient strength to carry loads. (See also Sec. 2.8).

At the other extreme, radiation shielding in nuclear reactor facilities requires high density. Lead is ideal for this purpose but expensive. Heavy-weight concrete can achieve the same purpose at less cost. Suitable aggregates are iron and titanium ores; scrap metal can be added as well.

2.3 ADMIXTURES

Most of the concrete produced in North America contains one or more admixtures that enhance certain properties of the concrete [1]. *Air-entraining agents* are added to improve the resistance of concrete against freeze-thaw cycles. *Chemical admixtures* are water-soluble components added to control the setting and early hardening of fresh concrete or to reduce its water requirements. *Mineral admixtures* are finely divided solids added to improve the concrete's workability or durability, or to provide additional cementing properties. There are also other admixtures that serve a variety of purposes and are not easily associated with any of the three main categories.

2.3.1 Air-entraining Agents

These are chemicals that cause the mixing water to foam and form small air bubbles, which stay intact during concrete hardening. These air voids need to be spaced very closely so that the free water does not have to migrate too far when under the hydraulic pressure caused by the incipient freezing. Once inside those voids, the water can expand harmlessly. The ACI Code [8] requires air entrainment for concrete exposed to freezing and thawing or deicer chemicals. The required amount varies between about three and seven percent volume, depending on the maximum aggregate size and severity of exposure.

2.3.2 Chemical Admixtures

This class of admixtures can be further categorized into water-reducing admixtures, retarding admixtures, amd accelerating admixtures.

Water-reducing admixtures lower the amount of water required by between 5 and 30% to attain a given slump (see next section). High-range, water-reducing admixtures are also referred to as *superplasticizers* and work as follows. In regular mixes, electrostatic surface charges on the cement and very fine aggregate particles cause these particles to flocculate, thereby tying up water that is then lost in making the paste and concrete flow. Molecules of the water reducers neutralize or reverse these surface charges, causing the particles to repel each other, thereby making most of the water available to reduce the concrete viscosity. Because these molecules increase the flow of the fresh concrete, it is possible to reduce the w/c ratio without reducing the workability. A lower w/c ratio leads to higher strength, lower permeability, and a general improvement of durability.

Retarding admixtures are added to delay the setting time. This may be necessary in situations where delays in the mixing and placing of concrete can be expected; or to avoid cold joints between successive lifts, especially in massive structures; or where strains in the freshly placed concrete could lead to cracks, unless it remains longer in its plastic state. The retarders slow down the rate of early hydration of C_3S by extending the dormant period (stage 2, Fig. 2.3).

Accelerating admixtures have the opposite effect of retarders by shortening the dormant period of C_3S hydration. Quick-setting admixtures (often used in combination with high-early cement) are necessary, for example, for emergency repair situations that call for the very rapid development of rigidity or strength.

2.3.3 Mineral Admixtures

Blast furnace slag is a waste product of iron production and is suitable both for use as aggregate or cementing material. Although widely used in North America as aggregate, it is more popular in Europe as a substitute for portland cement. To obtain the hydraulic (cementitious) properties, the molten slag is granulated (chilled rapidly) as it leaves the blast furnace. Subsequently, it is ground like cement clinker and blended with the portland cement during mixing, making up between 25 and 65% of the total cementitious material. Granulated blast furnace slag cement reduces the temperature rise in mass concrete; it lowers the concrete permeability by reducing pore sizes and it improves the concrete resistance to sulfate attack [9].

Fly ash is a major residue in coal-burning power plants. It is the fine dust trapped, for example, by electrostatic precipitators that such plants require for air pollution control. Of the approximately 50 million tons produced in 1987 in the United States, about 10% are used in concrete [10]. It is a *pozzolanic* material, that is, it contains amorphous silica, which reacts with calcium hydroxide, CH (one of the end products of portland cement hydration, see Eqs. 2.1 and 2.2), to produce more C-S-H,

$$CH + S + H \longrightarrow C\text{-}S\text{-}H \tag{2.6}$$

Thus, by continuing the hydration where the regular cement leaves off, it adds additional cementing power and can therefore be used either in addition to or as a substitute of part of regular portland cement. It generates less heat of hydration and is therefore suitable in high-strength and mass concrete applications, where excessive heat generation can be a problem. The addition of fly ash improves the workability of the fresh concrete, which means that the ratio of water to cementitious material can be reduced. The long-term reaction of fly ash reduces the pore sizes and thus the permeability, with various benefits for the concrete durability. Not the least of the advantages of fly ash is its low cost, since it is a waste byproduct of power generation.

Silica fume or *condensed silica fume* is another industrial waste product [11], the result of production of metallic silicon and various silicon alloys in submerged-arc electric furnaces. The industry used to have no interest or incentive to collect this waste material, and instead discharged it into the atmosphere. Clean air legislation compelled it to install dedusting systems, which act like big vacuum cleaners to collect the dust. Its extremely low weight of about 15 to 20 lb/ft^3 (only about 10% of its apparent volume is comprised of solid particles) makes it difficult to handle and transport. It can be mixed with water to form a slurry, or it can be densified by keeping it in slow motion in a silo, where it is transformed into micropellets.

The chemical makeup of silica fume is mostly SiO_2, but the detailed chemical composition depends on the specific silicon end product of which it is the byproduct. The fume is condensed vapor and therefore consists of extremely fine spherical particles of about 0.1 μm average size, that is, two orders of magnitude smaller than cement particles. This small size, or rather the large surface area, causes the material to be highly reactive and pozzolanic, that is, like fly ash it reacts with $Ca(OH)_2$, but it also affects the hydration of regular portland cement. Therefore, it can be used as a partial substitute of cement or a straight additive. The final hydrated cement paste appears to be much better consolidated if the cementitious material contains silica fume. The porosity is reduced, the permeability decreased, and as a result, the strength and durability are greatly enhanced. Silica fume also increases the bond between aggregate and paste, which is an important indicator of strength. Because of its large surface area, it binds more water than portland cement, reducing the effective w/c ratio. To maintain workability, a superplasticizer is usually added. Most high-strength concrete now produced in North America contains some silica fume.

2.4 PRODUCTION OF CONCRETE

The details of producing concrete with specified strength and other properties are not as important to the designer today as they were in earlier days, because the specialization of the concrete industry has given us producers who can readily satisfy most demands. For example, we now can simply order the necessary number of truck loads of concrete with a given strength and leave it up to the supplier to worry about the details. However, as a matter of principle, it is widely believed that a designer of concrete structures should be familiar with the main principles of concrete mix design.

Mix design.   Unfortunately, the state of the art in concrete materials science has not yet advanced to the point where we can rationally compute the expected strength of concrete, even if we were provided with all conceivable information about the constituent materials and their proportions. Concrete mix design therefore remains a highly empirical subject, so much so that it is necessary to produce *trial mixes*, from which the final mix design parameters can be determined.

The objective of mix design is to determine the mix proportions such that the resulting concrete has a specified strength and meets specific durability requirements. Actually, this objective involves three separate goals:

1. to achieve the specified strength and durability requirements;
2. to assure that the mix is workable;
3. to minimize the amount of cement, which is the most expensive ingredient of the mix.

For simplicity, we shall exclude here mineral admixtures, such as fly ash or silica fume, which are often added as supplements or substitutes for regular cement in order to achieve high strength or a durable end product. The remaining main variables at our disposal are the following:

1. The w/c ratio. As mentioned earlier, the concrete strength is inversely proportional to the w/c ratio, usually specified as a weight ratio (see Fig. 2.7). (For further details see Sec. 2.5.)
2. The cement-to-aggregate ratio.
3. The aggregate grading curve, including the ratio of coarse to fine aggregate, maximum aggregate size, fineness modulus, etc.

The cement-to-aggregate ratio, or for a given w/c ratio the cement paste-to-aggregate ratio, influences the workability of the mix. This is measured by the *slump test* (Fig. 2.8). A standard conical metal form is filled with fresh concrete, following a

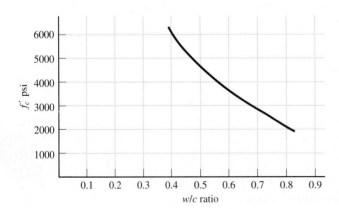

Figure 2.7   Typical relationship between concrete strength and water-cement ratio [12].

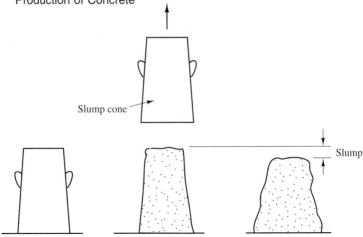

Figure 2.8   Slump test.

standard procedure. The mold is then lifted vertically, and the resulting loss in height of the concrete cone, or slump value, can be determined. This is a measure of workability. Low-slump concrete mixes require extensive labor to produce high-quality concrete and are therefore not very popular on construction sites. Increasing the paste-aggregate ratio would improve the situation (at an increase in cost and with other negative consequences of *rich* mixes), but this is not practical at a job site where concrete trucks may be waiting to make their deliveries. The time-honored "solution" of workers is to simply add water when the site engineer is looking the other way. This improves the flow of the mix and makes it easier to work with, but it also raises the w/c ratio and, according to Fig. 2.7, lowers the strength. Thus, this is *not* a solution. The most common way to resolve the old dilemma of strength versus workability is the addition of a superplasticizer, which increases the workability of the mix without lowering its strength. Such additives are relatively expensive but may be preferable over alternatives, such as using rich mixes with increased shrinkage and heat of hydration.

  With these preliminaries, we can now present the mix design procedure, which has been recommended by ACI Committee 211 [12]. As a prerequisite, basic information on the mix ingredients is needed, especially on the coarse and fine aggregate, that is, their grading curves, unit weights, moisture absorption capacities, etc.

Step 1. *Choice of slump.* The slump is often specified, taking into account the equipment available at the site. Alternatively, the values of Table 2.2 are recommended for various types of construction.

Step 2. *Choice of maximum aggregate size.* This choice is controlled by the requirement that no large voids or honeycombs be caused by large aggregate particles getting stuck during placement. This means the maximum aggregate size should not exceed $\frac{1}{5}$ of the narrowest dimension between sides of forms, $\frac{1}{3}$ the thickness of slabs, or $\frac{3}{4}$ of the minimum clear spacing between bars.

TABLE 2.2   RECOMMENDED SLUMP VALUES [12]

| | Slump (in) | |
Types of construction	Maximum	Minimum
Reinforced foundation walls and footings	3	1
Plain footings, caissons, and substructure walls	3	1
Beams and reinforced walls	4	1
Building columns	4	1
Pavements and slabs	3	1
Mass concrete	2	1

Authorized reprint from ACI Committee 211, Report 2.11.1-89, 1989.

Step 3.   *Estimation of mixing water and air content.* Table 2.3 provides estimates of required mixing water and air content for different slumps and nominal maximum sizes of aggregate with and without air entrainment.

Step 4.   *Selection of water-cement ratio.* Since strength does not depend solely on the w/c ratio, the relationship suggested in Table 2.4 (compare this with Fig. 2.7) is only approximate, aside from the fact that concerns about the concrete's durability also influence the choice of the w/c ratio.

Step 5.   *Calculation of cement content.* The amount of cement per unit volume of concrete is fixed by the determinations made in Steps 3 and 4. It is equal to the mixing-water content (Step 3) divided by the w/c ratio (Step 4).

Step 6.   *Estimation of coarse aggregate content.* The volume of coarse aggregate per unit volume of concrete is selected from Table 2.5.

Step 7.   *Estimation of fine aggregate content.* If the unit weight of the concrete is known (from experience or using Table 2.6), the required weight of fine aggregate is simply the difference between the weight of fresh concrete and the total weight of the other ingredients. Alternatively, the total volume displaced by the water, air, cement, and coarse aggregate is subtracted from the unit volume of concrete to obtain the required volume of fine aggregate. This is then multiplied with the unit weight to yield the weight of the fine aggregate.

Step 8.   *Adjustments for aggregate moisture.* The mixing water added to the batch must be reduced by an amount equal to the free moisture contributed by the aggregate. Conversely, if the aggregate is very dry, an amount of water equal to that lost to absorption must be added to the mixing water.

Step 9.   *Trial batch adjustments.* The calculated mix proportions should be checked by means of trial batches for unit weight and yield, air content, and workability, and adjusted if necessary.

TABLE 2.3 APPROXIMATE MIXING WATER AND AIR CONTENT
REQUIREMENTS FOR DIFFERENT SLUMPS AND AGGREGATE SIZES [12]

Water, lb/yd^3 of concrete for indicated nominal maximum sizes of aggregate

Slump (in)	3/8	1/2	3/4	1	$1\frac{1}{2}$	2	3	6
Non-air-entrained concrete								
1 to 2	350	335	315	300	275	260	220	190
3 to 4	385	365	340	325	300	285	245	210
6 to 7	410	385	360	340	315	300	270	—
More than 7	—	—	—	—	—	—	—	—
Approximate amount of entrapped air in non-air-entrained concrete (%)	3	2.5	2	1.5	1	0.5	0.3	0.2
Air-entrained concrete								
1 to 2	305	295	280	270	250	240	205	180
3 to 4	340	325	305	295	275	265	225	200
6 to 7	365	345	325	310	290	280	260	—
More than 7	—	—	—	—	—	—	—	—
Recommended averages total air content, % for level of exposure:								
Mild exposure	4.5	4.0	3.5	3.0	2.5	2.0	1.5	1.0
Moderate exposure	6.0	5.5	5.0	4.5	4.5	4.0	3.5	3.0
Severe exposure	7.5	7.0	6.0	6.0	5.5	5.0	4.5	4.0

Authorized reprint from ACI Committee 211, Report 2.11.1-89, 1989.

TABLE 2.4 RELATIONSHIP BETWEEN *w-c* RATIO AND
COMPRESSIVE STRENGTH [12]

Compressive strength at 28 days (psi)	Water-cement ratio (by weight)	
	Non-air-entrained concrete	Air-entrained concrete
6000	0.41	—
5000	0.48	0.40
4000	0.57	0.48
3000	0.68	0.59
2000	0.82	0.74

Authorized reprint from ACI Committee 211, Report 2.11.1-
89, 1989.

TABLE 2.5 VOLUME OF COARSE AGGREGATE
PER UNIT OF VOLUME OF CONCRETE [12]

Nominal maximum size of aggregate (in)	Volume of oven-dry-rodded coarse aggregate (per unit volume of concrete for different fineness moduli of fine aggregate)			
	2.40	2.60	2.80	3.00
$\frac{3}{8}$	0.50	0.48	0.46	0.44
$\frac{1}{2}$	0.59	0.57	0.55	0.53
$\frac{3}{4}$	0.66	0.64	0.62	0.60
1	0.71	0.69	0.67	0.65
$1\frac{1}{2}$	0.75	0.73	0.71	0.69
2	0.78	0.76	0.74	0.72
3	0.82	0.80	0.78	0.76
6	0.87	0.85	0.83	0.81

Authorized reprint from ACI Committee 211, Report 2.11.1-89, 1989.

TABLE 2.6 ESTIMATE OF WEIGHT OF FRESH CONCRETE [12]

Nominal maximum size of aggregate (in)	First estimate of concrete weight (lb/yd^3)	
	Non-air-entrained concrete	Air-entrained concrete
$\frac{3}{8}$	3840	3710
$\frac{1}{2}$	3890	3760
$\frac{3}{4}$	3960	3840
1	4010	3850
$1\frac{1}{2}$	4070	3910
2	4120	3950
3	4200	4040
6	4260	4110

Authorized reprint from ACI Committee 211, Report 2.11.1-89, 1989.

EXAMPLE 2.1 [12]

Design a concrete mix with a compressive strength of 3500 psi for a structure not exposed to severe weathering or sulfate attack. Given: Type I cement; coarse aggregate with a weight of 100 lb/ft^3, 2% total moisture and 0.5% water absorption; fine aggregate with 6% total moisture and 0.7% absorption; fineness modulus = 2.8; maximum aggregate size $1\frac{1}{2}$ in; required slump 3 to 4 in.

Solution

Step 1. Slump is specified as 3 to 4 in.

Step 2. Locally available aggregate with $1\frac{1}{2}$ in maximum size is indicated as suitable.

Step 3. Since the concrete will not be exposed to severe weathering, no air entrainment is necessary. The approximate amount of mixing water follows from Table 2.3 as 300 lb/yd^3. The same table shows the approximate amount of entrapped air to be 1%.

Step 4. The w/c ratio for non-air-entrained concrete with 3500 psi strength, according to Table 2.4, is about 0.62.

Step 5. The required cement content is found to be 300/0.62 = 484 lb/yd^3.

Step 6. The volumetric ratio for coarse aggregate, for a fineness modulus of 2.80 and $1\frac{1}{2}$ in maximum size, is 0.71 according to Table 2.5; this makes (0.71)(27) = 19.17 ft^3/yd^3, or (100)(19.17) = 1917 lb/yd^3 dry weight of coarse aggregate.

Step 7. With the quantities of water, cement, and coarse aggregate established, the required fine aggregate can be determined on the basis of weight. Estimating the concrete weight from Table 2.6 as 4070 lb/yd^3, we find: weight of fine aggregate (dry) = 4070 lb for concrete − 300 lb water − 484 lb cement − 1917 lb coarse aggregate = 1369 lb. The amount of absorbed water is neglected here.

Step 8. Coarse aggregate weight (wet) = (1.02)(1917) = 1955 lb

Fine aggregate weight (wet) = (1.06)(1369) = 1451 lb

Surface water contributed by coarse aggregate = 2 − 0.5 = 1.5%

Surface water contributed by fine aggregate = 6 − 0.7 = 5.3%

These amounts need to be subtracted from the mixing water to be added, that is, added water = 300 − (1917)(0.015) − (1369)(0.053) = 199 lb.

Estimated batch weights for one yd^3 of concrete:

Water to be added	= 199
Cement	= 484
Coarse aggregate (wet)	= 1955
Fine aggregate (wet)	= <u>1451</u>
Total	= 4089 lb/yd^3

Step 9. To produce 0.81 ft^3 of concrete for a trial batch, the weights are scaled down by a factor 0.81/27: cement = (484)(.81/27) = 14.52 lb; similarly, coarse aggregate (wet) = 58.65 lb (dry = 57.51 lb); fine aggregate (wet) = 43.53 lb (dry = 41.07 lb); but instead of (199)(.81/27) = 5.97 lb water, 7.0 lb was added in the trial batch. These figures add up to a total of 123.7 lb and a measured unit weight of 149.0 lb/ft^3. Thus, the trial batch yielded 123.7/149.0 = 0.83 ft^3 of concrete, with a measured slump of 2 in.

   The mixing water content was 7.0 lb (added) + 0.86 lb (1.5% of the 57.51 lb of dry coarse aggregate) + 2.18 lb (5.3% of the 41.07 lb of fine aggregate) = 10.04 lb. For 1 yd^3, this corresponds to (10.04)(27/0.83) = 327 lb mix water. This needs to be increased by 15 lb, because the slump was 1.5 in less than required (add 10 lb for each inch by which the slump falls short), that is, increase the mix water to 342 lb; and to keep the w/c ratio constant, required cement = 342/0.62 = 552 lb. Coarse aggregate = (58.65)(27/.83) = 1908 lb wet, or 1908/1.02 = 1871 lb dry, or (1871)(1.005) = 1880 lb saturated-surface-dry.

Total concrete weight = (149.0)(27) = 4023 lb; fine aggregate = 4023 lb (total) − 342 lb water − 552 lb cement − 1880 lb coarse aggregate saturated-surface-dry = 1249 lb saturated-surface-dry, or 1249/1.007 = 1240 lb dry.

Adjusted basic batch weights per yd^3 of concrete:

Water (net mixing)	= 342
Cement	= 552
Coarse aggregate (dry)	= 1871
Fine aggregate (dry)	= 1240
Total	= 4005 lb

Placing, compacting, and curing.   It is obvious that the concrete must be mixed well and placed carefully into the forms to avoid segregation. If dropped too far, heavy or big aggregate particles tend to settle down and lighter components such as water tend to rise. Just like the aggregate, forms and reinforcing bars should be clean and free from organic matter. As an exception, the formwork surfaces can be deliberately coated with a thin film of oil (release agent) to break the bond, allowing for easy removal of the formwork after the concrete has hardened.

During placement, large amounts of air get entrapped; this lowers the concrete strength unless the air is removed by compaction, which is achieved by vibration. High-frequency vibrators are either immersed into the fresh concrete or attached to the outside faces of the formwork. Care must be taken to ensure that the concrete does not vibrate excessively, lest the heavy aggregate settles down and the light mixing water rises to the surface.

Once the concrete has been placed and compacted, it is critical that none of the mixing water needed for cement hydration be lost. This is the objective of *curing*. For example, in hot or dry weather, large exposed surfaces will lose water by evaporation. This can be avoided by covering such surfaces with sheets of plastic or canvas, or period-ically spraying them with water [13]. High temperatures per se are not harmful, but high temperature gradients are (see Fig. 2.2). If this is a concern, mixing water can be pro-vided in the form of crushed ice. In precast concrete plants, concrete elements are often steam-cured, because the simultaneous application of hot steam and pressure accelerates the hydration process, which permits high turnover rates for the formwork installations.

If freshly poured concrete is frozen, the free water forms ice, and the increasing volume damages the young and weak concrete's internal structure. Therefore, concrete should not be produced at temperatures below 40°F, unless special precautions are taken, such as heating the mixing water and/or the aggregate, or enclosing and heating the space in which the concreting takes place [14].

Sometimes it helps to use air-entraining agents to produce several percent of air in the form of small bubbles. This lowers the strength of the concrete somewhat but increases its resistance to both seawater and freeze-thaw cycling. We shall return to questions of concrete durability in Sec. 2.6.

Quality control.   The quality of concrete determines how well it satisfies a number of requirements. First to come to mind is the compressive strength, but equally important is its durability. Both depend on numerous factors, not the least of which are the skill and care of the workers who produce the concrete. Many durability properties are directly related to strength. Thus, the single most important indicator of quality is the compressive strength, which is measured by testing standard cylinders of 6-in diameter and 12-in height, 28 days after mixing. This standard compressive strength is designated as f'_c. To assure proper quality control of the concrete, it is necessary to monitor its strength on a continual basis. The ACI Code [8] requires that samples consisting of two cylinders each be tested at least

a) once every day a given class of concrete is placed, but not less than

b) once for each 150 yd^3 of concrete placed, or

c) once for each 5000 ft^2 of slab or wall surface area.

Because of numerous influences, the sample strengths exhibit a significant amount of statistical scatter (Fig. 2.9). For this reason it is not enough to aim for a mean value equal to the required design strength f'_c, because half of the samples would have a strength less than f'_c. The mean strength should be higher than f'_c. Adequate strength can still be assured by requiring that only a small number of samples fall below a given minimum value, such as f'_c. This number of rejects is represented by the shaded area under the distribution curve of Fig. 2.9. The ACI Code specifies this requirement in the form of the following two criteria:

$$f'_{cr} = f'_c + 1.34s \qquad\qquad (2.7)[5\text{-}1]^*$$

$$f'_{cr} = f'_c + 2.33s - 500 \qquad\qquad (2.8)[5\text{-}2]$$

where f'_{cr} is the average strength, and s is the standard deviation, defined as

$$s = \sqrt{\sum_{i=1}^{n}(f'_{ci} - f'_{cr})^2/(n-1)} \qquad\qquad (2.9)$$

with f'_{ci} = measured strength of sample i, and n = number of samples. Equation 2.7 assures that the probability of the average of three consecutive sample strengths falling

---

* Numbers in brackets refer to the equation numbers in the ACI Code [8].

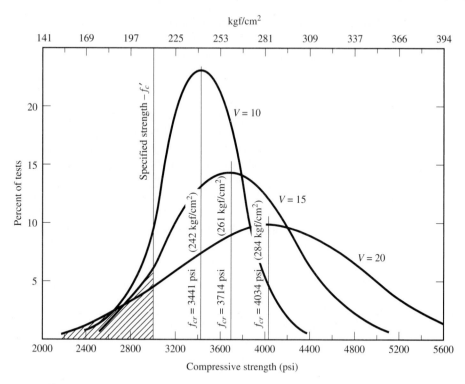

Figure 2.9 Effect of quality control on strength distribution curve [15]. (Authorized reprint from ACI Committee 214, Report 214.3R-88, 1988.)

below f'_c is less than 1 in 100. Similarly, Eq. 2.8 provides the same probability for an individual sample falling 500 psi below f'_c. Good quality control cannot guarantee that an individual test specimen meets the design strength, it can only assure that the probability for this to happen is sufficiently small. In addition, quality control makes economical sense, as illustrated in Fig. 2.9. For tighter quality control, the strength data tend to cluster around the mean with a smaller standard deviation s. Thus, for a prescribed maximum number of rejects, better quality control results in a smaller s, and a reduced necessary "overstrength," $(f'_{cr} - f'_c)$. This apparent conservatism is also needed because the actual strength of concrete in a structure (as measured with drilled core samples) is lower than that determined by control cylinders.

If the tests fall below target, that is, the concrete turns out to have insufficient strength, consequences can be dire. When the specimens are tested, 28 days after the concrete has been placed, a good amount of construction may have since been added, and the removal of substandard concrete can become a very costly proposition. For this reason, quality control does pay. This means tight supervision, inspection, field testing, and record keeping by competent engineers during the mixing, placing, and curing of the concrete.

2.5 MECHANICAL PROPERTIES OF CONCRETE

2.5.1 General Behavior in Compression

The mechanical behavior of concrete is largely controlled by the evolution of cracks, which start out as microcracks invisible to the eye and grow until they become large, open macrocracks. Microcracks and cracklike voids typically exist even prior to the application of any external load and are found mostly at the interfaces between coarse aggregate particles and the mortar matrix. There are several causes for these cracks; for example, differences in early volume changes between aggregate and the hardening cement paste, or the settlement of coarse aggregate particles during the placing process. Under monotonically applied uniaxial compression, several distinct phases can be observed (Fig. 2.10). Up to a stress of about 30% of f_c', the extent of cracking is very limited. As a result, the stress-strain response is almost linear and basically elastic. Under stresses between $0.3 f_c'$ and about $0.75 f_c'$, most new cracks are still limited to the aggregate-matrix interface, and the growth of such bond cracks is accompanied by a gradual softening of the stress-strain response. From $0.75 f_c'$ to about $0.9 f_c'$, cracks begin to extend noticeably into the mortar matrix and to form continuous patterns by connecting separate bond cracks along the larger aggregate particles. At this point, strain increases rapidly with increasing stress. Beyond the 90% load level, crack growth becomes unstable and causes rapidly increasing inelastic deformations, with failure strains between about 0.002 and 0.003. This final phase is accompanied by significant volume expansion. The actual material volume does not increase, of course, but the wholesale cracking and opening of relatively large voids within the material create that appearance.

Under a purely uniaxial state of compression, the failure mode is one of splitting (Fig. 2.11a). The material, before being crushed, develops large tensile cracks parallel to the direction of load application. This fact can be explained in a simplified manner: the stiff aggregate particles can be thought of as exerting a wedging action on the softer mortar matrix between them (Fig. 2.11b). However, before such

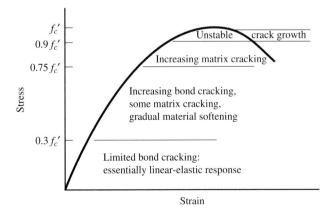

Figure 2.10 Phases of uniaxial stress-strain response.

(a)

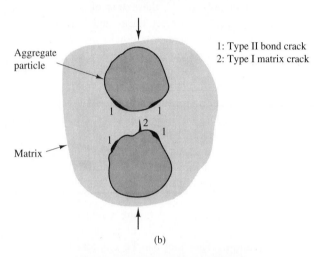

Aggregate
particle

Matrix

1: Type II bond crack
2: Type I matrix crack

(b)

(c)

Figure 2.11 Splitting failure under
uniaxial compression: (a) splitting failure
of cube (friction at ends eliminated);
(b) wedging action of coarse aggregate
particles; (c) shear failure of cube (friction
at ends not eliminated).

wedging action can be activated, the bond between aggregate particles and the mortar matrix needs to be broken and the opposing faces of the resulting cracks slide along each other. Then, these cracks can propagate away from the interface into the matrix. In fracture mechanics terminology, cracks that open up in a plane normal to the direction of a tensile stress are called *type I*, whereas cracks that don't quite open but have their opposing faces slide along each other are referred to as *type II* cracks.

During the standard test of a concrete cylinder, the frictional stresses between the top and bottom cylinder faces and the loading platens of the testing machine prevent the material from expanding laterally, thereby exerting a confining effect. As a result, such a cylinder does not fail by splitting but rather in shear, with fairly intact cones at the top and bottom (or pyramids in the case of cubes, as in Fig. 2.11c). Because of this restraint, the strength as measured in such a standard test is noticeably higher than that of a specimen in which the end friction is eliminated to produce a pure uniaxial state of stress.

The frictional stresses present in rough cracks of both type I and II give the material a limited amount of ductility, so that the stress-strain curve, after reaching the peak, does not drop abruptly. Still, a considerable amount of energy is stored in the material at this point, and once failure is initiated, cracking propagates very rapidly, especially in a load-controlled experiment (as opposed to a strain-controlled test), and failure can be very explosive. For higher-strength concretes, the ductility is even less, the post-peak stress-strain response steeper, and the failure more violent (Fig. 2.12).

Because of the highly nonlinear nature of the stress-strain response of concrete, the definition of a modulus of elasticity is by necessity arbitrary. Yet, as long as the material is not stressed beyond $0.5 f_c'$ or the level typically associated with service loads, the deviation from linearity is not serious, and the use of an average value is justified for practical purposes. The following commonly used empirical expression [8] relates the modulus of elasticity, E_c, to the strength, f_c', and the unit weight of the concrete, w_c.

$$E_c = w_c^{1.5} 33 \sqrt{f_c'} \qquad (2.10)$$

For normal-weight concrete weighing $w_c = 145$ pcf, Eq. 2.10 simplifies to

$$E_c = 57,000 \sqrt{f_c'} \qquad (2.11)$$

For example, for 3000 psi concrete, we find $E_c = 3,122,000$ psi, approximately one-tenth of the value for steel. It should be noted that the actual modulus of elasticity of concrete is a function of numerous other factors (mostly the type of aggregate) and can easily deviate 20% or more from the value computed with Eq. 2.10 or Eq. 2.11. In particular, these formulas become unreliable for concrete strengths in excess of 6000 or 7000 psi.

Poisson's ratio, the other elasticity constant used in solid mechanics, typically varies between 0.15 and 0.20, with 0.17 being a useful average value.

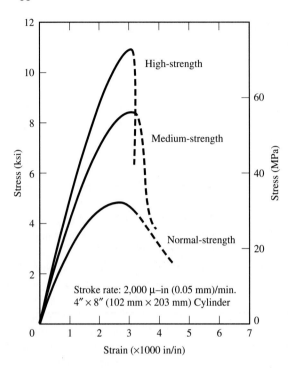

Figure 2.12 Stress-strain curves for different strength concretes [16]. (Authorized reprint from *ACI Journal*, vol. 78, 1981.)

2.5.2 Compressive Strength

Compressive strength, while not the only important property of concrete, is usually the most important. It depends on the properties of the constituent materials, the method of production and curing, and the method of testing. This last factor raises the puzzling question: if strength is dependent on the way we are measuring it, what, then, is real or intrinsic strength? If indeed there is something like intrinsic strength, we do not know how to measure it. We can resort only to a (rather arbitrary) standard test [17]: load a 6 × 12 in cylinder (because strength depends on the size and shape of the specimen), 28 days after casting (because concrete age is a major variable), at a rate between 20 and 50 psi/sec (because the loading rate also influences the strength) up to failure. The result of such a test, designated as standard cylinder strength f_c', provides information useful primarily for comparative purposes. Large differences between the results of standardized tests of laboratory specimens and tests performed on in-situ concrete are to be expected. In spite of this shortcoming, standard tests constitute a useful method of quality control: they verify that the concrete was properly batched and mixed, they can pinpoint a problem if substandard materials have been used, and they permit the determination of statistical scatter as a measure of quality control. Also, if the workers have evidence that their concrete is regularly tested, they are more likely to work carefully. Finally, the tests are instrumental in determining the time when formwork may be removed or prestress forces applied.

Since hydration is a continual process, concrete strength increases as a function of time, as shown in Fig. 2.13a. This graph illustrates the arbitrariness of the 28-day mark for defining the standard strength and also indicates that the strength increases almost indefinitely, albeit asymptotically.

For construction efficiency it is very important that data on concrete strength be available as early as possible. A 28-day waiting period may be unacceptable. For this purpose the concept of maturity (see Sec. 2.1) has been shown to be useful. An example of a strength-maturity relationship is [6],

$$S = \frac{M - M_0}{\frac{1}{A} + \frac{(M - M_0)}{S_\infty}} \tag{2.12}$$

where M = maturity [e.g., $M = t(T - T_0)$]; M_0 = offset maturity, prior to which strength gain is negligible; S_∞ = limiting strength; and A = initial slope of the strength-maturity curve, which is a function of the w/c ratio, type of cement, and other factors (Fig. 2.13b). An analytical expression such as Eq. 2.12 allows the estimation of the 28-day strength on the basis of 7-day or even 1-day strength data. The ratio between 28-day and 7-day strengths generally varies between 1.3 and 1.7. Obviously such a function depends on the type of cement; more specifically on the C_3S/C_2S ratio, fineness, and the method of curing. As an alternative, accelerated tests may be performed in which the concrete is cured under high-temperature (and possibly high-pressure) conditions to accelerate the cement hydration [18].

The single most important factor affecting the concrete strength is the w/c ratio (Fig. 2.7), or more precisely, the (water + air)/cement ratio, since the pores do not carry load, only the solids do. The inverse of porosity is the gel-to-space ratio, that is [1],

$$X = \frac{\text{solid products}}{\text{space available for these products}} = \frac{0.68\alpha}{0.32\alpha + w/c} \tag{2.13}$$

where α = degree of hydration. For example, for $w/c = 0.42$ and $\alpha = 1.0$, we get

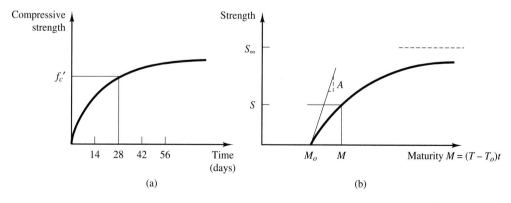

Figure 2.13   Strength gain and maturity.

$X = 0.92$ (8% porosity). If air is present (volume A), the w/c ratio should be replaced by $(w + A)/c$, and the strength can be determined approximately with the expression

$$S = S_i X^n \tag{2.14}$$

where S_i = intrinsic strength for a gel/space ratio of $X = 1.0$, and $n = 2.6$ to 3.0. The increase in strength is directly proportional to the increase in the gel/space ratio, regardless of age, original w/c ratio, or type of cement.

The strength of aggregate is important mostly for lightweight and high-strength concretes, because in regular concrete the aggregate strength is typically much higher than that of the cement paste, which in turn exceeds the bond strength between aggregate and cement paste. The aggregate shape and surface texture affect the concrete strength, because the surface texture influences the bond strength and the stress level at which microcracking begins. The aggregate size has only a slight effect on strength.

Standard test specimens in the United States are cylinders of 6-in diameter and 12-in height. In Europe, cubic specimens are often used. Strength test results are quite different. A cube exhibits a strength approximately 25% higher than that of a cylinder made of the same concrete. This fact can be explained by the friction forces between test specimen and loading platens mentioned earlier. In a cylinder with diameter $d = 6$ in, cones of about $d\sqrt{3}/2 = 5.2$ in height develop, so as long as the cylinder height h is at least twice that much, free lateral expansion of the concrete between the cones is possible. In a cube, no such unconstrained region exists between the top and bottom pyramids (see Fig. 2.11c). The end restraint has a confining effect that delays the development of cracks within the end cones or pyramids and therefore raises the strength over that of a specimen subjected to a purely uniaxial stress. A cylinder with $d/h = 1$ fails at a load 10% higher than one with $d/h = 0.5$.

The specimen size also affects the strength. A cylinder with more than an 18-in diameter fails at about 85% of the stress at which a standard size cylinder fails, partly because of the increased probability of the presence of weak spots. Also, smaller specimens are likely to be better compacted, (i.e., the concrete may indeed be stronger). Recent fracture mechanics theories can also explain the size effect on the basis of strain localization in the fracture zone [19].

The dependence of strength on the loading rate, also referred to as the strain-rate effect, is the reason for the prescribed loading rate in the standard test (Fig. 2.14). At high strain rates, propagating cracks do not have time to choose the path of least resistance and may have to pass through stronger or tougher aggregate particles rather than pass around them. At strain rates associated with traffic, wind, and earthquake-type loads, concrete strength appears to be 10 to 20% higher than the standard or "static" strength. Under blast and impact loading, the strength may even be twice as high.

At very slow strain rates, concrete strength is only about 75% of its short-term value. This *static fatigue strength* approximately correlates with the *discontinuity stress* level, at which microcracks start to depart from the aggregate-paste interface and propagate into the paste. At stresses higher than that, creep can cause isolated bond cracks to link up under sustained load and eventually cause failure. Failure strains due to such

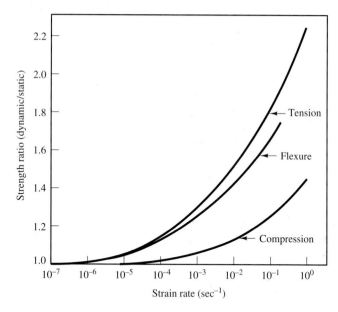

Figure 2.14   Strain-rate effect on compressive, tensile, and flexural strength [20]. (From W. Suaris and S. P. Shah, "Constitutive Model for Dynamic Loading of Concrete," *ASCE, J. of Struct. Eng.*, vol. 111, no. 3, March 1985. Reprinted by permission of The American Society of Civil Engineers.)

loading are larger than those caused by short-term loads, because more cracking and creep strains have time to develop. At lower sustained stresses, cracking that does occur may be compensated for by consolidation of concrete and its strengthening with time.

Finally, it should be pointed out that the common practice of producing concrete results in a slightly anisotropic material. During vibration, some air and free water rise and may get trapped below large aggregate particles. As a result, the compressive strength of concrete is larger and strains smaller in the direction of casting than in a direction normal to it.

2.5.3 Tensile Strength

The response of concrete to tensile loading is completely different than that to compression and can be characterized as brittle, that is, cracking is initiated without large prior plastic strains. As with compression, damage starts in the form of microcracks that grow around aggregate particles. However, once the cracks propagate into the matrix, macrocracks form perpendicular to the direction of load and quickly lead to failure—with much smaller strains compared to those observed in compression and hardly any ductility at all.

The tensile strength of concrete, f_t, is much smaller than its compressive strength. The f_t/f_c' ratio is of the order 0.1 and decreases with age. Because of its considerable statistical scatter, the tensile strength is unreliable and usually explicitly neglected in the design of concrete structures. Yet it is very important for the following reasons.

1. Because concrete in pure compression fails by splitting, the compressive strength is indirectly a function of the tensile strength.
2. The tensile strength controls the cracking and serviceability of concrete structures.
3. The stiffness of a reinforced concrete member depends to a large extent on the concrete's tensile strength.
4. The shear strength of concrete is directly proportional to its tensile strength.
5. The bond strength between steel reinforcement and the surrounding concrete is also largely proportional to the tensile strength.
6. In prestressed concrete design, the concrete's tensile strength is at times explicitly relied upon to resist load.

In sum, reinforced concrete design depends indirectly on the tensile strength of concrete. Reinforced concrete "works" because of its tensile strength, whether or not we consider it in design.

The tensile strength, f_t, can be measured directly using bone-shaped specimens (Fig. 2.15a). However, stress concentrations near the grips of the testing machine may cause premature failure, rendering the results unreliable. As an alternative, steel plates are glued directly to the ends of a prismatic specimen (Fig. 2.15b). Such tests are time-consuming and difficult to perform and therefore the tensile strength of concrete is usually measured by one of two other methods.

In the flexure test, a small prismatic beam with square cross section is subjected to a concentrated midspan load (Fig. 2.15c). The maximum flexural stress at failure is called *modulus of rupture* and can be determined easily from the failure load P,

$$f_r = \frac{PL}{4} \frac{6}{h^3} \tag{2.15}$$

where $L =$ span length and $h =$ size of the square cross section. f_r can be as much as 50% higher than f_t, because the maximum stress occurs only along the line "a," whereas in the direct tension test, each material point in the entire specimen is stressed to the same maximum value. The probability of encountering a weak spot or a large initial microcrack along line "a" is much smaller than that of encountering one within the entire volume of the specimen. Also, the stress gradient makes load sharing between neighboring material points possible. A widely used empirical relationship between the modulus of rupture and the compressive strength is

$$f_r = 7.5\sqrt{f_c'} \tag{2.16}$$

A second test for measuring the tensile strength is the *split cylinder test*, also known as the *Brazilian test*. Here a standard cylinder is subjected to longitudinal line loads (Fig. 2.15d), which except for small wedges under the loading plates, produce along the

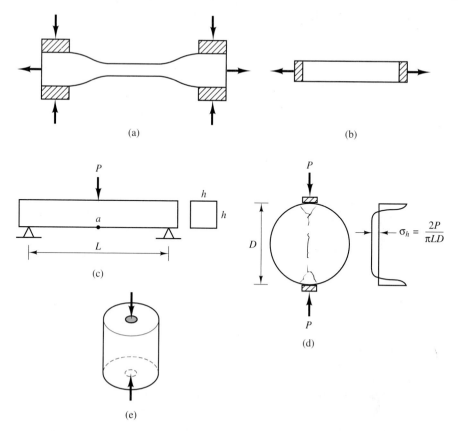

Figure 2.15 Tests to measure tensile strength: (a) direct tension test; (b) direct tension test; (c) flexure test; (d) split cylinder test; (e) double punch test.

center plane almost uniform horizontal tensile stresses of the magnitude

$$\sigma_h = \frac{2P}{\pi L D} \qquad (2.17)$$

where P = applied force, D = cylinder diameter (6 in), and L = cylinder length (12 in). The tensile strength measured in this test is considerably lower than f_r, because an entire vertical plane (as opposed to a single line) is subjected to the maximum tensile stress. Thus, the probability of encountering a weak spot is an order of magnitude larger than in the flexure test, but not as large as in the direct tension test. In addition, the compressive stress normal to the tensile stress lowers the tensile strength, as will be explained further below.

The split cylinder test concept can be extended by subjecting a standard cylinder to two near-concentrated loads (Fig. 2.15e). In this *double punch test* [21], failure also occurs by splitting. However, this time the failure plane can be any vertical plane, so that all material points, except for the small end regions, are subjected to the same maximum stress. Thus, the tensile stress measured by this test is very close to f_t.

2.5.4 Fatigue Behavior

A material can fail at a stress level below the static strength if the stress is applied repeatedly. This phenomenon is called *fatigue*. Each load application increases the amount of cracking or internal damage until the material can no longer resist load. The most common representation of fatigue behavior is the so-called *S-N* curve, which depicts the total number of load cycles to failure, N, as a function of the maximum stress level, S. Because N increases exponentially as S decreases, it is common to plot S as a function of log N (Fig. 2.16). Most metals exhibit an endurance limit, which is the maximum stress level that the material can withstand indefinitely. Such *S-N* curves are bilinear, with the second branch having zero slope (Fig. 2.16a).

Concrete does not seem to have such an endurance limit, that is, no matter how small the stress level, failure can be induced if the load is applied often enough. The *S-N* curve for concrete is also bilinear, except for a small transition range, but the second branch has a nonzero slope (Fig. 2.16b) [22]. The two branches can be associated with low- and high-cycle fatigue. Their point of intersection seems to correlate with the discontinuity stress, or the sustained strength limit of about $0.7 f_c'$ to $0.75 f_c'$.

A typical load-deformation diagram is shown in Fig. 2.17a for an experiment in which each maximum strain or deformation exceeds that of the previous load cycle. Note that the envelope to this cyclic response curve closely resembles the monotonic

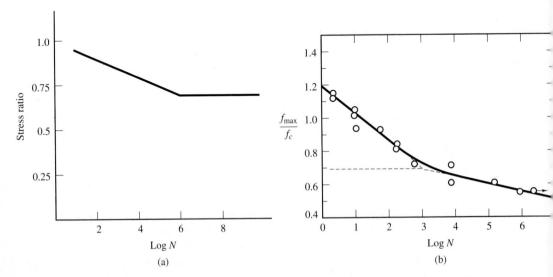

Figure 2.16 *S-N* curves for steel and concrete: (a) steel; (b) concrete [22] (authorized reprint from *ACI Materials Journal*, vol. 85, 1988).

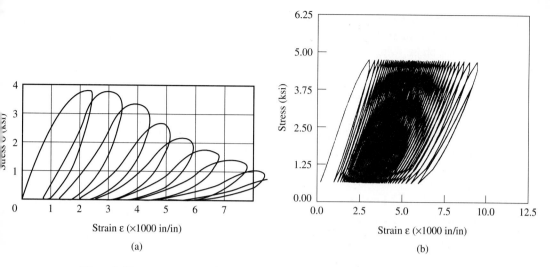

Figure 2.17 Response of concrete to cyclic load: (a) increasing strain amplitude [23] (authorized reprint from *ACI Journal*, vol. 61, 1964); (b) constant stress amplitude [24].

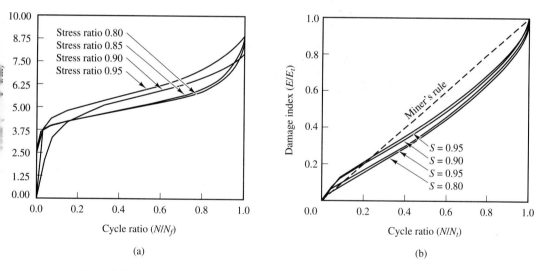

Figure 2.18 Damage accumulation in concrete [24]: (a) total strain accumulation; (b) dissipated energy.

stress-strain relationship [23]. The response curve of Fig. 2.17b was obtained from a test where the maximum load amplitude was kept constant throughout, and the strains increased as a result [24]. Plotting the total strain accumulation as a function of the number of load cycles results in the graph of Fig. 2.18a, with its three distinct phases of the material's fatigue life. The first phase, representing possibly 10% of the fatigue life, appears to be characterized by a relatively rapid increase of damage; this is deceptive, however. Most of this strain is elastic and recoverable. Also, the material undergoes a period of adaptation to the cyclic load, including some kind of consolidation. In the second phase, which covers almost 80% of the material's life, we note an almost steady-state increase in damage. In the third and final phase, damage accumulation accelerates all the way up to failure.

If, instead of strain accumulation, the dissipated energy (as defined by the area under the stress-strain diagram) is used as a measure of damage, then the test data of Fig. 2.18a appear as shown in Fig. 2.18b. Here the three phases of fatigue life are no longer clearly distinct; the first phase all but disappears. The results also indicate that the assumption of linear damage accumulation, which underlies the so-called Miner's Rule, does not quite apply.

2.5.5 Influence of State of Stress

From what we now know about the micromechanics basis for failure of concrete, it is easy to predict what happens if a compressive stress σ_1 is accompanied by an orthogonally applied compressive stress σ_2: the splitting cracks that would normally lead to failure are now being overstressed by the added stress component, and failure is delayed. The result is a considerable increase in ductility and strength. Figure 2.19 shows the resulting stress-strain relationships and strength envelope [25]. The strengths are quite different in the three different quadrants of the σ_1–σ_2 plane: In the

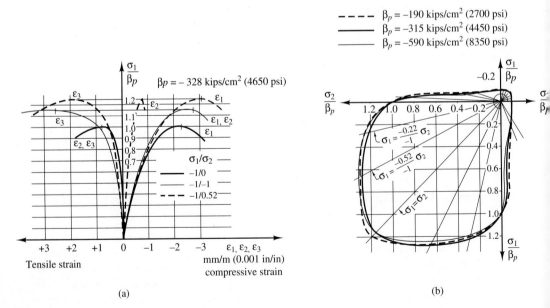

Figure 2.19 Constitutive behavior and failure envelope for biaxial stress states: (a) stress-strain response; (b) biaxial strength. (Authorized reprint from *ACI Journal*, vol. 66, 1969.)

compression-compression quadrant, a maximum strength increase of about 25% is realized for a stress ratio of $\sigma_1/\sigma_2 \approx 0.5$. In the tension-compression quadrant, the strength envelope is almost linear, that is, the compressive strength drops as a linear function of the tensile stress in the orthogonal direction. In the tension-tension quadrant, there is hardly any interaction, that is, the tensile strength is almost unaffected by tensile stresses applied in the orthogonal direction. Note that the strength envelope of Fig. 2.19b applies only to proportional loading, where the σ_1/σ_2 ratio is kept constant during loading. For nonproportional loading, the damage accumulation is strongly dependent on the history of stress application and failure is more difficult to predict.

Under triaxial states of compression, the concrete experiences an enormous increase in strength and ductility. The failure envelope in three-dimensional stress space is now a surface and is shown in Fig. 2.20a for proportional loading. Under pure hydrostatic pressure, it is not easy to visualize how the material can fail at all. The corresponding stress-strain curve does indeed maintain a positive slope, presumably indefinitely. In reality, however, the microporous structure of the mortar matrix collapses if the pressure is high enough and the concrete consolidates. This manifests itself in a loss of residual uniaxial compressive strength (Fig. 2.20b). For example, after applying a hydrostatic pressure of $6f_c'$, a 25% drop of the residual uniaxial strength has been reported [26].

A triaxial state of compression affects concrete behavior so profoundly that one may be tempted to call concrete under such a state of stress a different material. The basically brittle concrete has become ductile—a fact with considerable consequences for design practice. It is the challenge of the structural designer to reinforce the concrete with steel in such a way that it is *confined*, that is, the reinforcement constricts the concrete's lateral expansion and thereby creates a three-dimensional state of compression. We shall address this task in more detail in later chapters. Here, it suffices to consider the stress-strain curves of Fig. 2.21 to appreciate the profound effect of confinement reinforcement

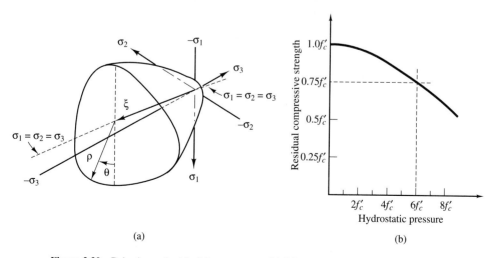

Figure 2.20 Behavior under triaxial stress states: (a) failure surface; (b) residual strength [26].

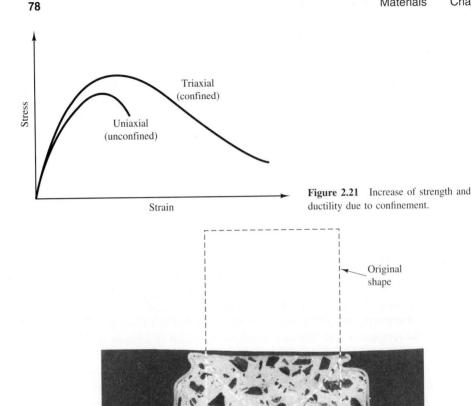

Figure 2.21 Increase of strength and ductility due to confinement.

Figure 2.22 Very large deformations of concrete confined by brass cylinder.

on concrete behavior. For design purposes, it is acceptable to estimate the enhanced strength f_c^* in the 1-direction due to compressive stresses f_h in the 2- and 3-directions by the expression,

$$f_c^* = f_c' + 4.1 f_h \qquad (2.18)$$

Figure 2.22 illustrates a related experiment: a standard sized brass cylinder was filled with concrete and tested 28 days later. The lateral expansion of the concrete was restrained by the brass shell, which introduced a triaxial state of compression. However, whereas in typical cylinder tests maximum strains of the order of 0.003 to 0.005 are achieved, the cylinder of Fig. 2.22 was reduced in height from 12 to 6 inches, that is, it experienced a strain of 1.0! We can assume that the bulk of the material did not undergo these extremely large deformations while remaining intact, but when the specimen was sawed open approximately one year later, the material had regained considerable strength, presumably through self-healing. The two end cones can clearly be noted in the photograph, as well as the flow lines of the material that had traveled several inches. Such behavior is possible under triaxial stress states, and the designer is well-advised to remember the remarkable qualities of well-confined concrete.

2.5.6 Creep

The cement gel has viscoelastic properties that cause deformations of the concrete under sustained load (Fig. 2.23a). Upon loading, the material experiences an immediate or instantaneous elastic deformation. Its gradual increase with time (while stress is constant) is referred to as *creep*. The rate of deformation decreases gradually with time but seems to never completely stop. If strain is held constant, the same viscoelastic nature of the material causes the stress to decrease with time. This is called *relaxation* (Fig. 2.23b). Upon unloading, the elastic deformation is recovered immediately and completely. The creep deformation, on the other hand, is recovered gradually and only partially. The creep deformations of concrete can be a multiple of the elastic deformations. Therefore, they play a significant role in practice, and we need to be familiar with the important factors that influence creep behavior [3].

Stress.   For most practical purposes (i.e., for stresses below about 0.5 f_c'), creep deformations can be assumed to be linearly proportional to applied stress.

Age.   Concrete age or degree of hydration has a very strong effect. Creep deformations of concrete loaded at a very young age can be a multiple of creep deformations of mature concrete. Therefore, it is crucial that formwork not be removed prematurely unless proper shoring is installed.

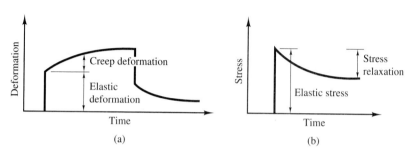

Figure 2.23   Creep and relaxation: (a) creep deformation (constant stress); (b) stress relaxation (constant strain).

Aggregate.   Coarse aggregate has a restraining effect on the creep of the cement paste. Therefore, creep deformations are affected by aggregate content, grading, shape, and maximum size. The elastic modulus of the aggregate also plays a role—lightweight aggregate typically causes higher creep rates.

Water and cement content.   Creep deformations increase almost linearly with the w/c ratio, that is, a high w/c ratio not only lowers the strength but also increases the long-term deformations. Creep is also proportional to cement content, in that rich mixes cause larger long-term deformations.

Humidity.   Creep is dependent on the extent of drying and the amount of evaporable water while the concrete is undergoing creep. Thus, low relative humidity in the surrounding air allows more free water to evaporate than in an atmosphere with high relative humidity. A specimen stored in water exhibits the least amount of creep. The size effect is of the same origin: a massive concrete member or one with a large volume-to-surface ratio will creep less than a small or thin member, where the free water has shorter distances to travel to escape.

Temperature.   Higher temperatures, too, facilitate free water transport and therefore are conducive to creep deformations.

For stresses in the service load range (i.e., those below about $0.5 f_c'$), elastic strains are almost proportional to stress (Fig. 2.12). Since, in this range, creep strains also are proportional to stress, we can define a *creep coefficient*

$$\phi = \frac{\epsilon_{\text{creep}}}{\epsilon_{\text{elastic}}} \tag{2.19}$$

which is a function of the various factors enumerated above. For concrete with strength 3000 to 4000 psi, it is approximately three.

Figure 2.24 shows the strain increase over time for different stress-strength ratios and clearly illustrates that the *static fatigue* mentioned earlier lies somewhere between $0.75 f_c'$ and $0.80 f_c'$ [27]. This is the highest stress level that the material can withstand indefinitely without failure.

In conclusion, it is in order to mention that creep not only has its downside (by increasing deformations) but also its advantages. In time, the viscoelastic action reduces stress concentrations, decreases the significance of secondary stresses and moments in indeterminate structures, and causes a long-term redistribution of loads. Thus, it plays a role for concrete quite similar to what plastic yielding does for steel, except on a different time scale.

2.5.7 Shrinkage

Shrinkage is a process of volume decrease due to the drying of concrete. It is affected by many of the same factors that influence creep, and it is therefore difficult to separate these two phenomena. Shrinkage is nonexistent in concrete stored under water and may reach

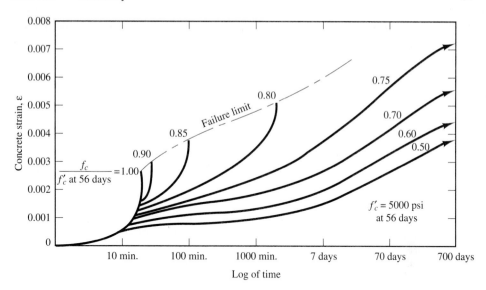

Figure 2.24 Strain-time relationship for different stress-strength ratios [27]. (Authorized reprint from *ACI Journal*, vol. 57, 1960.)

an order of 0.1% for low ambient humidity, high surface-to-volume ratio, rich mix, or high w/c ratio. Since the concrete near free surfaces shrinks more than the material near the center of a member, large tensile stresses can develop at member surfaces. When these exceed the tensile strength of the young concrete, they lead to cracks often seen in pavements, floor slabs, and walls. Light reinforcement is added to keep such cracks small and well-distributed, even if it is not needed for strength.

2.6 DURABILITY

2.6.1 General Comments

The mechanical properties of concrete, especially its strength, have traditionally been the properties of primary concern to engineers. In recent years, however, it has become increasingly obvious that a material's ability to maintain its qualities over long time spans is of equal importance. This property is generally referred to as *durability*. In fully developed countries like the United States, new construction is becoming less and less significant when compared to the requirements of maintaining, repairing, and rehabilitating existing facilities. Proof of this state of affairs can be encountered on a daily basis in the form of a "crumbling infrastructure." Questions of durability and the lifespan extension of structures and facilities are therefore of growing importance, as society must set priorities in allocating limited resources.

What then are the factors that influence the durability of concrete material? First, there is the mechanical effect of repeated load application that can lead to fatigue failure, as discussed earlier. Other mechanical sources of deterioration are surface wear and abrasion, as well as scour and cavitation in hydraulic structures exposed to rapidly flowing water. The damaging effects of repeated freeze-thaw cycles are of primary concern for structures and facilities exposed to harsh climates. Also, there are a number of sources of chemical attack, such as sulfates, salt spray, acids, and interaction with carbon dioxide in the atmosphere. Increasing attention is being directed toward the potential of damaging chemical reactions between the alkalis in portland cement and the silica present in certain types of aggregate. Corrosion of steel reinforcement has serious consequences regarding the serviceability and safety of structures. Moreover, there are other environmental conditions that may affect the durability of materials, such as the high temperatures of fires.

Many of these durability problems are related directly to the pore structure of the cement paste, because this determines the concrete permeability and the transport of water with various dissolved chemicals. The settlement of solid particles after the placing of fresh concrete, especially during vibration, can cause mixing water to rise or move to exterior faces—a process known as *bleeding*. The channels left behind after the evaporation of this water can form a continuous pore and capillary system that makes it easy for aggressive agents to penetrate the concrete and perform their destructive work. The best line of defense against this occurrence is to use a low w/c ratio, because less excess water will be available for bleeding. This is only one advantage; a low w/c ratio results in a stronger and less permeable concrete, which is also more likely to be durable.

Ignoring any one long-term effect can have catastrophic consequences—if not to life and limb, then at least in terms of financial losses. We shall now briefly summarize the main aspects and some of the preventive measures against these causes of durability problems.

2.6.2 Freeze-Thaw Cycling

Damage caused by repeated freezing and drying is part of what is often referred to as *weathering*. Weathering requires that the pore structure within the mortar matrix and/or in the coarse aggregate be saturated with free water. When freezing, this free water expands and exerts hydraulic pressure on the surrounding material, gradually destroying its structure until it literally crumbles. The most common preventive measure is the addition of an air-entraining agent mentioned in Sec. 2.3. This causes the mixing water to foam and form small air bubbles, into which the water can expand harmlessly when freezing. The ACI Code [8] requires a volume of between three and seven percent air entrainment for concrete exposed to freezing and thawing or deicer chemicals.

Concrete is more vulnerable to deterioration near free surfaces, where it is more exposed to wetting and drying cycles compared to the inside of solid members. The most important property associated with this vulnerability is its permeability, because this controls the migration of moisture and the transport of potentially aggressive chemicals. A low w/c ratio or (water + air)/cement ratio not only increases strength but also resistance to freeze-thaw cycling. A hypothetical concrete without any pores would be very strong and free of most types of durability problems. Likewise, a concrete with little or no free water to freeze is less likely to suffer frost damage. An effective sealant that keeps the moisture out of the concrete will preserve such frost resistance.

The choice of coarse aggregates also influences the concrete resistance to freeze-thaw cycles. If porous and saturated with water, individual aggregate particles experience the same problem as the cement matrix, but cannot benefit from the air bubbles produced by an air-entraining agent unless the maximum aggregate size is kept small. In this case, the average travel distance of the absorbed water is reduced and the hydraulic pressure may force enough of it into neighboring air voids to avoid serious deterioration of the material.

2.6.3 Chemical Attack

Sulfates are present in some soils, groundwaters, and seawater and are the result of all kinds of pollution. They penetrate the cement paste through its pore structure and react with the C_3A hydration products to form ettringite. This chemical reaction is accompanied by a considerable volume increase, which results in the cracking and deterioration of the concrete. This problem can be reduced by specifying a cement with low C_3A content, namely, portland cement types II and IV, and especially type V (see Table 2.1), which has been developed specifically for sulfate resistance. Also a high cement content and a low w/c ratio with the resulting low permeability help stem chemical attack. The addition of fly ash and silica fume increases the concrete resistance against sulfate attack for the same reasons, that is, because they reduce the permeability of the matrix.

The salinity and other chemicals present in seawater are less severe than one might think, because the chloride ions largely inhibit sulfate attack. Problems can be more severe in the splash or intertidal zones that are subjected to frequent wetting and drying cycles. The deleterious effects of salts are more serious where corrosion of steel reinforcement can occur (see Sec. 2.9). Fly ash or slag will help protect the concrete against seawater and acid attack. Brucite ($Mg(OH)_2$) tends to protect concrete against ingress of seawater.

Regarding attack by other chemicals, the alkaline nature of the cement paste protects the concrete against bases but not against acids. These are encountered in some groundwater, especially in the vicinity of mine tailings. Acid rain can be of concern, as can various industrial and agricultural operations. Special measures must be taken to protect concrete against such threats.

2.6.4 Alkali-Silica Reaction (ASR)

The alkalis present in the cement paste have various ways in which they chemically interact with reactive forms of silica that are found in certain aggregates, provided the moisture level is sufficiently high to sustain the reaction. These chemical processes can be quite complex and depend very much on the specific minerals within the aggregate. The result can be a volume increase, which may take years to build up before damaging the concrete. Alkali-silica reaction is, therefore, a truly long-term problem. One effective preventive measure is a thorough petrographic evaluation of the coarse aggregate. This includes a series of tests that attempt to simulate in a relatively short time the long-term deleterious effects [28, 29]. If the choice of aggregate is limited and alkali-aggregate reaction appears to be a possibility, the concrete durability can be improved by specifying a low-alkali cement or controlling the moisture content of the concrete. Also, the addition of mineral admixtures such as silica fume and granulated blast-furnace slag have been shown to provide some protection against ASR.

2.7 THERMAL PROPERTIES

The properties of concrete (like the properties of other materials) are functions of temperature. For certain applications, these functional relationships need to be known. If temperature is part of the service load condition, as it is in the case of nuclear reactor containment vessels or various furnace structures, then material properties specific to thermal analysis need to be given as well.

Thermal coefficient of expansion. This coefficient varies with the type of aggregate and moisture content, yet for many concretes it is not too different from that of steel ($\alpha_s \approx 6.5 \times 10^{-6}$ in/in/°F). This is fortunate because, otherwise, daily and seasonal temperature changes could suffice to weaken the bond between concrete and steel reinforcement and to cause extensive concrete cracking and strength deterioration. In fact, limestone-, granite-, and basalt-based aggregates have considerably smaller coefficients than the cement paste, so extensive cracking can be expected on the aggregate-paste interface. Since coarse aggregate constitutes the bulk of concrete, the coefficient of expansion of such a composite is bound to be small, compared with that of steel, so that large temperature variations can cause damage. As a reasonable average, one may assume a value of $\alpha_c = 5.5 \times 10^{-6}$ in/in/°F.

High-temperature effects. The susceptibility of steel to damage due to the high temperatures of a fire is well known and is the cause for the need for fireproofing. Concrete is also subject to high-temperature effects, but to a lesser degree. Because of the relatively low thermal conductivity and large heat capacity of massive members, concrete structures are well suited for fire-resistant construction. The large number of significant variables involved makes it difficult to predict the change of material properties under high temperatures. One quantitative relationship, reproduced from [1], is shown in Fig. 2.25.

Thermal conductivity. This is a measure of the amount of heat flow through a unit area for a unit temperature gradient, and is thus an indicator of the stresses that can develop in response to thermal loads. The conductivity is strongly dependent on material density. This explains why water is a much better heat conductor than air and much worse than most solids. The thermal conductivity of concrete varies from 0.9 to 2.0 Btu/ft/hr/°F (1.5 to 3.5 W/m/K) [1], compared with 0.5 W/m/K for water and 120 W/m/K for steel. The relatively low value for concrete is the reason for its low heat absorption and high fire resistance.

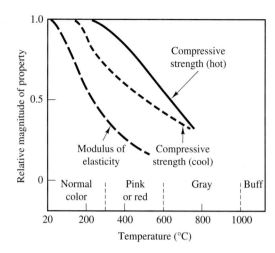

Figure 2.25   Effect of temperature on strength and elastic modulus of concrete. (From S. Mindess and J. F. Young, *Concrete*, 1981. Reprinted with permission of Prentice-Hall, Inc., Englewood Cliffs, NJ.)

Heat capacity.   Also known as specific heat, the heat capacity is a function of porosity, water content, and temperature. For concrete, heat capacity varies between 0.19 and 0.29 Btu/lb-°F (800 and 1200 J/kg-°C). Compared with steel (460 J/kg-°C), concrete can store about twice as much heat, which makes it an important material for the construction of heat sinks.

Diffusivity.   This thermal property is also referred to in heat conduction analyses. It is related to the other thermal properties through the following relationship [30],

$$\kappa = \frac{k}{\rho c} \tag{2.20}$$

where κ = diffusivity, k = thermal conductivity, ρ = mass density, and c = heat capacity.

2.8 SPECIAL CONCRETES

Special applications often call for special concretes. However, normal-type construction can also benefit from certain characteristics of nonstandard mixes.

Lightweight concrete.   As stated in Chapter 1, the heavy weight of concrete structures has its advantages and disadvantages. If we want to reduce the dead weight, we can achieve this by substituting lightweight for the standard stone aggregate. There is an entire spectrum of such aggregates available for various purposes, as was mentioned during our discussion of aggregates. In general, concrete strength increases with density, and density is proportional to specific weight. A reduction in weight is thus coupled with a reduction in strength, all other factors being equal, and the selection of an aggregate requires a compromise between weight and strength. For many applications, the concrete strength is not a controlling factor. For example, for building floor slabs, the concrete weight can be reduced and sufficient strength achieved without difficulty.

Mixes containing sintered fly ash, expanded clay or shale, or foamed slag as aggregate weigh as little as 100 lb/ft^3 and have strengths of at least 3000 psi, which qualifies them for structural applications [31]. Lighter concretes are more limited in their applications, and those weighing less than 70 lb/ft^3 (cellular or foam concretes can weigh as little as 30 pcf) are generally unsuitable for load carrying members. Because of their high porosity, these concretes are used primarily for insulation. We should note that creep deformations are generally higher for lightweight than for normal-weight concretes because of the higher porosity associated with lower density.

High-strength concrete.   The definition of high-strength concrete is somewhat arbitrary and subject to change with time. Not long ago, 5000 psi would have qualified a mix as high-strength. Nowadays, the borderline strength is more like 7000 psi, while strengths up to 10,000 psi are commercially available. In the laboratory, strengths in excess of 20,000 psi have been achieved without much difficulty. High strength has several advantages. In tall buildings, for example, the use of high-strength concrete permits a reduction in column sizes and thus an increase in rentable floor area. In precast concrete construction, higher strengths lead to a weight reduction of the elements to be shipped and erected, or, for a given weight limit, the element sizes can be increased or more elements included in a shipment. The cost/strength ratio decreases up to very high-strength mixes, which means the use of high-strength concrete makes economical sense for some applications.

As mentioned previously, the factor that influences strength more than anything else is porosity. Whatever reduces porosity increases strength. For example, a low water-cement ratio automatically yields high strength, up to the point of a mix with zero slump (*no-slump concrete*). In order to maintain workability, it is common to add a superplasticizer.

When a few links of a chain are strengthened, the strength of the chain increases, provided no other links are now the weak links. The same holds for high-strength concrete. The aggregate, mix design, concrete production, curing, and quality control all assume potentially controlling importance. With a strength increase of the cement paste, through a low w/c ratio, the strength of the coarse aggregate gains in importance. A failure crack can now pass right through coarse aggregate particles. Likewise, good aggregate grading and surface texture become more important, as crushed stone with rough surfaces is better than smooth gravel because of the better bond between paste and aggregate. A smaller maximum aggregate size increases the surface area for good bonding with the paste and thereby increases strength. It also requires an increased amount of paste, which together with a low w/c ratio, calls for more cement per unit volume of concrete. This raises the cost of the mix. w/c ratios as low as 0.35 (and even 0.30 for no-slump mixes) are common, and in some instances much smaller values are used [32]. Such low w/c ratios require particularly good compaction. If vibration is insufficient, high-pressure curing may be needed.

The addition of *silica fume* is now commonly used to increase the strength of concrete mixes (see Sec. 2.3). There are several reasons why the addition of this extremely fine compound increases the strength of concrete. First, it reacts with lime during the hydration of cement to form a stable cementitious compound and thus may replace a part of the cement. Because of its larger surface area, it binds more water than portland cement, thereby reducing the effective w/c ratio. As it also reduces the workability, silica fume should be used in conjunction with a superplasticizer. Silica fume also increases the aggregate-paste bond, an important indicator of concrete strength. Furthermore, it increases the mix density by reducing the pore sizes, resulting in a lower permeability.

The failure of high-strength concrete is much more brittle than that of normal-strength material (Fig. 2.12). The amount of microcracking is greatly reduced, especially on the paste-aggregate interface. As a result, only a few cracks develop, and the stress-strain response is very linear and steeper, that is, the modulus of elasticity is higher. When the few dominant cracks become unstable, less redundancy is left in the material structure. Failure is very brittle, with cracks typically passing through coarse aggregate particles. Because of the reduced porosity and lower w/c ratio, high-strength concretes exhibit less creep and shrinkage deformations than normal-strength mixes.

As an example, a 28-day strength of 16 ksi was achieved with the following mix proportions: 1682 lb/yd^3 Thornton limestone coarse aggregate, 905 lb/yd^3 Elgin sand, 1000 lb/yd^3 Type I cement, 266 lb/yd^3 water, and 200 lb/yd^3 silica fume. The w/c ratio was 0.22. (Note that the c stands for cementitious material, which includes the silica fume admixture.) The average air content was 1.5% and the average slump 4.7 in [33].

Fiber reinforced concrete.   The low tensile strength and brittle failure mode of concrete call for suitable reinforcing materials. Reinforcement in the form of steel bars, wires, or wire mesh has been the favored solution for more than a century, and has led to what is now called reinforced concrete—the subject of this book. However, a bar reinforces the concrete only in the direction in which it is oriented. Randomly distributed and oriented small fibers, on the other hand, strengthen the material in all directions more or less equally, literally changing the properties of the material itself.

The concept of reinforcing brittle materials is not new. For example, clay bricks have been reinforced with straw or horsehair since ancient times. At present, the most common types of fiber are made of steel, polymeric, glass, and organic materials [34]. Glass fibers require special attention because of the potential alkali-aggregate reactions and their related long-term durability effects.

The effectiveness of steel fibers is determined by their aspect ratio (length divided by equivalent diameter) and shape. To improve their bond with the surrounding cement paste matrix, they may be hooked or crimped. They are typically supplied in clips of many fibers joined with a special adhesive that releases the individual fibers during mixing. Typical lengths vary between 0.25 and 3 in, with diameters between 0.01 and 0.03 in.

If modest amounts of fibers are added to the mix (say up to 1% steel fiber by volume), they have little effect on the concrete strength, because when a crack opens up in the matrix, it is likely to be bridged by only a small number of fibers with a combined strength barely exceeding that of the paste. However, the large amount of work needed to open the cracks and to pull out the fibers from the matrix considerably increases the overall ductility and toughness of the material. This large increase in energy absorption capacity (Fig. 2.26), makes steel fiber reinforced concrete particularly suitable for structures subjected to dynamic loadings, such as blast and impact. It is widely used in conjunction with shotcrete, which is concrete that is sprayed onto difficult-to-form surfaces, such as unstable slopes, swimming pools, and tunnel linings.

Polymer-modified concrete.   The addition of polymers to concrete can dramatically improve its properties [35]. This can be achieved in various ways. A monomer, a low-viscosity but reactive organic liquid, is forced into the pore space of cured concrete and polymerized. The result is *polymer-impregnated concrete*, or PIC. The reduction in porosity, as polymer fills the pores, causes an increase in the strength of both the cement paste and the paste-aggregate bond. If a polymer latex (a dispersion of polymer particles of 0.1 to 5 microns in diameter in water), is added to wet portland cement, mixed with aggregate, and cured, a *latex modified concrete* or LMC results. The latex particles are moved by the growth of hydrate crystals and ultimately end up in those regions where pores would otherwise exist. The latex particles fuse with adjacent latex to form a continuous film. The result is reduced porosity and permeability and increased strength and durability.

The reduced permeability makes polymer-modified concrete particularly suitable for bridge decks. This includes rehabilitation and repair work, because frequent use of deicing salts can cause corrosion of the embedded steel reinforcement. Thin PMC overlays have been shown to serve as effective sealants that can protect the reinforcement against corrosion.

Refractory concrete.   This is a material designed specifically to resist high temperatures, such as occur in ovens and furnaces. The constituent materials should be chosen for their refractory properties. Some good materials in this regard are calcium aluminate, high alumina cement, and special aggregates such as silicon carbide or fired clay [36].

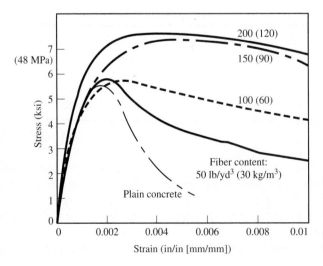

Figure 2.26 Stress-strain behavior of concrete without and with steel fibers. (From P. N. Balaguru and S. P. Shah, *Fiber-Reinforced Cement Composites*, 1992. Reprinted with permission of McGraw-Hill, Inc., New York.)

2.9 REINFORCING STEELS

General. Because of its low tensile strength, concrete is normally reinforced with steel. The relatively high cost of steel mandates its prudent use. This means that the concrete is assigned the task of resisting compressive stresses, while the steel carries primarily the tensile forces. In certain situations, however, the steel is used in compression as well. For example, in columns of tall buildings, a large amount of steel can reduce the required cross-sectional areas and increase the useable floor area. For reinforced concrete to work properly, it is necessary to effectively transfer stresses from the concrete to the steel and vice versa. This transfer is facilitated through bond. Chemical adhesion between the two dissimilar materials is not sufficient for this purpose, and surface deformations rolled onto the bars in the form of ribs are now used almost exclusively. Figure 2.27 shows some representative rib patterns of *deformed* reinforcing bars, called *rebars* for short. If the bond strength developed through the bar deformations is not sufficient to anchor a rebar, its ends may be bent into hooks for better anchorage.

Rebars are supplied in lengths of up to 60 ft and are available in standardized sizes, listed in Table B.1 of Appendix B. The size or number of a rebar refers to the nominal diameter expressed in eighths of an inch. For example, a number 6 rebar has a diameter of $\frac{6}{8} = 0.75$ in. Various markings are rolled into the rebars for identification. Figure 2.28 shows a rebar with: an initial to identify the producing mill; the size number of the rebar; a code for the type of steel (*S* for A615 steel, *R* for A616 rail steel, *A* for A617 axle steel, *W* for low-alloy A706 steel); and the steel grade, that is, the yield strength in ksi or a simple vertical line for grade 60.

Pavements, floor slabs, walls, shells, and thin webs of girders are often reinforced with *welded wire fabric* (or mesh). Either smooth or deformed cold-drawn wires are assembled at uniform spacing in two orthogonal directions and welded together at all

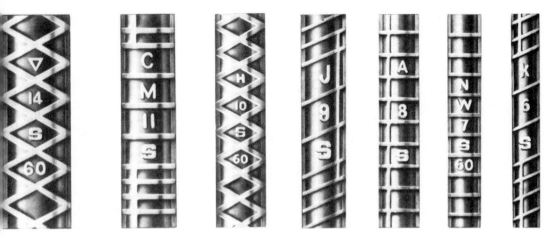

Figure 2.27 Typical deformed reinforcing bars (courtesy of the Concrete Reinforcing Steel Institute).

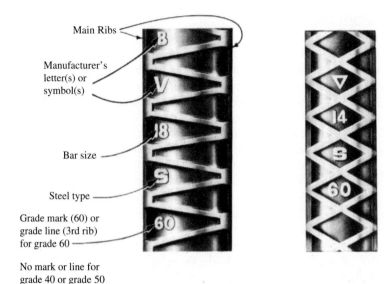

Main Ribs

Manufacturer's
letter(s) or
symbol(s)

Bar size

Steel type

Grade mark (60) or
grade line (3rd rib)
for grade 60

No mark or line for
grade 40 or grade 50

Figure 2.28   Rolled-in identification marks of reinforcing bars (courtesy of the Concrete
Reinforcing Steel Institute).

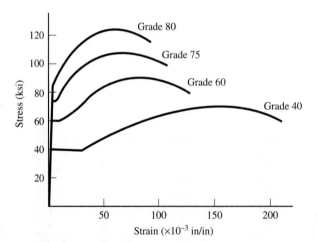

Figure 2.29   Typical stress-strain curves
for reinforcing steels.

points of intersection. Wire sizes vary from about $\frac{1}{8}$ to $\frac{5}{8}$ inches in diameter. ASTM
standard wire reinforcement is listed in Table B.2.

Mechanical Properties.   The important mechanical properties of reinforcing
steels are displayed in their stress-strain curves (Fig. 2.29). The yield plateau determines the *grade* of the steel. Until the 1970s, grade 40 steel was the most commonly
used. At present, it has been largely replaced by grade 60 steel. The higher strength

requires less steel and makes it easier to avoid problems of congestion in heavily rein-forced members and joints. However, according to Fig. 2.29, the higher-strength steels exhibit less ductility and shorter yield plateaus before the onset of the strain-hardening range. For analysis and design purposes, the stress-strain relationship is assumed to con-sist of a linear-elastic branch and a horizontal yield branch, that is, the strain-hardening branch is typically ignored. This is a conservative assumption in most cases, but not always. For example, when it is necessary to develop the full strength of a bar by bond to avoid pull-out, it is not sufficient to determine the anchorage requirements on the basis of the assumed yield strength. Also, actual yield and ultimate tensile strengths exhibit a certain amount of statistical scatter (though much less than concrete strength properties), and mean values may be considerably higher than specified minimum values.

The ACI Code permits the use of steel with yield strength up to 80 ksi. It inde-pendently requires a yield strain of 0.0035, which with the standard value of Young's modulus of $E_s = 29,000$ ksi translates to a conservative 100 ksi yield strength.

The use of higher-strength steels reduces not only the ductility capacity. In addition, larger strains are necessary to develop the full capacity of the bars, which means the concrete must undergo more extensive cracking. For steel of grade 60 or higher this can become a problem. This is the basis for the requirement that designs utilizing the higher-strength steels be checked for excessive cracking (see Sec. 3.3).

For prestressed concrete, steels with much higher strengths are used. Ultimate tensile strengths as high as 250 ksi and more are common. In this text, we will not cover prestressed concrete.

Corrosion.   Corrosion is an electrochemical process during which the steel reacts with its environment in the presence of dissolved oxygen, water, and in particular with salt and acids. The destructive nature of corrosion is based on two factors. First, the corroded part of the material is weak and therefore lost to load resistance, that is, the load-bearing cross-sectional area of a steel bar is gradually reduced. Second, the corrosion products occupy a larger volume than the base material. As a result, they exert considerable pressure on the protective concrete cover and tend to spall it off. The result is the potential total loss of bond between steel and concrete.

The alkalinity of the cement paste offers a natural protection of the embedded steel reinforcement against corrosion. This protection is lost if two conditions are met: (a) extensive cracking of the protective concrete cover; and (b) the presence of chloride ions in the pore solution, which are provided by salts and other chloride-bearing sub-stances. Together, these two conditions lead to corrosion and the above-mentioned conse-quences. The problem is particularly serious in regions with harsh climates, where deicing salts are used during the winter months to clear bridge decks and pavements of snow and ice.

In most other cases no special protective measures are required to guard against corrosion. Reinforcing bars, after having been stored in the open for some time, usually have a rusty appearance, the result of the formation of a thin protective oxide layer. Concern is appropriate if the corrosion has advanced sufficiently to produce flaking, in which case these products should be removed by brushing. If protective measures are in order, epoxy-coated bars (with their characteristic greenish color) are widely used. The smooth coating has an adverse effect on the bond strength, so longer development lengths are needed. Such bars also require careful handling during transport and placement. Even minor damage to the coating may defeat the purpose of the protection, since severe localized damage can have the same overall consequences as widespread corrosion damage. Also, their effectiveness is not universally accepted, especially in severe environments. Other methods of corrosion protection are based primarily on reducing the porosity of the concrete cover by making the concrete denser. The problem with these measures is that it is not easy to guarantee sufficiently small cracks, unless the entire concrete section is prestressed, thereby overstressing any potential cracks. This is a relatively expensive solution. Another method is the use of corrosion inhibitors. Finally, in *cathodic protection*, common in ocean-going ships, an external power supply is used to reverse the electrical potential and thereby prevent the electrical current associated with corrosion.

REFERENCES

1. Mindess, S., and Young, J. F., *Concrete*, Prentice-Hall, Inc., Englewood Cliffs, NJ, 1981.
2. Mehta, P. K., and Monteiro, P. J. M., *Concrete*, 2d ed., Prentice-Hall, Inc., Englewood Cliffs, NJ, 1993.
3. Neville, A. M., *Properties of Concrete*, 3d ed., Longman Scientific and Technical, New York, 1987.
4. *1989 Labor and Energy Report*, Portland Cement Association, Skokie, IL, 1989.
5. ASTM C150, *Standard Specifications for Portland Cement*, American Society for Testing and Materials, Philadelphia, PA, 1989.
6. Malhotra, V. M., and Carino, N. J., *CRC Handbook on Nondestructive Testing of Concrete*, CRC Press, Boca Raton, FL, 1991.
7. ASTM C33, *Standard Specification for Concrete Aggregate*, American Society for Testing and Materials, Philadelphia, PA, 1990.
8. ACI Committee 318, *Building Code Requirements for Reinforced Concrete*, Publication ACI 318-95, American Concrete Institute, Detroit, MI, 1995.
9. ACI Committee 226, *Ground Granulated Blast-Furnace Slag as a Cementitious Constituent in Concrete*, Report 226.1R-87, American Concrete Institute, Detroit, MI, 1987.
10. ACI Committee 226, *Use of Fly Ash in Concrete*, Report 226.3R-87, American Concrete Institute, Detroit, MI, 1987.
11. Malhotra, V. M., Ramachandran, V. S., Feldman, R. F., and Aitcin, P. C., *Condensed Silica Fume in Concrete*, CRC Press, Boca Raton, FL, 1987.
12. ACI Committee 211, *Standard Practice for Selecting Proportions for Normal, Heavyweight, and Mass Concrete*, Report ACI 211.1-89, American Concrete Institute, Detroit, MI, 1989.

13. ACI Committee 305, *Hot Weather Concreting*, Report ACI 305R-89, American Concrete Institute, Detroit, MI, 1989.

14. ACI Committee 306, *Cold Weather Concreting*, Report ACI 306R-88, American Concrete Institute, Detroit, MI, 1988.

15. ACI Committee 214, *Simplified Version of the Recommended Practice for Evaluation of Strength Test Results on Concrete*, Report ACI 214.3R-88, American Concrete Institute, Detroit, MI, 1988.

16. Carrasquillo, R. L., Nilson A. H., and Slate, F. O., "Properties of High Strength Concrete Subject to Short-Term Loads," *ACI Journal*, vol. 78, no. 3, May–June 1981.

17. ASTM C39, *Standard Test Method for Compressive Strength of Cylindrical Concrete Specimens*, American Society for Testing and Materials, Philadelphia, PA, 1986.

18. ACI Committee 214, *Use of Accelerated Strength Testing*, Report ACI 214.1R-81, American Concrete Institute, Detroit, MI, 1981.

19. Bazant, Z. P., ed., *Fracture Mechanics of Concrete Structures*, Elsevier Applied Science, London, 1992.

20. Suaris, W., and Shah, S. P., "Constitutive Model for Dynamic Loading of Concrete," *ASCE, J. of Struct. Eng.*, vol. 111, no. 3, March 1985.

21. Chen, W. F., and Yuan, R. L., "Tensile Strength of Concrete: Double-Punch Test," *ASCE, J. of the Struct. Div.*, vol. 106, no. ST8, Aug. 1980.

22. Su, E. C. M., and Hsu, T. T. C., "Biaxial Compression Fatigue and Discontinuity of Concrete," *ACI Materials Journal*, vol. 85, no. 3, May–June 1988.

23. Sinha, B. P., Gerstle, K. H., and Tulin L. G., "Stress-Strain Relations for Concrete Under Cyclic Loadings," *ACI Journal*, vol. 61, no. 2, 1964.

24. Paskova, T. I., "Low-Cycle Fatigue Behavior of Concrete With and Without Fiber Reinforcement," Ph.D. Thesis, Department of Civil Engineering and Engineering Mechanics, Columbia University, New York, 1994.

25. Kupfer, H., Hilsdorf, H. K., and Rüsch, H., "Behavior of Concrete Under Biaxial Stresses," *ACI Journal*, vol. 66, no. 8, Aug. 1969.

26. Schickert, G., and Danssmann, J., "Behavior of Concrete Stressed by High Hydrostatic Compression," International Conference on Concrete Under Multiaxial Conditions, Toulouse, France, May 22–24, 1984.

27. Rüsch, H., "Researches Toward a General Flexural Theory for Structural Concrete," *ACI Journal*, vol. 57, no. 1, July 1960.

28. Swamy, R. N., *The Alkali-Silica Reaction in Concrete*, Blackie, Glasgow and London, Van Nostrand Reinhold, New York, 1992.

29. Helmuth, R., *Alkali-Silica Reactivity: An Overview of Research*, Strategic Highway Research Program, Report SHRP-C-342, National Research Council, Washington, D.C., 1993.

30. Carslaw, H. S., and Jaeger, J. C., *Conduction of Heat in Solids*, 2d ed., Oxford University Press, 1959.

31. ACI Committee 213, *Guide to Structural Lightweight Aggregate Concrete*, Report ACI 213R-87, American Concrete Institute, Detroit, MI, 1987.

32. ACI Committee 363, *State-of-the-Art Report on High-Strength Concrete*, Report ACI 363R-84, American Concrete Institute, Detroit, MI, 1984.

33. Wolsiefer, J., "Ultra High-Strength Field Placeable Concrete With Silica Fume Admixture," *Concrete International*, April 1984.

34. Balaguru, P. N., and Shah, S. P., *Fiber-Reinforced Cement Composites*, McGraw-Hill, Inc., New York, 1992.

35. ACI Committee 548, *Guide for the Use of Polymers in Concrete*, Report ACI 548.1R-86, American Concrete Institute, Detroit, MI, 1986.

36. ACI Committee 547, *Refractory Concrete: State-of-the-Art Report*, Report 547R-79, American Concrete Institute, Detroit, MI, 1983.

PROBLEMS AND QUESTIONS

2.1 How can high-early strength of a cement be achieved?

2.2 Why does the rate of hydration decay asymptotically?

2.3 What are the raw materials for cement?

2.4 Why is it necessary to add small amounts of gypsum to portland cement?

2.5 List the consequences of using aggregate with only one size.

2.6 What would the consequence be if all aggregate particles were coated with a smooth sealant?

2.7 Why can very large aggregate particles reduce concrete strength?

2.8 Could a reinforced concrete beam carry load if the concrete tensile strength were zero? Explain your answer.

2.9 A standard cylinder and cube are made of the same concrete. Which one will fail at a higher stress in a standard test and why?

2.10 A 6×12 in standard cylinder and a 12×24 in cylinder are made of the same concrete. Which one will fail at a higher stress in a standard test and why?

2.11 How do you explain the fact that the long-term static strength is about 25% less than the short-term strength?

2.12 A clever concrete worker might want to improve the workability of the concrete mix by adding water; then, after placing and compacting the concrete, he would remove excess water with dry and hot air. Would you as an inspector accept his argument that the final w/c ratio is as specified? If not, why not?

2.13 The only difference between two mixes A and B is that mix B has more cement. List all the resulting differences of the properties of the end products, both mechanical and otherwise.

2.14 Will there be any difference between the properties of two identical concrete specimens, one cured under water and one cured in an atmosphere with 100% relative humidity?

2.15 A 4-in concrete cube is loaded in the x-direction up to $0.75 f_c'$, unloaded, then loaded in the y-direction up to failure. Will it fail at a stress other than f_c' and if so, why?

2.16 How does the result of a split cylinder test compare with that of a direct tension test and why?

2.17 Note that nothing in this chapter was said on the shear strength of concrete. Using your knowledge of the mechanics of solids and the behavior of concrete in tension and compression, what is the approximate strength in pure shear as a function of f_c'? (Hint: Use Fig. 2.19b.)

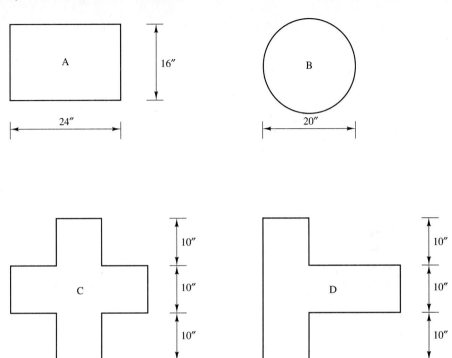

Figure P2.18

2.18 Four columns with the cross sections shown in Fig. P2.18 are made of the same concrete. Rank them from most to least in terms of the creep and shrinkage strains to be expected.

2.19 What are the advantages and disadvantages of a high-slump mix?

2.20 A concrete producer implements quality control measures that reduce the standard deviation of test samples from 400 to 280 psi. By how much can the concrete producer reduce the target strength f'_{cr} according to the ACI Code?

2.21 Enumerate the beneficial effects of lowering the porosity of concrete.

2.22 Enumerate the ways in which steel reinforcement can be protected against corrosion.

2.23 What is the definition of a pozzolanic material?

2.24 What are the beneficial effects of silica fume?

2.25 What is the best way to protect concrete against frost damage?

3

Behavior of Beams

3.1 INTRODUCTION

Before studying the behavior and design of concrete beams, it will be very helpful for the reader to review the computation of bending moment diagrams, some of which are shown in Fig. 3.1. We will be making use of them throughout this and subsequent chapters.

It will also prove useful to recall a few fundamental facts about bending. A simply supported beam carrying downward-acting loads such as those depicted in Fig. 3.2a deflects downwards, whereby the fibers in the upper beam portion are in compression and shorten, and the fibers in the beam's lower portion are in tension and lengthen. In elementary beam theory, it is assumed that cross sections that were plane before application of external loads remain plane in the deformed configuration, so that the stresses vary linearly over the depth of the beam, as shown in Fig. 3.2b. It is a common convention to call bending moments positive if they produce the curvature shown in Figs. 3.1a and 3.2a, with compression at the top and tension at the bottom. Conversely, negative moments produce curvature with tension at the top and compression at the bottom, such as in the cantilever overhangs of Fig. 3.1c and near the interior support of the continuous beam of Fig. 3.1b.

The skill in determining the sense of bending is extremely important in the design of reinforced concrete members. Since the tensile strength of concrete is very low compared with its compressive strength, we have to reinforce it with steel—always *on*

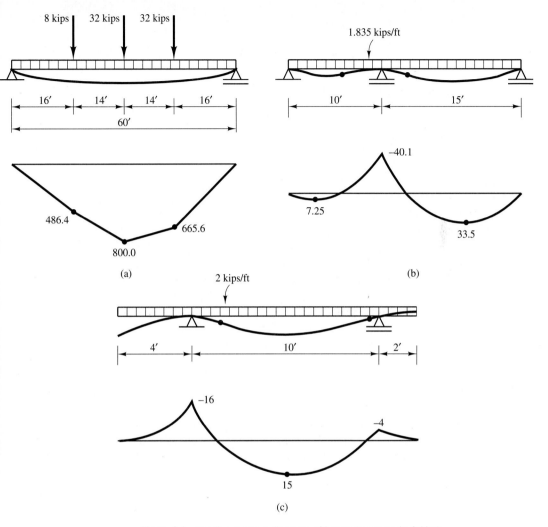

Figure 3.1 Bending moment diagrams of beams (moments in ft-kips).

the tension side. The absence of properly placed steel will most likely lead to structural failure. It cannot be overemphasized that students develop their skills to draw bending moment diagrams and the corresponding deflected shapes and qualitatively place the reinforcement in all regions subjected to tension. We shall discuss this subject in some detail in Chapter 6. In contrast to some texts on structural analysis, we shall consistently plot positive bending moments downwards and negative bending moments upwards. Such moment diagrams indicate immediately on which side the steel should be placed.

Using only the laws of statics, we can always make two general statements about stresses, regardless of their magnitudes or signs. These two statements are so important,

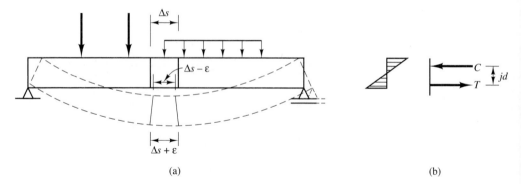

Figure 3.2 Bending of a beam: (a) loaded beam; (b) stresses.

we will use them repeatedly throughout this book:

1. The resultant of all compressive stresses on a section, designated by C, must be equal to the resultant of all tensile stresses, designated by T. In the presence of an external axial force, this statement of equilibrium must be modified accordingly.
2. The moment of these two forces, called the *internal couple*, must be equal to the bending moment due to the external loads at the section considered. The internal couple can be expressed in the following two forms:

$$M = C \cdot jd = T \cdot jd \tag{3.1}$$

where jd is the internal lever arm between the force resultants C and T (Fig. 3.2b).

Equation 3.1 shows that for a moment of given magnitude, equilibrium can be achieved with either a large lever arm jd and small stress resultants C and T, or with a small lever arm and large stress resultants. In other words, for a given moment, stresses in a deep beam are smaller than those in a shallow beam.

If the beam is subjected to loads that result in a varying bending moment, a constant depth will cause stresses to be relatively small where the moment is small (Fig. 3.3a). In such a case, the strength capacity of the material is not well utilized. If the depth of the beam were varied to approximate the bending moment diagram, the material could be utilized more efficiently throughout a large portion of the beam, as shown in Fig. 3.3b. There are two reasons why (in spite of this fact) most beams have constant depth: (1) the formwork is more expensive for nonprismatic beams; and (2) the moment diagram and therefore the optimum depth profile are functions of the loading, which can vary appreciably in time, if it is largely due to live load.

If a beam is made of a homogeneous material such as steel, it makes sense to concentrate the material near the top and bottom faces, where it can be utilized most effectively. The logical result is the I-beam (Fig. 3.4a), with a maximized lever arm between the compression and tension forces and a minimum amount of material needed to resist a given moment. In a concrete beam, the tensile force T must be supplied by the reinforcing steel, because of the insignificant tensile strength of concrete. The

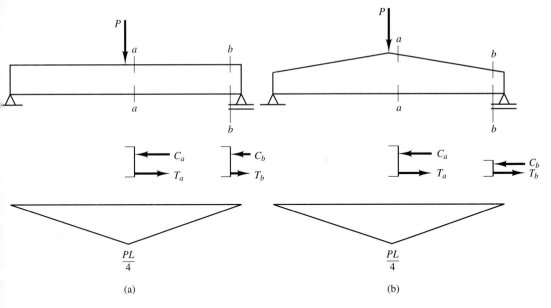

Figure 3.3 Prismatic vs. tapered beam: (a) prismatic beam; (b) tapered beam.

primary task of the concrete in the tension zone is to enclose the steel and to protect it against corrosion. Therefore, it would be logical to reduce the amount of concrete in the tension zone to the bare minimum. On the compression side, where the concrete is utilized very well, it pays to add material. In the web, material is needed to carry the shear forces. The logical result is the T-section of Fig. 3.4b, with the two materials concentrated where they are of most use. This is a very common section for reinforced concrete beams.

The situation is different if, for some reason, it is possible for the moment to change its direction so that the compression and tension stresses are reversed. In this case, it may be more appropriate to choose a symmetric cross section such as a rectangle-, I-, or box beam (Fig. 3.4c), with reinforcing steel provided in the top and bottom flanges, and with enough concrete available in both flanges to resist a moment of either direction.

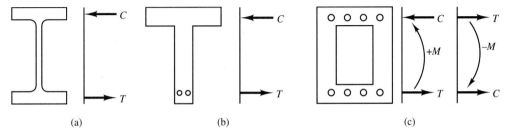

Figure 3.4 Efficient beam cross sections: (a) steel wideflange section; (b) concrete T-beam; (c) concrete box beam for moment reversal.

The I-girder is also used in concrete construction, for example, in multicell box girders, a popular cross section for highway bridge construction (Fig. 3.5).

If concrete beams are precast in a shop, especially if in conjunction with prestressing, formwork cost is less of a concern because it can be amortized over hundreds or thousands of precast elements. As a result, cross sections can be further optimized and still be produced economically. Some of these are shown in Fig. 3.6. In this chapter, we will consider primarily rectangular shapes. The design equations in Chapter 4, however, will be derived for nonrectangular sections as well.

In steel and timber structures, beams are distinct or separate and recognizable members. In concrete structures, beams are typically constructed as integral parts of continuous floor systems (Fig. 3.7). Tied together by steel reinforcement and poured as a unit, the beams and slabs form a structural system that is rigidly connected to the supporting columns or walls. In some cases, the beams are actually invisible from the outside, because they consist only of concentrations of reinforcing steel within a slab of constant thickness (Fig. 3.8). Such a slab system, also known as *flat slab*, requires more steel because of the smaller available depth. However, because of the architectural advantages and simplified formwork, this slab system is quite popular and commonly used for multistory apartment buildings.

When a beam such as the one shown in Fig. 3.7 deflects under load, a portion of the adjacent slab acts as if it were the flange of a T-shaped beam. As a result, the strength and stiffness of the "effective" beam is greatly increased and determined not only by the beam dimensions but also by the slab properties. The width of the slab on

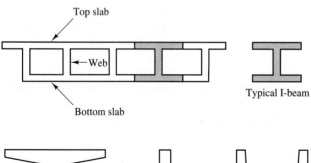

Figure 3.5 Box girder bridge idealized as series of I-beams.

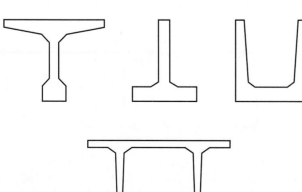

Figure 3.6 Sections for precast and/or prestressed concrete beams.

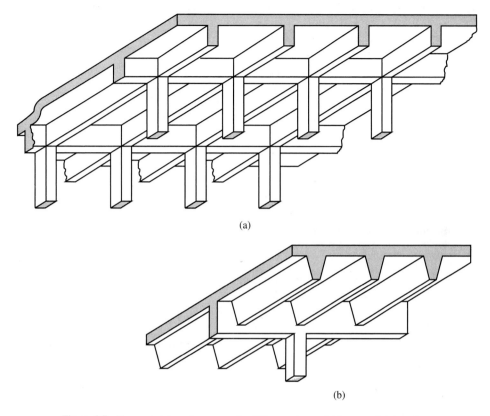

(a)

(b)

Figure 3.7   Beams as integral components of floor systems: (a) beam and slab system; (b) joist floor system.

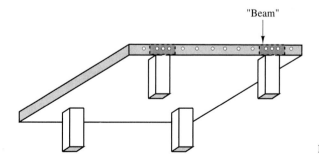

"Beam"

Figure 3.8   Flat slab with invisible beams.

either side of the stem that can be considered part of the beam is called the *effective flange width*.

Owing to their great width and length, slabs possess substantial stiffness in their own plane and are able to provide continuous lateral support for the beams. This support prevents them from displacing laterally and thus eliminates torsional buckling as a possible mode of failure. For an isolated beam not rigidly connected to a slab, such a failure mode should be investigated. In that case, the ACI Code [Sec. 10.4.1] requires that the spacing of lateral supports not exceed 50 times the width of the beam's compression flange. However, unlike slender steel members, concrete members are usually so massive that torsional buckling is seldom a possibility.

In many concrete buildings, the beams, slabs, columns, and walls form a highly indeterminate three-dimensional rigid-jointed structure. An exact analysis of so complex a structural system is almost impossible to carry out. In practice, the three-dimensional system is idealized as a series of two-dimensional frames in an approximate but conservative way. Historically, structures designed on the basis of such simplified analysis have behaved satisfactorily. In Chapter 8, we will describe in detail the simplified methods of analysis currently in use.

A continuous frame of beams and columns subjected to gravity loads deforms as shown (somewhat exaggeratedly) in Fig. 3.9. Because of its low tensile strength, concrete must be reinforced with steel near all tension faces if the beam is to develop significant bending capacity throughout. If reinforcing steel were omitted from the top of a beam where it passes over supports, the concrete alone could not produce the restraint necessary for continuity. The beam would deflect in each span as if it were simply supported. The resulting large rotations over the supports would be accompanied by extensive cracking. Steel bars bridging such potential cracks enforce continuity, thereby stiffening the beam considerably. For example, a uniformly loaded beam whose ends are simply supported deflects five times as much as an identical beam with both ends fixed against rotation (Fig. 3.10). Therefore, a properly reinforced continuous beam does not have to be as deep as a simply supported one if both are allowed to deflect the same amount.

Normally, a beam is defined as a structural member whose span length is large compared with its cross-sectional dimensions. It can be shown that the standard assumptions of ordinary beam theory are valid as long as the depth-span ratio does not exceed a value of approximately $\frac{1}{2}$. This is particularly true for the most basic assumption of all, namely, that plane sections remain plane, or that strains vary linearly over the cross

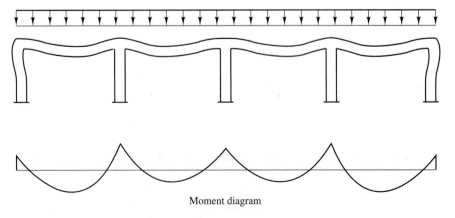

Moment diagram

Figure 3.9   Continuous beam on columns.

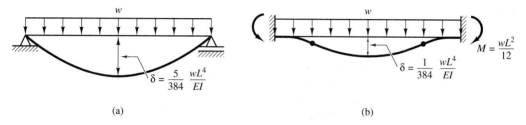

Figure 3.10   Effect of fixed ends on beam deflections: (a) simply supported beam; (b) fixed-end beam.

section. For beams with larger depth-span ratios, this assumption becomes invalid. The ACI Code [Sec. 10.2.2] therefore defines beams with depth-span ratios greater than $\frac{4}{5}$ for simple spans—and greater than $\frac{2}{5}$ for continuous spans—as *deep beams*. Strains in such beams vary nonlinearly, so that ordinary beam theory is not applicable and more advanced theories, such as the theory of plane elasticity, must be used to determine the state of stress and necessary steel reinforcement. In this book, the term *beam* shall always imply *shallow beam*, with the inherent assumption that ordinary beam theory and general ACI Code provisions apply.

The remainder of this chapter is devoted to the behavior of reinforced concrete beams at various stages of loading. A thorough understanding of this behavior is essential in designing beams to be safe and to behave satisfactorily. First, we shall consider loadings of such small magnitude that even the relatively low tensile strength of concrete is not exceeded. We refer to such a case as the *uncracked section*. Next, we shall concern ourselves with a beam subjected to *service loads*, that is, loads of such magnitude as a beam may actually experience during its service life. The designer will want to assure that the beam behaves well under such conditions, without excessive deflections or unsightly cracking. Finally, we shall study the behavior of a beam loaded to the verge of collapse, and the different modes in which such a seriously overloaded beam can fail. It is this stage of loading for which the beam dimensions and the amount of steel reinforcement are determined, as we shall see in subsequent chapters.

3.2 THE UNCRACKED SECTION

In order for a section to remain uncracked, the stresses must be so small that they do not exceed the low tensile strength of concrete. In this rather unusual case, a reinforced concrete beam behaves for all practical purposes like an idealized elastic beam, familiar from theory of strength of materials. However, the beam is not homogeneous because it is made up of two different materials—concrete and steel—with very different strengths and stiffnesses. If the concrete is reinforced with deformed steel bars having good bond characteristics, the strain in the steel, ϵ_s, can be assumed equal to the strain in the surrounding concrete, ϵ_c (Fig. 3.11), that is,

$$\epsilon_s = \frac{f_s}{E_s} = \epsilon_c = \frac{f_c}{E_c} \tag{3.2}$$

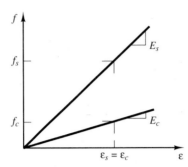

Figure 3.11   Stress-strain relationships for steel and concrete.

where E is Young's modulus, f is stress, and subscripts s and c designate, respectively, steel and concrete. From Eq. 3.2, the relationship between the stresses in steel and concrete follows as

$$f_s = \frac{E_s}{E_c} f_c = n f_c \tag{3.3}$$

where $n = E_s/E_c$ is called the *modular ratio*, which, for standard materials, typically varies between 6 and 10. It should be recalled that Young's modulus of concrete is not well defined because of the nonlinear stress-strain relationship (see Sec. 2.5.1). Values such as those based on Eq. 2.10 are only approximate.

For analysis and design purposes, it is convenient to convert the nonhomogeneous beam section to an equivalent homogeneous section. This can be achieved by replacing the steel area A_s by an equivalent concrete area $n A_s$ (Fig. 3.12). The factor n reflects the fact that steel is that much stiffer than concrete and therefore, for the same unit strain, develops n times as much stress as concrete. Defining A_g as the *gross section area*, where for a rectangular section $A_g = bh$, the net concrete area follows as

$$A_c = A_g - A_s \tag{3.4}$$

and the *transformed area* as

$$A_t = A_c + n A_s = A_g + (n - 1) A_s \tag{3.5}$$

Figure 3.12 represents a somewhat awkward attempt to illustrate the replacement of the steel bars of area A_s by an area $n A_s$ of concrete. By filling the "holes" of area A_s left after removing the steel bars, the excess concrete of area $(n - 1) A_s$ is shown by appendages. Actually, these added areas are located at the position of the steel bars.

As long as the concrete remains uncracked, it is now permissible to consider the beam as if it were made out of a perfectly elastic and homogeneous material. This means that stresses anywhere in the concrete can be computed using the standard beam equation

$$f_c = \frac{M y}{I_t} \tag{3.6}$$

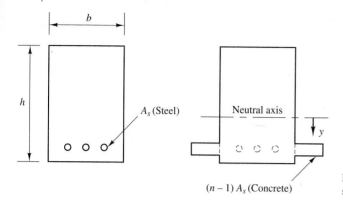

Figure 3.12   Transformed concrete section.

$(n - 1) A_s$ (Concrete)

where f_c = concrete stress at distance y from the neutral axis, M = bending moment on the section, and I_t = moment of inertia of the transformed section about its centroidal axis.

Recalling that for the same strain, steel stress is n times the concrete stress, the steel stress follows as

$$f_s = n \frac{M y_s}{I_t} \tag{3.7}$$

where y_s is the distance from the section centroid to the steel. Even though Eqs. 3.6 and 3.7 imply that the properties of the transformed section are used, the equivalent properties of the corresponding gross section often give results adequate for practical purposes.

To compute the neutral axis and moment of inertia of a section such as the one shown in Fig. 3.13, the section is subdivided into subareas A_i, preferably rectangles or triangles, for which the centroidal positions are known. The neutral axis position is then determined by

$$\bar{y} = \frac{\sum_i A_i y_i}{\sum_i A_i} \tag{3.8}$$

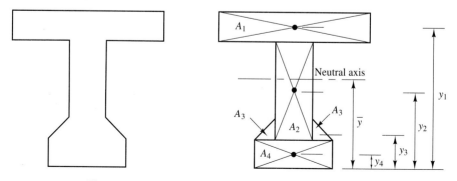

Figure 3.13   Determination of neutral axis and moment of inertia.

and the moment of inertia follows as

$$I_t = \sum_i \left[\frac{b_i h_i^3}{12} + b_i h_i (\bar{y} - y_i)^2 \right]$$

(3.9)

$$I_t = \sum_i b_i h_i \left[\frac{h_i^2}{12} + (\bar{y} - y_i)^2 \right]$$

if rectangular subareas only are involved. For triangles, $b_i h_i^3/36$ should be substituted for $b_i h_i^3/12$. Note that for areas with small depth, such as the $(n-1)A_s$ area add-ons in Fig. 3.12, the first term in Eq. 3.9 is generally negligible, because the moment of inertia of bars about their own axes is small compared with the section's entire moment of inertia.

EXAMPLE 3.1

Find the maximum concrete compression stresses in the two beams shown in Fig. 3.14 and determine the error incurred if gross section instead of transformed section properties are used. Given: $E_c = 3000$ ksi; $E_s = 29{,}000$ ksi; $n = E_s/E_c = 9.67$; area of #7 bar $= 0.6$ in^2; area of #10 bar $= 1.27$ in^2 (see Appendix B).

Solution   Using the gross section, the calculations are independent of the amount of steel, that is, stresses computed for both beams are equal.

Moment of inertia: $I = \dfrac{bh^3}{12} = \dfrac{(12)(20)^3}{12} = 8000$ in^4

Maximum moment: $M = \dfrac{PL}{4} = \dfrac{(5000)(20)(12)}{4} = 300{,}000$ lb-in

Maximum concrete stress: $f_c = \dfrac{Mc}{I} = \dfrac{(300{,}000)(10)}{8000} = 375$ psi

For the transformed section, the neutral axis must be located. In Beam 1, with $(n-1)A_s = (8.67)(4)(0.6) = 20.8$ in^2, the distance of the neutral axis from the bottom becomes

$$\bar{y} = \frac{(12)(20)(10) + (20.8)(2)}{(12)(20) + 20.8} = \frac{2441.6}{260.8} = 9.362 \text{ in}$$

$$I_t = (12)(20) \left[\frac{20^2}{12} + (10 - 9.362)^2 \right] + (20.8)(9.362 - 2.0)^2 = 9225 \text{ in}^4$$

$$f_c = \frac{(300{,}000)(10.638)}{9225} = 346 \text{ psi}$$

For Beam 2, with $(n-1)A_s = (8.67)(4)(1.27) = 44.0$ in^2, we find

$$\bar{y} = \frac{(12)(20)(10) + (44.0)(2)}{(12)(20) + 44.0} = \frac{2448.0}{284.0} = 8.761 \text{ in}$$

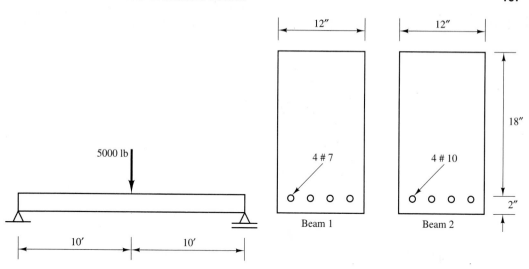

Figure 3.14 Example 3.1.

$$I_t = (12)(20)\left[\frac{20^2}{12} + (10 - 8.761)^2\right] + (44.0)(8.761 - 2.0)^2 = 10{,}380 \text{ in}^4$$

$$f_c = \frac{(300{,}000)(11.239)}{10{,}380} = 325 \text{ psi} \qquad\blacksquare$$

The use of gross instead of transformed section properties overestimates the concrete stresses by 8 and 15%, respectively, for the two beams. Moreover, by increasing the steel area by the factor $1.27/0.6 = 2.1$, or over 100%, maximum concrete stresses change by only 6%. This result illustrates the minor role that the reinforcing steel plays in an uncracked section. There are two reasons for this. First, the steel area, even if multiplied by n, is relatively small compared to the uncracked concrete area. Secondly, assuming concrete cracks at a strain of .0003 in/in, steel, strained by the same amount, experiences a stress of $(.0003)(29{,}000) = 8.7$ ksi, which is very small compared with its typical yield strength of 60 ksi. After the concrete cracks, the steel becomes much more effective.

At this point, it is appropriate to mention that an uncracked concrete section is very unusual. After concrete has been poured into forms, the hydration process generates heat, but the concrete cools down faster on the periphery than the interior. The resulting temperature gradients cause tensile stresses on the surface, which may exceed the tensile strength of the young concrete. Also, shrinkage often causes cracks in the concrete, in particular in the vicinity of reinforcing bars, which tend to resist such shrinkage. Finally, an unloaded beam might already have been subjected to loading that exceeded the beam's cracking strength. As a result, concrete beams are most likely to be cracked, even when no load is applied at all. To the layperson, the idea of cracked concrete may cause some concern, as if "cracked" were identical to "broken." In reinforced concrete design, we must learn to accept the fact that concrete is almost always cracked. In fact, for reinforced concrete to work properly, the steel must be strained to a sizeable fraction of its yield strength, and this is not possible without the cracking of the surrounding concrete. If the structure is designed properly, such cracks are hairline cracks and generally invisible to the eye under service loads. Only in prestressed concrete structures is it a frequent design objective to keep all concrete in compression and therefore crack-free.

In ordinary reinforced concrete design, the concept of an uncracked section is unrealistic. However, the bending moment at which a section can be expected to crack is needed in some cases. For example, at sections where the externally applied bending moment is smaller than the cracking moment, theoretically no reinforcing steel would be required. Because the tensile strength of concrete is not very reliable, however, and actual moments will change for different loads, a certain minimum amount of steel is always necessary in flexural members.

The *cracking moment*, M_{cr}, is defined to be that moment at which the maximum tensile stress in the concrete reaches the modulus of rupture, f_r. It can be determined from Eq. 3.6 by setting f_c equal to f_r, and, for simplicity, accepting I_g instead of I_t:

$$f_r = \frac{M_{cr} y_t}{I_g}$$

from which

$$M_{cr} = \frac{I_g f_r}{y_t} \qquad (3.10)[9\text{-}8]$$

where I_g = moment of inertia of gross section about centroidal axis, neglecting reinforcement; f_r = modulus of rupture of concrete; and y_t = distance from centroidal axis to extreme fiber in tension.

According to the ACI Code, the modulus of rupture for normal-weight concrete can be estimated as

$$f_r = 7.5\sqrt{f_c'} \qquad [9\text{-}9]$$

When lightweight concrete is used, and the splitting tensile strength f_{ct} is specified, then

$$f_r = 7.5\frac{f_{ct}}{6.7} \le 7.5\sqrt{f_c'}$$

should be used (see Chapter 2).

EXAMPLE 3.2

The beam of Fig. 3.15 is made of normal-weight concrete weighing 150 lb/ft³. Considering the dead weight of the beam, estimate the value of the concentrated midspan load P that will cause the beam to crack. Given: $f_r = 360$ psi.

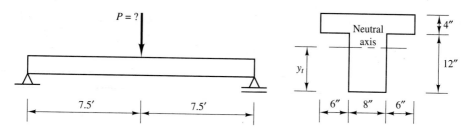

Figure 3.15   Example 3.2.

Solution

Gross concrete area: $A_g = (4)(20) + (8)(12) = 176$ in^2

Location of neutral axis: $y_t = \dfrac{(80)(14) + (96)(6)}{176} = 9.64$ in

Moment of inertia: $I_g = (80)\left(\dfrac{4^2}{12} + 4.36^2\right) + (96)\left(\dfrac{12^2}{12} + 3.64^2\right) = 4051$ in^4

Cracking moment: $M_{cr} = I_g f_r / y_t = \dfrac{(4051)(360)}{(9.64)(12)} = 12{,}607$ ft-lb

Beam dead weight: $w_D = \dfrac{(176)(150)}{144} = 183.3$ lb/ft $w_D = \dfrac{A_g \cdot w}{144}$

Dead weight moment: $M_D = \dfrac{w_D L^2}{8} = \dfrac{(183.3)(15)^2}{8} = 5156$ ft-lb

Reserve capacity: $M = M_{cr} - M_D = 12{,}607 - 5156 = 7451$ ft-lb

Cracking load: $P = \dfrac{4M}{L} = \dfrac{(4)(7451)}{15} = 1987$ lb   ∎

When the concrete cracks, it suddenly releases a major fraction of its tensile force. The upper bound for this force is given by the triangular stress block of Fig. 3.16:

$$T = \frac{f_r}{2}\frac{hb}{2} = \frac{hb}{4} f_r$$

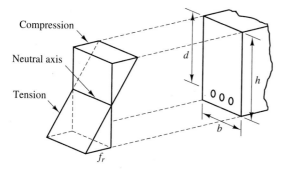

Figure 3.16   Stress distribution at cracking.

This would be the tensile force released if the concrete were to crack all the way to the neutral axis. If the beam is not reinforced with enough steel to absorb this tensile force, the steel could possibly break, resulting in a sudden failure of the beam. To prevent this type of failure, enough steel should be provided to resist the cracking moment M_{cr}. In Sec. 3.3 we shall determine how much steel is needed to resist a given moment after the concrete has cracked. Here, the following approximate estimate will suffice

$$M_{cr} = \frac{f_r I}{c} \approx A_s f_y d \tag{3.11}$$

that is, the steel is assumed to be stressed to its yield point f_y, and the internal moment arm is assumed to be equal to d. Since not all concrete below the neutral axis is cracked, and some of the cracked concrete can still carry tension, this assumption is not quite as unreasonable as it may seem. With $I/c \approx bd^2/6$, Eq. 3.11 can be solved for A_s:

$$A_s = \frac{f_r}{f_y} \frac{bd}{6} \tag{3.12}$$

This can be simplified by substituting $f_r = 7.5\sqrt{f_c'}$, and with a safety factor of 2.4 against this type of failure, it becomes

$$A_{s,\min} = \frac{3\sqrt{f_c'}}{f_y} bd \tag{3.13}$$

For a concrete of strength $f_c' = 4440$ psi, and introducing the reinforcement ratio

$$\rho = \frac{A_s}{bd} \tag{3.14}$$

Eq. 3.13 can also be written as

$$\rho_{\min} = \frac{200}{f_y} \qquad , \quad A_{s_{\min}} = \left(\frac{200}{f_y}\right)bd \tag{3.15}$$

Equation 3.13 constitutes the ACI Code requirement for minimum reinforcement for any section of a flexural member, where tensile reinforcement is required, with Eq. 3.15 being a lower bound [Sec. 10.5.1]. Alternatively, the steel area provided shall be at least one-third greater than required by analysis. In structural slabs of uniform thickness, overloads are distributed such that a sudden failure is less likely. For this reason, the minimum steel requirement for slabs is somewhat less than for beams [Sec. 7.12], namely, $\rho = .0020$ for steel of grade 40 or 50 ksi, and $\rho = .0018$ if grade 60 deformed bars or welded wire fabric are used.

EXAMPLE 3.3

Do the two beam sections of Fig. 3.17 satisfy the minimum steel requirement of the ACI Code if $f_y = 36,000$ psi and $f_c' = 4000$ psi?

ACI
CODE
CHECK

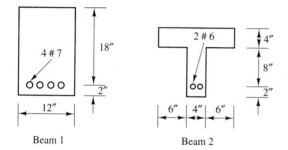

Beam 1   Beam 2   **Figure 3.17**   Example 3.3.

Solution

Beam 1: Actual steel, $A_s = (4)(0.6) = 2.4$ in^2

$$\text{Minimum steel, } A_{s,\min} = \frac{3\sqrt{f_c'}}{f_y} bd = \frac{3\sqrt{4000}}{36,000}(12)(18) = 1.14 \text{ in}^2$$

but not less than $\dfrac{200}{36,000}(12)(18) = 1.2$ in^2; $A_s > A_{s,\min}$ <u>OK</u>

Beam 2: Actual steel, $A_s = (2)(.44) = 0.88$ in^2

$$\text{Minimum steel, } A_{s,\min} = \frac{200}{36,000}(4)(12) = 0.27 \text{ in}^2 < A_s \quad \underline{OK}$$ ■

Note that f_y must be specified in psi when using Eqs. 3.14 or 3.16. In the T-beam of Example 3.3, the web width is used for b because this width determines the tension stress block (Fig. 3.16), not the compression flange width. We will return to this question in conjunction with the design of nonrectangular sections (Sec. 4.3).

In deep flexural members (about 3 ft or more), cracks appear to increase from the tension face. This phenomenon can be explained as follows. Near the tension face, there are many small and fairly closely spaced cracks. Only some of these propagate further toward the neutral axis, mainly as the result of several smaller cracks joining. In order to keep such crack widths small, it is necessary to add *skin reinforcement*, which is to be distributed uniformly over the two faces of the tension zone. For specific ACI Code requirements see [Sec. 10.6.7].

3.3 BEHAVIOR UNDER SERVICE LOAD

In this section we will study the behavior of beams under service loads. These are loads that a beam can be expected to actually experience during its lifetime. Examples are: its own weight, regular traffic loads on highway bridge girders, and normal occupancy loads on the floor beams of an office building. We can assume that a beam subjected to such service loads is cracked. If this were not the case, the beam would indeed be overdesigned. The reinforcing steel would be superfluous, although it can carry tension much more efficiently than concrete.

Once the beam section is cracked, the reinforcing steel must carry the entire tension if the tensile strength of the concrete below the neutral axis is ignored. Stresses and strains

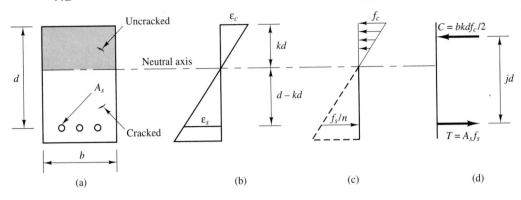

Figure 3.18   Cracked cross section: (a) cross section; (b) strains; (c) stresses; (d) resultant forces.

for this case are shown in Fig. 3.18. The neutral axis or line of zero strain is located by considering the equilibrium of forces acting on the cross section, $C = T$ (Fig. 3.18d), or

$$\frac{f_c}{2}(kd)b = A_s f_s \tag{3.16}$$

To eliminate the unknown stresses f_c and f_s, we utilize the relationship between the strains shown in Fig. 3.18b:

$$\epsilon_s = \frac{d - kd}{kd}\epsilon_c$$

With Hooke's Law ($\epsilon = f/E$), we can write this expression as

$$\frac{f_s}{E_s} = \frac{d - kd}{kd}\frac{f_c}{E_c}$$

or

$$f_s = \frac{d - kd}{kd}n f_c \tag{3.17}$$

where $n = E_s/E_c$ is the modular ratio, encountered previously in Eq. 3.3. Substitution of Eq. 3.17 into Eq. 3.16 leads to

$$\frac{kdb}{2} = A_s \frac{d - kd}{kd}n$$

or

$$kdb\frac{kd}{2} = nA_s(d - kd) \tag{3.18}$$

Before solving this quadratic equation for the only remaining unknown, (i.e., k or kd), which defines the neutral axis position, we may give it a physical interpretation. $(kd)b$ is the area of the uncracked compression zone (Fig. 3.18a), and $kd/2$ is the distance of its centroid from the neutral axis. nA_s represents the transformed area of steel and $(d - kd)$ is its lever arm to the neutral axis. Equation 3.18 is thus an equation of first moments about the neutral axis. It shows that a large steel area A_s must be balanced by a large compression area $(kd)b$, and an increase of kd shifts the neutral axis closer to the steel. Thus, the position of the neutral axis is a function of the steel area A_s. Equation 3.18 can be simplified to

$$k^2 = 2n\frac{A_s}{bd}(1-k)$$

or

$$k^2 + 2n\rho k - 2n\rho = 0 \tag{3.19}$$

where $\rho = A_s/bd$ is the reinforcement ratio, with b being the width of the compression face. The solution of Eq. 3.19 defines the position of the neutral axis:

$$k = -n\rho + \sqrt{(n\rho)^2 + 2n\rho} \tag{3.20}$$

With k and therefore kd known, the moment of inertia of the cracked section becomes

$$I_{cr} = \frac{b(kd)^3}{3} + nA_s(d-kd)^2 \tag{3.21}$$

For a given bending moment M, the maximum concrete stress follows as

$$f_c = \frac{M(kd)}{I_{cr}} \tag{3.22}$$

and the stress in the steel as

$$f_s = \frac{M(d-kd)}{I_{cr}}n \tag{3.23}$$

Stresses can also be computed differently. According to Fig. 3.18d, equilibrium between external and internal moments requires

$$M = C(jd) = T(jd) \tag{3.24}$$

where

$$jd = d - \frac{kd}{3} \tag{3.25}$$

is the lever arm for the internal couple. Substituting $T = A_s f_s$ into Eq. 3.24, we obtain for the steel stress:

$$f_s = \frac{M}{A_s\,jd} \tag{3.26}$$

Substituting $C = f_c bkd/2$ into Eq. 3.24, we find the maximum concrete stress:

$$f_c = \frac{M}{(jd)(kd)b/2} \tag{3.27}$$

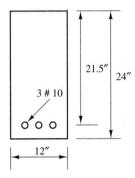

Figure 3.19   Example 3.4.

EXAMPLE 3.4

Assuming the section of Fig. 3.19 is cracked and subjected to the bending moment $M = 60,000$ ft-lb, compute the maximum concrete and steel stresses. Given: $n = 8.6$.

Solution

Reinforcing ratio: $\rho = \dfrac{A_s}{bd} = \dfrac{(3)(1.27)}{(12)(21.5)} = 0.0148$, $n\rho = (8.6)(0.0148) = 0.127$

Solution of Eq. 3.20: $k = -0.127 + \sqrt{(.127)^2 + (2)(.127)} = .393$

Neutral axis position: $kd = (.393)(21.5) = 8.44$ in

Internal lever arm: $jd = d - \dfrac{kd}{3} = 21.5 - \dfrac{8.44}{3} = 18.69$ in

Steel stress: $f_s = \dfrac{M}{A_s jd} = \dfrac{(60,000)(12)}{(3)(1.27)(18.69)} = 10,114$ psi

Concrete stress: $f_c = \dfrac{M}{\frac{1}{2} jdkdb} = \dfrac{(60,000)(12)}{(0.5)(18.69)(8.44)(12)} = 761$ psi

Using the cracked section moment of inertia:

Transformed steel area: $nA_s = (8.6)(3)(1.27) = 32.77$ in^2

Moment of inertia: $I_{cr} = \dfrac{(12)(8.44)^3}{3} + (32.77)(21.5 - 8.44)^2 = 7994$ in^4

Steel stress: $f_s = \dfrac{M(d - kd)n}{I_{cr}} = \dfrac{(60,000)(12)(21.5 - 8.44)(8.6)}{7994} = 10,114$ psi

Concrete stress: $f_c = \dfrac{M(kd)}{I_{cr}} = \dfrac{(60,000)(12)(8.44)}{7994} = 761$ psi   ∎

We must use Eq. 3.20 with caution. As Fig. 3.18 shows, it was derived on the assumption of a rectangular section, or more precisely, on the assumption of a rectangular uncracked portion of the section. If we are dealing with a nonrectangular section, such

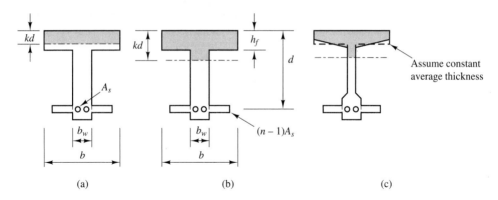

Figure 3.20 Neutral axis in nonrectangular sections.

as the T-beam of Fig. 3.20a whose uncracked part is confined to the top flange of width b, it is irrelevant that the cracked part of the section has variable width, and the neutral axis is correctly located by using Eq. 3.20. If the uncracked section is not rectangular, as in the case shown in Fig. 3.20b, then Eq. 3.20 is not applicable anymore. In this case, the first moment equation would read, with flange thickness h_f and web width b_w:

$$bh_f\left(kd - \frac{h_f}{2}\right) + b_w \frac{(kd - h_f)^2}{2} = nA_s(d - kd)$$

from which

$$k^2 + \left[\frac{2h_f}{db_w}(b - b_w) + 2n\rho\frac{b}{b_w}\right]k - \frac{h_f^2}{d^2b_w}(b - b_w) - 2n\rho\frac{b}{b_w} = 0$$

which is readily solved for k. Of course, for $b = b_w$, this equation reduces to Eq. 3.19. For a section with a tapered compression flange (Fig. 3.20c), it is acceptable to approximate the flange with a constant average thickness as indicated.

Working or allowable stress design.   The equations, derived for computing concrete and steel stresses (Eqs. 3.26 and 3.27), can be readily used to design and proportion flexural members. If the applied bending moment is known, the required beam dimensions and steel area are determined by limiting the stresses in the steel and concrete to certain allowable values, such as $f_{s,\text{all}} = 0.4f_y$ and $f_{c,\text{all}} = 0.45f_c'$. In this case, the apparent safety factors against steel and concrete being stressed to failure are 2.5 and 2.22, respectively. This design method is called *working stress design* (WSD) or *allowable stress design* (ASD) and has been widely used in the past.

EXAMPLE 3.5

Assuming the two concentrated loads are acting on the beam of Fig. 3.21 in addition to the beam's own weight, how large are they allowed to be according to WSD? Given: $n = 8$; $f_{s,\text{all}} = 20$ ksi; $f_{c,\text{all}} = 1.8$ ksi; 150 pcf concrete.

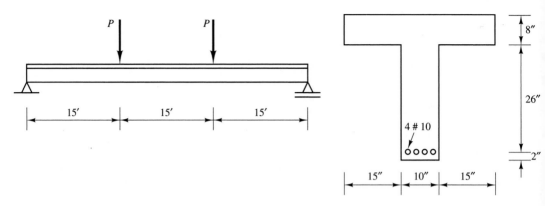

Figure 3.21   Example 3.5.

Solution

Dead load: $w_D = \dfrac{(8)(40) + (10)(28)}{144}(.15) = .625$ kips/ft

Maximum moment: $M = \dfrac{(.625)(45)^2}{8} + 15P = 158 + 15P$ ft-kips

Reinforcing ratio: $\rho = \dfrac{(4)(1.27)}{(34)(40)} = .00374$, $n\rho = (8)(.00374) = .0299$

Neutral axis position: $k = -.0299 + \sqrt{.0299^2 + (2)(.0299)} = .216$,
$\quad kd = (.216)(34) = 7.34 < 8$ in $\underline{\text{OK}}$

Internal moment arm: $jd = 34 - \dfrac{7.34}{3} = 31.6$ in

Capacity based on steel stress (Eq. 3.26): $M = A_s f_{s,\text{all}} jd = (4)(1.27)(20)\dfrac{31.6}{12} =$
$\quad 267.5$ ft-kips $\geq 158 + 15P$, from which $P = \dfrac{267.5 - 158}{15} = 7.3$ kips

Capacity based on concrete stress (Eq. 3.27): $M = f_{c,\text{all}}(jd)(kd)\dfrac{b}{2} =$
$\quad (1.8)(31.6)(7.34)\dfrac{40}{(2)(12)} = 695.8$ ft-kips > 267.5, that is, does not control.

Answer: $P_{\text{all}} = 7.3$ kips   ■

The advantage of WSD is the directness with which satisfactory beam behavior under service loads is achieved. Its main disadvantage lies in the fact that, due to the nonlinear material behavior, the real safety factors against failure are different from the apparent values 2.5 and 2.22. The recognition of the importance of structural safety has prompted a shift away from working stress design toward *ultimate strength design* (USD). This process started in the late 1950s and was completed in the early 1970s. Also responsible for this relatively rapid change of design practice was the fact that ultimate strength design generally leads to more economical structures (requiring less material), and that the design procedures are considerably simpler than in the case of allowable stress design. As a result, in present design practice, concrete structures are designed almost exclusively by the ultimate strength method. Only sanitary structures (e.g., water tanks) are still designed by working stress design, because in this case, serviceability (i.e., watertightness) is of equal if not greater importance than structural safety. Other structures may still be designed by the working stress method by following the provisions of the Code Appendix A, called *Alternate Design Method*.

When ultimate strength design is used to proportion members, the primary objective is to produce structures that can safely carry the loads to which they may be subjected during their lifetime. This objective is achieved by selecting the strength to be greater than or equal to that required to resist the effects of service loads multiplied with certain overload factors, as will be described in detail later. In reality, a structure is seldom if ever exposed to such overloads. Actual loads are most likely to be much smaller. In summary, then, the designer has two options:

1. When using working stress design, members are proportioned for satisfactory behavior under service loads. The actual factor of safety against failure must be checked separately.
2. When using ultimate strength design, members are proportioned for satisfactory strength, that is, for an adequate margin of safety against failure under overloads. The adequacy of the design under service loads must be checked separately. This means in particular that
   a) the concrete cracking should not be excessive, lest the steel reinforcement become exposed to corrosive agents, not to mention the unsightly appearance;
   b) the deflections should remain within certain limits to prevent damage to attached nonstructural components such as windows, partitions, or sensitive machinery, as well as to prevent a negative visual impact.

Since it is the second option that is typically followed in current practice, we will have to take a closer look at the two serviceability requirements.

Control of cracking.   Cracking of concrete is generally acceptable as long as the crack widths remain small. Of considerable concern is the cracking of the concrete cover of the reinforcing steel, because the primary purpose of this cover is to protect the steel against corrosion. Narrow cracks in the cover of high-quality concrete members will not significantly reduce its effectiveness, whereas wide open cracks will. The maximum acceptable crack width depends primarily on exposure conditions. In concrete exposed to seawater or numerous cycles of wetting and drying, the maximum crack width should not exceed .006 to .008 in. For members protected against weather, the ACI Code permits crack widths up to .016 in. For concrete surfaces exposed to view, especially architectural concrete, crack sizes should be controlled for esthetic reasons.

In the past, when working stress design was still the norm, reinforcing steels typically had yield strengths of 36 ksi, and stresses and strains due to service loads were rather low. Because of the low steel strains, the adjacent concrete was also strained only a little, causing minimal cracking. At present, steels with yield strengths of 60 ksi and higher are used regularly. To fully utilize these higher-strength steels, higher strain levels are necessary, which gives cause for concern about concrete cracking. This shift toward higher-strength steels, coupled with the introduction of ultimate strength design, was responsible for the introduction into the ACI Code of provisions to control the widths of cracks when steels with yield points over 40 ksi are used.

When the bending moment exceeds the cracking moment of a section, flexural microcracks develop, which are mostly invisible to the eye. This cracking increases in proportion to the steel stress, and under sustained load, creep deformations cause appreciable further increases. It has been shown that a large number of small bars, well distributed over the zone of maximum concrete tension, are more effective in keeping crack widths small than when the same steel area is supplied by a small number of large bars. Also, reducing the distance between the first row of bars and the surface reduces the crack widths.

An accurate prediction of crack widths is just about impossible. They depend on numerous variables and are subject to considerable statistical scatter. Recent experimental and analytical research using nonlinear finite element analysis and fracture mechanics has led to an improved understanding of many of the variables involved. For practical purposes, it is acceptable to use simple formulas, such as the one given in the ACI Code, which is based on a large number of laboratory experiments. Compliance with this formula assures that crack widths remain within acceptable limits:

$$Z = f_s \sqrt[3]{d_c A} \leq 145 \text{ for exterior exposure}$$

$$\leq 175 \text{ for interior exposure} \qquad (3.28)[10\text{-}4]$$

where

f_s = calculated stress in reinforcement at service loads (ksi); may be taken as $0.6 f_y$;

d_c = thickness of concrete cover measured from extreme tension fiber to center of bar located closest thereto;

A = effective tension area of concrete surrounding the flexural tension reinforcement, divided by the number of bars.

The quantity Z is directly related to the maximum crack width, w, in thousands of an inch, through the relationship $Z = w/.076\beta$, where β = ratio of distances to the neutral axis from the extreme tension fiber and from the centroid of the main reinforcement; an average value for beams is 1.2; for slabs, 1.35. If there is only one layer of reinforcing bars, the depth of the effective tension area is $2d_c$ (Fig. 3.22). In any case, the effective tension area should have the same centroid as the tension reinforcement. In Fig. 3.22, the distance of this centroid from the tension face is designated as y_s. When the flexural reinforcement consists of different bar sizes, the number of bars is computed as the total area of reinforcement divided by the area of the largest bar used:

$$\text{number of bars} = \frac{\text{total area of steel}}{\text{area of largest bar}}$$

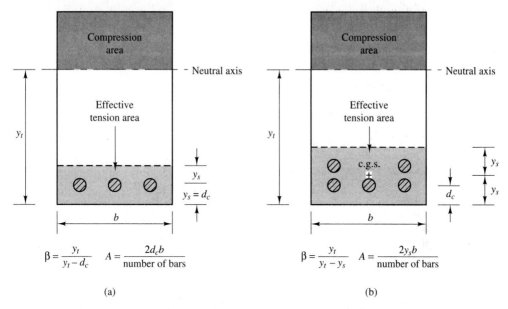

$$\beta = \frac{y_t}{y_t - d_c} \qquad A = \frac{2d_c b}{\text{number of bars}}$$

(a)

$$\beta = \frac{y_t}{y_t - y_s} \qquad A = \frac{2y_s b}{\text{number of bars}}$$

(b)

Figure 3.22 Definition of terms in Eq. 3.28: (a) single reinforcement layer; (b) multiple reinforcement layer.

The limits of Eq. 3.28 correspond to crack widths of .013 and .016 in, respectively. These may not be sufficient for structures subjected to very aggressive environments or designed to be watertight. For such structures, special investigations and precautions may be in order.

EXAMPLE 3.6

Determine if the reinforcement of the two beams shown in Fig. 3.23 satisfies the ACI Code requirement for crack control. Given: $f_y = 60$ ksi; exterior exposure.

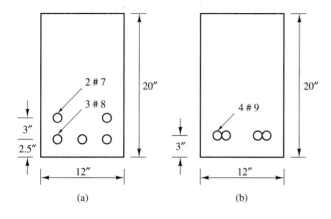

(a)

(b)

Figure 3.23 Example 3.6: (a) Beam 1; (b) Beam 2.

Solution

Beam 1:

Steel centroid from bottom: $y_s = \dfrac{(2)(0.6)(5.5) + (3)(0.79)(2.5)}{(2)(0.6) + (3)(0.79)} = 3.508$ in

Number of bars: $\dfrac{(2)(0.6) + (3)(0.79)}{0.79} = 4.52$

Effective tension area: $A = \dfrac{(2)(3.508)(12)}{4.52} = 18.63$

Crack check: $Z = (0.6)(60)\sqrt[3]{(2.5)(18.63)} = 130 < 145$ <u>OK</u>

Beam 2:

Number of bars (= number of bar bundles): 2

Effective tension area: $A = \dfrac{(2)(3)(12)}{2} = 36$

Crack check: $Z = (0.6)(60)\sqrt[3]{(3)(36)} = 171 > 145$ <u>NG</u>   ■

Note that the bundled bars are treated in a conservative manner, as if the area actually supplied by the two bars in a bundle were provided by a single bar. The bond properties of the bar bundle are better than those of the hypothetical single bar. Therefore, it is safe to compute A as illustrated in the above example.

Control of deflections.   The control of deflections is the other serviceability requirement to be satisfied by members designed by the ultimate strength method. Since the introduction of higher-strength materials and more efficient structural systems, structural members can often be held so slender that it is quite possible that their design is controlled by deflection criteria instead of strength requirements.

Excessive deflections of a floor system may cause damage to nonstructural attachments, such as ceilings, partitions, doors, windows, and piping. In the case of flat roof structures, the consequences can be disastrous. If the supporting members undergo large deflections, rainwater may accumulate near midspan, the weight of which aggravates the problem by increasing the deflections and permitting more water to accumulate. This phenomenon, known as *ponding*, is a serious one and has caused numerous structural collapses.

Floor beams that undergo noticeable deflections or cause ceilings to sag visibly can also cause discomfort or annoyance of occupants, who are not likely to have much confidence in the structural integrity of the building.

The control of deflections is further complicated by the creep behavior of concrete. When limiting deflections of concrete members, we should consider the total deflections, which include both instantaneous and long-term deflections due to creep. The integrity and serviceability of a structure must therefore be checked for different loading conditions and ages of the structure.

Deflection control rests on the ability to compute deflections: a formidable task. For any possible combination of service loads, both short- and long-term deflections of the structure need to be computed, taking into account the appropriate material properties (elastic modulus, creep constants), section properties (amount and extent of cracking), and boundary conditions. Add to this the tensile strength and stiffness of concrete between cracks (known as the tension-stiffening effect) and the change of both the elastic and creep properties of concrete due to a number of influences, and it becomes obvious that the accurate prediction of deflections is next to impossible. In tightly controlled laboratory experiments, it may be possible to obtain agreement between measured and predicted deflections within 20%. The predictability of deflections of structures built in the field is considerably lower. The premature stripping of formwork can, by itself, double or triple subsequent creep deflections. Loading a floor system with heavy construction equipment or materials, before the concrete has reached sufficient strength and stiffness, can have similar consequences. These and other similar influence factors depend on construction conditions, which may be beyond the designer's ability to control.

The single most important variable in this regard is the depth of the flexural member. In design practice, depths of beams and thicknesses of slabs are often predetermined. Architects generally wish to minimize floor construction depths. The advantage becomes obvious in multistory building construction: a two-inch reduction of each floor depth in a thirty-story building saves five feet of building height, which means five-feet savings for curtain walls, piping, stairs, elevator shafts, etc. These savings are likely to be a multiple of the cost for additional reinforcing steel that makes the two-inch depth reduction possible. In a sixty-story building, the same floor depth reduction may make it possible to add another story within a given height constraint. Likewise, if the depth of a highway overpass with 5% gradient is reduced by one foot, twenty feet of ramp length are saved on both sides. For multilevel freeway interchange structures, potential savings are multiplied accordingly. Thus, the designer is usually under pressure to minimize the construction depth, and the deflection constraints become more critical than strength requirements. If it is not possible to keep the deflections within prescribed limits, the use of prestressing may be necessary. This eliminates or minimizes cracking and renders the entire cross section effective in resisting bending moments. Moreover, by proper positioning of the prestressing steel, deflections opposite to those produced by the applied loads can be induced to result in small net deflections or none at all.

Practical methods of deflection control are by necessity of a highly approximate nature. The ACI Code distinguishes between those structures that, if undergoing excessive deflections, can damage attached partitions or other construction and those that do not.

1. If no partitions, windows, etc. can be damaged by deflections, it is sufficient to satisfy the lower limits on the depth-to-span ratios given in Table 3.1 (reproduced from ACI Code Table 9.5a). It will soon be shown that, in this case, deflections will remain below a certain fraction of the span.

2. If use of the first method is not permitted or not practical, for example, if deflection control is critical because of possible damage to attachments, then the actual deflections must be computed and checked against the maximum limits given in Table 3.2 (reproduced from ACI Code Table 9.5b).

The second method can always be used. For example, if the depth-to-span ratio limit of Table 3.1 is considered too restrictive, the designer has the option of reducing the depth, but needs to show that the computed deflections remain below the limits of Table 3.2.

Before describing practical ways of computing deflections, let us show how the first method of limiting the depth-to-span ratio places a limit on the deflection.

TABLE 3.1 (ACI CODE TABLE 9.5a) MINIMUM THICKNESS OF NONPRESTRESSED BEAMS OR ONE-WAY SLABS UNLESS DEFLECTIONS ARE COMPUTED

| Member | Minimum thickness, h | | | |
	Simply supported	One end continuous	Both ends continuous	Cantilever
	Members not supporting or attached to partitions or other construction likely to be damaged by large deflections.			
Solid one-way slabs	$\ell/20$	$\ell/24$	$\ell/28$	$\ell/10$
Beams or ribbed one-way slabs	$\ell/16$	$\ell/18.5$	$\ell/21$	$\ell/8$

TABLE 3.2 (ACI CODE TABLE 9.5b) MAXIMUM PERMISSIBLE COMPUTED DEFLECTIONS

Type of member	Deflection to be considered	Deflection limitation
Flat roofs not supporting or attached to nonstructural elements likely to be damaged by large deflections	Immediate deflection due to live load L	$\dfrac{\ell}{180}$
Floors not supporting or attached to nonstructural elements likely to be damaged by large deflections	Immediate deflection due to live load L	$\dfrac{\ell}{360}$
Roof or floor construction supporting or attached to nonstructural elements likely to be damaged by large deflections	That part of the total deflection occurring after attachment of nonstructural elements (sum of the long-time deflection due to all sustained loads and the immediate deflection due to any additional live load)	$\dfrac{\ell}{480}$
Roof or floor construction supporting or attached to nonstructural elements not likely to be damaged by large deflections		$\dfrac{\ell}{240}$

For any beam, the maximum deflection, δ, can be expressed as

$$\delta = k_1 \frac{ML^2}{EI} \tag{3.29}$$

where L is the beam span; E is the elastic modulus; I is the moment of inertia; and M is the maximum bending moment, for example, $wL^2/8$ for a simply supported beam with uniform load. k_1 is a constant, which depends on the boundary conditions and type of loading, for example, 5/48 for a simply supported beam with uniform load. The maximum bending moment M is related to the maximum stress f by

$$M = \frac{fI}{c} \tag{3.30}$$

where c is the distance from the neutral axis to the extreme compression fiber. This can be expressed as a fraction of the total beam depth, h, by

$$c = k_2 h \tag{3.31}$$

Substituting Eqs. 3.30 and 3.31 into Eq. 3.29 leads to

$$\delta = \frac{k_1}{k_2} \frac{fL^2}{Eh} \tag{3.32}$$

Both f and E are, within a limited range, approximately linear functions of the concrete strength f'_c, so that we may substitute $f/E = k_3$ into Eq. 3.32:

$$\delta = \frac{k_1 k_3}{k_2} \frac{L^2}{h} \tag{3.33}$$

It is a design objective to keep the δ/L ratio below a certain value, say k_4. Therefore, Eq. 3.33 becomes

$$\frac{\delta}{L} = \frac{k_1 k_3}{k_2} \frac{L}{h} \leq k_4 \tag{3.34}$$

This inequality is satisfied as long as the beam depth, h, meets the following condition

$$h \geq \frac{L}{k} \tag{3.35}$$

where $k = k_2 k_4 / k_1 k_3$ combines all constants involved. Representative values for k are listed in Table 3.1, for example, $k = 16$ for a simply supported beam. These minimum values were derived for reinforcing steel with $f_y = 60{,}000$ psi. For steel with different yield stresses, the L/k values of Table 3.1 are multiplied by the factor $(0.4 + f_y/100{,}000)$ to reflect the smaller deflections of members with higher-strength steels. In any case, it should be remembered that this first method of deflection control applies only to beams and one-way slabs that are not supporting or attached to partitions or other construction likely to be damaged by large deflections.

The actual computation of deflections is straightforward for beams made of linear elastic materials. Formulas for common support conditions and loadings are readily available in many handbooks and structural analysis texts. Some of the more frequent cases are listed in Table 3.3.

TABLE 3.3 BEAM DEFLECTION FORMULAS

$$\delta = \frac{PL^3}{3EI}$$

$$\theta = \frac{PL^2}{2EI}$$

$$\delta = \frac{ML^2}{2EI}$$

$$\theta = \frac{ML}{EI}$$

$$\delta = \frac{wL^4}{8EI}$$

$$\theta = \frac{wL^3}{6EI}$$

$$\delta = \frac{PL^3}{3EI}\left(\frac{a}{L}\right)^2\left(\frac{b}{L}\right)^2$$

$$\theta_a = \frac{PL^2}{6EI}\left[2\frac{a}{L} - 3\frac{a^2}{L^2} + \frac{a^3}{L^3}\right]$$

$$\theta_b = \frac{PL^2}{6EI}\left[\frac{a}{L} - \frac{a^3}{L^3}\right]$$

$$\delta = \frac{5}{384}\frac{wL^4}{EI}$$

$$\theta = \frac{wL^3}{24EI}$$

$$\delta = \frac{1}{384}\frac{wL^4}{EI}$$

In our case, the beam material (concrete) is neither elastic nor linear. First, the elastic modulus of concrete is not a constant, as can be observed in a typical stress-strain diagram, such as Fig. 2.12. For normal-weight concrete, it is common practice to approximate E using the ACI Code formula,

$$E = 57,000\sqrt{f_c'}$$

The use of a proper moment of inertia in any of the formulas of Table 3.3 is more problematic. If the entire beam were uncracked, the gross moment of inertia, I_g, could be substituted for I unless the more accurate moment of inertia of the transformed section were used. However, under service loads, the concrete is cracked wherever the bending moment exceeds the cracking moment. Thus, the real moment of inertia varies along the beam (Fig. 3.24). Large moments increase the extent of cracking and reduce the actual moment of inertia. For practical deflection computation, the ACI Code permits an average, effective moment of inertia, I_e, to be used in the deflection formulas, such as those listed in Table 3.3,

$$I_e = \left(\frac{M_{cr}}{M_a}\right)^3 I_g + \left[1 - \left(\frac{M_{cr}}{M_a}\right)^3\right] I_{cr} \leq I_g \qquad (3.36)[9\text{-}7]$$

where

M_{cr} = cracking moment = $f_r I_g / y_t$ (see Eq. 3.10);
M_a = maximum moment in member, due to service load;
I_g = moment of inertia of gross section;
I_{cr} = moment of inertia of cracked section transformed to concrete.

Equation 3.36 is based on numerous test data and considered sufficiently accurate for practical purposes. It is readily interpreted with the help of Fig. 3.25. For small moments ($M_a < M_{cr}$) the beam is uncracked, so $I_e = I_g$. For large moments (say $M_a > 3M_{cr}$) the beam is cracked so thoroughly that $I_e \approx I_{cr}$. For bending moments in between these two extremes, Eq. 3.36 represents an interpolation formula.

In continuous members, moments change sign, and near the inflection points, sections are theoretically uncracked. To simplify calculations, yet arrive at conservative results, the ACI Code suggests that the effective moment of inertia be taken as the

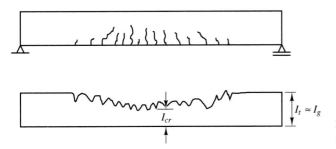

Figure 3.24   Variation of moment of inertia in cracked beam.

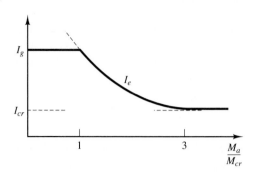

Figure 3.25   Effective moment of inertia as function of M_a/M_{cr} ratio.

average of values obtained from Eq. 3.36 for the critical positive and negative moment sections.

EXAMPLE 3.7

The beam shown in Fig. 3.26 spans a window opening of 20 ft, carrying a live load of 1 kip/ft in addition to its own weight. Disregarding long-term deflections, determine whether a $\frac{1}{2}$-in play in the window gaskets will be sufficient to protect the window from potential damage due to excessive beam deflection. Given: $f'_c = 4$ ksi, $f_y = 60$ ksi; normal-weight concrete weighing 150 lb/ft^3.

Solution

Gross concrete area: $A_g = (25)(5) + (7)(15) = 230$ in^2

Neutral axis location: $y_t = \dfrac{(25)(5)(17.5) + (7)(15)(7.5)}{230} = 12.9$ in, $y_c = 20 - 12.9 = 7.1$ in

Gross moment of inertia: $I_g = (25)(5)\left(\dfrac{5^2}{12} + 4.6^2\right) + (7)(15)\left(\dfrac{15^2}{12} + 5.4^2\right) = 7936$ in^4

Load on beam: $w = \dfrac{230}{144}(0.15) + 1.0 = 1.24$ kips/ft

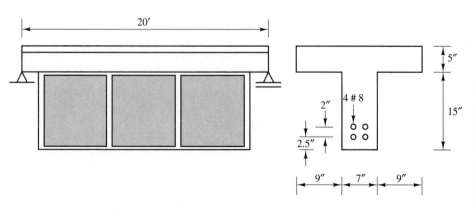

Figure 3.26   Example 3.7.

Maximum moment: $M_a = \dfrac{(1.24)(20)^2}{8} = 62$ ft-kips

Cracking moment: $M_{cr} = \dfrac{f_r I_g}{y_t} = \dfrac{7.5\sqrt{4000}(7936)}{(12.9)(12,000)} = 24.3$ ft-kips $< M_a$

Modular ratio: $n = \dfrac{E_s}{E_c} = \dfrac{29,000,000}{57,000\sqrt{4000}} = \dfrac{29,000,000}{3,605,000} = 8$

Reinforcing ratio: $\rho = \dfrac{(4)(.79)}{(25)(16.5)} = .00766$, $n\rho = (8)(.00766) = .0613$

Neutral axis of cracked section: $k = -.0613 + \sqrt{.0613^2 + (2)(.0613)} = .294$,
$\qquad kd = (.294)(16.5) = 4.85 < 5$ in (Neutral axis within top flange)

Cracked moment of inertia:

$$I_{cr} = \dfrac{b(kd)^3}{3} + nA_s(d-kd)^2 = \dfrac{(25)(4.85)^3}{3} + (8)(4)(.79)(16.5-4.85)^2 = 4382 \text{ in}^4$$

Effective moment of inertia: $I_e = \left(\dfrac{24.3}{62}\right)^3 (7936) + \left[1 - \left(\dfrac{24.3}{62}\right)^3\right](4382) = 4596 \text{ in}^4$

Midspan deflection: $\delta = \dfrac{5}{384}\dfrac{wL^4}{E_c I_e} = \dfrac{(5)(1.24)(20)^4(12)^3}{(384)(3605)(4596)} = 0.269$ in < 0.5 in

Answer: Considering instantaneous deflections only and trusting the ACI Code's limiting values, the window does not appear to be in danger of being damaged.   ∎

EXAMPLE 3.8

The beam shown in Fig. 3.27a supports a service load of 1.0 kip/ft, which includes its own weight. Using the provisions of the ACI Code, estimate the maximum instantaneous deflection of the cantilever tip. Given: $f_c' = 4$ ksi; $f_y = 60$ ksi; $n = 8$; $E_c = 3605$ ksi.

Solution   First, we compute the maximum and minimum bending moments as indicated in Fig. 3.27b. The cantilever deflection is the result of a combination of cantilever flexure and beam rotation over support B. Using the formulas of Table 3.3, we find:

$$\delta = \delta_{BC} - \theta_B L_{BC}$$

$$= \dfrac{wL_{BC}^4}{8E_c I_{e,BC}} - \left(\dfrac{wL_{AB}^3}{24E_c I_{e,AB}} - \dfrac{wL_{AB}L_{BC}^2}{6E_c I_{e,AB}}\right)L_{BC}$$

$$= \dfrac{wL_{BC}^4}{8E_c I_{e,BC}} + \dfrac{wL_{AB}L_{BC}}{24E_c I_{e,AB}}(4L_{BC}^2 - L_{AB}^2)$$

Positive effective moment of inertia for beam segment AB:

Gross moment of inertia: $I_g = \dfrac{(10)(18)^3}{12} = 4860 \text{ in}^4$

Cracking moment: $M_{cr} = \dfrac{(7.5)\sqrt{4000}(4860)}{(9)(12,000)} = 21.3$ ft-kips

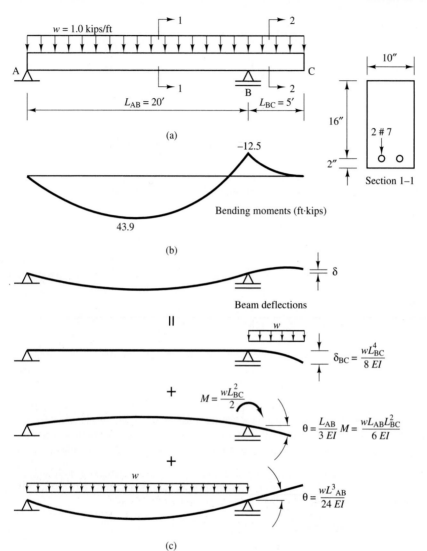

(a)

(b)

Bending moments (ft·kips)

Beam deflections

(c)

Figure 3.27 Example 3.8.

Maximum moment: $M_a = 43.9$ ft-kips $> M_{cr}$

Reinforcing ratio: $\rho = \dfrac{(2)(0.6)}{(16)(10)} = .0075$, $\rho n = (.0075)(8) = .06$

Neutral axis of cracked section: $k = -.06 + \sqrt{.06^2 + (2)(.06)} = .292$,
$\quad kd = (.292)(16) = 4.67$ in

Cracked moment of inertia: $I_{cr} = \dfrac{(10)(4.67)^3}{3} + (8)(1.2)(16 - 4.67)^2 = 1572$ in^4

Positive effective moment of inertia:

$$I_e^+ = \left(\frac{21.3}{43.9}\right)^3 (4860) + \left[1 - \left(\frac{21.3}{43.9}\right)^3\right](1572) = 1948 \text{ in}^4$$

Negative moment of inertia: $I_e^- = I_g = 4860 \text{ in}^4$, since $M_a = 12.5 < M_{cr}$

Effective moment of inertia of beam segment AB: $I_{e,\text{AB}} = \dfrac{1948 + 4860}{2} = 3404 \text{ in}^4$

Effective moment of inertia of cantilever BC: $I_{e,\text{BC}} = 4860 \text{ in}^4$

Cantilever deflection: $\delta = \dfrac{(1)(5)^4(12)^3}{(8)(3605)(4860)} + \dfrac{(1)(20)(5)(12)^3}{(24)(3605)(3404)}(4 \times 5^2 - 20^2) =$

$.0077 - .1760 = \underline{-.168 \text{ in}}$ (up)   ■

If the deflections in a statically indeterminate structure are to be computed, the moment diagram is dependent on the flexural stiffnesses of the members, EI, which in turn are functions of the amount of cracking and therefore of the bending moments. Thus, a solution can only be obtained by iteration. However, it may be recalled from structural theory that for a statically indeterminate analysis only the relative member stiffnesses are of concern. This means that if the members can be assumed to crack by similar degrees, it is permissible to use their uncracked moments of inertia.

The viscoplastic properties of concrete cause creep and shrinkage deformations, which increase under stress with time. Placing reinforcing bars in the compression zone is an effective way of reducing such long-time deformations, but there are many other factors that influence such deflections as well (see Sec. 2.5.6), most importantly: the strength and age of concrete at the time of loading; the level and duration of loading; and temperature, humidity, and curing conditions. According to the ACI Code, it is sufficient to consider only the influence of compression reinforcement. This implies that the concrete has been properly cured, the forms were not stripped prematurely, and the concrete had sufficient time to gain strength before load application. To estimate the long-term deflections due to creep and shrinkage of flexural members, instantaneous deflections caused by sustained load are then simply multiplied by the factor

$$\lambda = \frac{\xi}{1 + 50\rho'} \qquad\qquad (3.37)[9\text{-}10]$$

where ρ' is the compression steel reinforcing ratio and ξ is a time-dependent factor, specified as

Time	ξ
3 months	1.0
6 months	1.2
12 months	1.4
5 years or more	2.0

The total deflection is then a combination of the instantaneous deflection δ_{inst} and the long-term deflection δ_t:

$$\delta = \delta_{inst} + \delta_t = \delta_{inst} + \lambda\delta_p \tag{3.38}$$

where δ_{inst} is the instantaneous deflection due to simultaneous application of *all* service loads, and δ_p is the instantaneous deflection due to only *sustained* loading. Sustained loads include dead weight and that portion of live load that is applied long enough to cause significant time-dependent deflections, such as some storage loads. In the absence of compression reinforcement, the long-term deflections according to Eq. 3.37 can be as high as twice the instantaneous deflections, or according to Eq. 3.38, the combined total deflections can be three times as much.

EXAMPLE 3.9

The floor beam shown in Fig. 3.28 is reinforced with three #10 bars. It supports a dead load of 1.0 kip/ft and a live load of 2.0 kips/ft and is not attached to nonstructural elements likely to be damaged by large deflections. Compute the instantaneous and long-term deflections and check them against the Code-specified maximum value. Assume that only the dead weight is acting permanently. Given: $A_s = 3.81$ in^2; $f_c' = 3000$ psi; $f_y = 60$ ksi.

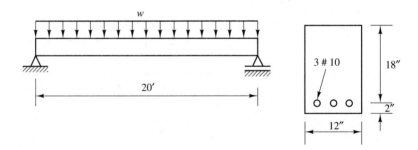

Figure 3.28 Example 3.9.

Solution

Gross moment of inertia: $I_g = \dfrac{bh^3}{12} = \dfrac{(12)(20)^3}{12} = 8000$ in^4

Modular ratio: $n = \dfrac{E_s}{E_c} = \dfrac{29{,}000{,}000}{57{,}000\sqrt{3000}} = \dfrac{29{,}000{,}000}{3{,}122{,}000} = 9.29$

Reinforcing ratio: $\rho = \dfrac{A_s}{bd} = \dfrac{3.81}{(12)(18)} = .0176$, $\rho n = (.0176)(9.29) = .164$

Neutral axis of cracked section: $k = -.164 + \sqrt{.164^2 + (2)(.164)} = .432$,
$\quad kd = (.432)(18) = 7.77$ in

Cracked moment of inertia: $I_{cr} = \dfrac{(12)(7.77)^3}{3} + (9.29)(3.81)(18 - 7.77)^2 = 5580$ in^4

Dead-load moment: $M_a = \dfrac{w_D L^2}{8} = \dfrac{(1)(20)^2}{8} = 50$ ft-kips

Cracking moment: $M_{cr} = \dfrac{f_r I_g}{y_t} = \dfrac{7.5\sqrt{3000}(8000)}{(10)(12,000)} = 27.39$ ft-kips $< M_a$

Effective moment of inertia: $I_e = \left(\dfrac{27.39}{50}\right)^3 (8000) + \left[1 - \left(\dfrac{27.39}{50}\right)^3\right](5580) = 5978$ in^4

Dead-load deflection: $\delta_D = \dfrac{5}{384}\dfrac{w_D L^4}{E_c I_e} = \dfrac{(5)(1)(20)^4(12)^3}{(384)(3122)(5978)} = .193$ in

Long-term deflection: $\delta_t = \dfrac{\xi}{1 + 50\rho'}\delta_D = \dfrac{(2)(.193)}{1 + (50)(0)} = .386$ in

Live-load moment: $M_L = \dfrac{(2)(20)^2}{8} = 100$ ft-kips

Total moment: $M_a = M_D + M_L = 50 + 100 = 150$ ft-kips

Effective moment of inertia (since $M_a \gg M_{cr}$): $I_e \approx I_{cr} = 5580$ in^4

Total instantaneous deflection: $\delta_{inst} = \dfrac{5(w_D + w_L)L^4}{384 E_c I_e} = \dfrac{(5)(3)(20)^4(12)^3}{(384)(3122)(5580)} = .620$ in

Total long-term deflection: $\delta = .620 + .386 = \underline{1.006 \text{ in}}$

According to Table 3.2: $\delta_{all} = \dfrac{L}{240} = \dfrac{(20)(12)}{240} = 1.0 \approx 1.006$ <u>OK</u>   ∎

It should be pointed out that the long-term deflections as estimated by the ACI Code method (Eq. 3.37) can be unconservative in the case of two-way slab systems, in which restrained shrinkage may increase the extent of cracking and reduce the effective moment of inertia. In addition, building slabs can often be observed to be subjected to very high construction loads when the concrete is still young (see Chapter 2). This combination of increased cracking and excessive loads applied to insufficiently cured slabs can cause large long-term deflections.

3.4 FAILURE MODES AND FLEXURAL STRENGTH

Continuing our discussion of beam behavior at different stages of loading, let us now consider loads that can cause the beam to fail. This final stage is of the highest importance for design because it addresses the question of structural safety. Suppose that M_u is the bending moment at which a beam fails and M_a is the largest moment to which it is likely to be subjected. The factor of safety against failure is defined as

$$\text{F.S.} = \frac{M_u}{M_a}$$

It indicates the margin of safety available for the unlikely but not impossible event that overloads are applied to the beam. In practice, this may be an exceptionally overweight vehicle passing over a highway bridge, or an unusual accumulation of heavy equipment in an office building that was designed only for normal occupancy loads.

The replacement of working stress design by ultimate strength design in engineering practice took place partially in recognition of the importance of structural safety. Therefore, we will now study the three modes in which a beam may fail in flexure under overloads. The mode the beam will actually fail in depends on the amount of tensile reinforcement.

First Mode: cracking of concrete followed by breaking of steel bars (insufficient steel area). When the tensile stress in the concrete exceeds its tensile strength, the concrete cracks and transfers the tensile force to the steel. If the provided steel has insufficient capacity to carry this additional force, it will break, and sudden failure of the section follows. We have seen earlier that this failure mode is possible when the bending moment exceeds the cracking moment M_{cr}.

Second Mode: yielding of steel followed by concrete compression failure (moderate steel area). If the steel is stressed to its yield point before the concrete is crushed in compression, the section will not fail. A small moment capacity reserve exists because the moment arm between the internal forces can still increase. This increase is small but significant because it is accompanied by large deflections and extensive cracking, which provide ample warning of impending failure. This capacity of a beam to undergo large deflections before failure is indicative of a member with high ductility. A beam failing in such a ductile mode is called *under-reinforced*.

Third Mode: crushing of concrete in compression before yielding of steel (large steel area). A large area of steel positions the neutral axis such that the strains in the concrete increase more rapidly than for small steel areas. As a result, the concrete may crush before the steel is stressed to its yield point. This crushing of concrete happens very suddenly and is the cause of a brittle failure of the section. A beam with such a large steel area is called *over-reinforced*.

The first mode of failure can be avoided by providing a certain *minimum* amount of steel reinforcement, as discussed earlier (see Eq. 3.13). The third mode can be prevented by observing a limit on the *maximum* amount of steel that can be used as reinforcement. Both of these failure modes are undesirable because they occur suddenly and without advance warning. On the other hand, the yielding of the steel in the second failure mode is by itself not tantamount to failure. The beam is in distress but can still support the applied loads, although with considerable cracking and large deflections.

At this point it is appropriate to clarify the term "desirable failure mode." Strictly speaking, no failure whatsoever is desirable. However, as designers we usually cannot control the loads to which our structures are exposed. We cannot prevent an owner or tenants of a building from foolishly overloading a building floor, but we can design the structure in such a way that it warns them of impending trouble by selecting the mode in which the structure should fail. By choosing the proper amount of reinforcement, we can cause large deflections and large-scale cracking to occur prior to failure, thereby forcing the structure to "talk" to its occupants. It will warn those responsible for the overload that something is amiss and suggest remedial action, or it may even warn the occupants to vacate the building. During a severe winter several years ago, more than a thousand roofs (of mostly garages and warehouses) in the Chicago area collapsed under heavy snow loads. Yet, no lives were lost, because in each case the occupants had received ample warning of impending failure. If a building were to fail in a brittle mode, its occupants would have little chance to escape harm. This is why modern design philosophy places great emphasis on ductile members with enough resilience to carry heavy overloads without failure, and if they do fail, then only after having undergone considerable inelastic deformations.

An under-reinforced beam as defined above is a beam reinforced with such an amount of steel that the concrete crushes in compression *after* the steel has started to yield. An over-reinforced beam has so much steel that the concrete crushes *before* the steel reaches its yield point. If a beam is reinforced with just the right amount of steel so that the concrete starts to crush *at exactly the same time* as when the steel starts to yield, the beam is said to be *balanced*.

We shall now determine this specific amount of steel, denoted as A_{sb}, that leads to balanced conditions, and investigate the failure of such a beam. If we wish to guarantee a ductile failure mode, we obviously must limit the amount of steel well below this value.

Failure of a balanced beam.   Consider the beam of Fig. 3.29, reinforced with the amount of steel, A_{sb}, which causes the concrete to reach the crushing strain (assumed to be $\epsilon_c = .003$) at the same time the steel reaches its yield strain, $\epsilon_s = \epsilon_y$. From similar strain triangles (Fig. 3.29b), the position of the neutral axis can readily be determined:

$$c = \frac{.003}{.003 + \epsilon_y} d \qquad (3.39)$$

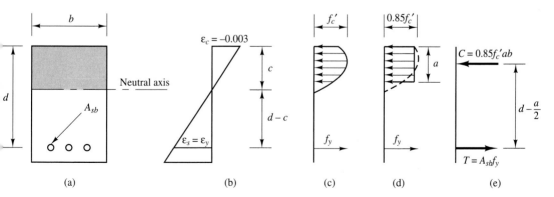

Figure 3.29   Failure of a balanced beam: (a) section; (b) strains; (c) actual stresses; (d) assumed stresses; (e) internal forces.

Substituting $\epsilon_y = f_y/E_s$ and multiplying numerator and denominator with $E_s = 29,000,000$ psi leads to

$$c = \frac{87,000}{87,000 + f_y}d \tag{3.40}$$

The variation of concrete stresses between the neutral axis and the compression face, at the time of failure, is similar to that of the stress-strain curve for concrete (compare Fig. 3.29c with Fig. 2.12). Since such a nonlinear stress distribution complicates the derivation of simple design equations, it is common practice to replace it by a uniform stress block. The equivalent uniform stress and the depth of the area over which it is acting must be determined such that the magnitude of the resultant force C and its centroid are identical to those for the actual stress distribution (Fig. 3.29d). The value of the equivalent uniform stress is normally taken as $0.85 f_c'$. The depth of the equivalent stress block, a, can be related to the depth of the actual stress block, c, by

$$a = \beta_1 c \tag{3.41}$$

where

$$\beta_1 = 0.85 \qquad\qquad\qquad \text{for } f_c' \le 4 \text{ ksi}$$
$$= 0.85 - 0.05(f_c' - 4) \ge 0.65 \quad \text{for } f_c' \ge 4 \text{ ksi}$$

(See ACI Code Sec. 10.2.7.3). Note that Eq. 3.41 is applicable to compression zones of any cross-sectional shape. The equivalent uniform stress block is also referred to as the *Whitney stress block*.

The resultant compression force in the concrete

$$C = 0.85 f_c' ab = 0.85 \beta_1 f_c' cb \tag{3.42}$$

acts at the centroid of the stress block. For a rectangular compression zone, this is located a distance $a/2$ from the compression face. With the resultant tension force in the steel given as

$$T = A_{sb} f_y \tag{3.43}$$

the steel area, A_{sb}, which leads to balanced conditions, can be determined from the equilibrium requirement, $C = T$, as

$$A_{sb} = \frac{0.85\beta_1 f_c' cb}{f_y}$$

$$= \frac{0.85\beta_1 f_c' b}{f_y}\left[\frac{87,000}{87,000 + f_y}\right]d$$

with the use of Eq. 3.40. Recalling the definition of the steel reinforcing ratio, $\rho = A_s/bd$, the balanced steel reinforcing ratio becomes

$$\rho_b = \frac{A_{sb}}{bd} = 0.85\beta_1 \frac{f_c'}{f_y}\left[\frac{87,000}{87,000 + f_y}\right] \tag{3.44}[8-1]$$

(Note that in these formulas, both f_c' and f_y are to be specified in psi). To ensure a ductile mode of failure, the ACI Code (Sec. 10.3.3) requires that no flexural member be reinforced with more than 75% of this amount:

$$\rho \le \rho_{max} = \frac{3}{4}\rho_b \tag{3.45}$$

Failure of an over-reinforced beam.   By definition, a beam is over-reinforced if $\rho \ge \rho_b$. The load-deflection curve for such a beam is almost linear up to failure (Fig. 3.30). Failure is typically accompanied by the explosive disintegration of the concrete. No large deflections or severe cracking take place to warn of impending failure. The more steel reinforcement the beam has, the lower the neutral axis of the transformed section, and the lower the stress in the steel at failure. Such large amounts of reinforcement are undesirable for two reasons. First, they lead to brittle failure. Second, steel added in excess of ρ_b is wasted, since the steel is not fully utilized at failure. The strength of the member is controlled by the concrete. Therefore, adding steel does not do any good, a situation similar to strengthening the strongest link in a chain.

Failure of an under-reinforced beam.   Consider the beam of Fig. 3.31a, which shall be reinforced with an amount of steel well below A_{sb}. Under service loads, the stress and strain variations tend to be linear (Fig. 3.31b). These are quite similar to those in an over-reinforced beam, except that the neutral axis is now closer to the compression face (because of less steel), and therefore the steel stresses are higher. Upon increasing the bending moment on the section, the steel will start to yield at some point (Fig. 3.31c). Because the concrete has not yet reached the crushing strain, the beam has a small strength reserve, which can be explained as follows. The yield plateau in the steel stress-strain curve limits the force in the steel to $T = A_s f_y$. Because of the equilibrium requirement (i.e., $C = T$), the concrete compression force is limited to the same value. This means the *area* under the concrete stress diagram cannot increase, but its *shape* can change (Fig. 3.31d). With the changing shape of the concrete stress triangle, its centroid moves up and the internal lever arm increases, thus activating the moment

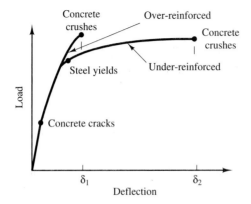

Figure 3.30   Load-deflection curves for over- and under-reinforced beam.

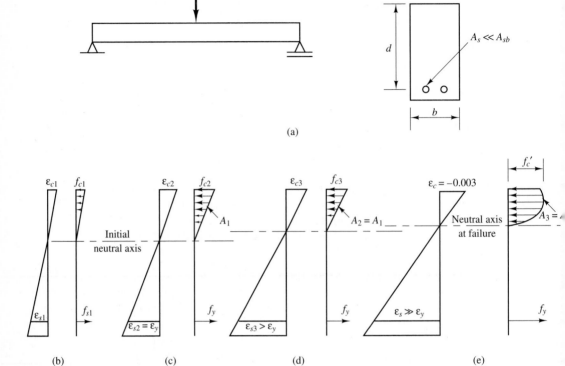

Figure 3.31 Behavior of under-reinforced beam: (a) beam loading and section; (b) strains and stresses at service load; (c) strains and stresses at steel yield; (d) strains and stresses after steel yield; (e) strains and stresses at failure.

strength reserve. In other words, the neutral axis moves toward the compression face until the concrete crushing strain of 0.003 is reached (Fig. 3.31e). When that happens failure occurs rapidly. It is important to note the similarities and differences between this failure mode and the failure of a balanced or over-reinforced beam. The final failure of concrete is equally brittle and sudden in both cases. However, in the under-reinforced beam, this happens only after the steel has yielded a considerable amount. The steel strain at failure is typically a multiple of the yield strain ϵ_y. As a result, the beam undergoes large deflections before collapse (Fig. 3.30), accompanied by severe cracking, thus providing ample warning of impending collapse.

Computation of failure moment. A comparison between Figs. 3.31e and 3.29d suggests that it should be straightforward to compute the bending moment that causes an under-reinforced beam with steel area A_s to fail. In fact, if the actual stress block is again replaced by the Whitney stress block, $C = .85 f'_c ab$ (Eq. 3.42), the equilibrium condition ($C = T$) allows the determination of the depth of the concrete compression block:

$$a = \frac{A_s f_y}{.85 f_c' b} \tag{3.46}$$

With the C-force located the distance $a/2$ from the compression face, the *nominal moment capacity* of the beam, M_n, is defined by

$$M_n = C\left(d - \frac{a}{2}\right) = .85 f_c' ab \left(d - \frac{a}{2}\right) \tag{3.47}$$

or, using $C = T$, by

$$M_n = T\left(d - \frac{a}{2}\right) = A_s f_y \left(d - \frac{a}{2}\right) \tag{3.48}$$

Substitution of Eq. 3.46 into Eq. 3.48 leads to

$$M_n = A_s f_y \left(d - \frac{A_s f_y}{1.7 f_c' b}\right)$$

and, introducing $\rho = A_s / bd$

$$M_n = \rho bd^2 f_y \left(1 - \frac{\rho f_y}{1.7 f_c'}\right) \tag{3.49}$$

Equation 3.49 forms the basis for flexural strength design, and we shall return to it in Chapter 4.

EXAMPLE 3.10

How much superimposed load w_L will collapse the beam shown in Fig. 3.32? Does the beam satisfy the Code's maximum reinforcement limitation? Given: Concrete weighing 150 pcf, with $f_c' = 3000$ psi; $f_y = 60$ ksi.

Solution

Tensile force at failure: $T = A_s f_y = (2)(.79)(60) = 94.8$ kips $= C$

Depth of compression block: $a = \dfrac{C}{.85 f_c' b} = \dfrac{94.8}{(.85)(3)(12)} = 3.1$ in

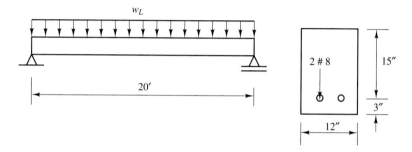

Figure 3.32 Example 3.10.

Internal lever arm: $d - \dfrac{a}{2} = 15 - \dfrac{3.1}{2} = 13.45$ in

Nominal moment capacity: $M_n = T \left(d - \dfrac{a}{2} \right) = \dfrac{(94.8)(13.45)}{12} = 106.3$ ft-kips

Beam dead weight: $w_D = \dfrac{(18)(12)}{144}(.15) = .225$ kips/ft

Maximum moment due to load:

$$M = \dfrac{(w_D + w_L)L^2}{8} = \dfrac{(.225 + w_L)20^2}{8} = 106.3 \text{ ft-kips}$$

Failure load: $w_L = \dfrac{(106.3)(8)}{20^2} - .225 = \underline{1.9 \text{ kips/ft}}$

Beam reinforcing ratio: $\rho = \dfrac{A_s}{bd} = \dfrac{(2)(.79)}{(15)(12)} = .00878$

Balanced reinforcement:

$$\rho_b = .85\beta_1 \dfrac{f'_c}{f_y} \left[\dfrac{87,000}{87,000 + f_y} \right] = (.85)(.85) \dfrac{3,000}{60,000} \dfrac{87,000}{87,000 + 60,000} = .0214$$

Maximum reinforcing ratio: $\rho_{\max} = \dfrac{3}{4}\rho_b = (.75)(.0214) = .0160 > .00878 \text{ OK}$ ∎

It should be pointed out that the moment capacity of an under-reinforced section is controlled by the amount and strength of the steel and therefore barely affected by the concrete strength. This is easily shown with the help of Eq. 3.48. The only quantity affected by the concrete strength f'_c is the depth of the concrete compression block, a (Eq. 3.46), which is somewhat smaller for higher-strength concrete, thus slightly increasing the internal moment arm.

EXAMPLE 3.11

By how much would the nominal moment capacity of the beam of Example 3.10 increase if the concrete strength were increased from 3000 to 4000 psi?

Solution

New depth of concrete compression block: $a = \dfrac{94.8}{(.85)(4)(12)} = 2.32$ in

New moment capacity: $M_n = T \left(d - \dfrac{a}{2} \right) = \dfrac{94.8}{12} \left(15 - \dfrac{2.32}{2} \right) = \underline{109.3 \text{ ft-kips}}$ ∎

In other words, a 33% increase in concrete strength leads to a flexural strength increase of barely 3%. However, there is a more significant benefit of the higher-strength concrete. For $f'_c = 3$ ksi, we found $a = 3.1$ in, so that $c = 3.1/.85 = 3.65$ in. From similar strain triangles the steel strain at failure follows to be $\epsilon_s = 11.35/3.45(.003) = .0093$. For $f'_c = 4$ ksi, we found $a = 2.32$ in, so that $c = 2.32/.85 = 2.73$ in, and $\epsilon_s = 12.27/2.73(.003) = .0135$. This indicates that the steel strain at failure is 45% higher, with much more cracking and deflection to warn of impending failure. This post-yield deformation capacity, called *ductility*, will be the topic of the next section.

The beam width b can be shown to have as little effect on the flexural capacity of a member as concrete strength.

EXAMPLE 3.12

By how much would the nominal moment capacity of the beam of Example 3.10 increase if the beam width were doubled from 12 to 24 in?

Solution

New depth of concrete compression block: $a = \dfrac{94.8}{(.85)(3)(24)} = 1.55$ in

New moment capacity: $M_n = \dfrac{94.8}{12}\left(15 - \dfrac{1.55}{2}\right) = 112.4$ ft-kips   ■

Thus, a 100% increase in beam width (and dead weight!) effects a flexural capacity increase of less than 6%.

3.5 THE PLASTIC HINGE

The telltale sign of an under-reinforced beam is the capability to undergo large deflections before failure (Fig. 3.30). The large yield deformations of the steel make this possible. These are quite localized in the regions of maximum moment, for example, near midspan of a uniformly loaded simply supported beam. When the steel starts to yield, large cracks open up, and the beam curvature increases dramatically in this relatively small region with a large rotation θ (Fig. 3.33). This is reminiscent of the behavior of a hinge, and therefore such a plastic region is generally referred to as a *plastic hinge*. It is important to realize the difference between a plastic hinge and a real hinge, such as a simple door hinge. With a real hinge, an infinitesimally small moment can cause large hinge rotations. In a plastic hinge, only a moment equal to the yield moment M_y can cause large hinge rotations. However, with such a moment acting in the hinge, it will be possible to displace the beam with very small *additional* force, that is, the beam's bending stiffness at this stage, with respect to additional load, has been reduced basically to zero.

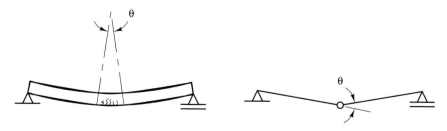

Figure 3.33   Plastic hinge.

The total rotation that a plastic hinge can undergo is obviously much more limited than that of a real hinge. In fact, in Fig. 3.31 it was shown that the plastic deformation capacity of the beam depends on the concrete's strain reserve beyond the point at which the steel starts to yield. For a balanced beam, this strain reserve is zero, and thus there is no plastic rotation capacity. The smaller the amount of steel reinforcement in the beam, the lower the maximum concrete strain at the instant when the beam's yield moment capacity is reached, and the larger the rotational capacity of the plastic hinge. When all other possible failure modes are eliminated by proper design details, the beam is assumed to fail when the concrete strain finally reaches the crushing value $\epsilon_c = 0.003$. Let δ_u be the deflection of the beam just before failure and δ_y the deflection of the beam at the instant when the yield moment capacity of the beam is reached (Fig. 3.34). The ratio

$$\mu = \frac{\delta_u}{\delta_y} \tag{3.50}$$

is defined as the beam *ductility*, which is an extremely important structural property. It is a measure of a beam's capacity to undergo large inelastic deformations without failure. It is this property that allows plastic moment redistribution in indeterminate structures and gives structures the energy absorption capacity and toughness needed to withstand severe dynamic and impulsive loads. Ductility is a property that causes similarity between steel and reinforced concrete members. But while steel is an inherently ductile material, concrete must be designed properly in order to become ductile. For $\rho = \rho_b$, $\mu = 1$, because the beam fails when $\delta_u = \delta_y$. For smaller values of ρ, the ductility μ increases. This is the reason for the ACI Code requirement

$$\rho \le \frac{1}{2}\rho_b \tag{3.51}$$

if use is to be made of plastic moment redistribution, discussed in Chapter 6.

According to Fig. 3.34, a plastic hinge forms gradually after the steel yield strain has been exceeded. For analysis purposes, the load-deformation curve is often replaced by a bilinear curve, with an imaginary point of slope discontinuity defined as the intersection

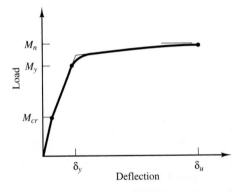

Figure 3.34   Load-deflection curve for under-reinforced beam.

of the two tangents shown. For convenience, we will henceforth assume that the bending moment M_n is acting in a plastic hinge instead of M_y. The actual moment is usually somewhere in between, depending on the amount of plastic rotation (or steel yield) already experienced.

The significance of ductility in structures cannot be overemphasized. For example, compared with structural steel and its excellent ductility, concrete, until not too long ago, was considered to be an inferior material for construction in areas with strong seismic activity. However, since the mid-1960s, structural engineers have succeeded in improving the designs of ductile concrete structures to such an extent that even very tall buildings can now be designed to withstand strong earthquake ground motions.

3.6 MOMENT-CURVATURE RELATIONSHIP

It is instructive to construct a function that relates the moment M at a section to the curvature ϕ. We recall that in regular flexural beam theory, moment and curvature are related through the beam equation

$$M = EI\phi \approx EIy'' \tag{3.52}$$

where for small beam deflections, $y = y(x)$, the second derivative y'' is a sufficiently close approximation of the curvature ϕ. For a linear elastic material, $E = $ const, and as long as the geometric properties of the section do not change, $I = $ const, that is, the sectional rigidity or stiffness, EI, remains constant. For a reinforced concrete section, these conditions exist only for very small moments. For more practical moment levels, the M-ϕ relationship is highly nonlinear. Constructing it for the entire range from zero to failure can serve as a useful review of reinforced concrete mechanics principles and some of the concepts introduced in this chapter.

We shall construct the M-ϕ relationship for the simple rectangular section of Fig. 3.35a. The stress-strain diagrams for both steel and concrete must be given or assumed. We shall assume the curve of Fig. 3.35b for steel (i.e., $f_y = 60$ ksi, $E_s = 30,000$ ksi) and for concrete we assume the curve depicted in Fig. 3.36c, which is subdivided into three segments as follows:

$$f_c = 4000\epsilon \qquad\qquad\qquad\qquad\qquad \epsilon \leq .0005 \tag{3.53a}$$

$$= -\frac{1}{2} + 6000\epsilon - 2 \times 10^6\epsilon^2 \quad .0005 \leq \epsilon \leq .0015 \tag{3.53b}$$

$$= 3 + \frac{4000}{3}\epsilon - \frac{4}{9} \times 10^6\epsilon^2 \quad .0015 \leq \epsilon \leq .003 \tag{3.53c}$$

that is, we approximate the lowly stressed part with a straight line and $E_c = 4000$ ksi, which will also be valid for tension up to $f_c = f_r = 0.4$ ksi. The other two branches are approximated by quadratic functions, and it can easily be verified that the three functions satisfy slope continuity at their transition points.

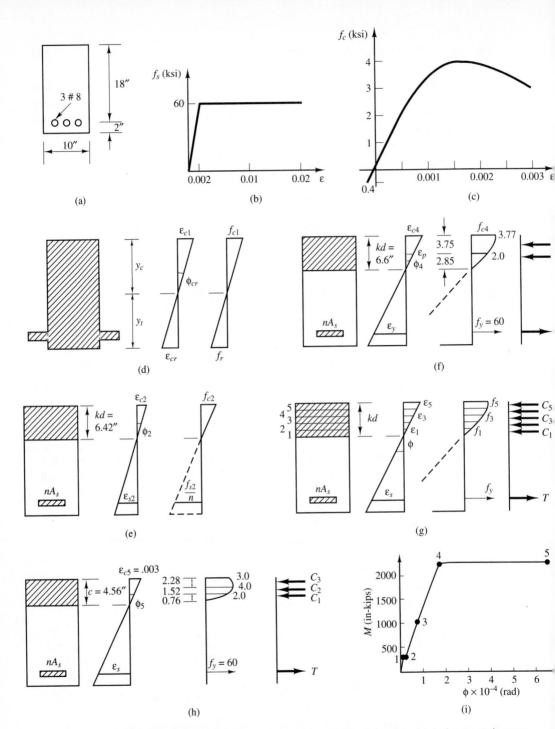

Figure 3.35 Constructing a moment-curvature curve: (a) cross section; (b) steel stress-strain curve; (c) concrete stress-strain curve; (d) stresses and strains at cracking; (e) stresses and strains after cracking; (f) stresses and strains at first steel yield; (g) numerical determination of M and ϕ; (h) stresses and strains at failure; (i) moment-curvature curve.

Point 1: precracking.   As long as the section remains uncracked and both steel and concrete are linear-elastic, EI remains constant. Assuming the first deviation from linearity occurs when the cracking moment is reached, we may use Eq. 3.10, but instead choose to use transformed section properties (see Fig. 3.35d):

Modular ratio: $n = \dfrac{30,000}{4,000} = 7.5$

Excess transformed steel area: $(n - 1)A_s = (3)(.79)(7.5 - 1) = 15.4 \text{ in}^2$

Neutral axis position: $y_t = \dfrac{(10)(20)(10) + (15.4)(2)}{200 + 15.4} = 9.43 \text{ in}, \ y_c = 10.57 \text{ in}$

Moment of inertia: $I_t = \dfrac{(10)(20)^3}{12} + (200)(.57)^2 + (15.4)(7.43)^2 = 7582 \text{ in}^4$

Cracking moment: $M_1 = M_{cr} = \dfrac{f_r I_t}{y_t} = \dfrac{(0.4)(7582)}{9.43} = 322 \text{ in-kips}$

Cracking curvature: $\phi_1 = \phi_{cr} = \dfrac{\epsilon_{cr}}{y_t} = \dfrac{.0001}{9.43} = 0.0106 \times 10^{-3}$

Check concrete compressive strain: $\epsilon_{c1} = \phi_1 y_c = (0.0106 \times 10^{-3})(10.57) = .00011 < .0005 \text{ \underline{OK}}$.

With this information we have the coordinates for point 1 in Fig. 3.35i.

Point 2: postcracking.   After cracking, only part of the concrete is effective, but we assume that both concrete and steel remain within the elastic domain, that is, Eqs. 3.19 through 3.27 apply (see Fig. 3.35e).

Reinforcing ratio: $\rho = \dfrac{(3)(.79)}{(10)(18)} = .013, \ n\rho = (7.5)(.013) = .099$

Neutral axis position: $k = -.099 + \sqrt{.099^2 + (2)(.099)} = .356,$
$kd = (.356)(18) = 6.42 \text{ in}$

Cracked moment of inertia: $I_{cr} = \dfrac{(10)(6.42)^3}{3} + (3)(.79)(7.5)(11.58)^2 = 3266 \text{ in}^4$

Maximum concrete stress: $f_{c2} = \dfrac{M_{cr}(kd)}{I_{cr}} = \dfrac{(322)(6.42)}{3266} = 0.633 < 2.0 \text{ ksi \underline{OK}}$

Concrete strain: $\epsilon_{c2} = \dfrac{f_{c2}}{E_c} = \dfrac{0.633}{4000} = .000158$

Curvature: $\phi_2 = \dfrac{\epsilon_{c2}}{kd} = \dfrac{.000158}{6.42} = 0.0247 \times 10^{-3}$

The cracking reduces the section stiffness by more than a factor of 2, that is, for the same moment, the curvature more than doubles (see point 2 in Fig. 3.35i).

Point 3: concrete proportionality. After cracking, the flexural stiffness remains constant until the concrete reaches the proportionality limit ($f_{c3} = 2$ ksi, $\epsilon_{c3} = .0005$; see Fig. 3.35c). The corresponding moment and curvature are readily determined.

Moment: $M_3 = \dfrac{f_{c3} I_{cr}}{kd} = \dfrac{(2)(3266)}{6.42} = 1017$ in-kips

Curvature: $\phi_3 = \dfrac{\epsilon_{c3}}{kd} = \dfrac{.0005}{6.42} = 0.0779 \times 10^{-3}$

Check steel stress: $f_{s3} = \dfrac{M_3(d - kd)}{I_{cr}} n = \dfrac{(1017)(11.58)(7.5)}{3266} = 27 < 60$ ksi <u>OK</u>

Point 4: yield moment. For $f_c > 2$ ksi, the concrete stress-strain relationship is nonlinear; therefore, the neutral axis starts to shift and needs to be located iteratively. After a few trials we find (see Fig. 3.35f)

Neutral axis location: $kd = 6.6$ in

Curvature: $\phi_4 = \dfrac{\epsilon_y}{d - kd} = \dfrac{.002}{11.4} = .1754 \times 10^{-3}$

Maximum concrete strain: $\epsilon_{c4} = \phi_4 kd = .00116$

Maximum concrete stress (Eq. 3.53b): $f_{c4} = 3.77$ ksi

Distance of proportional point above neutral axis: $y_p = \dfrac{\epsilon_p}{\phi_4} = \dfrac{.0005}{.0001754} = 2.85$ in

Integration of concrete stress block:

Linear part: $C_1 = \dfrac{1}{2}(2)(10)(2.85) = 28.5$ kips

For the nonlinear part, integrate Eq. 3.53b, with $\epsilon = \phi y$:

$$C_2 = \int_{2.85}^{6.6} f_c b \, dy = (10)\left[-\frac{1}{2}y + 6000\phi\frac{y^2}{2} - 2 \times 10^6 \frac{\phi^2 y^3}{3} \right]_{2.85}^{6.6} =$$

113.6 kips; $C = C_1 + C_2 = 142.1$ kips

Check axial force equilibrium: $T = (2.37)(60) = 142.2 \approx C$ <u>OK</u>

The bending moment is the sum of two concrete contributions (found by integration) and the steel contribution:

$$M_{c1} = (28.5)\frac{2}{3}(2.85) = 54 \text{ in-kips}$$

$$M_{c2} = \int_{2.85}^{6.6} f_c y b \, dy = \int_{2.85}^{6.6} \left(-\frac{y}{2} + 6000\phi y^2 - 2 \times 10^6 \phi^2 y^3 \right) b \, dy$$

$$= (10)\left[-\frac{y^2}{4} + \frac{6000\phi y^3}{3} - 2 \times 10^6 \phi^2 \frac{y^4}{4} \right]_{2.85}^{6.6} = 557 \text{ in-kips}$$

$$M_{st} = (142.2)(11.4) = 1621 \text{ in-kips}$$

$$M_4 = M_y = 54 + 557 + 1621 = 2232 \text{ in-kips}$$

If analytical expressions for the concrete stress-strain relationship, such as Eq. 3.53, are considered too cumbersome to work with, the concrete compression region can be subdivided into a number of layers, in each of which the concrete stress is then assumed to be constant (Fig. 3.35g). The total C-force and the bending moment about the neutral axis are then computed numerically.

Point 5: ultimate moment.   Failure is assumed to take place when the maximum concrete strain is 0.003. The stress distribution for that case is shown in Fig. 3.35h. The position of the neutral axis is again located by trial and error and found to be 4.56 in from the top.

Curvature: $\phi_5 = \dfrac{.003}{4.56} = 0.658 \times 10^{-3}$

Distance of proportional point above neutral axis: $y_p = \dfrac{\epsilon_p}{\phi_5} = \dfrac{.0005}{.000658} = 0.76 \text{ in}$

Concrete force C_1: $\dfrac{1}{2}(2)(10)(.76) = 7.6 \text{ kips}$

Concrete force C_2:

$$\int_{.76}^{2.28} f_c b dy = (10) \left[-\frac{1}{2}y + 6000\phi\frac{y^2}{2} - 2 \times 10^6 \frac{\phi^2 y^3}{3} \right]_{.76}^{2.28} = 50.7 \text{ kips}$$

Concrete force C_3:

$$\int_{2.28}^{4.56} f_c b dy = (10) \left[3y + 4000\phi\frac{y^2}{6} - 4 \times 10^6 \frac{\phi^2 y^3}{27} \right]_{2.28}^{4.56} = 83.6 \text{ kips}$$

Total concrete force:

$$C = 7.6 + 50.7 + 83.6 = 141.9 \text{ kips versus } T = A_s f_y = 142.2 \text{ kips} \underline{\text{ Say OK}}$$

Moment contribution of C_1: $M_{c1} = (7.6)\dfrac{2}{3}(.76) = 3.9 \text{ in-kips}$

Moment contribution of C_2:

$$M_{c2} = \int_{.76}^{2.28} f_c b y dy = (10) \left[-\frac{y}{4} + 6000\phi\frac{y^3}{3} - 2 \times 10^6 \phi^2 \frac{y^4}{4} \right]_{.76}^{2.28} =$$
$$81.0 \text{ in-kips}$$

Moment contribution of C_3:

$$M_{c3} = \int_{2.28}^{4.56} f_c b y dy =$$
$$(10) \left[-\frac{3y^2}{2} + 4000\phi\frac{y^3}{9} - 10^6 \phi^2 \frac{y^4}{9} \right]_{2.28}^{4.56} = 281.1 \text{ in-kips}$$

Moment contribution of steel: $M_{st} = (142.2)(18 - 4.56) = 1911.2$ in-kips

Total moment: $M_5 = M_n = 4 + 81 + 281 + 1911 = 2277$ in-kips

Comparing this with the failure moment based on the simplified Whitney stress block:

Depth of concrete compression block: $a = \dfrac{A_s f_y}{.85 f_c' b} = \dfrac{142.2}{(.85)(4)(10)} = 4.18$ in

Moment capacity: $M_n = T\left(d - \dfrac{a}{2}\right) = (142.2)(18 - 2.09) = 2262$ in-kips

The error in this case is less than 0.7%.

Note that the curvature ductility that this beam is capable of developing is $\mu = \phi_5/\phi_4 = .658/.1754 = 3.7$.

3.7 MOMENT CAPACITIES OF OTHER SECTIONS

The equations of reinforced concrete mechanics used in the previous pages suffice to determine the moment capacities of other sections, for example, *composite sections*, in which the tension is carried by structural steel shapes and sections with several layers of reinforcing bars, including those with bars in the compression zone (Fig. 3.36). Both kinds of sections will be dealt with in more detail in subsequent chapters. To determine their ultimate moment capacities requires no new information beyond that used to deal with the simple rectangular section considered so far. However, care needs

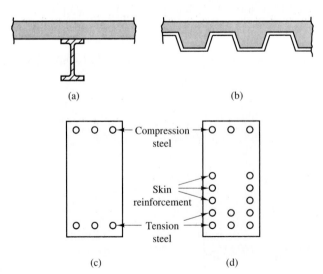

(a)   (b)

(c)   (d)

Figure 3.36 Composite sections and beams with compression reinforcement: (a) composite section; (b) corrugated steel deck with concrete topping; (c) section with compression reinforcement; (d) section with several layers of reinforcement.

to be taken that, because of the bond between steel and concrete, strain compatibility between these two materials is maintained consistently. Because of the general nature of the problem, the solution procedure is iterative and consists of the following five steps:

1. Assume a position of the neutral axis, c, measured from the compression face, and determine the depth of the concrete compression block, $a = \beta_1 c$.
2. Draw the strain diagram for the section. With $\epsilon = 0$ at the neutral axis and $\epsilon_c = .003$ at the compression face, strains can be determined for any layer of steel bars or structural steel shapes from similar triangles.
3. Using the known stress-strain diagram for steel, compute the stresses and resultant forces for each layer of reinforcing bars or structural steel shapes.
4. If the sum of the tensile forces is not equal to the sum of the compression forces (in the absence of an applied axial force), modify the neutral axis position and go back to step 1.
5. Once axial force equilibrium is achieved, take moments about the neutral axis (or any other point) to determine the moment capacity of the section.

Two examples shall serve to illustrate this method.

EXAMPLE 3.13

A steel metal deck of $\frac{1}{16}$-in thickness is topped with concrete, as shown in Fig. 3.37a. The metal deck has small deformations that prevent relative slip between steel and concrete and assure full composite action. The floor deck is also reinforced with W4 welded wire mesh spaced 6 in in both directions. Determine the nominal moment capacity of a one-ft-wide strip of the deck. Given: Concrete with $f_c' = 3000$ psi; metal deck with $f_y = 50$ ksi; welded wire mesh with $A_s = 0.08$ in^2/ft and $f_y = 60$ ksi.

Solution

Step 1. We start by assuming the entire metal deck to yield at failure and ignore the welded wire mesh contribution.

Total tension force: $T = A_s f_y = (12 + 8)\dfrac{1}{16}(50) = 62.5$ kips

Depth of concrete compression block: $a = \dfrac{T}{.85 f_c' b} = \dfrac{62.5}{(.85)(3)(12)} = 2.04$ in

Neutral axis position, with $\beta_1 = .85$: $c = \dfrac{2.04}{.85} = 2.4$ in

Step 2. Strain diagram (Fig. 3.37b).

Maximum steel strain: $\epsilon = \dfrac{2.6}{2.4}(.003) = .00325$

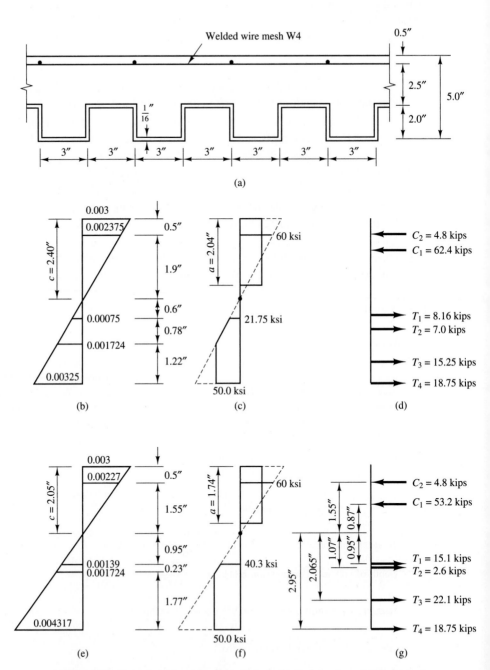

Figure 3.37 Example 3.13: composite metal deck. (a) Corrugated metal deck; (b) strains (for $c = 2.4''$); (c) stresses (for $c = 2.4''$); (d) internal forces; (e) strains (for $c = 2.05''$); (f) stresses (for $c = 2.05''$); (g) internal forces.

Strain at which metal deck yields: $\epsilon_y = \dfrac{50}{29,000} = .001724$

Strain at top of metal deck: $\epsilon = \dfrac{0.6}{2.4}(.003) = .00075$

Distance from neutral axis to point at which steel yields: $y = \dfrac{.001724}{.003}(2.4) = 1.38$ in

Strain in welded wire mesh: $\epsilon = \dfrac{1.9}{2.4}(.003) = .002375$

Step 3.   Stresses (Fig. 3.37c) and forces (Fig. 3.37d).

Stress in welded wire mesh: $f_s = (.002375)(29,000) = 68.9 > 60$ ksi

Concrete compression force: $C_1 = (.85)(3)(2.04)(12) = 62.4$ kips

Force in welded wire mesh: $C_2 = (.08)(60) = 4.8$ kips

Stress in top flange of metal deck: $f = (.00075)(29,000) = 21.75$ ksi

Force in top flange: $T_1 = (6)\left(\dfrac{1}{16}\right)(21.75) = 8.16$ kips

Force in nonyielding part of web: $T_2 = (.78)(4)\left(\dfrac{1}{16}\right)\dfrac{21.75 + 50}{2} = 7.0$ kips

Force in yielding part of web: $T_3 = (1.22)(4)\left(\dfrac{1}{16}\right)(50) = 15.25$ kips

Force in bottom flange: $T_4 = (6)\left(\dfrac{1}{16}\right)(50) = 18.75$ kips

Step 4.   Check equilibrium.

$$\sum C = 67.2 \neq \sum T = 49.2 \text{ kips}$$

Horizontal equilibrium is not satisfied, because the neutral axis was located on the assumption that the entire steel deck yields at failure. Since this is not the case, a smaller concrete stress block is needed to balance the smaller tension force. Try $c = 2.05$ in (Fig. 3.37e, f, and g).

Depth of concrete compression block: $a = (.85)(2.05) = 1.74$ in

Strain in top flange: $\epsilon = \dfrac{.95}{2.05}(.003) = .00139$

Distance from neutral axis to point at which steel yields: $y = \dfrac{.001724}{.003}(2.05) = 1.18$ in

Strain in welded wire mesh: $\epsilon = \dfrac{1.55}{2.05}(.003) = .00227$

Concrete compression force: $C_1 = (.85)(3)(1.74)(12) = 53.2$ kips

Force in welded wire mesh: $f = (.00227)(29,000) = 65.8 > 60$ ksi,
$\quad C_2 = (.08)(60) = 4.8$ kips

Force in top flange: $f = (.00139)(29,000) = 40.3$ ksi, $T_1 = (6)\left(\dfrac{1}{16}\right)(40.3) = 15.1$ kips

Force in nonyielding part of web: $T_2 = (.23)(4)\left(\dfrac{1}{16}\right)\dfrac{40.3 + 50}{2} = 2.6$ kips

Force in yielding part of web: $T_3 = (1.77)(4)\left(\dfrac{1}{16}\right)(50) = 22.1$ kips

Force in bottom flange: $T_4 = (6)\left(\dfrac{1}{16}\right)(50) = 18.75$ kips

Equilibrium check: $\sum C = 58.0 \approx \sum T = 58.5$ kips

Step 5.   Take moments.

$$M_n = (4.8)(1.55) + (53.2)\left(\dfrac{1.74}{2}\right) + (15.1)(0.95) + (2.6)(1.07) + (22.1)(2.065)$$
$$+ (18.75)(2.95) = \underline{172 \text{ in-kips}} \qquad \blacksquare$$

EXAMPLE 3.14

Determine the nominal flexural capacity of the composite floor system shown in Fig. 3.38a.
A W18 × 71 flange section is acting together with a 6-in-thick and 96-in-wide wide concrete
slab, which is also reinforced with 8 #7 bars as shown. (The mechanics of such composite
sections shall be described in more detail later). Given: $f'_c = 4$ ksi; for rebars, $f_y = 60$ ksi;
for the wide flange steel shape, $A = 20.8$ in^2, $I = 1170$ in^4, $f_y = 36$ ksi.

Solution   At failure, the entire W18 × 71 shall be assumed to yield in tension, and the
rebars are assumed to yield in compression.

Tension force: $T = (20.8)(36) = 748.8$ kips

Rebar compression force: $C_2 = (8)(0.6)(60 - .85 \times 4) = 271.7$ kips (see Sec. 4.4)

Concrete compression force: $C_1 = T - C_2 = 748.8 - 271.7 = 477.1$ kips

Depth of concrete compression block: $a = \dfrac{477.1}{(.85)(4)(96)} = 1.46$ in

Neutral axis position: $c = \dfrac{1.46}{.85} = 1.72$ in

Strain and stress at top of steel shape (see strain diagram, Fig. 3.38b):

$$\epsilon = \dfrac{4.28}{1.72}(.003) = .00746; \quad f = (.00746)(29,000) = 216 > 36 \text{ ksi}$$

(i.e., the entire wide flange section yields as assumed)

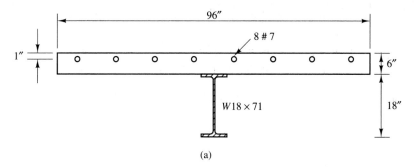

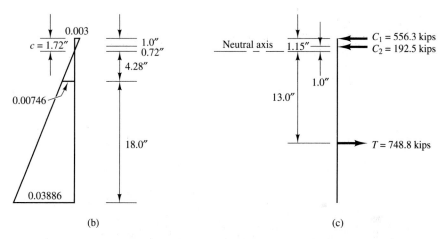

Figure 3.38 Example 3.14: Composite floor system. (a) Composite section; (b) strains (for $c = 1.72''$); (c) final forces (for $c = 2.0''$).

Rebar strain and stress: $\epsilon = \dfrac{.72}{1.72}(.003) = .00125; \ f_s = (.00125)(29{,}000) =$
$\quad 36.4 < 60$ ksi <u>NG</u>. Try $f_s = (36 + 60)/2 = 48$ ksi

$C_2 = (8)(0.6)(48 - .85 \times 4) = 214.1$ kips

$C_1 = 748.8 - 214.1 = 534.7$ kips

$\qquad a = \dfrac{534.7}{(.85)(4)(96)} = 1.64$ in; $\ c = \dfrac{1.64}{.85} = 1.93$ in

New rebar stress: $f_s = \dfrac{.93}{1.93}(.003)(29{,}000) = 41.9$ ksi

After two more iterations, find $f_s = 43.5$ ksi:

$\qquad C_2 = (4.8)(43.5 - .85 \times 4) = 192.5$ kips
$\qquad C_1 = 748.8 - 192.5 = 556.3$ kips; $\ a = 1.70$ in; $\ c = 2.0$ in

Nominal moment capacity (see Fig. 3.38c):

$$M_n = (748.8)(13.0) + (556.3)\left(2 - \dfrac{1.7}{2}\right) + (192.5)(1.0) = \underline{10{,}567 \text{ in-kips}} \qquad \blacksquare$$

SUMMARY

1. In the absence of external axial loads, the resultants of all compressive and tensile stresses must always be equal:

$$C = T$$

2. The external bending moment must always be equal to the internal couple:

$$M = Cjd = Tjd$$

3. To design a beam efficiently, place the reinforcing steel in the tensile zone and concentrate the concrete in the compression zone, reducing the amount of concrete in the tensile zone to that required to enclose the steel.

4. The transformed section is defined as

$$A_t = A_c + nA_s = A_g + (n-1)A_s$$

where $n = E_s/E_c$ is the modular ratio.

5. Concrete stresses in an uncracked section are given by

$$f_c = \frac{My}{I}$$

The amount of steel has relatively little effect on concrete stresses in an uncracked section.

6. The cracking moment is given by

$$M_{cr} = \frac{I_g f_r}{y_t}$$

where $f_r \approx 7.5\sqrt{f_c'}$ is the rupture modulus.

7. The minimum amount of steel to be provided is

$$A_{s,\min} = \frac{3\sqrt{f_c'}}{f_y}bd \geq 200\frac{bd}{f_y}$$

This follows from the requirement that the steel be able to take all tension released by the concrete upon cracking.

8. The neutral axis in a cracked rectangular section subjected to service loads is located a distance kd from the compression face, where

$$k = -n\rho + \sqrt{(n\rho)^2 + 2n\rho}$$

For nonrectangular sections, similar equations can be derived by taking area moments about the neutral axis. This is necessary only if the concrete compression area is nonrectangular.

9. Steel and concrete stresses for a rectangular section under service loads are

$$f_s = \frac{M}{A_s jd} \qquad f_c = \frac{M}{jkbd^2/2}$$

where $jd = d - kd/3$.

10. According to the ACI Code, the preferred method of design is the Ultimate Strength Design (USD). The Working Stress Design is still permissible, as described in the Code Appendix A.

11. When designing beams using the Ultimate Strength Design with steel yield strengths over 40 ksi, the ACI Code requires that crack widths be controlled by

$$Z = f_s \sqrt[3]{d_c A} \le 145 \quad \text{for exterior exposure}$$

$$\le 175 \quad \text{for interior exposure}$$

where $Z = w/0.076\beta$, with $\beta = 1.2$ for beams, and $f_s =$ steel stress due to service load, may be assumed to be equal to $0.6 f_y$. A is the effective tension area divided by the number of bars.

12. Deflections are controlled by keeping the span-to-depth ratio below the values specified in Table 3.1 if no attached construction can be damaged by excessive deflections. Otherwise, deflection values must be shown to remain below those specified in Table 3.2.

13. To compute deflections, the ACI Code permits one to estimate the effective moment of inertia according to

$$I_e = \left(\frac{M_{cr}}{M_a}\right)^3 I_g + \left[1 - \left(\frac{M_{cr}}{M_a}\right)^3\right] I_{cr} \le I_g$$

14. Long-term deflections may be estimated according to

$$\delta_t = \lambda \delta_{\text{inst}}$$

where $\lambda = \xi/(1 + 50\rho')$, and $1.0 \le \xi \le 2.0$.

15. A beam may fail in any one of three flexural modes:
   a) Cracking of concrete followed by breakage of steel, because of an insufficient amount of steel for the tension released by the concrete (sudden failure);
   b) Yielding of steel followed by crushing of concrete in compression after considerable cracking and large displacements (ductile failure);
   c) Crushing of concrete before steel reaches yield point, because too much steel is provided (brittle failure).

16. Failure mode **a** shall be prevented by observing the minimum steel requirement

$$A_{s,\min} = \frac{3\sqrt{f_c'}}{f_y} bd \ge 200 \frac{bd}{f_y}$$

Failure mode **c** (over-reinforced beam) shall be prevented by observing the maximum steel limitation

$$\rho_{max} = \frac{3}{4}\rho_b = .6375\beta_1 \frac{f'_c}{f_y}\left[\frac{87,000}{87,000 + f_y}\right]$$

where

$$\beta_1 = 0.85 \qquad\qquad\qquad\qquad \text{for } f'_c \leq 4 \text{ ksi}$$

$$= 0.85 - 0.05(f'_c - 4) \geq 0.65 \qquad \text{for } f'_c \geq 4 \text{ ksi}$$

ρ_b is the balanced reinforcing ratio, which causes the concrete to reach its crushing strain at the same instant the steel is stressed to its yield point.

17. The only acceptable failure mode is the second one, with sufficient ductility to safely carry overloads and to give ample warning (through cracking and large deflections) of impending failure.

18. The nominal moment capacity of a beam is given by

$$M_n = .85 f'_c ab\left(d - \frac{a}{2}\right)$$

$$= A_s f_y\left(d - \frac{a}{2}\right)$$

where $a = A_s f_y/.85 f'_c b$, provided the reinforcement limits are observed.

19. Ductility is needed for
   a) assurance of a nonbrittle failure mode;
   b) inelastic redistribution of moments in continuous structures;
   c) design against dynamic loads, such as those due to earthquakes.

20. The flexural design of reinforced concrete members is based on the following assumptions:
   a) plain sections remain plain;
   b) perfect bond between steel and concrete assure strain compatibility;
   c) all concrete below the neutral axis is cracked and ineffective;
   d) concrete crushes at the maximum strain value of 0.003.

PROBLEMS

3.1 For each of the structures shown (see Fig. P3.1) and the loadings indicated,
   (a) draw the shear diagram;
   (b) draw the moment diagram;
   (c) qualitatively draw the deflected shape.

3.2 A standard AASHTO truck is positioned on an 80-ft simple span bridge as shown in Fig. P3.2.
   (a) Draw the moment and shear diagrams.
   (b) Show that the truck is in a position that produces the largest moment possible. (Hint: Designate the distance from the left support to the left wheel load as x and use the rules of calculus to solve this problem.)

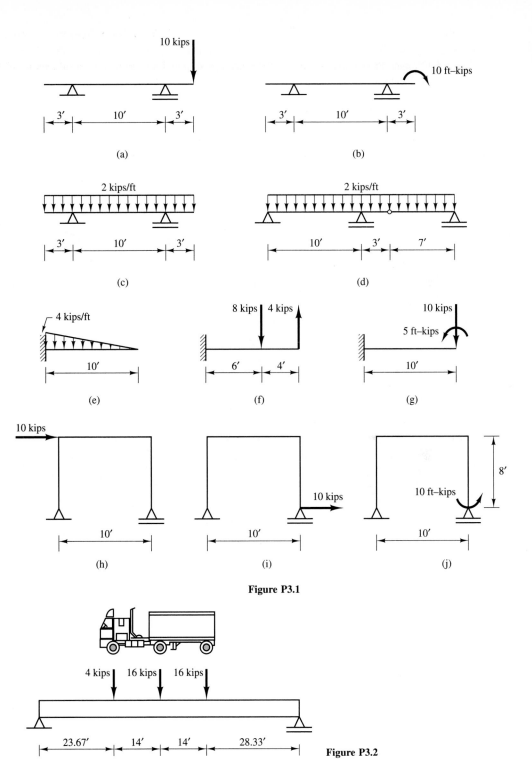

10 kips

3′ | 10′ | 3′

(a)

10 ft–kips

3′ | 10′ | 3′

(b)

2 kips/ft

3′ | 10′ | 3′

(c)

2 kips/ft

10′ | 3′ | 7′

(d)

4 kips/ft

10′

(e)

8 kips | 4 kips

6′ | 4′

(f)

10 kips

5 ft–kips

10′

(g)

10 kips

10′

(h)

10 kips

10′

(i)

10 ft–kips

8′

10′

(j)

Figure P3.1

4 kips | 16 kips | 16 kips

23.67′ | 14′ | 14′ | 28.33′

Figure P3.2

155

3.3 Compute the moment of inertia and both section moduli about the horizontal axis for each of the cross sections (see Fig. P3.3).

3.4 Compute for each of the sections shown (see Fig. P3.4) the moment of inertia and both section moduli, considering their transformed section properties. Note that we normally transform steel to an equivalent area of concrete. The section of Fig. P3.4b is made up of two hypothetical materials. Compute its moment of inertia after transforming material 2 to an equivalent area of material 1, then after transforming material 1 to an equivalent area of

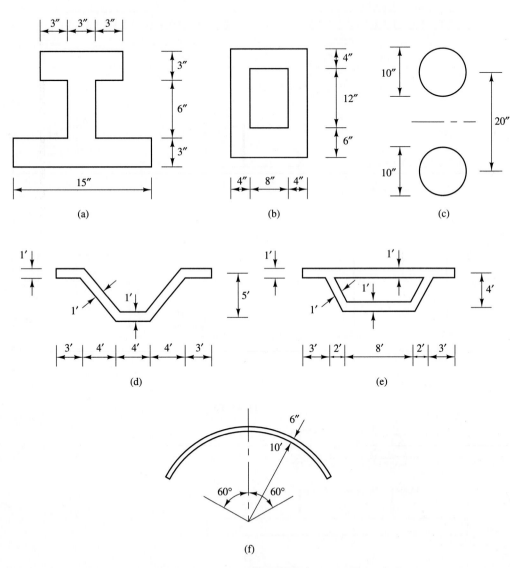

Figure P3.3

material 2, and compare the results. What is your conclusion? (Except for the section of Fig. P3.4b, use $E_s = 29,000$ ksi, $E_c = 3000$ ksi.)

3.5 For the section shown in Fig. P3.4c and an externally applied moment of 50 ft-kips,
   (a) compute the stresses at the top and bottom of the concrete slab;
   (b) compute the stresses in the top and bottom fibers of the steel beam.

3.6 Assuming the section of Fig. P3.4d is uncracked, compute for an externally applied moment of 20 ft-kips,
   (a) the concrete stress in the top fiber;
   (b) the steel stress in the reinforcing bars.

3.7. The two column cross sections shown in Fig. P3.7 are both subjected to an axial compression force of 500 kips, a bending moment about its x-axis of $+1500$ in-kips, and a bending moment about its y-axis of $+1000$ in-kips. Compute the maximum compressive stress in the concrete and the maximum tensile stress in the steel
   (a) using the gross section properties,
   (b) using the transformed section properties.
   Note: Ignore the fact that the concrete may be cracked. Use $n = 9$.

3.8 Using gross section properties, determine the cracking moment for each of the three sections shown in Fig. P3.8. Given: $f_r = 450$ psi.

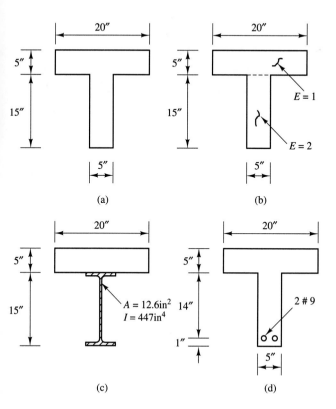

(a)   (b)

(c)   (d)   **Figure P3.4**

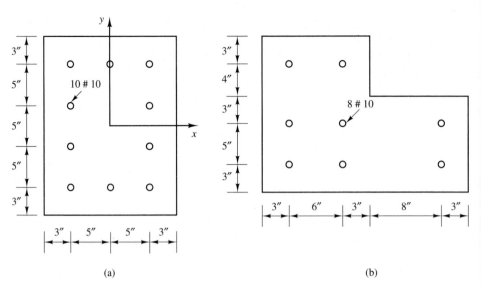

Figure P3.7

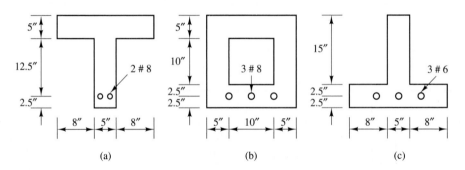

Figure P3.8

3.9 The beam shown in Fig. P3.9 carries only its own dead weight.

   (a) Assuming the concrete weighs 150 lb/ft^3 and is not cracked, compute stresses in the steel and in the extreme compression fiber of the concrete. Use transformed section properties and $n = 8$.

   (b) Assuming the modulus of rupture of the concrete is 450 psi, determine the magnitude of a concentrated midspan load applied to the beam in addition to its dead weight, which will cause the concrete to crack.

   (c) After having cracked the concrete, the concentrated midspan load of problem **(b)** is removed. Compute stresses in the steel and in the extreme compression fiber of the concrete, due to dead weight alone, this time using the cracked section properties. ($E_c = 3600$ ksi).

   (d) Does the section satisfy the ACI Code's minimum reinforcement requirement? ($f_y = 60,000$ psi).

   (e) Does the section satisfy the ACI Code's requirement for crack control?

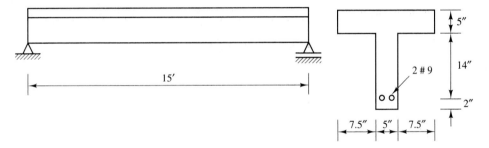

Figure P3.9

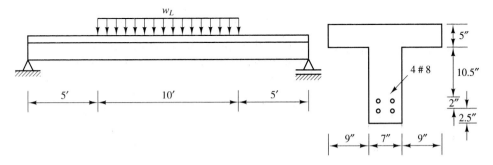

Figure P3.10

3.10 The T-beam shown in Fig. P3.10 is loaded by its own dead weight and live load distributed only over the center half of its 20-ft simple span.

(a) Determine the amount of live load w_L that will cause the concrete to crack. Given: $f'_c = 4000$ psi, $f_r = 7.5\sqrt{f'_c}$, 150 pcf concrete. Note: You may use the gross section properties.

(b) For the service load $w_L = 2$ kips/ft, compute the maximum stresses in the steel and concrete. Given: $E_s = 29,000$ ksi, $E_c = 57,000\sqrt{f'_c}$.

(c) Assuming external exposure, does the design satisfy the ACI Code requirement for crack control? Given: $f_y = 60$ ksi.

3.11 Suppose the beam of Problem 3.10 spans a window opening. Will a $\frac{1}{2}$-in play in the window gaskets be sufficient if the live load is rather temporary? (Hint: to compute live-load deflections, you may take the average of the deflection due to the load distributed over the entire beam, and the deflection due to the same load concentrated at midspan.)

3.12 Check whether the total long-term deflection of the cantilever ends of the beam shown in Fig. P3.12 satisfies the limit specified in the ACI Code. Given: $f'_c = 4000$ psi; $n = 8$.

3.13. The tank wall shown in Fig. P3.13 is designed to resist water pressure when the tank is full. Compute the long-term horizontal displacement of the top of the wall, assuming the wall is perfectly fixed at the base. Given: $f'_c = 4$ ksi; note that water weighs 62.5 pcf.

3.14. Compute the instantaneous deflection of the bridge girder subjected to the truck load shown in Fig. P3.14. Given: $f'_c = 4000$ psi; $n = 8$.

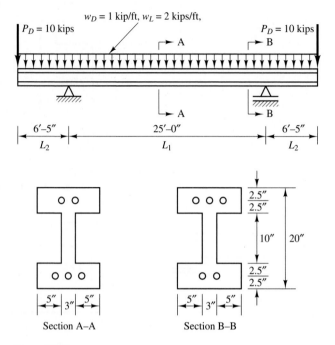

Figure P3.12

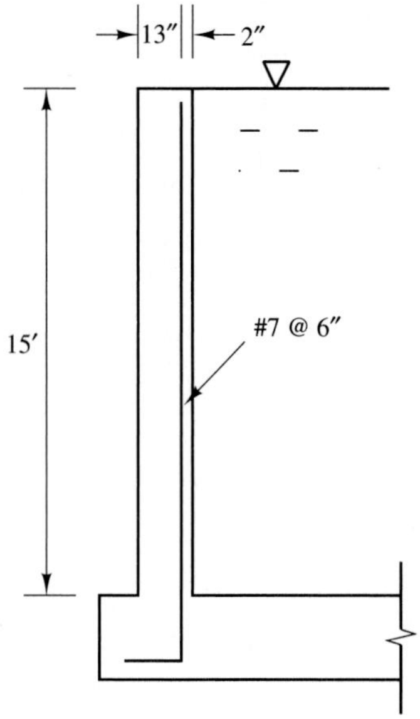

Figure P3.13

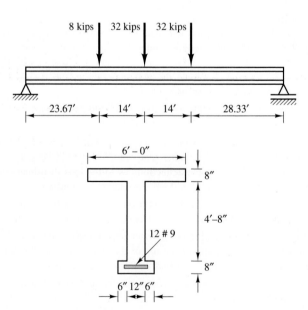

Figure P3.14

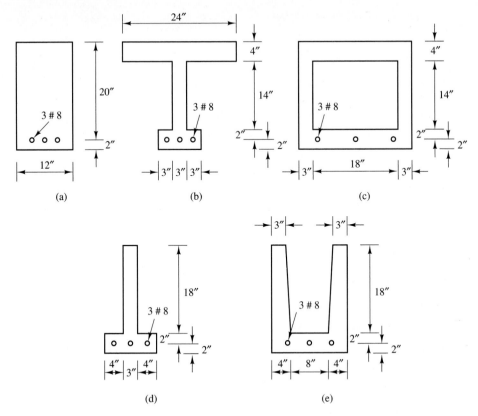

Figure P3.15

3.15. Compute the nominal moment capacities of the cross sections shown in Fig. P3.15. Given: $f'_c = 4000$ psi; $f_y = 60,000$ psi.

3.16. For each of the sections of Problem 3.15, check whether the reinforcement satisfies the minimum and maximum limits of the ACI Code.

3.17. A building is to be constructed with high-strength concrete, which in a laboratory test fails at a strain of 0.002. Derive the balanced steel formula for this case.

3.18. For the problem of Sec. 3.6 (Fig. 3.35), find the moment that corresponds to the curvature $\phi = .0001$. Assume the neutral axis is located 6.44 in from the top of the beam.

Design of Beams for Flexure

4.1 INTRODUCTION

In the preceding chapter we studied in detail the behavior of beams subjected to loads of different intensities. In this chapter, we will apply these findings to design beams for flexure. The design against shear as well as reinforcing details dictated by bond and other requirements will be dealt with thereafter. It should be noted that the beam design procedures are directly applicable to the design of slabs, which may simply be considered as a series of one-foot-wide rectangular beams, placed side by side (see Chapter 8).

Following a design principle stated earlier, structural material should be concentrated where it is most effective. For concrete in a flexural member, this is the compression flange. On the tension side, only a minimum amount of concrete is needed to enclose the steel. In the web area between the two flanges, theoretically no concrete would be needed to resist flexure. However, since bending moments are almost always coupled with shear forces, the width of the beam web is normally dimensioned on the basis of shear requirements, the subject of Chapter 5. The result is a T-section, which can be thought of as the ideal section for a reinforced concrete beam (Fig. 4.1), just as the I-beam is ideal for structural steel.

All ultimate strength design is based on equations of the form

$$R > U \tag{4.1}$$

where R is the structural strength or resistance, and U is the load effect. Equation 4.1 simply expresses the obvious requirement that to avoid failure, structural strength should

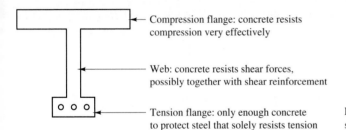

Compression flange: concrete resists compression very effectively

Web: concrete resists shear forces, possibly together with shear reinforcement

Tension flange: only enough concrete to protect steel that solely resists tension

Figure 4.1   Ideal reinforced concrete beam section.

always exceed the effects of loads. For example, the effect of a uniform load w on the maximum moment in a simply supported beam is $wL^2/8$, so that Eq. 4.1 would take the form

$$M > \frac{wL^2}{8}$$

where M is the beam's moment capacity, which should exceed the bending moments produced by the load.

The $>$ sign in Eq. 4.1 signifies the need for a safety factor. It is not sufficient that the strength of a structural member merely equals the effect of loads that have a high probability of being applied to the member. The ACI Code calls for two sets of safety factors (as do many other design codes).

1. It specifies multipliers, or *load factors* (or even better, *overload* factors), for the service loads for which the member is to be designed; these are generally larger than unity.

2. It specifies factors by which the nominal member strength is to be reduced. Designated by the symbol ϕ, such *strength reduction factors* are always less than unity.

The load factors depend on the type of load, their probability of occurrence, and on the load combination under consideration. For example, the ACI Code [Sec. 9.2] specifies the following requirements

$$U = 1.4D + 1.7L \tag{4.2}$$

$$U = 0.75(1.4D + 1.7L + 1.7W) \tag{4.3}$$

$$U = 0.9D + 1.3W \tag{4.4}$$

where D is the dead-load effect, L the live-load effect, and W the wind-load effect. All load effects are those due to actual service loads as prescribed in the governing code, such as the Uniform Building Code (see Chapter 1). The factor 1.4 for dead load in Eq. 4.2 is quite small, because dead weight can typically be determined with high accuracy; and the probability that the actual dead weight exceeds the assumed value is relatively small. Live loads, on the other hand, are often very unpredictable, and the probability that the maximum value prescribed in some code will be exceeded at least once during the lifetime of the member can be very high. This fact is reflected by the overload factor 1.7.

When wind is considered in addition to dead and live load (Eq. 4.3), the Code allows a 25% reduction of the load factors, because of the small likelihood that the maximum wind will be acting at the same time a full live load is present. However, to guard against situations where live loads are counteracting wind effects, the Code requires that, in Eq. 4.3, both the full live load L and no live load shall be considered to determine the more severe condition. Similarly, Eq. 4.4 applies to situations where the dead load counteracts the wind effects. This is one of the few cases where safety requires a load factor less than unity.

The second type of safety factor, the strength reduction factor ϕ, is introduced to account for uncertainties inherent in the computed (or *nominal*) strength of a member. For example, material properties always have a certain statistical scatter. Also, because of construction tolerances, the internal moment arm in a beam may be less than assumed. For flexural design, the ACI Code specifies $\phi = 0.9$. Equation 4.2 can thus be written as

$$M_u = 0.9M_n \geq 1.4M_D + 1.7M_L \tag{4.5}$$

where M_n is the nominal moment capacity of the section (see Eq. 3.49), M_u is the ultimate or reduced moment capacity, M_D is the dead-load moment, and M_L the live-load moment. Equation 4.5 simply requires that the ultimate moment capacity of the section exceed the combined factored moments due to dead and live load. Note that by using the $=$ of the $\geq$ sign in Eq. 4.5, the symbol M_u assumes a dual meaning; it can be both the ultimate moment capacity and the moment due to factored loads. We will distinguish between these two different meanings whenever necessary.

Even though the nominal moment capacity of a section has been derived in detail in Chapter 3, it is appropriate to review the major steps. First, we recall the two fundamental conditions of equilibrium:

$$T = C \tag{4.6}$$

and

$$\begin{aligned} M_n &= T\left(d - \frac{a}{2}\right) \\ &= C\left(d - \frac{a}{2}\right) \end{aligned} \tag{4.7}$$

Figure 4.2 shows strains and stresses in a section of an under-reinforced beam at failure. With $C = 0.85f_c'ab$ and $T = A_s f_y$, the ultimate moment capacity follows as

$$\begin{aligned} M_u &= \phi 0.85 f_c'ab\left(d - \frac{a'}{2}\right) \\ &= \phi A_s f_y\left(d - \frac{a}{2}\right) \end{aligned} \tag{4.8}$$

where $\phi = 0.9$. Equation 4.6, or

$$0.85 f_c'ab = A_s f_y$$

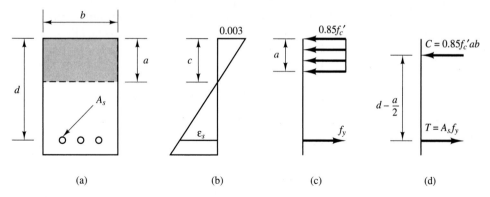

Figure 4.2 Ultimate strains and stresses in an under-reinforced beam: (a) section; (b) strains; (c) stresses; (d) internal forces.

yields the depth of the stress block a as

$$a = \frac{A_s f_y}{0.85 f_c' b} \qquad (4.9)$$

Substitution of Eq. 4.9 into 4.8 leads to

$$M_u = \phi A_s f_y \left(d - \frac{A_s f_y}{1.7 f_c' b} \right)$$

and with the reinforcing ratio, $\rho = A_s / bd$,

$$M_u = \phi \rho f_y bd^2 \left(1 - \frac{\rho f_y}{1.7 f_c'} \right) \qquad (4.10)$$

M_u is the ultimate moment capacity required to resist the moments due to factored loads. If the loads are known, the moments they produce can be determined by structural analysis. In that case, Eq. 4.10 represents a design problem with the objective of providing a beam with moment capacity M_u at least equal to the factored moments due to load. This capacity is a function of five design variables: b, d, ρ, f_y, and f_c'. Theoretically, there are infinite solutions to this problem in *five-dimensional design space*. In practice, it is common to preselect the material properties f_y and f_c' and the beam cross-sectional dimensions b and d, based on practical experience or other constraints. Then, the only remaining unknown on the right-hand side of Eq. 4.10 is the reinforcing ratio ρ, so that Eq. 4.10 can be solved directly for ρ,

$$\rho = \beta - \sqrt{\beta^2 - \frac{2 M_u \beta}{\phi f_y bd^2}} \qquad (4.11)$$

where

$$\beta = \frac{0.85 f_c'}{f_y} \qquad (4.12)$$

As an alternative, the required steel area may be determined iteratively. We start by assuming a value for a, the depth of the concrete compression block. This is typically a small value, maybe a few inches. In the absence of any idea for a starting value, even $a = 0$ will do. Equation 4.8 can then be solved for a first estimate of the steel area:

$$A_s = \frac{M_u}{\phi f_y (d - \frac{a}{2})} \tag{4.13}$$

Next, Eq. 4.9, that is, our second equation of equilibrium, is used to compute an improved estimate for a. This is substituted into Eq. 4.13, and iteration continues until A_s converges.

Why would we want to replace a direct solution (Eq. 4.11) with a fixed number of calculation steps by an iterative procedure with an unknown number of iterations? There are basically three reasons. First, the iteration converges very rapidly, because $a/2$ is always small compared with d as long as the beam is under-reinforced. Seldom will more than two or three iterations be required. With experience, even a single iteration is typically sufficient, in which case the computational effort is less than that required to solve Eq. 4.11. Also, the required accuracy is not high, underscored by the discreteness of the available bar sizes. A second reason why the iterative solution might be preferable is that the two equations of equilibrium, Eqs. 4.9 and 4.13, have clear physical significance and give the designer a good feeling for the design process. Third, the iterative procedure is readily adopted to more complex designs involving nonrectangular and composite sections and sections with compression reinforcement, as we shall see later in this chapter.

Instead of preselecting cross-sectional dimensions b and d, we can choose a reasonable value for ρ and solve Eq. 4.10 for bd^2:

$$bd^2 = \frac{M_u}{\phi \rho f_y (1 - \frac{\rho f_y}{1.7 f_c'})} \tag{4.14}$$

Equation 4.14 also has an unlimited number of cross sections as solutions. However, if the depth d is limited or even prescribed, Eq. 4.14 can readily be solved for the beam width b.

Before outlining practical design procedures in detail, some comments on the five design parameters are appropriate.

For concrete strength f_c', 3000 or 4000 psi are normally adequate. Concrete mixes with such strengths are readily available, cost-effective, and easily placed and compacted. For longer spans and heavily loaded members, slightly more expensive and stronger concrete may be justified, especially if this results in smaller member dimensions and dead weight. However, as Example 3.10 showed, the flexural strength is only marginally controlled by f_c'; therefore, the use of lightweight concrete is more cost-effective in such cases. Exposure to aggressive environments such as seawater may dictate the use of denser and higher-strength concretes. For prestressed concrete members, the minimum concrete strength is usually 5000 psi or more. The choice of a concrete strength is usually made beforehand for an entire structure or structural segment. For in-situ construction, the choice of a uniform concrete strength simplifies both the design and construction.

For highly stressed members like columns in tall buildings, higher-strength concrete is frequently specified.

In the past, most regular reinforcing steel had a yield point of 40,000 psi. In recent years, however, standard practice has shifted toward the use of steels with 60,000 psi yield strength or more. Higher strength reduces the required steel area, which in turn simplifies some reinforcing details such as clearance requirements, where large numbers of lower-strength bars might lead to congestion. However, the problem of crack control becomes more acute for higher-strength steels. Thus, if crack control turns out to be a problem, lower-strength steel may have to be used. Again, the use of only one grade of steel for a job simplifies construction, although for highly stressed members, the specification of higher-grade steel is often justified.

The upper and lower bounds on the reinforcing ratio ρ have been discussed at length in the previous chapter. These bounds still permit a considerable latitude in the selection of ρ. However, higher ρ values lead to reduced ductility and often cause congestion. Although the use of more than one layer of reinforcing bars is not uncommon, the feasibility of alternative designs should be explored, because rebars are most effective if placed as close as possible to the tension face, and the unit cost of placing several layers of rebars is higher than of single-layer arrangements.

The selection of bars to furnish a required steel area is restricted by the limited number of available sizes. These are listed in Table B.1 of Appendix B. The Code also specifies *cover requirements*. These are minimum concrete thicknesses meant primarily as protection for the steel against corrosion. For nonprestressed, cast-in-place beams the required minimum cover requirements are listed in Table 4.1 (ACI Code Sec. 7.7). Similar limits are prescribed by the Code for precast and for prestressed concrete members.

The spacing of reinforcing bars is also subject to Code restrictions. If spaced too closely, it can be difficult to place concrete between and below the bars without leaving voids. Also, closely spaced bars, when stressed, have a tendency to split off the concrete cover. Therefore, the clear distance between parallel bars in a layer shall not be less than either the nominal diameter of the bar nor one inch. In addition, the maximum size of coarse aggregate shall not be larger than three-fourths the minimum clear spacing between bars. When parallel reinforcement is placed in two or more layers, the bars in the upper layers shall be placed directly above those in the bottom layer, with the clear distance between layers not less than one inch.

Beam dimensions, in particular the beam depth h, are in many instances predetermined by architectural, economic, or construction constraints. Smaller beam depths reduce the overall construction height, with potential economic benefits, for example, in multistory buildings and multilevel freeway interchange structures. On the other hand, beams that are too shallow may deflect excessively and require too much steel unless recourse is taken to prestressing. Also, if the lateral stiffness of a building depends entirely on rigid frame action, exceedingly shallow beams will lead to unacceptable lateral displacements called *drift*. In Table 3.1, we saw a set of limiting depth/span ratios born out of the necessity to control deflections. These may serve as initial rules of thumb. As designers accumulate experience, they will refine these rules, but they will also find that choices are rather limited.

TABLE 4.1 MINIMUM COVER REQUIREMENTS FOR NONPRESTRESSED CAST-IN-PLACE MEMBERS

Concrete exposure	Minimum cover* (in)
Cast against and permanently exposed to earth	3
Exposed to earth or weather	
#6 through #18 bars	2
#5 bars, W31 or D31 wire, and smaller	$1\frac{1}{2}$
Not exposed to weather or in contact with ground	
Slabs, walls, joists:	
#14 and #18 bars	$1\frac{1}{2}$
#11 bar and smaller	$\frac{3}{4}$
Beams, columns:	
primary reinforcement, ties, stirrups, spirals	$1\frac{1}{2}$
Shells, folded plate members:	
#6 bar and larger	$\frac{3}{4}$
#5 bar, W31 or D31 wire, and smaller	$\frac{1}{2}$

*Measured from exposed concrete surface to outermost steel surface

    The beam width b is not subject to constraints, such as the beam depth h. Also, it contributes little to flexural strength. The web width of nonrectangular sections is controlled by shear, not flexure. Therefore, the h/b ratio for rectangular beams is quite arbitary, but normally kept around two (or somewhat larger for longer spans).

    To conclude this introduction, it may be worthwhile to repeat an earlier statement: the art of design cannot be taught entirely. It must emerge from practical experience through numerous trial-and-error designs and the study of other designers' work. Through experience, the designer learns to preselect almost all free design parameters so that only one parameter remains (usually the steel area) to satisfy all design requirements. The only way for the student to gain such insight is by working out many examples and exercises and not getting discouraged if the first trial-and-error attempts turn out to be rather lengthy affairs.

4.2 DESIGN OF RECTANGULAR BEAMS

With the preliminaries now behind us, we are in a position to design beams with rectangular cross sections for flexure, following the standard step-by-step procedure below.

    Step 1.   *Select material properties f'_c and f_y.* As stated earlier, this choice is usually made beforehand for the entire structure or structure segment.

    Step 2.   *Establish minimum beam depth.* The minimum beam depth requirements of Table 3.1 (ACI Code Table 9.5a) apply strictly only to members not supporting or attached to partitions or other construction likely to be damaged by large deflections. However, they can serve as a rule of thumb for reasonable minimum beam depths.

Step 3. *Select beam cross-sectional dimensions b and d.* If the beam depth is not fixed by architectural or other requirements, start out with a reasonable depth/span ratio, say $\frac{1}{12}$, guided by the minimum depth requirement of Step 2. Draw the beam elevation to scale and judge if the selected depth/span ratio looks reasonable. Then choose a beam width b, possibly equal to $d/2$, or less for longer spans. Beam dimensions are usually expressed in full inches, larger beams in even inches, and slabs in half inches. Note that the overall beam depth h is equal to d plus an allowance for concrete cover and an estimated half-bar diameter (say $\frac{1}{2}$ in for a #8 bar).

Step 4. *Determine factored moments.* Establish the service live loads and dead load on the basis of the assumed beam dimensions, multiply them with their respective load factors (Eqs. 4.2 through 4.4, or ACI Code Sec. 9.2), then compute the combined design or factored bending moment, which establishes the required ultimate moment capacity M_u. Although concrete is an inelastic material, experience has shown that designs based on moments computed by elastic theory are generally satisfactory.

Step 5a. *Determine required steel area A_s by iteration*
   1. Assume a value for a, the depth of the concrete compression area.
   2. Compute the required steel area:

$$A_s = \frac{M_u}{\phi f_y \left(d - \frac{a}{2}\right)} \tag{4.15}$$

   3. Update the depth of the compression area:

$$a = \frac{A_s f_y}{.85 f_c' b} \tag{4.16}$$

   4. Repeat until convergence has been achieved.

Step 5b. *Determine required steel area A_s by direct solution, using Eq. 4.11.*

$$\rho = \beta - \sqrt{\beta^2 - \frac{2 M_u \beta}{\phi f_y b d^2}} \tag{4.17}$$

where

$$\beta = \frac{0.85 f_c'}{f_y} \tag{4.18}$$

The steel area follows as

$$A_s = \rho b d$$

Note that, as a rule, A_s is almost proportional to M_u.

Step 6. *Check steel area for minimum and maximum bounds.* The steel ratio ρ must satisfy the requirements of Eqs. 3.15, 3.17, and 3.45.

$$\rho_{min} = \max \left\{ \frac{3\sqrt{f'_c}}{f_y}; \frac{200}{f_y} \right\} \text{ or } \frac{4}{3}\rho \qquad (4.19)[10\text{-}3]$$

$$\frac{A_s}{b_{comp}d} \leq \rho_{max} = .6375\beta_1 \frac{f'_c}{f_y} \left(\frac{87,000}{87,000 + f_y} \right) \qquad (4.20)$$

If Eq. 4.19 is violated, it is likely that a smaller beam section is adequate; if Eq. 4.20 is violated, a larger beam or one with compression reinforcement may be necessary. In either case, go back to Step 3 to revise the beam dimensions. The factored moment needs to be updated only if the dead weight is changed considerably and constitutes a large fraction of the total load.

Step 7. *Select and arrange reinforcing bars.* The selected rebars must have a combined area of at least A_s. If these bars can be placed in the beam with proper clearances and cover (Table 4.1, or ACI Code Secs. 7.6 and 7.7), the design is almost complete. For ease of construction, it may be preferable to choose bars of only one size, although bars of more than one size are acceptable. For obvious reasons, bars in symmetric beams are always arranged symmetrically. For large steel areas, the bars may have to be bundled or placed in more than one layer. If, even with these measures, the required clearances and minimum cover cannot be maintained, go back to Step 3 and increase the beam dimensions. Another alternative is the use of compression reinforcement (see Sec. 4.4).

Step 8. *Check maximum crack width.* In order to avoid excessive cracking, Eq. 3.28 must be satisfied if $f_y > 40$ ksi, that is,

$$Z = f_s\sqrt[3]{d_c A} \leq 145 \text{ for exterior exposure}$$
$$\leq 175 \text{ for interior exposure} \qquad (4.21)$$

where $f_s = 0.6f_y$ may be taken.

Step 9. *Check maximum deflections.* If attached partitions, windows, etc. can be damaged through excessive deflections, compute elastic as well as long-term deflections (as described in Chapter 3) and compare these with the maximum permissible deflections of Table 3.2. If these limits are exceeded, the beam depth will probably have to be increased, although small improvements are possible by increasing the beam width and steel area.

EXAMPLE 4.1

Design a beam with a simple span of 24 ft to support a live load of 0.20 kips/ft. Assume that the beam will not be attached to other construction. Given: $f'_c = 3000$ psi; $f_y = 50,000$ psi; normal weight concrete weighing 150 lb/ft^3.

Solution

Step 1. Materials are predetermined.

Step 2. Minimum beam depth follows from Table 3.1 as $L/16 = (24)(12)/16 = 18$ in.

Step 3. Start with $L/h = 12$ and $h/b = 2$, from which $h = (24)(12)/12 = 24$ in, $d = 24 - 2 = 22$ in, $b = h/2 = 12$ in.

Step 4. Dead weight: $w_D = \dfrac{(24)(12)(.15)}{144} = 0.3$ kips/ft

Factored load: $w_u = 1.4w_D + 1.7w_L = (1.4)(0.3) + (1.7)(0.2) = 0.76$ kips/ft

Factored moment: $M_u = \dfrac{w_u L^2}{8} = \dfrac{(.76)(24)^2}{8} = 54.7$ ft-kips

Step 5. Start with $a = 2$ in, then

Steel area: $A_s = \dfrac{M_u}{\phi f_y (d - \frac{a}{2})} = \dfrac{(54.7)(12)}{(0.9)(50)(22 - \frac{2}{2})} = 0.69$ in^2

Improved compression block depth: $a = \dfrac{A_s f_y}{.85 f'_c b} = \dfrac{(.69)(50)}{(.85)(3)(12)} = 1.14$ in

Improved steel area: $A_s = \dfrac{(54.7)(12)}{(0.9)(50)(22 - \frac{1.14}{2})} = 0.68$ in^2.

Even though a changed by almost 100%, A_s hardly changed.

Steel reinforcing ratio: $\rho = \dfrac{A_s}{bd} = \dfrac{0.68}{(12)(22)} = 0.0026$

Alternatively, we could have solved Eq. 4.17 directly for ρ: $\beta = \dfrac{0.85 f'_c}{f_y} =$

$\dfrac{(.85)(3)}{50} = 0.051; \quad \rho = .051 - \sqrt{.051^2 - \dfrac{(2)(54.7)(12)(.051)}{(0.9)(50)(12)(22)^2}} = .0026$

Step 6. Minimum steel requirement: $\rho_{min} = \max\left\{\dfrac{3\sqrt{3000}}{50,000}; \dfrac{200}{50,000}\right\} = .004 > .0026$

NG. A smaller beam is justified. Reduce h to 20 in, that is, $d = 18$ in, $b = 9$ in, and repeat calculations.

$w_D = \dfrac{(20)(9)(.15)}{144} = .1875$ kips/ft

$w_u = (1.4)(.1875) + (1.7)(0.2) = .6025$ kips/ft

$M_u = \dfrac{(.6025)(24)^2}{8} = 43.4$ ft-kips

Let $a = 1$, $A_s = \dfrac{(43.4)(12)}{(0.9)(50)(18 - \frac{1}{2})} = 0.66$, $a = \dfrac{(.66)(50)}{(.85)(3)(9)} = 1.44$, close enough.

$\rho = \dfrac{.66}{(9)(18)} = .0041 > \rho_{min}$ OK

$\rho_{max} = (.6375)(.85)\dfrac{3}{50}\dfrac{87,000}{87,000 + 50,000} = .0206 > .0041$ OK

Step 7. Use 2 #6 bars with $A_s = .88$ in^2 and $\rho = .88/(9)(18) = .0054$. Cover and clearance requirements are easily satisfied (see Fig. 4.3).

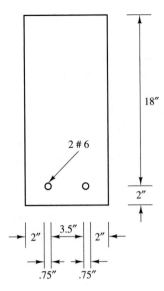

Figure 4.3 Section for Example 4.1.

Step 8. Crack control, with $d_c = 2$ in, $A = (2)(2)(9)/2 = 18$.

$$Z = f_s \sqrt[3]{d_c A} = (0.6)(50)\sqrt[3]{(2)(18)} = 99 < 145 \ \underline{OK}$$

Step 9. Not necessary because of stated assumption and minimum depth requirement.

∎

It is worth noting that the smallest bars available to satisfy $A_s = .66$ in^2 supply a steel area 33% in excess of the requirement. Although this is somewhat unusual, this example illustrates that excessive accuracy in design calculations is not warranted.

EXAMPLE 4.2

The window opening of Fig. 4.4 is to be spanned by a beam that supports a uniform live load of 500 lb/ft and a concentrated midspan live load of 7000 lb in addition to its own weight. The beam depth is set by the architect as $h = 18$ in. Given: $f'_c = 3000$ psi; $f_y = 60,000$ psi; $f_r = 350$ psi.

Solution

Step 1. Material properties are given.

Step 2. $h_{min} = \dfrac{L}{16} = \dfrac{(20)(12)}{16} = 15$ in < 18 in $\underline{OK}$.
The architect's requirement appears to be feasible.

Step 3. Choose $h = 18$ in, $d = 18 - 2.5 = 15.5$ in, $b = 9$ in.

Step 4. Dead weight: $w_D = \dfrac{(18)(9)(.15)}{144} = .169$ kips/ft

Dead-weight moment: $M_D = \dfrac{(.169)(20)^2}{8} = 8.45$ ft-kips

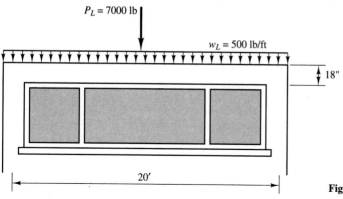

$P_L = 7000$ lb

$w_L = 500$ lb/ft

18"

20′

Figure 4.4   Example 4.2: Beam layout.

Live-load moment:

$$M_L = \frac{w_L L^2}{8} + \frac{P_L L}{4} = \frac{(0.5)(20)^2}{8} + \frac{(7)(20)}{4} = 60 \text{ ft-kips}$$

Factored moment: $M_u = (1.4)(8.45) + (1.7)(60) = 114$ ft-kips

Step 5. Assume $a = 2$.

Steel area: $A_s = \dfrac{(114)(12)}{(0.9)(60)(15.5 - 1)} = 1.75$ in^2

Revise a: $a = \dfrac{(1.75)(60)}{(.85)(3)(9)} = 4.57$ in

Revise steel area: $A_s = \dfrac{(114)(12)}{(0.9)(60)(15.5 - 2.28)} = 1.92$ in^2

Check a again: $a = (4.57)\dfrac{1.92}{1.75} = 5.0 \approx 4.57$ in OK

Steel ratio: $\rho = \dfrac{1.92}{(9)(15.5)} = .0138$

Step 6. $\rho_{\min} = \max \left\{ \dfrac{3\sqrt{3000}}{60,000}; \dfrac{200}{60,000} \right\} = .00333 < .0138$ OK

$\rho_{\max} = (.6375)(.85)\dfrac{3}{60}\dfrac{87}{87 + 60} = .0160 > .0138$ OK.

Step 7. Use 2 #9 bars with $A_s = 2.0$ in$^2 > 1.92$ in^2 and $\rho = \dfrac{2.0}{(9)(15.5)} = .0143 < .0160$

OK; vertical cover, $2.5 - \dfrac{1.125}{2} = 1.9375 > 1.5$ in OK.

Step 8. $d_c = 2.5$ in, $A = \dfrac{(2)(2.5)(9)}{2} = 22.5$;

$Z = (0.6)(60)\sqrt[3]{(2.5)(22.5)} = 138 < 145$ OK.

Step 9. Uncracked gross moment of inertia: $I_g = \dfrac{(9)(18)^3}{12} = 4374$ in^4

Cracking moment: $M_{cr} = \dfrac{I_g f_r}{y_t} = \dfrac{(4374)(.350)}{(9)(12)} = 14.17$ ft-kips

Maximum service load moment:

$$M_a = \frac{(0.169 + 0.5)(20)^2}{8} + \frac{(7)(20)}{4} = 68.4 \text{ ft-kips}$$

Young's modulus: $E_c = 57,000\sqrt{3000} = 3,122,000 \text{ psi}$

Modular ratio: $n = \dfrac{29,000}{3,122} = 9.29, \; \rho n = (9.29)(.0143) = 0.133$

Neutral axis:

$$k = -.133 + \sqrt{(.133)^2 + (2)(.133)} = 0.40; \; kd = (0.4)(15.5) = 6.2 \text{ in}$$

Cracked moment of inertia: $I_{cr} = \dfrac{(9)(6.2)^3}{3} + (9.29)(2)(15.5 - 6.2)^2 = 2322 \text{ in}^4$

Effective moment of inertia:

$$I_e = \left(\frac{14.17}{68.4}\right)^3 (4374) + \left[1 - \left(\frac{14.17}{68.4}\right)^3\right](2322) = 2340 \text{ in}^4$$

Dead-load deflection: $\delta_D = \dfrac{5}{384}\dfrac{w_D L^4}{E_c I_e} = \dfrac{(5)(.169)(20)^4(12)^3}{(384)(3122)(2340)} = .0833 \text{ in}$

Live-load deflection: $\delta_L = (.0833)\dfrac{0.5}{0.169} + \dfrac{(7)(20)^3(12)^3}{(48)(3122)(2340)} = .522 \text{ in}$

Total instantaneous deflection: $\delta_{\text{inst}} = .0833 + .522 = .605 \text{ in}$

Long-term deflection: $\delta_t = (2)(.0833) = .167 \text{ in}$

Total deflection: $\delta = .605 + .167 = .772 \text{ in}$

Maximum allowable deflection: $\dfrac{L}{480} = \dfrac{(20)(12)}{480} = 0.5 < .772 \text{ in NG}$

Discussion: The deflection exceeds the allowable value. However, a part of the deflection will have taken place at the time of window installation. If the window manufacturer can assure a play of at least 0.7 in, the design might be acceptable. It would be preferable, however, if the architect were to ease the depth requirement, which is quite stringent. (Try a 12-in-wide beam!) ∎

As mentioned earlier, the design procedure can be modified such that a reinforcing ratio ρ is assumed, together with f_c' and f_y. Beam dimensions b and d are then determined from Eq. 4.14. The drawback of this procedure is the fact that for a dead-load estimate, values for b and d must be assumed anyway. The step-by-step procedure is as follows.

Step 1. Select material properties f_c' and f_y.

Step 2. Establish minimum beam depth.

Step 3. Determine required ultimate moment capacity M_u, using assumed cross-sectional dimensions to estimate the dead weight.

Step 4. Establish upper and lower bounds for reinforcing ratio.

Step 5. Select reinforcing ratio ρ.

Step 6. Determine beam dimensions b and d from Eq. 4.14, that is,

$$bd^2 = \frac{M_u}{\phi \rho f_y \left(1 - \frac{\rho f_y}{1.7 f_c'}\right)}$$

Step 7. Select and arrange reinforcing bars.

Step 8. Check maximum crack width.

Step 9. Check maximum deflection.

EXAMPLE 4.3

Design the beam of Example 4.1 using the modified step-by-step procedure.

Solution

Step 1. Material properties were given: $f_c' = 3000$ psi, $f_y = 50$ ksi.

Step 2. Minimum beam depth is 18 in, as before.

Step 3. Assuming again a 9-by-20-in cross section:

Dead weight: $w_D = \dfrac{(20)(9)(.15)}{144} = .1875$ kips/ft

Factored load: $w_u = (1.4)(.1875) + (1.7)(0.2) = .6025$ kips/ft

Factored moment: $M_u = \dfrac{(.6025)(24)^2}{8} = 43.4$ ft-kips

Step 4. $\rho_{min} = .004$, $\rho_{max} = .0206$, as before.

Step 5. Select reinforcing ratio, $\rho = .005$.

Step 6. $bd^2 = \dfrac{(43.4)(12)}{(0.9)(.005)(50)\left[1 - \frac{(.005)(50)}{(1.7)(3)}\right]} = 2433$ in^3.

For $d = 20$ and $b = \dfrac{2433}{20^2} = 6$, the d/b ratio is high.

Try $d = 18$, $b = \dfrac{2433}{18^2} = 7.5$ (say $b = 8$ in).

Step 7. $A_s = (.005)(18)(8) = .72$ in^2. Use 2 #6 bars with $A_s = .88$ in^2.

The remainder of the example is as in Example 4.1.   ∎

The standard design procedure outlined earlier is applicable not only to regular beams, but also to slabs. These carry load in either one or two directions. The design moment M_u is computed for a typical one-foot-wide strip of the slab, and the required reinforcement is determined for this slab strip. When reinforcing bars are selected, it becomes apparent that the bar spacing introduces a new variable at the designer's disposal. Thus, a requirement of 0.66 in^2 of steel per foot of slab width can be satisfied with #6 bars spaced 8 in apart, #5 bars spaced $5\frac{1}{2}$ in apart, or with #4 bars spaced $3\frac{1}{2}$ in apart. The provided steel area can be selected to be closer to the required steel area than is possible for beams, because the discrepancies caused by the discreteness of available bar sizes can all but be removed. For convenience, Table B.2 in Appendix B summarizes the steel areas (in in²/ft of slab width) for different bar sizes and spacings.

EXAMPLE 4.4

Design a one-way slab spanning 15 ft and supporting a live load of 100 psf. Given: $f'_c = 3500$ psi; $f_y = 60,000$ psi; normal-weight concrete @ 150 pcf.

Solution

Step 1.   Material properties are given.

Step 2.   Minimum slab depth: $\dfrac{L}{20} = \dfrac{(15)(12)}{20} = 9$ in.

Step 3.   Try $h = 9$ in, with $\frac{3}{4}$ in cover, and estimated bar diameter of $\frac{1}{2}$ in; $d = 9 - 0.75 - 0.25 = 8$ in.

Step 4.   Dead weight: $w_D = \dfrac{(9)(12)(.15)}{144} = .1125$ ksf

Live load: $w_L = (0.10)(1) = 0.1$ ksf

Factored moment: $M_u = [(1.4)(.1125) + (1.7)(.10)]\dfrac{15^2}{8} = 9.21$ ft-kips/ft

Step 5.   Let $a = 1$ in, $A_s = \dfrac{(9.21)(12)}{(0.9)(60)(8 - 0.5)} = 0.27$ in²/ft.

Check $a = \dfrac{(0.27)(60)}{(.85)(3.5)(12)} = .46$ in, $A_s = \dfrac{(9.21)(12)}{(0.9)(60)(8 - \frac{.46}{2})} = .26$ in²/ft

Step 6. $\rho_{min} = .0018$ (ACI Code Sec. 7.12), $\rho = \dfrac{.26}{(8)(12)} = .0027 > \rho_{min}$ OK.

Step 7.   Use #4 bars @ 9 in with $A_s = .26$ in²/ft   ■

The American Concrete Institute and the Concrete Reinforcing Steel Institute have developed design aids [1, 2] to speed up the design process. For instance, we may rewrite Eq. 4.14 in the form

$$bd^2 = \frac{M_u}{K_u} \qquad (4.22)$$

where

$$K_u = \phi \rho f_y \left(1 - \frac{\rho f_y}{1.7 f'_c}\right) = \phi q f'_c (1 - 0.59q)$$

and

$$q = \frac{\rho f_y}{f'_c}$$

K_u is a function of only ρ, f_y, and f'_c and has been tabulated as such [1]. Multiplying Eq. 4.22 with 12,000 converts M_u from ft-kips to in-lb. The resulting function

$$F = \frac{M_u}{K_u} = \frac{bd^2}{12,000} \qquad (4.23)$$

has also been tabulated for b and d combinations that satisfy Eq. 4.23. F has the characteristics of a section modulus required for a rectangular beam to carry a given moment.

In the past, design aids such as the quoted ACI and CRSI tables have been in wide use. With the availability of pocket calculators it seems that design practice is shifting away from tabulated design aids, because a typical design can be performed just as easily with a few keystrokes on a pocket calculator, using the step-by-step procedures outlined above. Also, these procedures can be programmed for programmable calculators and spreadsheets.

As a further step, the entire design process can be programmed for a digital computer. Here is not the place to discuss such automated design procedures, which are in fact finding increased use in design offices. However, a word of caution is in order. The student should first become familiar with all details of the design process and feel confident of mastering it before turning to design programs. Otherwise, such use will be based on blind reliance on the computer. The use of the computer without full understanding of the theoretical basis for the program bears the constant danger of program misuse. There are a number of reasons why program users should be intimately familiar with the design procedures of the programs they are using:

1. Users should be aware of all assumptions and limitations of the theory and design procedure used, which often are not clearly spelled out in the program documentation.
2. Users should be able to check and verify by hand the computer output to prevent the use of faulty output, which is usually caused by faulty input.
3. Users should be able to service, maintain, or extend a design program, for example, in case the ACI Code provisions are changed. (With design practice shifting away from programs written in-house to commercial software, this requirement is getting less important.)
4. Users should be able to produce designs even if the computer is temporarily unavailable.
5. Finally, and most important of all, engineers should never forget that they bear the responsibility for their work; this responsibility cannot be delegated to the computer or the program supplier. The use of computers only appears to offer an escape from professional liability, but the engineer remains responsible for his or her work, no matter how the design was arrived at. This implies a healthy skepticism toward all computer-generated results and the capability of checking these for soundness.

4.3 DESIGN OF BEAMS WITH NONRECTANGULAR SECTIONS

Beams with nonrectangular sections play an important role in practice, whether cast in place or precast in factories, where the use of specially designed forms allows greater flexibility toward more efficient cross sections. T-beams in particular are frequently parts of floor slab systems, discussed in further detail in Chapter 8. The principles governing the design of such sections are the same as those for rectangular sections. There are basically two differences. First, the shape of the concrete area in compression, A_c, may now be irregular, thus somewhat complicating the design. Second, the computation of the maximum allowable amount of steel is not as straightforward as for rectangular sections.

Before proceeding with actual design details for beams with nonrectangular sections, two important observations need to be made. First, since the entire concrete area below the neutral axis is assumed to be cracked, there is no difference between the design of an I- and a T-beam, except that it may be more difficult to place the same amount of steel in the stem of a T-beam instead of spreading it over the bottom flange of an I-beam. Secondly, if the depth of the concrete compression zone is less than or equal to the thickness of the top flange, then the design of either a T-, I-, or box-beam is no different than that of a rectangular beam having a width equal to that of the top flange (Fig. 4.5). Even if the compression zone should extend somewhat into the stem or webs, the designs of the various sections are not likely to be much different, that is, the steel required to carry a given moment will be about the same as for a beam with a rectangular section.

Usually, the dimensions of irregular sections are predetermined, reducing the design problem to the determination of the steel area required for a given moment. Sometimes, the amount of steel is known or assumed so that only the computation of the ultimate bending moment capacity is needed. In either case, the amount of steel should be less than the maximum allowed by the Code to assure a ductile mode of failure. In other words, the moment capacity associated with the maximum permissible amount of steel must be at least equal to the moment due to factored loads. Otherwise, the assumed section is inadequate for the given loads.

First, we will derive ρ_{max} and the corresponding maximum design capacity of a beam with a nonrectangular cross section, such as the I-beam of Fig. 4.6. If reinforced with the "balanced" amount of steel, A_{sb}, the steel starts to yield just when the concrete is strained to its crushing point. As in the case of a rectangular beam, the neutral axis position is determined from similar strain triangles (Fig. 4.6b):

$$c = \frac{.003}{.003 + \epsilon_y}d$$

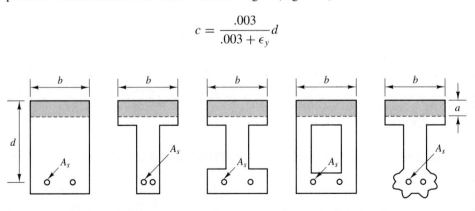

Figure 4.5   Beams with equal ultimate moment capacities.

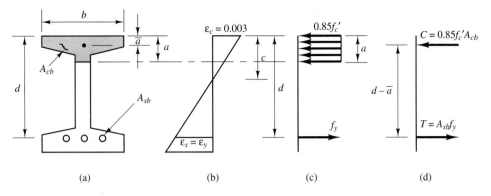

Figure 4.6   Beam at balanced conditions: (a) section; (b) strains; (c) stresses; (d) internal forces.

With $\epsilon_y = f_y/E_s$ and $E_s = 29,000,000$

$$c = \frac{87,000}{87,000 + f_y} d \tag{4.24}$$

and

$$a = \beta_1 c \tag{4.25}$$

where, as before

$$\beta_1 = 0.85 \qquad\qquad\qquad \text{for } f'_c \le 4 \text{ ksi}$$
$$= 0.85 - 0.05(f'_c - 4) \ge 0.65 \quad \text{for } f'_c \ge 4 \text{ ksi}$$

Equilibrium of internal forces (Fig. 4.6d) requires

$$A_{sb} = \frac{.85 f'_c A_{cb}}{f_y} \tag{4.26}$$

where A_{sb} is the balanced steel area and A_{cb} the corresponding concrete area under compression, with a total depth a. Since this depth a is known from Eq. 4.25, A_{cb} can be determined by inspection (Fig. 4.6a), and the balanced steel area follows from Eq. 4.26. The maximum allowable steel area is then given by

$$A_{s,\text{max}} = \frac{3}{4} A_{sb} = .6275 \frac{f'_c}{f_y} A_{cb} \tag{4.27}$$

or, in the form of reinforcing ratio,

$$\rho_{\text{max}} = \frac{A_{s,\text{max}}}{bd} = .6275 \frac{f'_c}{f_y} \frac{A_{cb}}{bd} \tag{4.28}$$

The bending capacity of a beam that contains the maximum permissible amount of reinforcement, $A_{s,\text{max}}$, can now be computed. With the total tensile force being $T = A_{s,\text{max}} f_y$, the corresponding concrete area under compression follows as

$$A_{c,\max} = \frac{A_{s,\max} f_y}{.85 f_c'} = \frac{3}{4} A_{cb} \tag{4.29}$$

for which the total depth and the position of the centroid, $\bar{a}$, are readily determined, knowing the geometry of the section. The maximum moment capacity is then given by

$$M_{u,\max} = \phi A_{s,\max} f_y (d - \bar{a}) \tag{4.30}$$

There is an alternative to computing the maximum allowable amount of steel. In this method, which is outlined in the ACI Code Commentary, the balanced steel A_{sb} is divided into a part A_{sw} needed to balance the concrete compression in the web only, and a part A_{sf} needed to balance the concrete compression in the flange overhangs. Figure 4.7 shows the strains, stresses, and internal forces at balanced conditions. The web components of the forces involved are

$$T_w = A_{sw} f_y = C_w = 0.85 f_c' a b_w \tag{4.31}$$

Similarly, for the flange overhang components,

$$T_f = A_{sf} f_y = C_f = 0.85 f_c' h_f (b - b_w) \tag{4.32}$$

Defining

$$\bar{\rho}_b = \frac{A_{sw}}{b_w d} = .85\beta_1 \frac{f_c'}{f_y} \left(\frac{87,000}{87,000 + f_y} \right) \tag{4.33}$$

as the balanced reinforcing ratio of the section without flange overhangs, and

$$\rho_f = \frac{A_{sf}}{b_w d} = \frac{.85 f_c' h_f (b - b_w)}{f_y b_w d} \tag{4.34}$$

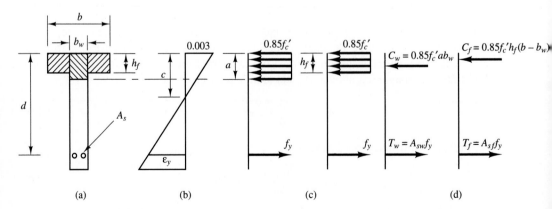

Figure 4.7 Separation of force components for balanced conditions: (a) section; (b) strains; (c) stresses; (d) internal forces.

as the reinforcing ratio needed to develop the compression strength of the flange overhangs, $h_f(b - b_w)$, then the combined balanced reinforcing ratio follows from the definition

$$T = T_w + T_f = A_s f_y = \rho_b b d f_y \tag{4.35}$$

With Eqs. 4.33 and 4.34, this becomes

$$T = \overline{\rho}_b b_w d f_y + \rho_f b_w d f_y = \rho_b b d f_y$$

so that

$$\rho_b = \frac{b_w}{b}(\overline{\rho}_b + \rho_f) \tag{4.36}$$

and

$$\rho_{max} = \frac{3}{4}\rho_b \tag{4.37}$$

EXAMPLE 4.5

Determine the maximum amount of steel with which the beam of Fig. 4.8 may be reinforced. What is the corresponding maximum moment capacity? Given: $f_c' = 3000$ psi; $f_y = 60,000$ psi.

Solution

Neutral axis position at balanced conditions: $c = \dfrac{87}{87 + 60}(27.5) = 16.3$ in

Depth of concrete compression block: $a = (.85)(16.3) = 13.8$ in

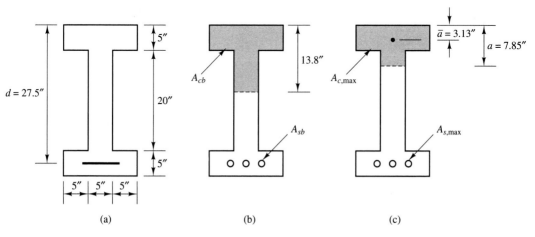

(a)   (b)   (c)

Figure 4.8   Example 4.5 and 4.6: Beam section.

Concrete compression area (see Fig. 4.8b): $A_{cb} = (5)(10) + (13.8)(5) = 119 \text{ in}^2$

Maximum permissible steel: $A_{s,\max} = (.6375)\dfrac{3}{60}(119) = \underline{3.79 \text{ in}^2}$

Corresponding concrete area in compression: $A_{c,\max} = \dfrac{3}{4}(119) = 89.25 \text{ in}^2$

Depth of compression area $A_{c,\max}$ (see Fig. 4.8c): $a = \dfrac{89.25 - (10)(5)}{5} = 7.85 \text{ in}$

Centroid of this area: $\bar{a} = \dfrac{(50)(2.5) + (5)(7.85)^2/2}{89.25} = 3.13 \text{ in}$

Ultimate moment capacity: $M_{u,\max} = (0.9)(3.79)(60)\dfrac{27.5 - 3.13}{12} = \underline{415.6 \text{ ft-kips}}$

We can also compute $A_{s,\max}$ using the alternative method:

Web balancing reinforcing ratio: $\bar{\rho}_b = (.85)(.85)\dfrac{3}{60}\dfrac{87}{87 + 60} = .02138$

Flange overhang balancing steel: $\rho_f = \dfrac{(.85)(3)(5)(15 - 5)}{(60)(5)(27.5)} = .01545$

Maximum combined reinforcing ratio: $\rho_{\max} = (.75)\dfrac{5}{15}(.02138 + .01545) = .00921$

Maximum permissible steel: $A_{s,\max} = (.00921)(15)(27.5) = \underline{3.80 \text{ in}^2}$ ∎

Determining the maximum amount of steel permitted for nonrectangular sections is the more time-consuming task. Computing the amount of steel necessary to carry a given moment is comparatively easy. It is best to start by assuming a value for $\bar{a}$, the position of the centroid of the concrete area under compression. For I- and T-beams, the half-thickness of the top flange is usually a good first estimate. The initial answer for A_s is then simply

$$A_s = \frac{M_u}{\phi f_y(d - \bar{a})} \tag{4.38}$$

and the concrete area under compression follows as

$$A_c = \frac{A_s f_y}{.85 f_c'} \tag{4.39}$$

The overall depth a and the revised position $\bar{a}$ of the centroid for this area A_c are determined, taking the area's specific geometry into account. If the new centroidal position is markedly different from our first estimate, repeated use of Eqs. 4.38 and 4.39 will refine the answer until satisfactory convergence is reached.

Turning now to the minimum reinforcement requirement (Eq. 4.19), we must recall the basis on which it was derived. The minimum steel requirement ensures that the tension force carried by the concrete before cracking can be resisted by the steel after cracking. This tension force is proportional to the section width on the tension side.   For a rectangular section, the width b is constant and unambiguous (Fig. 4.9a). For a T-section with the web of width b_w in tension, $b = b_w$ should be used (Fig. 4.9b). In the case of I-, box-, and inverted T-sections with very wide tension flanges (Fig. 4.9c), the minimum steel requirement of $\rho_{min} = \max\left\{3\sqrt{f_c'}/f_y; 200/f_y\right\}$, multiplied by $b_{tens}d$ will be so conservative as to be inappropriate. In fact, for a slender T-section with the flange in tension, we might get $A_{s,min} > A_{s,max}$, which is indeed a quandary!

Figure 4.9c may serve to derive an alternative minimum steel requirement for sections with wide tension flanges. Retracing the derivation of Eq. 3.15 for this case, we ignore the web contribution to the stress block, that is,

$$T \approx b_{tens}h_f f_r$$

where b_{tens} and h_f are the width and thickness of the tension flange. The steel needed to carry this force is given by

$$A_s = b_{tens}h_f\frac{f_r}{f_y} \tag{4.40}$$

Assuming again for the rupture modulus $f_r = 7.5\sqrt{4440} = 500$ psi and a safety factor of 2.4, we get

$$A_{s,min} = b_{tens}h_f\frac{1200}{f_y} \tag{4.41}$$

or

$$\rho_{min} = \frac{1200}{f_y} \tag{4.42}$$

where ρ_{min} is now the steel ratio relative to the tension flange area. The ACI Code as well as the AASHTO Specifications for Highway Bridges [3] are silent on the inconsistency of

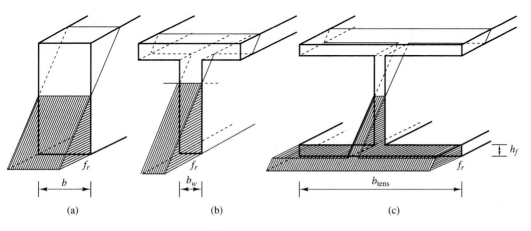

Figure 4.9 Tension stress block.

$\rho_{\min}$ with respect to wide tension flanges. The alternative of using $\frac{4}{3}$ of the steel required by analysis implies an unwarranted penalty for massive or lowly stressed members. Equation 4.42 appears to be more rational.

We are now in the position to formulate the design procedure for nonrectangular sections, assuming that the material properties and cross-sectional dimensions have been specified or assumed.

Step 1.   Assume the location of the centroid of the concrete area in compression, $\bar{a}$, such as the half-thickness of the compression flange.

Step 2.   Compute the required steel area (Eq. 4.38):

$$A_s = \frac{M_u}{\phi f_y (d - \bar{a})}$$

The strength reduction factor ϕ is still 0.9, as before.

Step 3.   Compute the corresponding concrete area under compression (Eq. 4.39):

$$A_c = \frac{A_s f_y}{.85 f_c'}$$

Step 4.   Compute the centroid of this revised area by inspection. If it is markedly different from the previous estimate, repeat Steps 2 through 4 until convergence is achieved.

Step 5.   Check A_s against the minimum steel requirement (Eq. 4.19 or 4.42 in case of a section with wide tension flange):

$$A_{s,\min} = \max \left\{ \frac{3\sqrt{f_c'}}{f_y}; \frac{200}{f_y} \right\} b_{\text{tens}} d \qquad \text{or} \qquad \frac{1200}{f_y} b_{\text{tens}} h_f$$

The following steps are needed to check A_s against the maximum steel requirement.

Step 6.   Locate the neutral axis for balanced conditions (Eq. 4.24):

$$c = \frac{87,000}{87,000 + f_y} d$$

Step 7.   Compute the corresponding depth of concrete area under compression (Eq. 4.25):

$$a = \beta_1 c$$

where β_1 is computed as before.

Step 8.   Determine by inspection the concrete area under compression, A_{cb}.

Step 9.   Compute the maximum permissible steel (Eq. 4.27):

$$A_{s,\max} = .6275 \frac{f_c'}{f_y} A_{cb}$$

If $A_s > A_{s,max}$, the section dimensions must be increased or compression steel is needed. In this case, it is necessary to determine the capacity of the section reinforced with $A_{s,max}$.

Step 10. Compute the concrete area under compression needed to balance $A_{s,max}$ (Eq. 4.29):

$$A_{c,max} = \frac{3}{4} A_{cb}$$

Step 11. Determine $\bar{a}$, the centroidal position of compression area $A_{c,max}$.

Step 12. Compute the maximum moment capacity (Eq. 4.30):

$$M_{u,max} = \phi A_{s,max} f_y (d - \bar{a})$$

EXAMPLE 4.6

Select for the beam of Fig. 4.8 the amount of steel needed to resist a factored moment of 400 ft-kips.

Solution

Step 1. Assume for $\bar{a}$ the half-thickness of top flange: $\bar{a} = 2.5$ in.

Step 2. Required steel area: $A_s = \dfrac{(400)(12)}{(.9)(760)(27.5 - 2.5)} = 3.56$ in.

Step 3. Check concrete area: $A_c = \dfrac{(3.56)(60)}{(.85)(3)} = 83.8$ in$^2 > (5)(15) = 75$ in^2.

Step 4. Check assumption for $\bar{a}$: $a = \dfrac{83.8 - (5)(10)}{5} = 6.76$ in,

$$\bar{a} = \frac{(50)(2.5) + (6.76)^2(5)/2}{83.8} = 2.85 \text{ in.}$$

Revise A_s: $A_s = \dfrac{4800}{(.9)(60)(27.5 - 2.85)} = 3.61$ in^2.

This is close enough to the previous value of 3.56 in^2, so $\bar{a}$ will hardly change.

Answer: Use 2 #8 and 2 #9 bars, with $A_s = 3.58$ in^2.

Check minimum reinforcement: $\rho_{min} = \max\left\{ \dfrac{3\sqrt{3000}}{60,000}; \dfrac{200}{60,000} \right\} = .00333$

$A_{s,min} = (.00333)(15)(27.5) = 1.37 < 3.58$ OK

Check maximum reinforcement (see Example 4.5):

$A_s = 3.58 < A_{s,max} = 3.80$ in^2 OK   ∎

Note that if the flange overhangs are tapered, it is usually permissible to ignore the taper and to assume a constant average flange thickness.

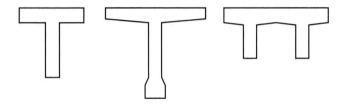

Figure 4.10 Isolated T-beams.

The most important category of nonrectangular sections are T-beams. These are often encountered as isolated, that is, as separately cast and often precast members (Fig. 4.10). More commonly, however, they appear as integral parts of slab-and-beam and joist floor systems, which are ubiquitous in concrete construction (Fig. 4.11).

The structural behavior of a beam cast integrally with a slab needs to be well understood, because some part of the slab is more effective in stiffening the beam than other parts. For such an understanding, let us consider a typical slab supported by beams spaced a distance s_b apart (Fig. 4.12a), and isolate a single beam together with its tributary slab by separating it from the remaining slab halfway between the beams (Fig. 4.12b). If we also cut away the slab overhangs from the beam proper, it is easy to see that the beam with its larger moment of inertia is much stiffer in bending than the slab strips and therefore attracts a much larger share of the load. When the isolated beam bends under this load, its top compression fibers shorten and the bottom fibers lengthen in tension, whereas the fibers of the separated unloaded slab do not change their length. This strain incompatibility between slab and beam can be removed only by applying corrective shear stresses, as shown in Fig. 4.12c. These shorten the slab overhangs and lengthen the top fibers of the beam until they fit together again. In effect,

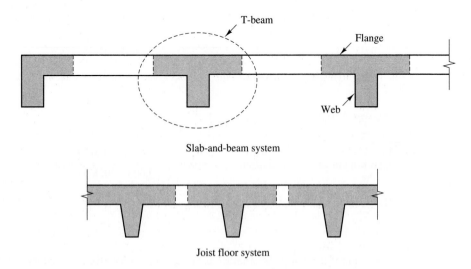

Figure 4.11 T-beams in slab-and-beam and joist floor systems.

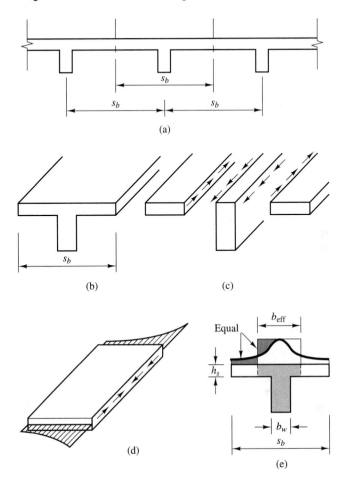

Figure 4.12 Shear lag effect in T-beam: (b) typical beam with tributary slab; (c) shear stresses to restore strain compatibility between beam and slab; (d) longitudinal stresses in slab overhangs due to corrective shear stresses; (e) qualitative variation of actual and assumed longitudinal stresses along slab centerline.

the slab overhangs are now loaded by these corrective edge shear stresses, which cause the normal stress distribution shown qualitatively in Fig. 4.12d, so that the maximum compression stress varies across the entire section as shown in Fig. 4.12e. The stress decrease as a function of distance from the beam is known as the *shear-lag* effect, because it is caused by the shear stresses and shear deformations in the slab. It is more pronounced for very thin slabs and widely spaced beams, in which case the slab portions far removed from the beam are virtually ineffective in stiffening the beam, (i.e., in carrying the load). In general, it is rather difficult to accurately determine the stress variation. For practical purposes it is therefore common to replace the actual T-beam with flange width s_b and variable stress distribution by a beam with an *effective* flange width

b_{eff} and constant stress distribution having about the same flexural stiffness (Fig. 4.12e). The ACI Code [Sec. 8.10] permits the use of the following simplified criteria to determine b_{eff},

$$b_{\text{eff}} = \min\left\{\frac{L}{4};\ 16h_s + b_w;\ s_b\right\} \tag{4.43}$$

where L is the beam span; h_s is the slab thickness; b_w is the web width; and s_b is the average beam spacing. Obviously, b_{eff} cannot be larger than s_b. We also see that the slab overhang of $8h_s$ is small for thin slabs. The $L/4$ criterion can be explained with the help of Fig. 4.12d. In a short span, the corrective shear stresses cannot easily activate the slab part that is far removed, that is, the load cannot spread as well laterally as in a longer beam, and therefore its effective width will be smaller.

For beams with a slab overhang on one side only, the effective flange width may be estimated as

$$b_{\text{eff}} = \min\left\{\frac{L}{12} + b_w;\ 6h_s + b_w;\ \frac{s_b + b_w}{2}\right\} \tag{4.44}$$

EXAMPLE 4.7

Determine the flexural reinforcement for an exterior beam and a typical interior beam of the floor system spanning 15 ft, shown in Fig. 4.13, which carries a live load of 50 psf in addition to its own dead weight. Given: $f_c' = 3000$ psi; $f_y = 60,000$ psi; lightweight concrete @ 120 pcf; $d = 10$ in.

Solution

Interior beam:

Dead weight: $w_d = \dfrac{(4)(72) + (5)(8)}{144} 0.12 = .27$ kips/ft

Factored load: $w_u = (1.4)(.27) + (1.7)(0.05)(6) = .89$ kips/ft

Factored moment: $M_u = \dfrac{(.89)(15)^2(12)}{8} = 300$ in-kips

Effective flange width:

$$b_{\text{eff}} = \min\left\{\frac{L}{4} = \frac{(15)(12)}{4} = 45;\ 16h_s + b_w = (16)(4) + 5 = 69;\ s_b = 72\right\} = 45\ \text{in}$$

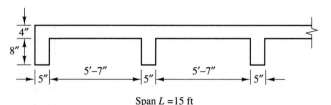

4"

8"

5"   5'-7"   5"   5'-7"   5"

Span L =15 ft   **Figure 4.13** Example 4.7.

Required steel: $\bar{a} = 2$ in, $A_s = \dfrac{300}{(0.9)(60)(10-2)} = .69$ in^2,

$$a = \dfrac{(.69)(60)}{(.85)(3)(45)} = .36 \text{ in}; A_s = \dfrac{300}{(0.9)(60)(10-.18)} = .57 \text{ in}^2.$$

Use 2 #5 bars with $A_s = .61$ in^2.

Check minimum steel: $\dfrac{200}{60,000}(5)(10) = .167$ in$^2 < .61$ in^2 <u>OK</u>

Exterior beam:

Dead weight: $w_d = \dfrac{(4)(38.5)+(5)(8)}{144}0.12 = .162$ kips/ft

Factored load: $w_u = (1.4)(.162) + (1.7)(0.05)\left(\dfrac{38.5}{12}\right) = .50$ kips/ft

Factored moment: $M_u = \dfrac{(.50)(15)^2(12)}{8} = 169$ in-kips

Effective slab overhang:

$$\min \left\{ \dfrac{L}{12} = \dfrac{(15)(12)}{12} = 15;\ 6h_s = (6)(4) = 24;\ \dfrac{s_b - b_w}{2} = \dfrac{72-5}{2} = 33.5 \right\} = 15 \text{ in}$$

Effective flange width: $b_{\text{eff}} = 15 + 5 = 20$ in

Required steel: $\bar{a} = 0.2$ in, $A_s = \dfrac{169}{(0.9)(60)(10-.2)} = .32$ in^2,

$$a = \dfrac{(.32)(60)}{(.85)(3)(20)} = .37 \text{ in } \underline{\text{OK}}.$$

Use 2 #4 bars with $A_s = 0.4$ in$^2 > A_{s,\text{min}} = .167$ in^2.

Note: The part of the slab that is not effective does not contribute to the load resistance of the floor system, but it still contributes dead load. ∎

EXAMPLE 4.8

Determine the flexural steel required for a typical interior girder of the multicell box girder bridge of span $L = 60$ ft shown in Fig. 4.14. Given: $f'_c = 5000$ psi; $f_y = 60,000$ psi; maximum positive factored moment: 1372 ft-kips.

Solution

Assume $\bar{a} = \dfrac{h_f}{2} = 3.5$ in; $d = 45$ in

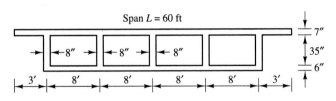

Span $L = 60$ ft

Figure 4.14 Example 4.8: Box girder bridge.

Required steel area: $A_s = \dfrac{(1372)(12)}{(.9)(60)(45-3.5)} = 7.35 \text{ in}^2$

Effective flange width:

$$b_{\text{eff}} = \min\left\{\dfrac{(60)(12)}{4} = 180;\ (16)(7)+8 = 120;\ (8)(12) = 96\right\} = 96 \text{ in}$$

Concrete compression block depth: $a = \dfrac{(7.35)(60)}{(.85)(5)(96)} = 1.08 \text{ in} \ll 7 \text{ in}$

Revise steel area: $A_s = \dfrac{(1372)(12)}{(.9)(60)(45-.54)} = 6.86 \text{ in}^2$

Minimum steel area (Eq. 4.19), with $b_{\text{tens}} = 96$ in:

$$\rho_{\min} = \max\left\{\dfrac{3\sqrt{5000}}{60,000};\ \dfrac{200}{60,000}\right\} = .00354$$

$A_{s,\min} = (.00354)(96)(45) = 15.3 \text{ in}^2$

Using $\dfrac{4}{3}A_{s,\text{req}}$ instead: $A_{s,\min} = \dfrac{4}{3}(6.86) = 9.15 \text{ in}^2$

Using Eq. 4.41: $A_{s,\min} = \dfrac{(1200)(96)(6)}{60,000} = 11.52 \text{ in}^2$

In either case, the very large effective width of the girder causes the minimum steel requirement to control the design, that is, $A_s = 9.15 \text{ in}^2$. This is provided by 12 #8 bars per girder,
or #8 bars spaced @ 8 in throughout the bottom slab.   ∎

It should be noted that any load acting on a floor system, such as shown in Fig. 4.11,
except line loads positioned exactly along the beam centerlines, produce transverse bending moments in the slab. To resist these moments, the slab must be reinforced in the
transverse direction as a one-way slab, continuous over the various beams acting as
supports.

4.4 DESIGN OF DOUBLY REINFORCED BEAMS

It is quite common that architectural, economical, or other requirements place limitations on the beam dimensions, especially the depth. If the maximum permissible amount
of tension reinforcement permitted by the ACI Code is not sufficient to carry a given
design moment, then the beam capacity may have to be increased by placing steel in
the compression zone. As we have seen earlier, the flexural capacity of a section is
limited by the maximum amount of steel to assure a ductile failure. If steel is placed
near the compression face, it delays the crushing failure of concrete, so that it is now
possible to add an equal amount of steel on the tension side without the threat of a
brittle failure (Fig. 4.15). In other words, the addition of compression reinforcement
alone does not change the flexural capacity of a beam. However, it helps the concrete
carry the compression force and thus makes it possible to add tension steel beyond the
maximum otherwise permitted. It is this added *tensile* reinforcement that increases the

flexural strength of a beam. If the same margin of safety against brittle failure for a singly reinforced beam is to be maintained, the maximum steel limitation, $\rho \leq .75\rho_b$, should be modified as follows.

$$\rho - \rho' \leq .75\rho_b \qquad (4.45)$$

where $\rho' = A'_s/bd$ is the compression reinforcing ratio, ρ is the total tension steel ratio, and ρ_b is the balanced steel ratio (see Eq. 3.44).

Compression steel has other beneficial effects as well. It is quite effective in reducing the long-term deflections of flexural members, because due to the creep of concrete over long time spans, the compression steel is forced to assume a disproportionate share of the total compression force. This effect was accounted for in the formula for long-term deflections (Eq. 3.37). Top reinforcing bars also provide a minimum negative moment capacity for unforeseen loading conditions and moment reversals. It improves the ductility of a section by raising the neutral axis, that is, for the same reason that an increase in concrete strength increases ductility (see Example 3.11). Finally, the addition of such bars makes it easy to tie up the vertical stirrups, which are needed for shear reinforcement (see Chapter 5). For efficient construction, *reinforcement cages* (Fig. 4.16) are often assembled on the ground and then lifted by crane into position inside the formwork.

For design, it is helpful to divide the total bending capacity M_u of a doubly reinforced beam into two components, M_1 and M_2. M_1 represents the moment furnished by the concrete compression block and the portion of the tension steel needed to balance it. This moment is equal to the maximum capacity of the beam if it were only singly reinforced. M_2 is the additional capacity gained by the compression steel A'_s and the corresponding amount of tension steel (Fig. 4.17). Both tension and compression steel are assumed to have (but do not necessarily have) the same yield strength f_y. Using again the capacity reduction factor ϕ, we have

$$M_u = \phi(M_1 + M_2) \qquad (4.46)$$

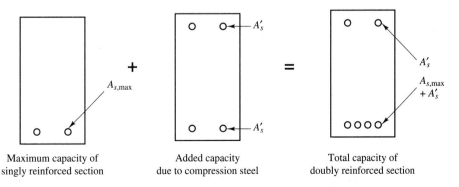

Maximum capacity of   Added capacity   Total capacity of
singly reinforced section   due to compression steel   doubly reinforced section

Figure 4.15   Doubly reinforced beam.

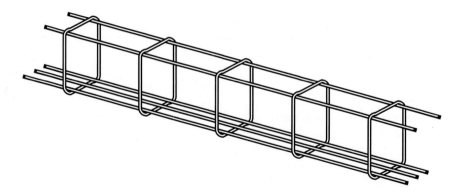

Figure 4.16 Reinforcement cage.

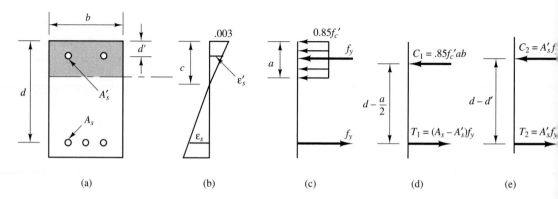

Figure 4.17 Doubly reinforced beam at failure: (a) section; (b) strains; (c) stresses; (d) moment M_1; (e) moment M_2.

where

$$M_1 = (A_s - A'_s) f_y \left(d - \frac{a}{2} \right) \qquad (4.47)$$

$$M_2 = A'_s f_y (d - d') \qquad (4.48)$$

and A_s is the total tensile reinforcement. The distance of the compression steel from the top is designated by d'. The depth of the concrete stress block follows from the requirement that $T_1 = C_1$:

$$a = \frac{(A_s - A'_s) f_y}{.85 f'_c b} \qquad (4.49)$$

Strictly speaking, the above equations are not correct because the area occupied by the compression bars is counted twice. This error, which can become nonnegligible when A'_s is large, can be corrected by adjusting the stress of the compression bars:

$$\overline{f'_y} = f'_y - 0.85 f'_c \qquad (4.50)$$

Equations 4.47 through 4.50 are based on the assumption that the compression steel yields at failure. This assumption can easily be verified with the strain diagram of Fig. 4.17b. The strain in the compression steel is given by

$$\epsilon_s' = \frac{c - d'}{c}(.003) \tag{4.51}$$

where

$$c = \frac{a}{\beta_1} \tag{4.52}$$

The compression steel stress then follows simply as

$$f_s' = E_s \epsilon_s' = E_s \frac{c - d'}{c}.003$$

$$= \frac{c - d'}{c} 87,000 \tag{4.53}$$

If $f_s' \geq f_y$, the compression steel yields; that is, our initial assumption was correct, and so are Eqs. 4.47 through 4.49. If $f_s' < f_y$, the compression steel does not yield. In this case, the C_2 force of Fig. 4.17e is smaller than assumed, and so is its moment contribution:

$$M_2 = A_s' f_s'(d - d') \tag{4.54}$$

The tension steel will still yield at failure, that is, $T_1 + T_2 = A_s f_y$. Thus, Eq. 4.49 must be replaced by

$$a = \frac{A_s f_y - A_s' f_s'}{.85 f_c' b} \tag{4.55}$$

and the first moment component (Eq. 4.47) now becomes

$$M_1 = (A_s f_y - A_s' f_s') \left(d - \frac{a}{2} \right) \tag{4.56}$$

Again, to be precise, f_s' should be replaced in the above equations, according to Eq. 4.50, by $\overline{f_s'} = f_s' - 0.85 f_c'$.

This updated solution is based on the linear strain diagram of Fig. 4.17b. In reality, however, concrete undergoes creep deformations, whereas the compression steel does not (or barely does). As a result, the concrete unloads some of its share of the compression force onto the steel, raising its stress above f_s'. Thus, if f_s' is close to f_y, it is safe to assume that the compression steel yields at failure, even though the strain diagram may indicate otherwise, so the original solution (Eqs. 4.46 through 4.49) can be assumed to be correct. It is interesting to note that in working stress design (or alternate design), the Code requires the compression steel to be transformed, not to nA_s', but to $2nA_s'$, because of the very reason just mentioned.

We can now summarize the procedure to determine the ultimate capacity of a doubly reinforced section.

Step 1.   Determine the moment M_2: $M_2 = A'_s f_y(d - d')$.

Step 2.   Compute the depth of the concrete stress block: $a = \dfrac{A_s f_y - A'_s f_y}{.85 f'_c b}$.

Step 3.   Compute the moment M_1: $M_1 = (A_s - A'_s) f_y \left(d - \dfrac{a}{2}\right)$.

Step 4.   Sum the total moment capacity: $M_u = \phi(M_1 + M_2)$.

Step 5.   To verify the assumption of compression steel yielding, determine the neutral axis position: $c = a/\beta_1$, with β_1 as before (see Eq. 4.25).

Step 6.   Compute the compression steel stress: $f'_s = \dfrac{c - d'}{c} 87,000$.

Step 7.   If $f'_s \geq f_y$, the steel yields, and the problem is solved. If $f'_s \ll f_y$, revise the concrete stress block depth, $a = (A_s f_y - A'_s f'_s)/.85 f'_c b$.

Step 8.   Repeat Steps 5, 6, and 7 until the value of a does not change appreciably.

Step 9.   Compute the revised ultimate moment capacity:

$$M_1 = (A_s f_y - A'_s f'_s)\left(d - \frac{a}{2}\right)$$

$$M_2 = A'_s f'_s(d - d')$$

$$M_u = \phi(M_1 + M_2)$$

EXAMPLE 4.9

Find the ultimate capacity of the beam section shown in Fig. 4.18. Given: $f'_c = 4000$ psi; $f_y = 60,000$ psi; $A_s = 6.0$ in²; $A'_s = 2.0$ in².

Solution

Step 1. $M_2 = (2)(60)(18 - 2.5) = 1860$ in-kips.

Step 2. $a = \dfrac{(6.0 - 2.0)(60)}{(.85)(4)(12)} = 5.88$ in.

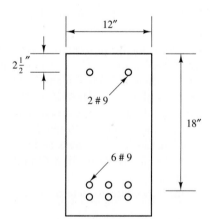

Figure 4.18   Section for Example 4.9.

Step 3. $M_1 = (6.0 - 2.0)(60)\left(18 - \dfrac{5.88}{2}\right) = 3614$ in-kips.

Step 4. $M_u = (0.9)(1860 + 3614) = 4927$ in-kips.

Step 5. $\beta_1 = 0.85; c = \dfrac{5.88}{.85} = 6.92$ in.

Step 6. $f'_s = \dfrac{6.92 - 2.5}{6.92} 87 = 55.57$ ksi < 60 ksi $\underline{\text{NG}}$.

Linear analysis indicates that the compression steel does not yield at failure. However, since f'_s is within 7% of f_y, it can be assumed that concrete creep will increase the steel stress in time so that it will yield at failure under factored loads. For illustrative purposes, we shall compute the correct moment capacity, including the more precise compression steel stress according to Eq. 4.50, that is, $\overline{f'_s} = 55.6 - 0.85 \times 4 = 52.2$ ksi.

Step 7. $a = \dfrac{(6.0)(60) - (2.0)(52.2)}{(.85)(4)(12)} = 6.26$ in, $c = \dfrac{6.26}{.85} = 7.37$ in.

Step 8. $f'_s = \dfrac{7.37 - 2.5}{7.37} 87 = 57.5$ ksi ≈ 55.6 ksi $\underline{\text{OK}}$.

$\overline{f'_s} = 57.5 - 0.85 \times 4 = 54.1$ ksi.

Step 9. $M_1 = [(6.0)(60) - (2.0)(54.1)]\left(18 - \dfrac{6.26}{2}\right) = 3744$ in-kips.

$M_2 = (2.0)(54.1)(18 - 2.5) = 1677$ in-kips.

$M_u = (0.9)(3744 + 1677) = \underline{4879}$ in-kips.

This result is just 1% less than the result obtained in Step 4.   ■

To determine the reinforcement needed to carry a given moment, the same basic steps are slightly rearranged.

Step 1.   Allowing for sufficient concrete cover, establish d and d'.

Step 2.   Determine the maximum allowable steel ratio (Eq. 4.20):

$$\rho_{\max} = \frac{3}{4}\rho_b = .6375\beta_1 \frac{f'_c}{f_y} \frac{87,000}{87,000 + f_y} \qquad (4.57)$$

Step 3.   Determine the depth of the concrete stress block:

$$a = \frac{A_{s,\max} f_y}{.85 f'_c b} = \frac{\rho_{\max} f_y d}{.85 f'_c} \qquad (4.58)$$

Step 4.   Compute the maximum moment capacity of the section with only tension steel:

$$M_{\max} = \phi \rho_{\max} b d f_y \left(d - \frac{a}{2}\right) \qquad (4.59)$$

Step 5.   If $M_{max} \geq M_u$, the beam needs to be only singly reinforced, as described earlier. If $M_{max} < M_u$, compute the added moment capacity needed:

$$M_2 = \frac{M_u - M_{max}}{\phi} \tag{4.60}$$

Step 6.   Compute the required compression steel:

$$A_s' = \frac{M_2}{f_y(d - d')} \tag{4.61}$$

and the tension steel:

$$A_s = \rho_{max}bd + A_s' \tag{4.62}$$

Step 7.   To check the yield condition of the compression steel, find the neutral axis position, $c = a/\beta_1$, and the compression steel stress, $f_s' = \frac{c-d'}{c} 87,000$. If $f_s' \ll f_y$, revise the required steel areas:

$$A_s' = \frac{M_2}{f_s'(d - d')}$$

$$A_s = \rho_{max}bd + A_s' \tag{4.63}$$

To be accurate, revise a again and repeat calculations. (This is seldom justified.)

EXAMPLE 4.10

A 10-by-20-inch beam shall be designed to carry a moment $M_u = 3500$ in-kips. Determine the required steel, with $f_c' = 4000$ psi and $f_y = 60,000$ psi.

Solution

Step 1.   Estimate 1-in bar diameter and 2-in cover, then $d' = 2.5$ in and $d = 17.5$ in.

Step 2. $\rho_{max} = (.6375)(.85)\dfrac{4}{60}\dfrac{87}{87 + 60} = .0214;$

$A_{s,max} = (.0214)(10)(17.5) = 3.74$ in^2.

Step 3. $a = \dfrac{(3.74)(60)}{(.85)(4)(10)} = 6.6$ in

Step 4. $M_{max} = (0.9)(3.74)(60)\left(17.5 - \dfrac{6.6}{2}\right) = 2868$ in-kips

Step 5. $M_2 = \dfrac{3500 - 2868}{0.9} = 702$ in-kips

Step 6. $A_s' = \dfrac{702}{(60)(17.5 - 2.5)} = 0.78$ in^2; $A_s = 3.74 + 0.78 = 4.52$ in^2.

Step 7. $c = \dfrac{6.6}{0.85} = 7.76$ in^2; $f_s' = \dfrac{(87)(7.76 - 2.5)}{7.76} = 59.0 \approx 60$ ksi <u>OK</u>.

Use 3 #11 bars for tensile reinforcement and 2 #6 bars for compression reinforcement.

To assess the error caused by counting the compression steel area twice, let us repeat the design, using $\overline{f}_s' = 59.0 - 0.85 \times 4 = 55.6$ ksi.

$$A_s' = \frac{702}{(55.6)(17.5 - 2.5)} = .84 \text{ in}^2, \quad A_s = 3.74 + 0.84 = 4.58 \text{ in}^2$$

The previously selected bars are still adequate.   ■

Note that the compression bars must be prevented from buckling by providing a sufficient number of transverse ties or stirrups. The minimum spacing requirements of the ACI Code are similar to those for column reinforcement (see Chapter 7). Such requirements are often satisfied by shear reinforcement (see Chapter 5).

If $\rho < \frac{3}{4}\rho_b$, that is, if the amount of tension reinforcement is less than the maximum allowable for singly reinforced beams, then the effect of compression bars is usually so small that it can be neglected.

EXAMPLE 4.11

Compute the moment capacity of the beam shown in Fig. 4.19, both including and neglecting the compression bars. Given: $f_c' = 3000$ psi; $f_y = 40,000$ psi.

Solution

Steel areas: $A_s = (4)(.79) = 3.16 \text{ in}^2$, $A_s' = (2)(.60) = 1.2 \text{ in}^2$,

$$\rho = \frac{3.16}{(10)(20)} = .0158, \quad \rho_{max} = (.6375)(.85)\frac{3}{40}\frac{87}{87+40} = .0278 > .0158 \text{ OK}$$

We first neglect the compression bars and use Eq. 4.10 to compute the moment capacity:

$$M_u = (.9)(.0158)(40)(10)(20)^2\left[1 - \frac{(.0158)(40)}{(1.7)(3)}\right] = \underline{1993 \text{ in-kips}}$$

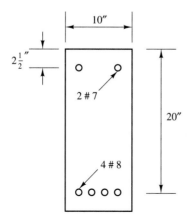

Figure 4.19   Section for Example 4.11.

When including the compression bars, we start by assuming these bars to yield, then we use Eqs. 4.49, 4.52, and 4.53 to find

$$a = \frac{(3.16 - 1.2)(40)}{(.85)(3)(10)} = 3.074, \, c = \frac{3.074}{.85} = 3.616 \text{ in}$$

$$f_s' = \frac{3.616 - 2.5}{3.616} 87 = 26.86 \text{ ksi} \ll 40 \text{ ksi NG}$$

Try $f_s' = \dfrac{40 + 26}{2} = 33$ ksi, then, with Eq. 4.55 replacing Eq. 4.49,

$$a = \frac{(3.16)(40) - (1.2)(33)}{(.85)(3)(10)} = 3.404, \, c = \frac{3.404}{.85} = 4.005 \text{ in}$$

$$f_s' = \frac{4.005 - 2.5}{4.005} 87 = 32.7 \text{ ksi} \approx 33 \text{ ksi OK}$$

Now we can compute the individual moments using Eqs. 4.56 and 4.54.

$$M_1 = [(3.16)(40) - (1.2)(32.7)]\left(20 - \frac{3.404}{2}\right) = 1595 \text{ in-kips}$$

$$M_2 = (1.2)(32.7)(20 - 2.5) = 687 \text{ in-kips}$$
$$M_u = (0.9)(1595 + 687) = 2054 \text{ in-kips}$$

The addition of 2 #7 compression bars increases the total amount of steel in the beam by 38%, but the moment capacity gains only 3% because the role of the compression steel is basically to strengthen the concrete, not to increase the beam's flexural capacity. As we have seen earlier, an increase in concrete strength is a similarly inefficient way of increasing the flexural strength of a beam, but both have a beneficial effect on the section ductility. ∎

SUMMARY

Procedure for the design of a singly reinforced rectangular beam:

Step 1. Select material properties f_c' and f_y.
Step 2. Establish a minimum depth h from Table 3.1 (ACI Code Table 9.5a).
Step 3. Select beam cross-sectional dimensions b and d (if not predetermined).
Step 4. Determine by structural analysis the factored design moment M_u.
Step 5. Compute the required steel ratio:

$$\rho = \beta - \sqrt{\beta^2 - \frac{2M_u\beta}{\phi f_y b d^2}} \qquad \text{with} \qquad \beta = \frac{0.85 f_c'}{f_y},$$

or determine the steel area by iteration:

a) Assume a value for a;

b) $A_s = \dfrac{M_u}{\phi f_y (d - \frac{a}{2})}$;

c) $a = \dfrac{A_s f_y}{.85 f'_c b}$;

d) Recompute A_s and a if necessary.

Step 6. Check steel ratio for upper and lower bounds:

$$\rho_{min} = \max \left\{ \frac{3\sqrt{f'_c}}{f_y}; \frac{200}{f_y} \right\} \qquad \text{or} \qquad \frac{4}{3}\rho$$

$$\rho_{max} = .6375\beta_1 \frac{f'_c}{f_y} \left(\frac{87,000}{87,000 + f_y} \right)$$

$$\beta_1 = 0.85 \qquad\qquad\qquad \text{for } f'_c \le 4 \text{ ksi}$$

$$= 0.85 - 0.05(f'_c - 4) \quad \text{for } f'_c \ge 4 \text{ ksi}$$

Step 7. Select reinforcing bars to furnish $A_s = \rho b d$ and arrange them to satisfy the clearance and minimum cover requirements of the ACI Code (Secs. 7.6 and 7.7).

Step 8. Check maximum crack width:

$$Z = f_s \sqrt[3]{d_c A} \le 145 \quad \text{for exterior exposure}$$

$$\le 175 \quad \text{for interior exposure}$$

Step 9. Check maximum deflections using the procedure and equations of Chapter 3.

Procedure for the design of singly reinforced nonrectangular beams: The procedure is similar to that for beams with rectangular sections, except that Steps 5 and 6 are replaced by the following procedure.

Step 1. Assume a location for the centroid of the concrete compression area, $\bar{a}$, such as the half-thickness of the compression flange.

Step 2. Compute the required steel area:

$$A_s = \frac{M_u}{\phi f_y (d - \bar{a})}$$

Step 3. Compute the corresponding concrete compression area:

$$A_c = \frac{A_s f_y}{.85 f'_c}$$

Step 4. Determine the centroid of the revised compression area and go back to Step 2 if necessary.

Step 5.   Check steel area against the minimum requirement:

$$A_{s,\min} = \max \left\{ \frac{3\sqrt{f_c'}}{f_y}; \frac{200}{f_y} \right\} b_{\text{tens}} d \quad \text{or} \quad \frac{1200}{f_y} b_{\text{tens}} h_f$$

The next four steps are needed only to check the steel area A_s against the upper bound permitted by the Code.

Step 6.   Locate the neutral axis for balanced conditions:

$$c = \frac{87,000}{87,000 + f_y} d$$

Step 7.   Compute the depth of the concrete compression block:

$$a = \beta_1 c$$

Step 8.   Determine by inspection the concrete area under compression, A_{cb}.

Step 9.   Compute the maximum permissible amount of steel:

$$A_{s,\max} = .6375 A_{cb} \frac{f_c'}{f_y}$$

The next three steps are needed only if A_s (from Step 2) exceeds $A_{s,\max}$.

Step 10.   Compute concrete compression area to balance $A_{s,\max}$:

$$A_c = \frac{A_{s,\max} f_y}{.85 f_c'} = \frac{3}{4} A_{cb}$$

Step 11.   Locate the centroid of the new compression area $\bar{a}$.

Step 12.   Compute the maximum moment capacity of the singly reinforced section:

$$M_{\max} = \phi A_{s,\max} f_y (d - \bar{a})$$

Procedure for the design of a doubly reinforced beam: If the maximum permissible capacity of a singly reinforced beam as computed in Step 12 above is not sufficient, add the following steps, assuming (again) for simplicity a rectangular section.

Step 13.   Compute the additional moment capacity needed:

$$M_2 = \frac{M_u - M_{\max}}{\phi}$$

Step 14.   Compute the required reinforcing steel:

$$A_s' = \frac{M_2}{f_y(d - d')} \quad \text{and} \quad A_s = A_{s,\max} + A_s'$$

Step 15. Check the yield condition of the compression steel.

Depth of concrete compression block: $a = \dfrac{A_{s,\max} f_y}{.85 f'_c b}$

Neutral axis position: $c = \dfrac{a}{\beta_1}$

Compression steel stress; $f'_c = \dfrac{c - d'}{c} 87{,}000$

If $f'_s \ll f_y$, then correct a: $a = \dfrac{A_s f_y - A'_s f'_s}{.85 f'_c b}$

Recompute c and f'_s until convergence is obtained.

Step 16. Compute the final moment capacity:

$$M_1 = (A_s f_y - A'_s f'_s)\left(d - \frac{a}{2}\right)$$

$$M_2 = A'_s f'_s(d - d')$$

$$M_u = \phi(M_1 + M_2)$$

Note that the effective flange width of a T-beam is defined as

$$b_{\text{eff}} = \min\left\{\frac{L}{4}; \ 16h_s + b_w; \ s_b\right\}$$

REFERENCES

1. ACE Committee 318, *Design Handbook in Accordance with the Strength Design Method of ACI 318*, vol. 1, *Beams, One-Way Slabs, Brackets, Footings, and Pile Caps*, American Concrete Institute, Detroit, MI, 1991.
2. CRSI, *CRSI Handbook*, 7th ed., Concrete Reinforcing Steel Institute, Schaumburg, IL, 1992.
3. AASHTO, *Standard Specifications for Highway Bridges*, 14th ed., American Association of State Highway and Transportation Officials, Washington, D.C., 1989.

PROBLEMS

Unless specified otherwise, assume for all problems normal-weight concrete weighing 150 pcf, with $f'_c = 4000$ psi and steel with $f_y = 60{,}000$ psi. Concrete cover $= 2$ in, $\beta_1 = 0.85$, $E_s = 29{,}000$ ksi.

4.1 A 30-ft long and 25-in-deep precast concrete girder with an I cross section is to be designed so that it can safely carry: its own weight; a uniform live load of 700 lb/ft; and a concentrated midspan live load of 10 kips, for which, in addition to the standard overload factor, an impact factor of 1.2 is to be considered.

   (a) Select the flange widths and the amount of longitudinal reinforcing steel needed to carry the above loads. Check the steel against the ACI Code's minimum and maximum limits.

   (b) Just in case the beam is stored upside down, determine the required minimum negative reinforcement. Use an impact factor of 1.2 with the dead weight.

(c) Show all positive and negative steel in a neatly drawn sketch of the beam cross section, complete with all spacing and clearance dimensions.

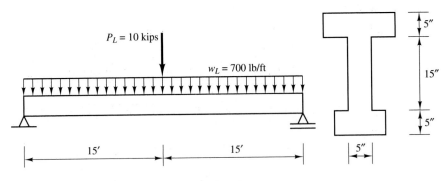

Figure P4.1

4.2 A 6-in-thick floor slab carries a 40-psf live load in addition to its own weight and is supported by 18-in-deep beams 5 ft on center and spanning 30 ft. Determine the required width of a typical beam and the reinforcement. Check against maximum and minimum reinforcement limits given in the ACI Code.

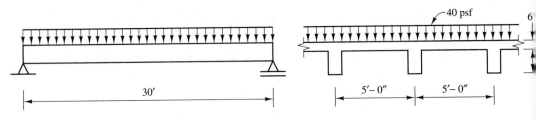

Figure P4.2

4.3 The beam shown in Fig. P4.3 is typical for a floor system in an existing building. It has been designed conservatively to carry a live load of 300 lb/ft and a dead weight of 400 lb/ft, includ-

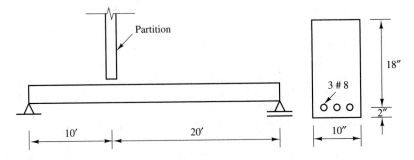

Figure P4.3

ing its own weight. The owner wants to add a partition as shown. Assuming the added load shares dead and live load equally, what is the total load you would allow the owner to add?

4.4 For the beam shown in Fig. P4.4, determine the amount of steel needed to carry its own weight, in addition to the two 60-kip live loads.

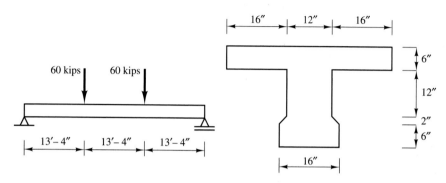

Figure P4.4

4.5 The beam shown in Fig. P4.5 carries a uniformly distributed live load of 1325 lb/ft, in addition to its own weight. (Use $1\frac{1}{2}$-in concrete cover.)

(a) Determine the amount of steel needed to resist the positive and negative moments. Select actual bars and show where in the cross section you would place them.

(b) Check whether the steel determined in problem (a) satisfies the Code's minimum and maximum steel requirements.

(c) If the section were reinforced with the maximum amount of steel permitted by the Code, estimate the beam's ultimate live-load capacity.

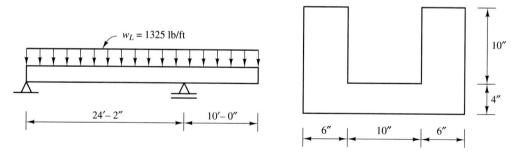

Figure P4.5

4.6 For a solid one-way slab of 9-in thickness and 15-ft span, determine the required reinforcement for the following live loads (use a 1-in concrete cover):

(a) 50 psf;

(b) 100 psf;

(c) 150 psf.

4.7 An 8-in-thick one-way concrete slab is reinforced with #6 bars @ $5\frac{1}{2}$ in ($d = 7$ in). Determine the allowable simple span length for a live load of 60 psf.

4.8 A one-way joist floor system has a simple span of 25 ft and is reinforced with 2 #6 bars per joist, as shown in Fig. P4.8.

    (a) Based on its flexural capacity, how much live load (in pounds per square foot) is this joist floor system permitted to carry in addition to its own weight?

    (b) Check if the amount of reinforcement satisfies the upper and lower limits of the ACI Code.

    (c) Determine the amount of reinforcement needed if the live load were raised to 200 psf. Indicate the number and size of bars per joist.

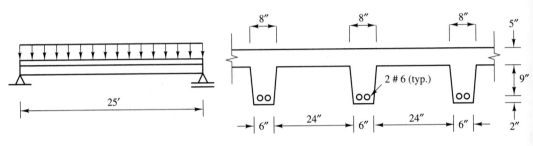

Figure P4.8

4.9 A parking garage deck is composed of precast double T-girders as shown in Fig. P4.9. In addition to its own weight, each typical girder shall be able to safely carry a live load of 100 psf. (Concrete cover 2 in to the edge of bars.)

    (a) Determine the amount of flexural reinforcement for each web.

    (b) Check whether the steel of problem (a) satisfies the maximum limit.

    (c) During erection, it appears practical to lift the girder as shown. For this purpose, four lifting lugs are concreted 5 ft away from the girder ends as shown. Allowing for an impact factor of 1.5, determine the maximum negative moment for which the girder will have to be designed to be safe during erection, and determine the necessary reinforcement. Make sure to satisfy the Code's minimum limit.

    (d) In a neatly drawn sketch, show the reinforcing bar arrangement, both for the negative and positive steel. Show the cross section and beam elevation. Bar cutoffs may be qualitative.

4.10 Determine the amount of flexural reinforcement needed for the beam shown to safely carry its own weight in addition to the two concentrated live loads. Check the steel against the Code's upper and lower limit. (see Fig. P4.10).

4.11 A cantilever beam with channel section carries its own weight, a uniform live load of 300 lb/ft, and a concentrated live load of 3000 lb at its tip (Fig. P4.11). Determine the required reinforcement and show the bar arrangement in a cross-sectional sketch and in a beam elevation.

4.12 A tank wall is to be designed against water pressure as shown. Determine the flexural reinforcement needed (per foot of wall length) and show the steel arrangement in a sketch, including a qualitative anchorage detail.

4.13 A warehouse floor system consists of precast inverted T-girders, carrying an 8-in cast-in-place slab as shown.

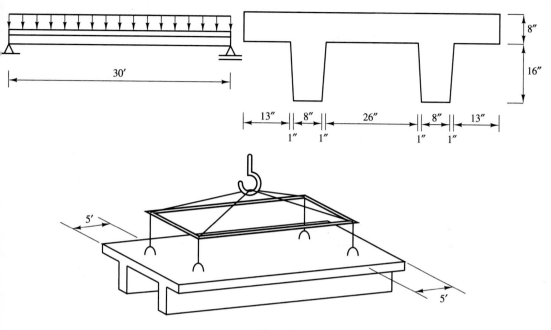

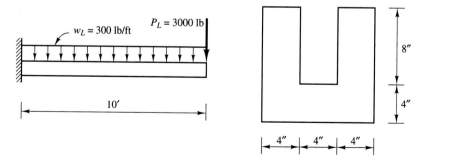

Figure P4.9

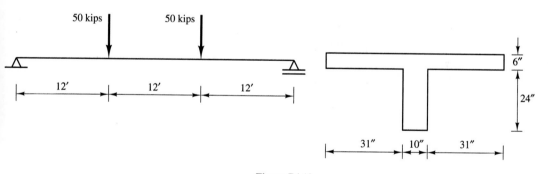

Figure P4.10

Figure P4.11

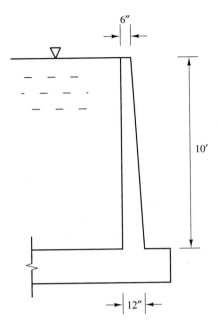

Figure P4.12

(a) Determine the amount of steel required for a typical precast girder to carry its own weight, plus its share of the weight of the (wet) concrete slab, plus a construction load of 10 psf (use the same overload factor as for live load), if no shoring of the 60-ft span is used during construction. Check for maximum steel limit.

(b) Assuming live load is applied only after the concrete deck has hardened and can serve as the compression flange for the precast girders, determine how much live load (in pounds per square foot) the same girder can carry. Check for the minimum steel limit.

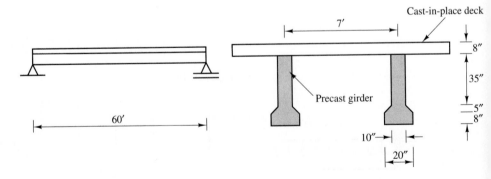

Figure P4.13

4.14 A one-way floor slab system is to be designed for a 15-ft simple span and 100-psf live load. Observe the Code's deflection control requirement, assuming no other construction is attached that is likely to be damaged by large deflections.

(a) Design a solid slab.

(b) Design a slab supported by beams spaced 8 ft apart.

(c) Design a slab supported by beams spaced 6 ft apart.

(d) Design a joist floor system.

Compare and discuss the various designs.

4.15 A 6-inch-thick one-way slab supporting a 20-ft simple span had originally been designed to carry a 100 psf live load in addition to its own weight. It is reinforced with #6 bars spaced 6 in apart. After years of service, the concrete cover spalled off at several places due to severe corrosion of the steel. An investigation leads to the estimate that on the average 20% of the steel area has been lost. Can the slab still safely carry its design load? (Assume $d = 5$ in.)

4.16 Write a subroutine in Fortran or any other programming language to automatically design a singly reinforced rectangular beam. Input provided by the user: b, d, f_c', f_y, and M_u. Output provided by the program: A_s, which satisfies the upper and lower bounds on the reinforcing ratio.

4.17 Write a subroutine in Fortran or any other programming language to determine the steel reinforcement in a one-way slab. Input provided by the user: slab thickness (in), slab span (ft), live load (psf), f_c' (ksi), f_y (ksi), and concrete weight (pcf). Output provided by the program: bar size and spacing, satisfying the upper and lower bounds on the reinforcing ratio.

Shear, Bond, and Torsion

Even if a beam has been designed properly to resist the effects of bending, it does not necessarily follow that it can safely carry its loads. First, shear stresses can crack the concrete and eventually cause failure. Second, bond stresses between the rebars and their surrounding concrete must remain below the capacity of the material, lest the rebars begin to slip relative to the concrete. Finally, if the beam is subjected to torsional loads, it must be designed properly in order to safely resist them as well.

In this chapter, we will discuss these three items in detail. Chapter 6 will deal specifically with the aspects of continuity, which will conclude our discussion of beams.

5.1 SHEAR AND DIAGONAL TENSION IN HOMOGENEOUS ELASTIC BEAMS

In courses on elementary strength of materials, it is taught that elastic homogeneous beams under load experience not only direct or *normal* stresses like tension and compression, but also shearing or *shear* stresses. The nature of these stresses can be visualized by considering loadings that tend to "shear off" part of a beam (Fig. 5.1). If we take a

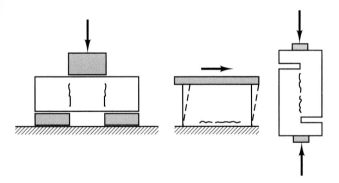

Figure 5.1 Basic shear loading.

free body of a beam at some section A (Fig. 5.2), shear stresses v are needed for vertical equilibrium. When integrating these stresses over the entire cross section, they must equal the shear force at that section, that is,

$$V = \int_A v \, dA$$

For an isolated infinitesimal element at some point P (Fig. 5.2b), vertical equilibrium requires that a shear stress equal and opposite to v act on the opposite face of the element as well (Fig. 5.2c). These two stresses form a couple that tends to rotate the element. For moment equilibrium, an equal moment of opposite sense is needed, which can be supplied by another two shear stresses acting horizontally on the top and bottom faces of the element, as shown in Fig. 5.2c. It is these horizontal shear stress components that would cause adjacent beam fibers to slip relative to each other if they could. Two beams placed on top of each other, without an adhesive or bond between them that could transmit shear stresses, will deflect as shown in Fig. 5.3a. However, if the two beams were glued together such that the glue could resist the shear stresses, the beam would be forced into the deflection pattern of Fig. 5.3b. The moment of inertia would then quadruple, from $I_a = 2(bh^3/12)$ to $I_b = [b(2h)^3]/12 = 4I_a$.

In many types of structures, it is common to place a concrete slab on top of either steel or precast concrete beams. To prevent the slab from sliding relative to the beams

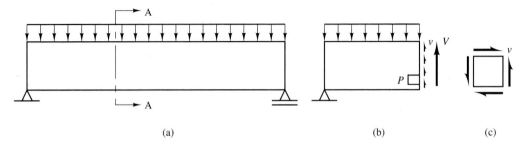

(a)   (b)   (c)

Figure 5.2   Shear stresses at a section.

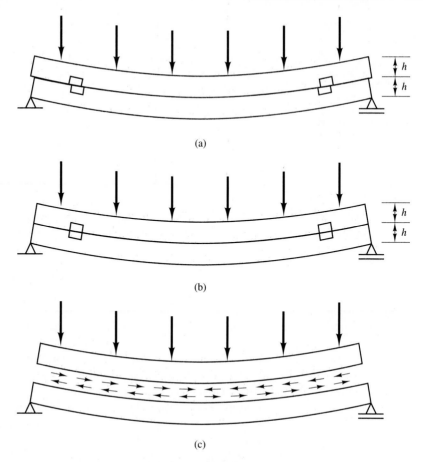

Figure 5.3   Effect of shear stresses.

as shown in Fig. 5.3a, *dowels* (also called *shear connectors* or *shear keys*), are used (Fig. 5.4). Such structures are referred to as *composite* systems. Their composite action relies on the shear connectors' capacity to transmit all shear forces from the beams to the slab and vice versa.

In an elastic homogeneous beam, the bending or direct stresses are given by

$$f = \frac{My}{I} \tag{5.1}$$

while shear stresses are computed as

$$v = \frac{VQ}{Ib} \tag{5.2}$$

where V is the shear force at the section considered, and b and Q are, respectively, the width and statical moment for the point where the shear stress is to be computed, a distance y away from the centroid. The quantity $q = vb = VQ/I$ is referred to as *shear*

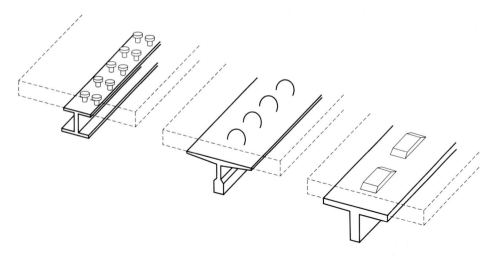

Figure 5.4 Composite beams.

flow with units force per unit length. The statical moment Q is a function of y and may
be computed as

$$Q = \sum_i A_i y_i \tag{5.3}$$

where it is assumed that the cross section above y is divided into a number of partial
areas A_i (Fig. 5.5). y_i is then the distance of the centroid of the ith partial area A_i from
the neutral axis, and the summation extends over all partial areas i for which $y_i > y$.

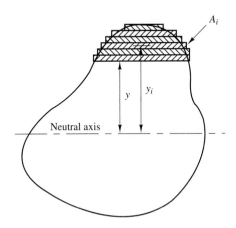

Figure 5.5 Definition of statical moment.

EXAMPLE 5.1

A precast concrete girder carries a cast-in-place slab as shown in Fig. 5.6. To tie the slab to
the girder, shear connectors are provided every 6 in. Compute the shear force that the first

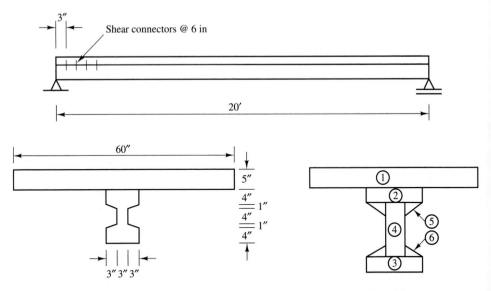

Figure 5.6 Example 5.1.

shear connector 3 in away from the support must resist, if the loading consists of dead load (150-pcf concrete) and a 50-psf live load.

Solution It is advantageous to calculate sectional properties in tabular form.

Area i	b_i	d_i	A_i	y_i	$A_i y_i$	$A_i(d_i^2/12)$	$A_i(y_i - \overline{y})^2$
1	60	5	300	16.5	4950	625	1587
2	9	4	36	12	432	48	174
3	9	4	36	2	72	48	5358
4	3	6	18	7	126	54	933
5	3	1	3	9.67	29	~.25	62
6	3	1	3	4.33	13	~.25	292
Total			396		5622	775.5	8406

Neutral axis: $\overline{y} = \dfrac{5622}{396} = 14.2$ in

Moment of inertia: $I = 775 + 8406 = 9181$ in^4

Statical moment: $Q = (300)(16.5 - 14.2) = 690$ in^3

Dead weight: $w_D = \dfrac{396}{144}(.15) = .4125$ kips/ft

Shear force at 3 in: $V = [.4125 + (.05)(5)]\left(10 - \dfrac{3}{12}\right) = 6.46$ kips

Shear flow: $q = \dfrac{VQ}{I} = \dfrac{(6.46)(690)}{9181} = 0.485$ kips/in

Shear to be resisted by first shear connector: $S = (.485)(6) = 2.91$ kips   ■

For reinforced concrete members, it is not so much the shear stresses that are of concern as the tension stresses, which result as a combination of shear *and* direct stresses. Considering some point A in a beam (Fig. 5.7a), the state of stress depends on the angle at which the reference element is isolated (Fig. 5.7b). The relationship between the two sets of stresses is found by using equations of statics (Fig. 5.7c), such that

$$\sigma = \sigma_x \cos^2 \alpha + \sigma_y \sin^2 \alpha + 2\tau_{xy} \sin \alpha \cos \alpha$$
$$\tau = \tau_{xy}(\cos^2 \alpha - \sin^2 \alpha) + (\sigma_y - \sigma_x) \sin \alpha \cos \alpha \tag{5.4}$$

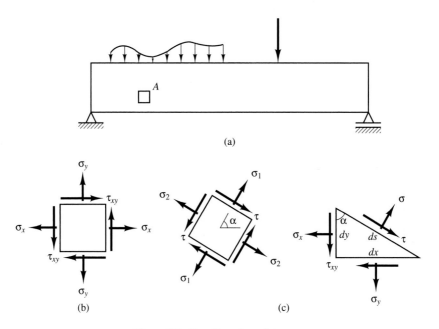

Figure 5.7 Transformation of stresses.

In texts on strength of materials, it is shown that the direct stresses σ reach a maximum or minimum for that angle α at which the shear stresses τ vanish. Hence, when setting $\tau = 0$ in Eq. 5.4, and neglecting the vertical stresses (i.e., $\sigma_y = 0$), the corresponding angle α follows as

$$\alpha = \frac{1}{2} \tan^{-1} \left(2\frac{\tau_{xy}}{\sigma_x} \right) \tag{5.5}$$

at which the direct stresses become

$$\sigma_{1,2} = \frac{\sigma_x}{2} \pm \sqrt{\frac{\sigma_x^2}{4} + \tau_{xy}^2} \tag{5.6}$$

These stresses are known as *principal stresses* and can be illustrated by Mohr's Circle (Fig. 5.8). For any given pair of stresses σ_x and τ_{xy}, we can construct the circle as shown and readily obtain the principal stresses σ_1 and σ_2, as well as the maximum shear stress τ_{max}. In reinforced concrete terminology, the principal tension σ_1 is normally referred to as *diagonal tension*.

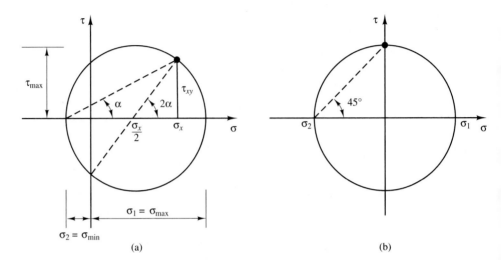

Figure 5.8   Mohr's circle: (a) general case ($\sigma_y = 0$); (b) pure shear ($\sigma_x = \sigma_y = 0$).

EXAMPLE 5.2

Compute the diagonal tension at point P at section A–A of the T-beam shown in Fig. 5.9 for the loading indicated.

Solution

Left reaction: $R_L = \dfrac{(10)(10) + (2)(10)(15)}{20} = 20$ kips

Moment at section A–A: $M = (20)(5) - \dfrac{(2)(5)^2}{2} = 75$ ft-kips

Shear at section A–A: $V = 20 - (2)(5) = 10$ kips $= 10,000$ lb

Neutral axis from bottom: $y_t = \dfrac{(5)(16)(17.5) + (4)(15)(7.5)}{(5)(16) + (4)(15)} = 13.2$ in

Neutral axis from top: $y_c = 20 - 13.2 = 6.8$ in

Moment of inertia: $I = (5)(16)\left(\dfrac{5^2}{12} + 4.3^2\right) + (4)(15)\left(\dfrac{15^2}{12} + 5.7^2\right) = 4720$ in^4

Statical moment at point P: $Q = (5)(16)(6.8 - 2.5) + (4)(1)(6.8 - 5.5) = 349.2$ in^3

Bending stress at P: $\sigma_x = \dfrac{(75,000)(12)(-0.8)}{4720} = -152.5$ psi (compression)

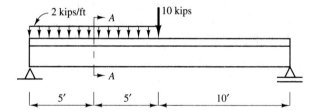

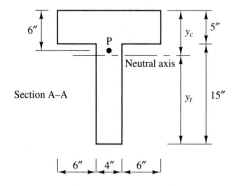

Section A–A

Neutral axis

Figure 5.9 Example 5.2.

Shear stress at P: $\tau_{xy} = \dfrac{(10,000)(349.2)}{(4720)(4)} = 185$ psi

Diagonal tension at P: $\sigma_1 = \dfrac{-152.5}{2} + \sqrt{\left(\dfrac{-152.5}{2}\right)^2 + 185^2} = \underline{124 \text{ psi}}$

Even in the "compression zone" of the beam, we encounter tensile stresses, depending on the angle α under which we decide to look at the stresses. In this example, the angle is $\alpha = \frac{1}{2} \tan^{-1}[(2)(185)/(-152.5)] = -33.8°$.   ∎

In the absence of bending stresses, such as along the neutral axis of a beam, the material is subjected to pure shear (Fig. 5.10a). However, viewing the same element at an angle of 45°, equivalent tension and compression stresses can be noticed (Fig. 5.10b). If these stresses are high enough, they can crack the concrete along a 45° line as shown,

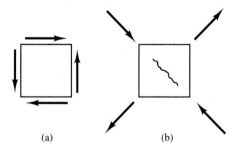

(a)

(b)

Figure 5.10 Pure shear and diagonal crack.

even though tensile stresses due to flexure are zero. In Fig. 5.11, the stress trajectories are shown for a uniformly loaded beam. These curves indicate for each point in the beam the direction of the minimum and maximum principal stresses and can be interpreted as sets of "cables" and "arches" that transmit the loads to the supports. The figure illustrates clearly that, theoretically, the beam is nowhere free of tensile stresses (except at the boundaries), not even in what is normally called the compression zone, as we saw in Example 5.2. Note that the stress trajectories cross the neutral axis at 45° angles, that is, in the absence of direct stresses, the principal stresses occur at a 45° angle (Fig. 5.10b).

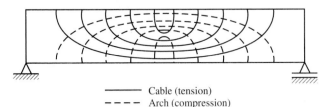

——————— Cable (tension)
– – – – Arch (compression)

Figure 5.11   Stress trajectories for uniformly loaded beam.

It can be argued that the ideal reinforcement then consists of steel bars that follow the tensile stress trajectories. Theoretically, this is correct, because the steel would resist all tension in the most direct way, while the concrete would be stressed exclusively in compression. There are two practical arguments against such a reinforcing scheme. First, it would be difficult and expensive to bend and place the bars appropriately, and this would require specially skilled labor. Second, the reinforcement would be optimal for one stress state, such as the one shown in Fig. 5.11, which is valid for one load case only. The live loads acting on real beams are movable and can be responsible for a considerable share of the total load. Thus, with each shift in live load, the stresses in the beam vary, and the concrete will be subjected to tensile stresses after all, for which it is not reinforced.

A concrete beam reinforced against only tensile stresses due to flexure can still crack under the combined action of shear and direct stresses if the diagonal tension exceeds the tensile strength of the concrete. Such diagonal cracks may lead to *shear failures*, which can occur rather suddenly and before the beam has a chance to fail in the ductile mode emphasized in the design for flexure. It is the objective of *shear design* to sufficiently reinforce the concrete for diagonal tension so that the beam will always reach its flexural capacity sooner than its shear capacity and, therefore, fail in a gradual and ductile mode. Such reinforcement is usually provided in the form of ties or closed stirrups, or by longitudinal bars bent upwards where they are no longer needed to carry flexural tension (Fig. 5.12).

It is extremely difficult to accurately compute by rational means the actual stresses in a cracked, reinforced concrete beam. Therefore, shear design is based largely on empirical knowledge. In the next sections, the behavior of beams with and without shear reinforcement will be discussed, followed by a practical design procedure based on the current ACI Code provisions.

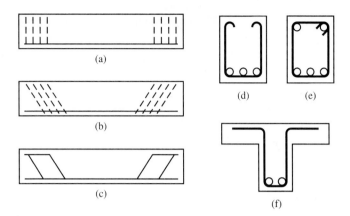

Figure 5.12   Shear reinforcement.

5.2 BEHAVIOR OF BEAMS WITHOUT SHEAR REINFORCEMENT

If uncracked, a reinforced concrete beam behaves very much like a homogeneous beam. Shear stresses follow closely the variation predicted by Eq. 5.2, which for a rectangular section is parabolic (Fig. 5.13a). Replacing the parabolic stress block with a rectangular stress block of average stress intensity, that is,

$$v = \frac{V}{bh} \tag{5.7}$$

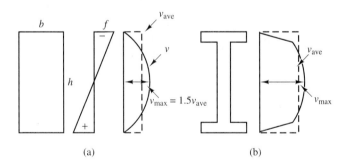

Figure 5.13   Shear stresses in rectangular and I-beam.

seems to be a gross oversimplification, but practical design methods are concerned primarily with overall behavior. In this regard, we ignore the fact that the maximum shear is 50% higher than the average stress. Note that in an I-beam the maximum shear stress is much closer to the average stress (Fig. 5.13b).

Once the loads acting on a beam are increased to such a value that the principal tension exceeds the concrete's tensile strength, cracks will appear. The flexural reinforcement is not of much help in this situation, because it has been placed in areas of maximum flexural tension, not maximum shear or diagonal tension. Such diagonal cracks change the state of stress in a beam much more than do flexural cracks, which are neutralized by the tensile reinforcement. We distinguish two different cases.

a) Consider a section of a beam where the bending moment is small and the shear force is high, such as near the supports of a simply supported beam. Flexural cracks in the tension zone are not likely to be significant, if they exist at all. The maximum principal tension occurs near the neutral axis at a 45° angle, causing diagonal cracks as shown in Fig. 5.14a. Since they originate in the web, such cracks are referred to as *web-shear cracks*. The shear force that causes such cracks has been determined experimentally to be approximately

$$V_c = 3.5\sqrt{f_c'}bd \tag{5.8}$$

Since the tensile strength of concrete is approximately proportional to $\sqrt{f_c'}$, the shear strength, or rather the principal tension resistance, is also proportional to $\sqrt{f_c'}$. The value indicated by Eq. 5.8 is considerably less than the one given in Chapter 2. This is a result of the biaxial stress state, with the coexisting principal compressive stress effectively lowering the tensile strength (Fig. 2.19b). After a web-shear crack appears, it tends to propagate rapidly toward the tension reinforcement, and from there to the supports, causing the bars to pull loose in a bond failure. Such failures initiated by web-shear cracks are most likely to occur in I- and T-beams with thin webs and less so in members with solid sections or thick webs.

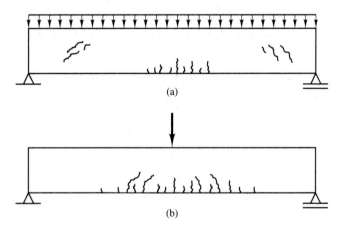

(a)

(b)

Figure 5.14 Diagonal cracks in beams without shear reinforcement: (a) web shear cracks for large V, small M; (b) flexure-shear cracks for large V, large M.

b) In a region in which the shear force V as well as the bending moment M are large, such as near a concentrated midspan load, vertical cracks will appear in the tension zone as soon as the external moment exceeds the cracking moment. As the load increases, these flexural cracks propagate toward the neutral axis region, where they tend to assume a 45° inclination (Fig. 5.14b). Because it is merely an extension of a flexure crack, such a crack is called a *flexure-shear* crack, and the flexural crack that caused it is referred to as the *initiating flexure* crack. The concrete area available to resist shear is reduced by the flexure cracks. Also, according to our assumption, the large M and large V combine for a large principal stress, so the concrete can be expected to carry less shear than in case a. This expectation has been confirmed by numerous tests, and the shear force that the concrete can carry in this case is approximately equal to

$$V_c = 1.9\sqrt{f'_c}bd \tag{5.9}$$

This is almost half of the capacity given by Eq. 5.8.

The transition between the two cases just described is obviously a function of M, or rather the ratio M/V, which is also known as the *shear span*, that is, the distance from the support to a concentrated load (Fig. 5.15). Using the dimensionless form M/Vd, the nominal (unreduced) shear strength of concrete allowed by the ACI Code is

$$V_c = \left(1.9\sqrt{f'_c} + 2500\rho_w \frac{V_u d}{M_u}\right) b_w d \leq 3.5\sqrt{f'_c}b_w d \tag{5.10}[11-5]$$

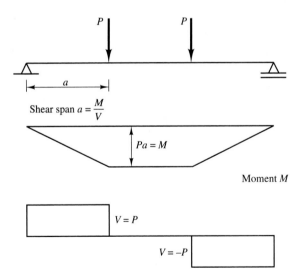

Shear span $a = \dfrac{M}{V}$

$Pa = M$

Moment M

$V = P$

$V = -P$

Shear V **Figure 5.15**   Definition of shear span.

where b_w is the web width and $\rho_w = A_s/b_w d$ the longitudinal steel area relative to the web area. In Fig. 5.16, Eq. 5.10 is plotted in dimensionless form as a function of $\rho_w V_u d/\sqrt{f'_c}M_u$, together with some test data [1]. For large $V_u d/M_u$ ratios, Eq. 5.10 gives capacities that are close to those of Eq. 5.8, while for small $V_u d/M_u$ ratios, they are closer to the limit of Eq. 5.9. To limit the value for V_c near points of inflection,

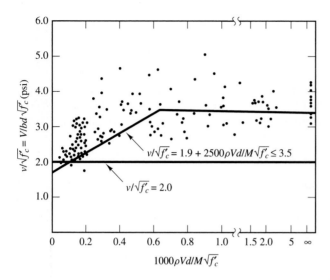

Figure 5.16 Shear strength of concrete [1]. (Authorized reprint from *ACI Journal*, vol. 59, 1962.)

where M_u is zero, a minimum value of $M_u = V_u d$ shall be assumed, or $V_u d/M_u \leq 1$. In effect, this requirement states that V_c be evaluated at least a distance d away from simple supports and inflection points. The shear strength of concrete according to Eq. 5.10 is also a function of ρ_w, because a large amount of tension reinforcement will keep the flexure cracks small and the shear-resisting concrete area large. Note the remarkable scatter of the test data, which explains why it is so difficult to establish a rational theory of shear strength, the lack of which needs to be compensated for with relatively large safety factors.

For simplicity, the ACI Code also permits the use of the formula

$$V_c = 2\sqrt{f_c'}b_w d \qquad\qquad (5.11)[11\text{-}3]$$

which is more conservative and independent of both ρ_w and $V_u d/M_u$ (Fig. 5.16).

EXAMPLE 5.3

Compute the shear strength of concrete at the supports and just outside the concentrated loads for the beam of Fig. 5.17. Given: $f_c' = 3000$ psi.

Solution

Using Eq. 5.11, $V_c = 2\sqrt{3000}(8)(10) = 8.76$ kips throughout the beam.

Using Eq. 5.10, at the supports, $V_u = 20$ kips, $M_u = 0$; therefore use $V_u d/M_u = 1$, that is, consider a section a distance d from the supports.

Reinforcing ratio: $\rho_w = \dfrac{1.58}{(10)(8)} = .0197$

Concrete shear strength: $V_c = [1.9\sqrt{3000} + (2500)(.0197)(1)](8)(10) = 12.27$ kips

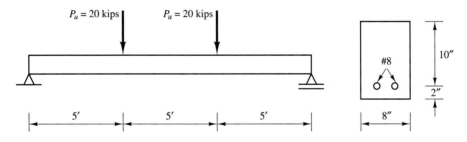

Figure 5.17   Example 5.3.

Check: $3.5\sqrt{3000}(8)(10) = 15.3$ kips <u>OK</u>

Under the loads, $V_u = 20$ kips, $M_u = (5)(20) = 100$ ft-kips $= 1200$ in-kips.

$$V_u d/M_u = \frac{(20)(10)}{1200} = .167$$

Concrete shear strength: $V_c = [1.9\sqrt{3000} + (2500)(.0197)(.167)](80) = 9.0$ kips.   ∎

The presence of an axial force in the beam, in addition to shear and the bending moment, has a definite influence on the amount of shear that the concrete can carry. An axial tension force will use up part of the tensile capacity of the concrete available for diagonal tension. Axial compression, on the other hand, reduces the diagonal tension in the beam and therefore increases the concrete's shear carrying capacity, which is then approximated by the ACI Code as

$$V_c = \left(1.9\sqrt{f'_c} + 2500\rho_w \frac{V_u d}{M_m}\right) b_w d \le 3.5\sqrt{f'_c}b_w d \sqrt{1 + \frac{N_u}{500A_g}} \qquad (5.12)[11\text{-}7]$$

where

$$M_m = M_u - N_u \frac{4h - d}{8} \qquad (5.13)[11\text{-}6]$$

and the ratio $V_u d/M_m$ is not limited to 1.0. N_u is the factored compression force (positive for compression), A_g is the gross section area, and h its total depth. Note that N_u/A_g is to be expressed in psi. Alternatively, the Code allows use of the simpler and more conservative formula

$$V_c = 2\left(1 + \frac{N_u}{2000A_g}\right)\sqrt{f'_c}b_w d \qquad (5.14)[11\text{-}4]$$

If the member is subject to significant tension, the capacity

$$V_c = 2\left(1 + \frac{N_u}{500A_g}\right)\sqrt{f'_c}b_w d \qquad (5.15)[11\text{-}8]$$

shall be used, where N_u is negative for tension. According to Eq. 5.15, a concrete section subjected to a tensile stress of $N_u/A_g = 500$ psi loses its capacity to carry any shear.

EXAMPLE 5.4

For the beam shown in Fig. 5.18, subjected to the factored load indicated, compute the concrete shear capacity 17.5 in to the right of both supports. Use $f'_c = 4000$ psi.

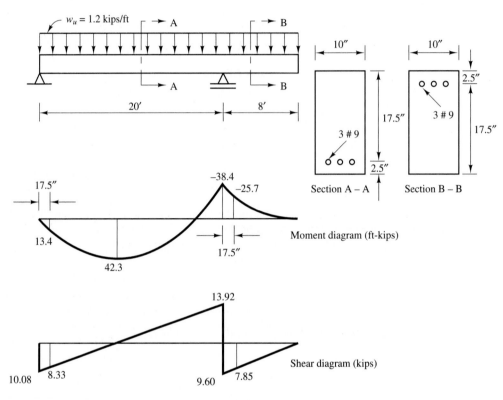

Figure 5.18   Example 5.4.

Solution   Using the refined formula of Eq. 5.10 requires the computation of bending moments and shears. These are shown in Fig. 5.18.

Reinforcing ratio: $\rho_w = \dfrac{3.0}{(10)(17.5)} = .0171$

Shear capacity near left support, using Eq. 5.10: $\dfrac{V_u d}{M_u} = \dfrac{(8.33)(17.5)}{(13.4)(12)} = .907 < 1.0$ OK;

$V_c = [1.9\sqrt{4000} + (2500)(.0171)(.907)](10)(17.5) = 27.8$ kips

Check: $3.5\sqrt{4000}(10)(17.5) = 38.7$ kips OK.

Shear capacity near right support, using Eq. 5.10: $\dfrac{V_u d}{M_u} = \dfrac{(7.85)(17.5)}{(25.7)(12)} = .445 < 1.0$ OK;

$V_c = [1.9\sqrt{4000} + (2500)(.0171)(.445)](10)(17.5) = 24.4$ kips

Using the simple formula of Eq. 5.11 gives a constant capacity throughout the beam:

$$V_c = 2\sqrt{4000}(10)(17.5) = 22.1 \text{ kips}$$ ∎

EXAMPLE 5.5

How does either an axial compression or a tension force of 10,000 lb affect the concrete shear capacity of the previous example near the beam's right support?

Solution   Using Eqs. 5.12 and 5.13 for compression, we have

$$M_m = (25{,}700)(12) - (10{,}000)\frac{(4)(20) - 17.5}{8} = 230{,}275 \text{ in-lb}$$

$$\frac{V_u d}{M_m} = \frac{(7.85)(17.5)}{230.3} = .597$$

$$V_c = [1.9\sqrt{4000} + (2500)(.0171)(.597)](10)(17.5) = 25.5 \text{ kips}$$

$$\text{Check: } 3.5\sqrt{4000}(10)(17.5)\sqrt{1 + \frac{10{,}000}{(500)(200)}} = 40.6 \text{ kips} > 25.5 \text{ OK.}$$

Using Eq. 5.14, we have $V_c = 2\left[1 + \dfrac{10{,}000}{(2000)(200)}\right]\sqrt{4000}(10)(17.5) = 22.7 \text{ kips}$

For tension, we use Eq. 5.15: $V_c = 2\left[1 - \dfrac{10{,}000}{(500)(200)}\right]\sqrt{4000}(10)(17.5) = 19.9 \text{ kips}$ ∎

After a diagonal crack has opened up, a number of different things can happen. The worst event is that a crack propagates all the way through the compression zone and causes a sudden and complete failure. This is possible in shallow beams without shear reinforcement. In deeper beams, it is more likely that strength reserves exist to force alternative stress paths that arrest the cracks before they penetrate too far into the compression zone. In this case, we can make several observations (Fig. 5.19).

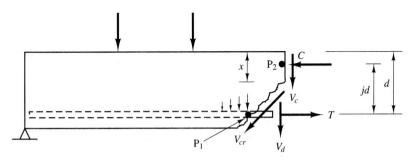

Figure 5.19   Cracked beam without shear reinforcement.

1. Since little shear can be transmitted directly across a crack, the shear force resisted by the concrete in the compression zone is confined to a considerably smaller area

than before cracking. Whereas the average shear stress before crack formation was V_c/bd, it is now closer to V_c/xb.

2. The concrete compression force C may also be acting on a slightly reduced area xb, resulting in a larger compressive stress.

3. A small amount of shear force, V_{cr}, can still be transferred across a crack. Its vertical component adds to the beam's shear strength.

4. The tensile reinforcement carries a shear force, V_d, through dowel action. This force tends to split the concrete along the bars (with subsequent bond failure) and therefore cannot be expected to contribute much to the section's shear capacity.

5. Taking moments about point P_1 (Fig. 5.19), the external moment M is effectively resisted by the internal moment Tjd, where $T = M/jd$ is the force in the steel at point P_2, which may be larger, depending on the loading. Thus, the formation of a diagonal crack may increase the stress in the reinforcing bars.

6. Some of the load can be carried by arch action (Fig. 5.20). Therefore, the share of the load to be carried by other mechanisms is reduced. The horizontal reaction for the arch base is supplied by the force in the longitudinal reinforcement, provided this is well anchored by bond.

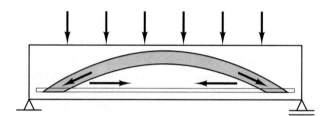

Figure 5.20   Arch action in beam.

In summary, a beam can survive after diagonal crack initiation only if the activated strength reserves exceed the losses caused by the crack. In general, because of the large number of factors involved, it is barely possible to predict whether a beam will fail after diagonal cracking or not. As a result, it has been accepted practice, as documented in the ACI Code, to assume that the capacity of concrete to carry shear (V_c) is the same, whether the section is cracked or not. This means any strength reserves that may exist are ignored, and so are any strength reductions caused by diagonal cracks. However, any shear force in excess of V_c will have to be carried by shear reinforcement.

5.3 BEHAVIOR OF BEAMS WITH SHEAR REINFORCEMENT

Typical shear reinforcement arrangements were illustrated in Fig. 5.12. In most cases, the stirrups cannot develop their full design forces over the short available development lengths by bond alone and therefore must be anchored at their ends. Whereas the stirrup shape of Fig. 5.12f may be appropriate for T-beams, closed hoops like those shown in

Fig. 5.12e are most common, especially for doubly reinforced sections, because the upper bars can be wired directly to the stirrups.

Most stirrups are placed vertically. If placed at an angle, they should form an angle of at least 45° with the longitudinal reinforcement and be securely anchored against slipping. Since the purpose of the stirrups is to tie the longitudinal bars into the main bulk of the concrete, both vertical and inclined stirrups shall extend a distance d from the extreme compression fiber.

If the longitudinal reinforcement is bent upwards to carry tension resulting from negative moments, such as over interior supports of continuous beams (Fig. 5.12c), then the inclined portion of this flexural reinforcement can be utilized as shear reinforcement. However, such bars are less effective than vertical stirrups, because they are loaded by a larger tributary area; they tend to cause longitudinal or inclined cracks and crushing near the bend points; they do not effectively confine the concrete in the shear region; and in nonrectangular sections such as I- or T-beams, they are less effective in tying the compression flange and web together. For these reasons, the Code [Sec. 11.5.6.6] considers only the center three-fourths of the inclined portion as effective.

A beam with shear reinforcement will experience diagonal cracks at approximately the same load as a beam without such reinforcement. Just as flexural steel becomes effective only after flexural cracking, the presence of shear reinforcement makes itself felt only after diagonal cracking. However, once cracking does occur, sufficient shear or web reinforcement is needed to carry the shear force not resisted by the concrete.

The nominal (i.e., unreduced) shear strength of a beam at a given section is defined as

$$V_n = V_c + V_s \qquad (5.16)[11\text{-}2]$$

where V_c is the amount of shear carried by concrete, as discussed in the previous section, and V_s is the share of the load carried by the shear reinforcement. Whereas in flexural design the steel is designed to carry all load after cracking and the concrete none, Eq. 5.16 implies that the concrete carries the same amount of shear after cracking as it does before. In view of the earlier discussion, this assumption is debatable, but we will soon consider additional strength reserves that can be thought of as compensation for this possibly unconservative assumption.

Since it is the task of shear reinforcement to bridge diagonal cracks, the ties should be spaced no farther apart than is needed to cross each potential diagonal crack with at least one stirrup. Assuming the crack to be inclined at 45°, it follows that the stirrup spacing s should not exceed half the beam depth, $d/2$ (Fig. 5.21). For some arbitrary stirrup spacing s, the number of stirrups crossing a crack is about $n = d/s$ (Fig. 5.22), and the shear force carried by the stirrups at failure follows as

$$V_s = n A_v f_y = \frac{A_v f_y d}{s} \qquad (5.17)[11\text{-}16]$$

where A_v is the area of the tie bar (or rather twice the bar area, if the tie is a hoop with two legs as is normally the case). By using the yield stress f_y, it is assumed that at failure, the diagonal crack has widened sufficiently to stress the web reinforcement to the

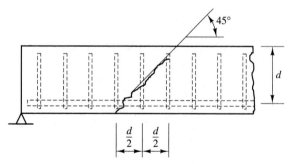

Figure 5.21   Maximum stirrup spacing.

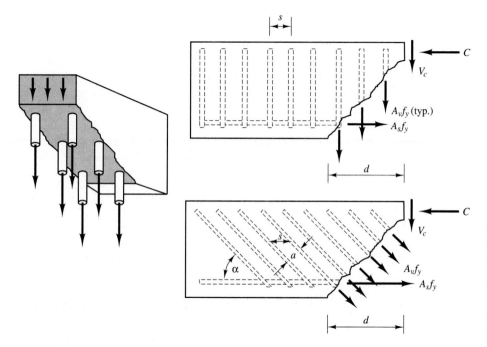

Figure 5.22   Action of shear reinforcement.

yield point. To avoid excessive crack widths prior to such failure, the Code [Sec. 11.5.2] limits the design yield strength of shear reinforcement to 60 ksi.

If the bars are inclined at some angle α (Fig. 5.22), then the spacing between two bars along a 45° crack is given by

$$a = \frac{s\sqrt{2}}{1 + \cot \alpha}$$

so that the number of bars crossing the crack becomes

$$n = \frac{d}{s}(1 + \cot \alpha)$$

and the total vertical force supplied by those inclined bars is

$$V_s = nA_v f_y \sin \alpha = \frac{d}{s} A_v f_y (\sin \alpha + \cos \alpha) \qquad (5.18)[11\text{-}17]$$

As can be seen, Eq. 5.17 is contained in Eq. 5.18 for the special case $\alpha = 90°$.

The action of the stirrups can be illustrated by visualizing the bars acting as vertical tension struts in an imaginary truss (Fig. 5.23). The tension members in the lower chord are furnished by the longitudinal reinforcement, while the diagonal compression members as well as the upper chord members are supplied by the concrete.

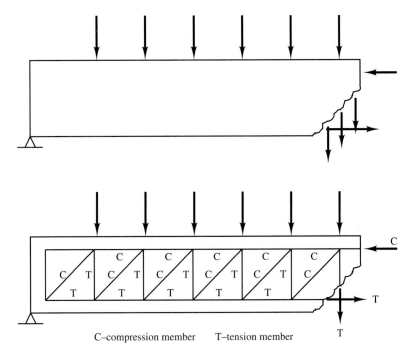

C–compression member   T–tension member

Figure 5.23   Truss analogy.

The shear reinforcement also has other effects of major importance. Once a crack opens, little shear can be transmitted by the concrete across the crack if no normal restraint keeps the two concrete blocks from separating. However, if steel reinforcement bridges the gap, it develops normal forces as the two faces of the crack ride up and down on each other (Fig. 5.24). The steel bars and compression struts between the diagonal cracks form little trusses, which can resist a shear force. Failure occurs when the compression diagonals crush under the combined axial and shear forces. The shear capacity is approximately proportional to the average restraining force, $T = A_v f_y$, supplied by the steel, which is somewhat analogous to ordinary friction. Therefore, this mechanism

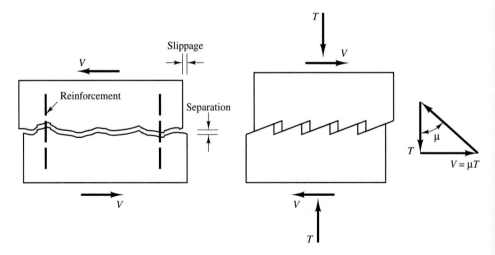

Figure 5.24   Shear friction analogy.

has been called *shear-friction*, but it is better known as *aggregate interlock* or *interface shear transfer*.

To summarize the various beneficial effects of shear reinforcement:

1. Web reinforcement that traverses a diagonal crack carries shear directly, as demonstrated by the truss analogy.
2. Web reinforcement prevents diagonal cracks from opening, so the concrete is capable of carrying shear across the crack through aggregate interlock.
3. By preventing the diagonal cracks from opening, shear reinforcement also keeps them from propagating too far into the concrete compression zone, thus leaving more concrete area available to carry shear. To assure this, the steel yield strength is limited to 60 ksi.
4. The stirrups tie the longitudinal bars into the main bulk of the concrete, thus keeping them from splitting off the concrete cover. This increases the dowel shear effect and provides support reactions for imaginary local arches to develop between diagonal cracks (Fig. 5.25).

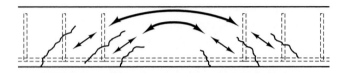

Figure 5.25   Local arch action.

5. Hoop reinforcement effectively confines the concrete, thus increasing its strength and especially its ductility.

Only the first item is directly accounted for when designing shear reinforcement. The other four beneficial effects are indirectly lumped together to compensate for the assumption that the shear carried by the concrete after cracking is equal to the shear that caused the diagonal crack—a possibly unconservative assumption. This qualitative tradeoff between unaccounted beneficial and detrimental effects is symptomatic of current ACI shear design. It cannot be sustained by rational means. Yet, the fundamental mechanisms of shear transfer are understood well enough from a qualitative point of view, and the large number of available test data provide a credible support for the highly empirical design method. The utterly erratic nature of crack initiation and propagation poses a major hurdle against solving the problem on a purely rational basis. Since concrete knows no geographical or political boundaries, it is no surprise that other major foreign design codes employ different but similarly empirical design methods. There are now efforts underway to approach the design for shear on a more rational basis [2]. These are likely to find their way into future editions of the ACI Code.

5.4 ACI CODE DESIGN FOR SHEAR AND DIAGONAL TENSION

Before presenting a detailed step-by-step procedure for shear design, it may be worthwhile to briefly recall some basic principles of flexural design (Fig. 5.26). On the compression side, the concrete is very effective. It is here where as much of it is concentrated as needed for strength and stiffness. On the tension side, the concrete is cracked and rendered ineffective as far as strength is concerned, even though between cracks it preserves its tensile capacity, thereby maintaining much of its flexural stiffness. (This is known as the *tension-stiffening* effect.) Thus, on the tension side, generally no more concrete is provided than is needed to enclose the reinforcement. In the central region (the web), flexural stresses are comparably low, but shear stresses high. When dimensioning the web width to carry shear, it should be kept in mind that steel is more effective than concrete in resisting diagonal tension. Therefore, it may be worthwhile to check whether a thin web with more shear reinforcement is more economical than a thick web with less or no shear reinforcement. The reduced dead weight usually speaks in favor of the former.

The nominal shear strength of a beam was given in Eq. 5.16. The basic shear design requires that the factored shear force not exceed the "ultimate," that is, the reduced shear

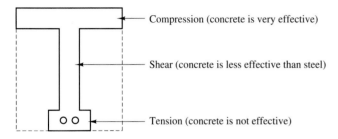

Compression (concrete is very effective)

Shear (concrete is less effective than steel)

Tension (concrete is not effective)

Figure 5.26   Design concepts for flexural members.

strength

$$V_u \leq \phi V_n = \phi(V_c + V_s) \qquad (5.19)[11\text{-}1]$$

where the strength reduction factor $\phi = 0.85$ for shear design. This is more conservative than the value for flexural design because of the larger statistical scatter of concrete shear strength. According to Eq. 5.19, no shear reinforcement would be needed if $V_u \leq \phi V_c$. In practice, however, the brittle shear failure of a flexural member without shear reinforcement is highly undesirable. The ACI Code [Sec. 11.5.5.1] translates this concern into the design requirement that only if

$$V_u \leq \frac{\phi V_c}{2} \quad \left(\text{doesn't need rebar}\right) \qquad (5.20)$$

does the member not need any shear reinforcement. To better understand this requirement, we can compare the strength of a beam with that of a chain consisting of two links, the first representing flexural strength and the second shear strength. The first link is guaranteed to fail in a ductile mode (if the member is under-reinforced); the other link is guaranteed to fail in a brittle mode if no shear reinforcement is present. To assure a ductile failure, the shear strength obviously must exceed the flexural strength. Moreover, because of the undesirability of a brittle failure, the second link is penalized with the additional safety factor of 2 in Eq. 5.20. This penalty is waived in the case of very wide or shallow beams and slabs, in which case there is the opportunity of load sharing between regions of under- and overstrength. A wide or shallow beam is defined to be one with a total depth h no larger than either 10 in, 2.5 times the flange thickness, or one-half the web width.

Another special case is that of *joist construction*. In order to qualify, the limitations indicated in Fig. 5.27 must be satisfied (ACI Code, Sec. 8.11.8). Such floor systems are quite popular because of their efficient load-carrying system, easy formwork, and good track record. Since the closely spaced ribs permit a certain amount of load redistribution, the ACI Code permits a 10% increase of the shear capacity of concrete, V_c.

Equations 5.19 and 5.20 imply that in the case of

$$\frac{\phi V_c}{2} \leq V_u \leq \phi V_c$$

shear reinforcement is not required for strength, but only to avoid the brittle failure of a section without shear reinforcement. In this situation, the minimum amount of shear reinforcement given by

$$A_v = \frac{50 b_w s}{f_y} \qquad (5.21)[11\text{-}13]$$

shall be provided. For a given or selected stirrup of size A_v, this requirement is equivalent to an upper limit on the stirrup spacing, that is,

$$s_{max} = \frac{A_v f_y}{50 b_w} \qquad (5.22)$$

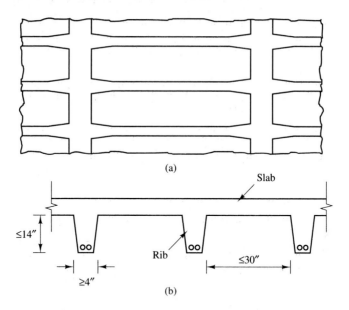

Figure 5.27   Joist construction: (a) plan view (seen from below); (b) section.

This minimum steel requirement is somewhat analogous to the minimum flexural reinforcement requirement. In both cases, care is taken that there is enough steel to carry the tension force released by the concrete after cracking.

If $V_u > \phi V_c$, Eq. 5.19 indicates that reinforcement is needed to carry whatever shear the concrete cannot, which is

$$V_s = \frac{V_u}{\phi} - V_c \tag{5.23}$$

Shear design generally proceeds by choosing a stirrup size, A_v. The required spacing s follows from Eq. 5.17 as

$$s = \frac{A_v f_y d}{V_s} \tag{5.24}$$

Aside from the maximum spacing limit resulting from the minimum steel requirement of Eq. 5.22, the ACI Code [Sec. 11.5.4] sets an independent upper limit for the stirrup spacing s by

$$s_{\max} = \min\left\{\frac{d}{2}; 24''\right\} \quad \text{if} \quad V_s \leq 4\sqrt{f_c'}b_w d$$

$$\tag{5.25}$$

$$= \min\left\{\frac{d}{4}; 12''\right\} \quad \text{if} \quad V_s > 4\sqrt{f_c'}b_w d$$

to assure that potential diagonal cracks are crossed by at least one line of shear reinforcement (Fig. 5.21). Designs controlled by Eq. 5.25 are likely to be less economical, so it is advisable to use smaller stirrups, such as #3 or #4 ties with close spacing, rather than heavier ties with wide spacing.

The ACI Code [Sec. 11.5.6.8] also places an upper limit on the amount of shear that the web reinforcement is permitted to carry, with

$$V_s \leq 8\sqrt{f_c'}b_w d \tag{5.26}$$

in order to keep the width of the cracks small and also to prevent crushing of the concrete between diagonal cracks prior to the yielding of the stirrups. Equation 5.26 can be rewritten, with the help of Eq. 5.24, to read

$$\frac{A_v}{b_w d} \leq 8\frac{\sqrt{f_c'}}{f_y}\frac{s}{d} \tag{5.27}$$

which corresponds to a maximum shear reinforcing ratio. In flexural design, the upper bound on ρ ensures yielding of the flexural steel before crushing of the concrete in compression. In shear design, the upper limit ensures that the crack widths remain small.

The limitation of Eq. 5.26, together with the nominal shear capacity of the concrete (Sec. 5.2), places an upper limit on the total amount of shear that a beam is allowed to carry. If this is exceeded by the factored shear force, the beam web must be widened. Simply adding steel will not be sufficient.

In actual beams, cracks seldom appear right at the supports but rather some distance away. The explanation for this can be found in the shear strength increase near concentrated loads and reactions that cause multiaxial compression (see Chapter 2). Where the support reaction introduces compression stresses into the end regions of a member, such as in most beam supports, the Code allows the maximum shear force V_u to be computed a distance d away from the face of the support (Fig. 5.28). This provision does not apply

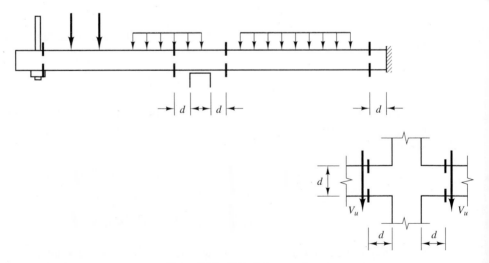

Figure 5.28   Critical shear sections.

to very short or deep beams or for beams with heavy concentrated loads applied close to the supports. Also, in beams suspended on hangers or other construction, the critical shear section is to be taken right at the reaction.

Design practice calls for the placing of stirrups a distance d beyond the point where they are no longer needed in order to restrain inclined cracks that may start at that point. This measure is also a safeguard against cracking due to unusual shifting of live loads. The minimum shear reinforcing requirement generally satisfies this provision.

Having listed all major Code requirements, we can now combine them in a concise step-by-step procedure for systematic shear design.

Step 1. Draw the nominal shear diagram for the factored beam loading $V_n = V_u/\phi$. Note that $V_{u,\max}$ is usually evaluated a distance d away from the face of a support.

Step 2. Draw the moment diagram for the factored beam loading M_u. This is needed only if the refined shear design method is used.

Step 3. Superimpose on the shear diagram the concrete shear capacity diagram by computing V_c at selected points, using Eq. 5.10 or 5.11. If axial forces are present, use Eq. 5.12, 5.14, or 5.15.

Step 4. Select a reasonable stirrup size, normally #3 or #4, unless the shear forces are very large.

Step 5. Check whether $V_s = (V_u/\phi) - V_c > 8\sqrt{f'_c}b_w d$. If this is the case anywhere, the web width must be increased.

Step 6. Compare the V_n versus V_c curves to determine the required stirrup spacing in the following three regions:

Region I, defined by $V_u/\phi \le V_c/2$: No stirrups are needed.

Region II, defined by $V_c/2 < V_u/\phi \le V_c + V_{s,\min}$: Minimum shear reinforcement (or maximum stirrup spacing) is sufficient, according to Eq. 5.22 or 5.25.

Region III, defined by $V_u/\phi > V_c + V_{s,\min}$: Determine the required stirrup spacing according to

$$s = \frac{A_v f_y d}{V_s} \quad \text{where} \quad V_s = \frac{V_u}{\phi} - V_c$$

If the web reinforcement is inclined, solve Eq. 5.18 for s.

Step 7. Select a reasonable and practical spacing pattern, considering the spacing limits of Step 6 as well as ease of placement. For example, varying the stirrup spacings too frequently will only confuse the workers.

In practice, this graphic method is likely to be used in a more simplified form. For instructive reasons, however, it can serve to develop a feeling for shear reinforcement design. Experience will later permit the designer to determine required stirrup spacings at a few selected points, coupled with engineering common sense.

EXAMPLE 5.6

The beam of Fig. 5.29 carries a factored load of $w_u = 7.9$ kips/ft. Determine the amount of shear reinforcement required, using a) the simplified, and b) the detailed ACI formula to calculate the shear carried by concrete. Given: $f_c' = 3000$ psi, $f_y = 60$ ksi.

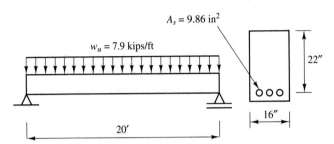

$A_s = 9.86$ in^2

$w_u = 7.9$ kips/ft

22"

16"

20'

Figure 5.29 Example 5.6.

Solution

Left reaction: $R_L = (7.9)(10) = 79$ kips

Shear at distance $d = 22''$ from support: $V_u = 79 - (7.9)\dfrac{22}{12} = 64.5$ kips

Maximum required nominal shear strength: $V_n = \dfrac{V_u}{\phi} = \dfrac{64.5}{.85} = 75.9$ kips

a) *Simplified Code formula* (see Fig. 5.30a)

Concrete shear capacity: $V_c = 2\sqrt{3000}(16)(22) = 38,560$ lb.

To determine the point beyond which no shear reinforcement is needed (Region I), find intersection of $V = V_c/2$ and V_n curves: $\dfrac{92.9 - 19.3}{92.9}(10) = 7.92$ ft.

Region I: No stirrups needed beyond $x = 7.92$ ft.

Region II: Maximum stirrup spacing for #3 stirrups @ 60 ksi:

$$s_{max} = \min\left\{\frac{d}{2}; 24; \frac{A_v f_y}{50 b_w}\right\} = \min\left\{11; 24; \frac{(.22)(60,000)}{(50)(16)} = 16.5\right\} = 11 \text{ in}$$

At maximum spacing, steel carries a minimum amount of

$$V_{s,min} = \frac{A_v f_y d}{s} = \frac{(.22)(60,000)(22)}{11} = 26,400 \text{ lb}$$

Nominal shear strength: $V_n = 38.6 + 26.4 = 65.0$ kips

This value intersects V_n diagram at $x = \dfrac{92.9 - 65.0}{92.9}(10) = 3.0$ ft

Maximum spacing is sufficient for $3.0 < x \leq 7.92$ ft.

Region III: Minimum stirrup spacing required for $V_{s,max} = 75.9 - 38.6 = 37.3$ kips

is $s = \dfrac{A_v f_y d}{V_s} = \dfrac{(.22)(60,000)(22)}{37,300} = 7.79$ in, say 7 in.

Final design: 11 stirrups #3; first stirrup at 3 in away from support; next 5 stirrups @ 7 in; final 5 stirrups @ 11 in.

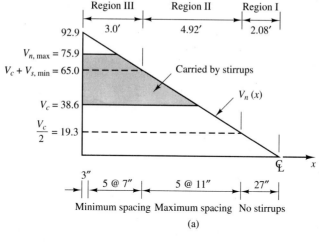

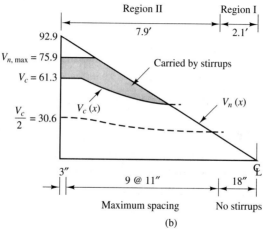

Figure 5.30 Example 5.6: Shear design. (a) Simplified design (Eq. 5.11); (b) detailed code formula (Eq. 5.10).

b) *Detailed Code formula* (see Fig. 5.30b)

We evaluate the concrete shear capacity in tabular form as shown on the following page. The curve for $V_c/2$ still intersects the V_u/ϕ curve at $x = 7.9$ ft, but now the shear carried by stirrups at maximum spacing is sufficient throughout, so only 10 stirrups are required.

Final design: 10 stirrups #3; first stirrup at 3 in away from support; next 9 stirrups @ 11 in.

In this example, use of the refined formula is obviously not justified, as it leads to the savings of just two stirrups over the length of the beam. In practice, one would probably ignore Region I and space stirrups uniformly along the entire length of the beam.

x	M_u	V_u	$\dfrac{V_u d}{M_u}$	V_c	$\dfrac{V_u}{\phi} - V_c$
0	0.0	79.0	1.000	61.28	14.6
1	75.0	71.1	1.000	61.28	14.6
2	142.2	63.2	0.815	56.72	17.6
3	201.4	55.3	0.503	49.03	16.0
4	252.8	47.4	0.344	45.11	10.7
5	296.2	39.5	0.244	42.65	3.8
6	331.8	31.6	0.175	40.95	0.0
7	359.4	23.7	0.121	39.61	0.0
8	379.2	15.8	0.076	38.51	0.0
9	391.0	7.9	0.037	37.54	0.0
10	395.0	0.0	0.000	36.63	0.0

∎

EXAMPLE 5.7

The one-way joist floor system of Fig. 5.31 supports a live load of 100 psf in addition to its own weight. Determine whether shear reinforcement is required or not. Given: Lightweight concrete at 110 pcf; $f'_c = 3500$ psi; $f_y = 60$ ksi

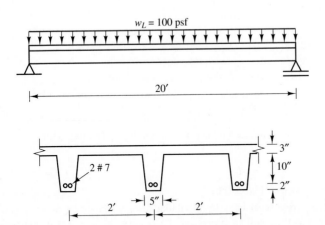

Figure 5.31   Example 5.7: One-way joist floor system.

Solution

Dead load: $w_D = [(5)(12) + (3)(24)]\dfrac{110}{144} = 100.8$ lb/ft

Factored load per joist: $w_u = (1.4)(100.8) + (1.7)(2)(100) = 481$ lb/ft

Left reaction: $R_L = (10)(481) = 4810$ lb

Shear at $d = 13''$ from support: $V_u = 4810 - (481)\dfrac{13}{12} = 4291$ lb

Moment at $d = 13''$ from support: $M_u = (4810)\dfrac{13}{12} - \dfrac{481}{2}\left(\dfrac{13}{12}\right)^2 = 4929$ ft-lb.

Check: $\dfrac{V_u d}{M_u} = \dfrac{(4291)(13)}{(4929)(12)} = .943 < 1.0$ OK

Reinforcing ratio: $\rho_w = \dfrac{(2)(0.6)}{(5)(13)} = .01846$

Nominal shear strength of concrete:

$$V_c = [1.9\sqrt{3500} + (2500)(.01846)(.943)](13)(5) = 10{,}135 \text{ lb}$$

Check: $3.5\sqrt{3500}(13)(5) = 13{,}459 > 10{,}135$ OK

Ultimate shear capacity of concrete:

$$V_u = 1.1\phi V_c = (1.1)(.85)(10{,}135) = 9476 \text{ lb} > 4291 \text{ lb}$$

No shear reinforcement is needed.

Use of the simplified Code formula would have led to the same conclusion. However, there may be some cases where only the detailed formula leads to this conclusion. In such a borderline situation, of course, it may be worthwhile to use it. ∎

EXAMPLE 5.8

In a laboratory experiment, two beams are to be tested to failure under concentrated third-point loads (Fig. 5.32). Both beams are identical except that one is reinforced with stirrups as shown, whereas the other is not. Predict the failure loads for both beams. Given: $f_c' = 3608$ psi, $f_y = 61{,}700$ psi; as measured in the laboratory.

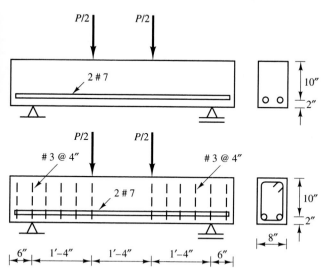

Figure 5.32   Example 5.8: Beams.

Solution   The failure load controlled by flexural capacity is the same for both beams:

Depth of compression block: $a = \dfrac{A_s f_y}{.85 f_c' b} = \dfrac{(2)(0.6)(61.7)}{(.85)(3.61)(8)} = 3.02$ in

Nominal moment capacity:

$$M_n = A_s f_y \left(d - \frac{a}{2}\right) = (1.2)(61.7)\left(10 - \frac{3.02}{2}\right) = 629 \text{ in-kips}$$

Failure load, neglecting dead weight of beam:

$$M_n = \frac{P_n L}{6} \text{ or } P_n = \frac{(6)(629)}{48} = \underline{78.6 \text{ kips.}}$$

Shear strength of first beam at third point: $\dfrac{V_u d}{M_u} = \dfrac{Pd}{(PL/3)} = \dfrac{3d}{L} = \dfrac{(3)(10)}{48} = .625$

Reinforcing ratio: $\rho = \dfrac{1.2}{(8)(10)} = .015$;

$$V_c = [1.9\sqrt{3608} + (2500)(.015)(.625)](80) = 11,005 \text{ lb} = \frac{P_n}{2}$$

Predicted failure load: $\dfrac{P_n}{2} = V_c$, that is, $P_n = (2)(11) = \underline{22.0 \text{ kips}}$

Contribution of shear reinforcement in second beam:

$$V_s = \frac{A_v f_y d}{s} = \frac{(2)(.11)(61.7)(10)}{4} = 33.9 \text{ kips}$$

Nominal shear strength of second beam: $V_n = V_c + V_s = 11.0 + 33.9 = 44.9$ kips
Predicted failure load: $P_n = 2V_n = (2)(44.9) = \underline{89.8 \text{ kips}}$

It is predicted that the first beam fails at $P = 22$ kips in shear, whereas the second beam fails at $P = 78.6$ kips in flexure. Note that no strength reduction factors ϕ are used when the actual rather than the design strength of a beam is to be estimated. ∎

5.5 BOND

The design for flexure described in Chapter 4 was based on the assumption (among others) that the strain in a reinforcing bar is the same as the strain in the concrete surrounding it. This assumption is equivalent to saying that perfect bond exists between concrete and steel, which does not permit any relative slippage to occur. In fact, it is fair to say that reinforced concrete would not work were it not for the transfer of bond stresses between steel and concrete. Consider a concrete beam with pipes embedded where we would normally place the steel bars (Fig. 5.33). Inserting well-greased steel

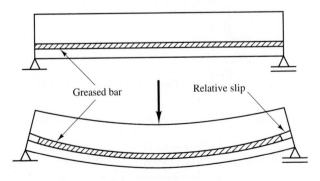

Greased bar   Relative slip

Figure 5.33 Concrete beam with unbonded reinforcement.

bars in these pipes would do the concrete no good, because when the concrete elongates under flexure, the bars would not stretch but simply slide relative to the concrete, pick up no stress, and not be able to assist the concrete in carrying any load.

If steel bars with smooth surfaces were embedded in the concrete, the bond between steel and concrete would depend entirely on chemical adhesion or friction, neither of which are very strong, and offer only moderate resistance in a pull-out test (Fig. 5.34). Under tension, the bar cross section contracts due to Poisson's ratio effect, thereby actually breaking whatever bond might exist. As a result, the bond strength of plain bars is so low that special anchorage devices (such as standard hooks at the ends) are necessary for all practical applications.

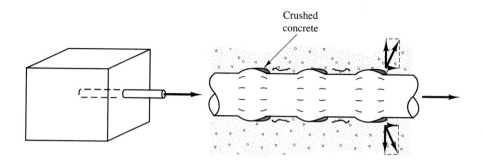

Figure 5.34   Pull-out test.

Nowadays, almost all reinforcement consists of *deformed* bars, that is, bars with ribs or lugs (Figs. 5.34 and 2.27). Breaking the bond between such bars and the concrete is obviously much harder, because rather than a relatively weak adhesion, the shear strength of the concrete itself must be overcome, as the steel ribs try to shear off small layers of concrete. In actual pull-out tests, small wedges of concrete can be observed sticking to the bar in front of the ribs (Fig. 5.34). These exert a high radial stress on the surrounding concrete, which tends to split it in its weakest direction. This radial pressure can spall off the concrete cover (Fig. 5.35), especially when the bars are closely spaced or the cover thickness is inadequate. If this happens, the bond strength is lost and slippage occurs. The minimum concrete cover and bar spacing requirements of the ACI Code

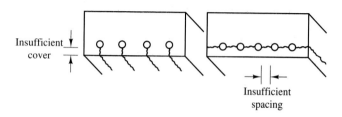

Figure 5.35   Concrete splitting caused by bond stresses.

are based partially on the necessity of preventing such bond failures. In Sec. 5.3, we discussed the contribution of dowel action of longitudinal bars to shear resistance. This dowel action also contributes to the tendency of the radial bond stress component to split off the concrete cover. Thus, shear and bond failures are often interrelated.

The bond strength of a flexural member can be considered the third link in a chain, the other two links being the flexural and shear strengths. Since bond failures are almost always sudden and brittle and therefore to be avoided, bar cutoffs must be detailed with great care to assure that the flexural capacity of the beam can be developed.

Let us start with the cantilever beam of Fig. 5.36a. The top reinforcing bars must provide the full tensile force necessary to equilibrate the bending moment PL. They obviously must be anchored securely in the support to develop this force. The length l_d necessary to accomplish this through bond alone is called *development length*. If the piece of steel of length l_d within the anchorage zone is isolated as a free body (Fig. 5.36b), bond stresses are evidently required to equilibrate the force T, that is,

$$T = \int_0^{l_d} u\pi d_b \, dx \tag{5.28}$$

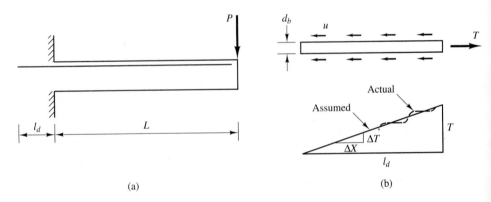

Figure 5.36   Bond stresses in cantilever beam.

where u is the bond stress and d_b the bar diameter. For convenience, let us assume bond stresses to be constant over the development length, that is,

$$u = u_{ave} = \text{const} \tag{5.29}$$

which is equivalent to saying that the tensile force in the bar increases linearly (Fig. 5.36b). Even though the actual bond stresses may vary considerably, it is the average stress u_{ave} that is of concern for design purposes, and this can be determined by tests. Substitution of Eq. 5.29 into Eq. 5.28 leads to

$$T = \pi d_b u_{ave} l_d \tag{5.30}$$

Equating this force to the bar capacity, $T = A_b f_y = (\pi/4)d_b^2 f_y$, the required development length follows as

$$l_d = \frac{d_b f_y}{4u_{ave}} \tag{5.31}$$

The ACI Code [Sec. 12.2] requires the following minimum development length,

$$\frac{l_d}{d_b} = \frac{3}{40} \frac{f_y}{\sqrt{f_c'}} \frac{\alpha\beta\gamma\lambda}{\left(\frac{c+k_{tr}}{d_b}\right)} \quad \text{with} \quad \left(\frac{c+k_{tr}}{d_b}\right) \le 2.5 \tag{5.32)[12-1]}$$

which recognizes the influence of several factors:

1. Experience has shown that during concrete placing and compaction, some excess water and air rise toward the top surface of the finished concrete and can be trapped below top reinforcing bars, thereby reducing their bond strength. For horizontal reinforcement so placed that more than 12 in of fresh concrete is cast below, the *bar location factor* α shall be 1.3.

2. The *coating factor* β recognizes the reduced bond properties of epoxy-coated bars. For all such bars with less than $3d_b$ cover or clear spacing of less than $6d_b$, $\beta = 1.5$; for all other epoxy-coated bars, $\beta = 1.2$. Note that the product $\alpha\beta$ need not be taken greater than 1.7.

3. Smaller bars do not have the tendency of larger ones to split the concrete. This fact is captured by the *bar size factor* γ, which is 0.8 for #6 and smaller bars.

4. To account for the generally lower strength of lightweight concrete, the *lightweight aggregate concrete factor* $\lambda = 1.3$ applies. However, when the splitting tensile strength f_{ct} is specified, $\lambda = 6.7\sqrt{f_c'}/f_{ct} \ge 1.0$ may be used.

5. If the concrete cover d_c or clear spacing d_s between bars is insufficient, concrete splitting may occur as shown in Fig. 5.35. This phenomenon is captured in Eq. 5.32 by the factor $(c + k_{tr})/d_b$, where c is the smallest of the side cover, cover over the bar (both measured to the center of the bar), or one-half the center-to-center spacing of the bars, and

$$k_{tr} = \frac{A_{tr} f_{yt}}{1500sn} \tag{5.33}$$

is a factor that represents the restraining effect of confining reinforcement across potential splitting planes, that is, A_{tr} is the total cross-sectional area of all transverse reinforcement having the spacing s and the yield strength f_{yt}, and n is the number of bars being developed along the plane of splitting.

6. If more reinforcement is provided than needed for strength, the full tensile capacity is not required, so the development length can be reduced by a factor equal to the ratio of required A_s divided by provided A_s.

As in many other instances of the ACI Code, Eq. 5.32 represents a relatively accurate estimate of required development length. A simpler (and more conservative) approach uses the formula

$$\frac{l_d}{d_b} = \frac{3}{40}\frac{f_y}{\sqrt{f_c'}}\alpha\beta\gamma\lambda \tag{5.34}$$

that is, it assumes $(c + k_{tr})/d_b = 1.0$. For the two cases,

   a) clear spacing of bars and clear cover at least d_b, and stirrups or ties through l_d not
   less than the Code minimum, or
   b) clear spacing of bars of at least $2d_b$ and clear cover of at least d_b;

the development length of Eq. 5.34 may be reduced by factor 2/3.

EXAMPLE 5.9

The design of a cantilever beam calls for a steel area of 2.7 in^2. Three #9 bars are selected
as shown in Fig. 5.37, together with #4 stirrups spaced 6 in apart. Determine the required
development length, using the refined and simple formulas of the ACI Code. Given: $f_y = 60$
ksi; lightweight concrete with $f_c' = 3500$ psi and $f_{ct} = 350$ psi.

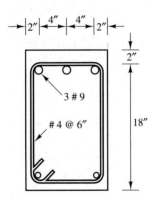

Figure 5.37   Example 5.9.

   Solution

   a) Simple formula, Eq. 5.34
      Bar location factor: $\alpha = 1.3$
      Coating factor: $\beta = 1.0$
      Bar size factor: $\gamma = 1.0$
      Lightweight aggregate concrete factor: $\lambda = \dfrac{6.7\sqrt{3500}}{350} = 1.13$
      Excess reinforcement factor: $\dfrac{2.7}{3.0} = 0.9$
      Check for 2/3 reduction factor:
         clear spacing $= 4 - 1.128 = 2.872 > 2d_b$ <u>OK</u>, cover $= 2 > d_b$ <u>OK</u>
      Required development length:
      $$l_d = \frac{2}{3}\frac{3}{40}\frac{60,000}{\sqrt{3500}}(1.3)(1.0)(1.0)(1.13)(0.9) = \underline{67\ in}$$

b) Refined formula, Eq. 5.32

Transverse reinforcement index: $k_{tr} = \dfrac{(2)(0.2)(60,000)}{(1500)(6)(3)} = 0.89$

Least spacing/cover: $c = 2 - \dfrac{1.128}{2} = 1.436 \text{ in}$

Spacing/transverse reinforcement factor: $\dfrac{c + k_{tr}}{d_b} = \dfrac{1.436 + .89}{1.128} = 2.06$

Required development length:

$$l_d = \frac{3}{40} \frac{60,000}{\sqrt{3500}} \frac{(1.3)(1.0)(1.0)(1.13)(0.9)}{2.06} = \underline{49} \text{ in}$$ ■

Until now we have dealt with the length required to develop the full yield force of a bar at its end anchorage. Bond stresses are also required to facilitate the variation of tensile forces in reinforcing bars at any section, not only near their ends. Consider the small segment of length Δx of a beam (Fig. 5.38). Moment equilibrium requires that $\Delta M = V \Delta x$. Since the internal lever arm jd is fairly constant, the tensile force in the steel is approximately proportional to the bending moment, that is,

$$\Delta T = \frac{\Delta M}{jd} = \frac{V \Delta x}{jd}$$

Force equilibrium for a small steel bar segment calls for bond stresses u acting over the contact surface area $\pi d_b \Delta x$:

$$\Delta T = \frac{V \Delta x}{jd} = u \pi d_b \Delta x$$

from which

$$u = \frac{V}{\pi d_b jd}$$

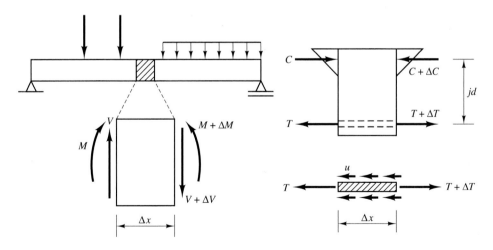

Figure 5.38 Bond action at arbitrary beam section.

or, if the symbol Σ_o represents the sum of all bar circumferences,

$$u = \frac{V}{\Sigma_o jd} \tag{5.35}$$

This bond stress u is proportional to the shear force and therefore referred to as *flexural bond*.

In reality, bond stresses are not directly proportional to the shear force because the concrete is cracked in the tensile zone. Between cracks, the uncracked concrete maintains its small tensile strength, which reduces the demand on bond stresses to develop tension in the bar. Therefore the bond stresses actually vary somewhat, as shown in Fig. 5.39. At a crack, the bond stress is zero; but in the vicinity, bond stresses can be quite large, especially in regions of high shear. These two effects can add up and cause local bond slip.

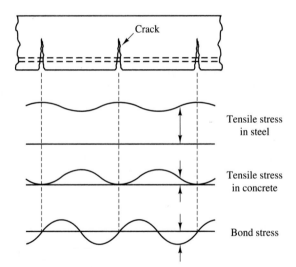

Figure 5.39 Crack-induced stress variations in steel and concrete.

Flexural bond requires that the length of a bar from any point with fully developed yield strength to its nearest free end must be at least equal to its development length. If the available length is insufficient to develop the full design force, special anchorage devices such as hooks or anchorage plates are required. This point is illustrated with the help of the simply supported beam of Fig. 5.40. At midspan, the full bar force $A_s f_y$ has half of the span length, $L/2$, available for development. At the quarterspan point, the bending moment is $\frac{3}{4}$ of the midspan moment. That is, the bar force is $\frac{3}{4} A_s f_y$, for the development of which only $L/4$ is available, so that 50% higher bond stresses are needed to develop this bar force compared to the full bar force at midspan. This increase is due to the higher shear force at the quarterspan point. Now consider a small segment of the beam of length Δx next to the support, where the shear force is a maximum (Fig. 5.41a). The bending moment at the right face is given by $M = (wL/2)\Delta x - (\Delta x^2/2)w$. The

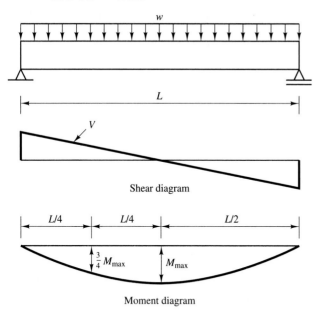

Shear diagram

Moment diagram

Figure 5.40 Simply supported beam.

term $(\Delta x^2/2)w$ is of a higher order and can be ignored, so that $M = (wL/2)\Delta x$, or $\Delta x = M/V$. Since the shear force equals the slope of the moment diagram, the quantity Δx indicates the length available for developing the steel force associated with the moment M. The higher the shear force at the support, the larger the demand on bond strength to facilitate this transfer of stress. In particular, if the nominal moment capacity $M_n = A_s f_y [d - (a/2)]$ is substituted for M, then the length M_n/V_n is needed for full bar force development, which should not be exceeded by the development length required for a specific bar. The ACI Code relaxes this requirement by substituting V_u for V_n and by giving credit for an end anchorage, l_a (Fig. 5.41c), defined as

$$l_a = \max\{12d_b; d\} \tag{5.36}$$

where d_b is the bar diameter and d the effective beam depth. With these two modifications, the Code requirement near simple supports and inflection points becomes

$$l_d \le \frac{M_n}{V_u} + l_a \tag{5.37}[12\text{-}2]$$

When the ends of the reinforcement are confined by a compressive reaction, Eq. 5.37 can be replaced by

$$l_d \le \frac{1.3 M_n}{V_u} + l_a \tag{5.38}$$

to reflect the improved bond properties in such cases.

Negative reinforcement can theoretically be cut off at an inflection point, beyond which it is no longer needed. In order to be on the conservative side and to accommodate shifting inflection points due to shifting live loads, the ACI Code requires that at least

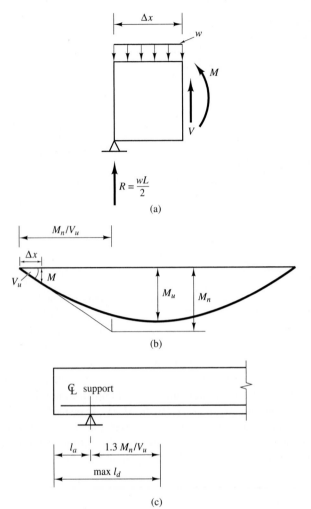

Figure 5.41 Bar development requirement near simple support.

one-third of the negative reinforcement extend the distance l_a beyond the point at which it is no longer required to resist flexure, or one-sixteenth of the clear span, whichever is greater.

EXAMPLE 5.10

A simple beam with cantilever overhang is reinforced with 4 #8 bars to resist both the positive and negative moments due to a factored uniform load of $w_u = 2.5$ kips/ft (Fig. 5.42). Shear reinforcement of 2 #3 closed stirrups spaced 5 in throughout the length of the beam is provided.

a) Determine the point at which the negative reinforcement can be terminated.

b) Check whether the bar forces can be developed at the left support.

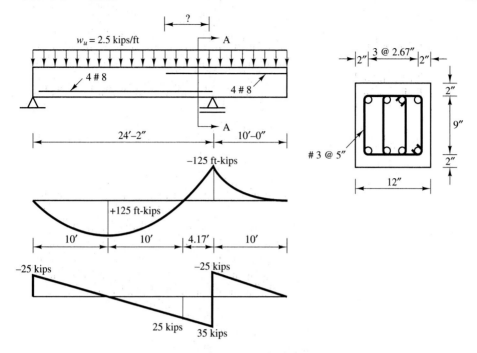

Figure 5.42   Example 5.10.

Given: Normal weight concrete with $f_c' = 4000$ psi; $f_y = 60,000$ psi. The moment and shear diagrams are as shown.

Solution

a) Bar location factor: $\alpha = 1.3$

Coating factor: $\beta = 1.0$

Bar size factor: $\gamma = 1.0$

Lightweight aggregate concrete factor: $\lambda = 1.0$

Check for 2/3 reduction factor:

clear spacing $= 2.67 - 1.0 = 1.67 < 2d_b$, but $1.67 > d_b$

and $A_v = (4)(.11) = .44 > \dfrac{(50)(12)(5)}{60,000} = 0.05$ __OK__

Required development length:

$$l_d = \frac{2}{3}\frac{3}{40}\frac{60,000}{\sqrt{4000}}(1.3)(1)(1)(1) = 62 \text{ in}$$

Length extension: $l_a = \max\{(12)(1) = 12; d = 11\} = 12$ in,

but $\max\left\{ l_a = 12; \dfrac{L_n}{16} = \dfrac{290}{16} = 18 \right\} = 18$ in

Available bar length 18 in beyond inflection point:

$50 + 18 = 68$ in > 62 in __OK__.

Let 2 #8 bars extend 50 in and 2 #8 bars extend 68 in from the support.

b) Depth of concrete compression block: $a = \dfrac{A_s f_y}{.85 f'_c b} = \dfrac{(4)(.79)(60)}{(.85)(4)(12)} = 4.64$ in

Nominal moment capacity: $M_n = (4)(.79)(60)\left(11 - \dfrac{4.64}{2}\right) = 1645.7$ in-kips

Required development length: $l_d = \dfrac{2}{3}\dfrac{3}{40}\dfrac{60{,}000}{\sqrt{4000}} = 48$ in

Development length check: $\dfrac{1.3 M_n}{V_u} + l_a = \dfrac{(1.3)(1645.7)}{25} + 12 = 97.6 \; > \; 48$ in OK.

∎

The design of flexural reinforcement is always based on the critical section or sections. Theoretically, at any other point, the beam is "overdesigned" because more steel is provided than necessary. For economy, it would be desirable to keep the amount of steel everywhere to the necessary minimum, by keeping the steel area approximately proportional to the factored bending moment. This can be achieved by either discontinuing some of the bars where they are no longer needed or by bending them up or down where the moment diagram changes sign. In this way the steel can do double duty, namely, resisting tension at the bottom in positive moment regions and resisting tension at the top in negative moment regions. Moreover, bond requirements are most likely satisfied if the bars are well anchored in the adjacent moment region.

To determine the point where it is permissible and feasible to discontinue some of the steel bars, consider a simply supported beam subjected to uniform load, which requires 4 #7 bars at midspan (Fig. 5.43a). If two of the four bars are discontinued, the remaining two bars provide only half the midspan capacity. From the moment diagram, the theoretical cutoff point can thus be determined, beyond which the two bars are no longer needed. The remaining two bars are then stressed at this point to their full capacity. The two discontinued bars contribute in this region only if they are properly anchored, for example, by extending them beyond this point for development. However, the actual loading on the beam is subject to fluctuations so that the theoretical cutoff points may shift somewhat in either direction. For these reasons, discontinued bars should extend beyond the point at which they are no longer needed to resist flexure, for a distance equal to l_a. In addition, the computed peak steel stress should at any section be developed by adequate length, that is, the bar length measured from a critical section such as midspan must be at least equal to l_d.

For the reasons stated above it is standard practice to add the length l_a beyond the point where bars are no longer needed. Figure 5.43 shows the theoretical variation of steel stress and a more realistic variation that takes into account the effect of bond. As can be seen, the steel is better utilized than if all steel would be continued through to the supports, as it is stressed to near capacity over a larger portion of the beam. However, terminating flexural reinforcement in a tension zone has been shown to reduce the shear strength and cause loss of ductility. Therefore, it is permitted only in areas where the shear force does not exceed two-thirds of the shear strength provided, and additional stirrups are provided to prevent the opening of cracks that are typically observed at points of bar cutoff.

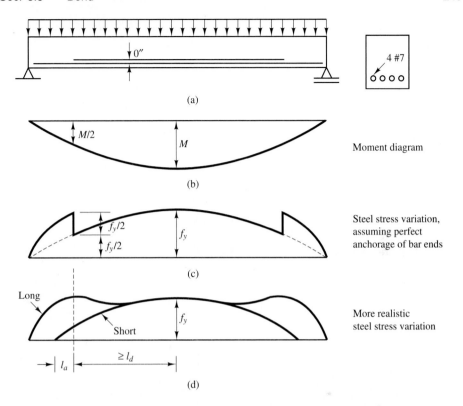

Figure 5.43   Qualitative steel stress variation in beam with discontinued bars.

EXAMPLE 5.11

For the positive reinforcement of the beam in Example 5.10, determine at what point two of the four #8 bars can be cut off.

Solution

For 2 #8 bars, depth of compression block: $a = \dfrac{(2)(.79)(60)}{(.85)(4)(12)} = 2.32$ in

Ultimate moment capacity:

$$M_u = (.9)(2)(.79)(60)\left(11 - \frac{2.32}{2}\right) = 839 \text{ in-kips} = 70 \text{ ft-kips}$$

Theoretical cutoff point x follows from $\dfrac{wx^2}{8} = 125 - 70 = 55$ ft-kips,

that is, $x = \sqrt{\dfrac{(8)(55)}{2.5}} = 13.3$ ft or $10 - \dfrac{13.3}{2} = 3.37$ ft $= 40$ in from the left support or the inflection point.

Add safety margin: $l_a = \max\{12d_b = 12;\, d = 11\} = 12$ in.

Answer: 2 #8 bars can be cut off at $40 - 12 = 28$ in to the right of the left support and to the left of the inflection point, provided the ACI Code's provisions for sufficient shear reinforcement at these cutoff points are satisfied. In practical situations, it is debatable whether the savings of a few feet of bar length justify the added design effort and the increased bar placement effort. ∎

5.6 TORSION

A moment that tends to twist a member about its longitudinal axis is referred to as a *twisting moment*, a *torsional moment*, or simply *torque*. Examples are encountered frequently in everyday life (Fig. 5.44a and b). Also, in concrete construction, members

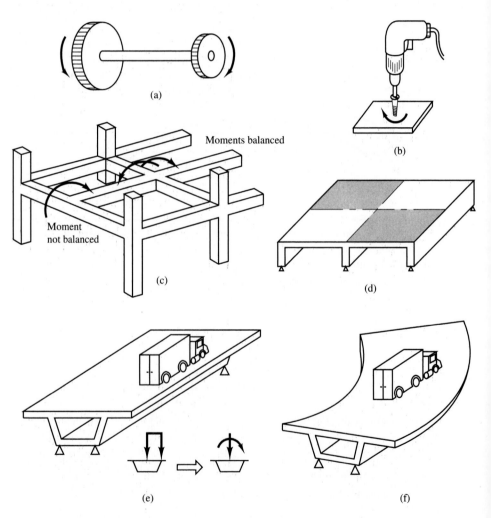

Figure 5.44   Structures subjected to torsion: (a) shaft; (b) drill or screwdriver; (c) spandrel beam; (d) checkerboard loading on floor slab; (e) eccentrically loaded straight bridge; (f) curved bridge.

are frequently subjected to torsion. A spandrel beam experiences a twisting moment if it supports a floor beam whose end moment is not balanced like that of a beam supported in the interior of a building (Fig. 5.44c). A slab loaded by a checkerboard-type live load (Fig. 5.44d) tends to rotate one-half of the supporting beam clockwise and the other half counterclockwise, thereby subjecting the beam to torsion. In bridge design, torsional moments are very common. Consider a truck placed eccentrically with respect to the centerline of a box girder (Fig. 5.44f). If the bridge ends are properly supported to prevent bridge rotation, the box girder will be subjected to torsion, as is a curved bridge, even under a concentrically placed truck (Fig. 5.44f).

In mechanics of solids, two types of torsion are distinguished—St. Venant (pure torsion) and warping torsion. In pure torsion, the member is assumed to be free to warp, as can be demonstrated with a solid piece of an eraser (Fig. 5.45a). Warping torsion, on the other hand, is generally associated with thin-walled sections, such as steel members (Fig. 5.45b), in which the two thin flanges are subjected to opposite bending stresses and the associated deformations. In most structural members, both types of torsion are present, but as stresses associated with warping torsion are significant primarily in thin-walled sections, only pure torsion shall be considered here.

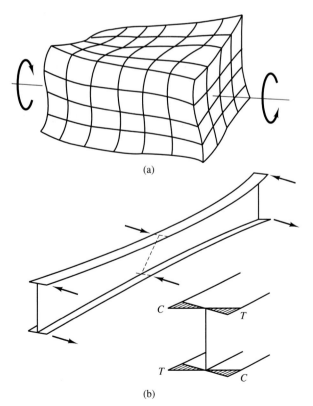

(a)

(b)

Figure 5.45 St. Venant and warping torsion.

The St. Venant torsion formula, which relates the externally applied torque T to the angle of twist, Φ (Fig. 5.46a), can be derived using the laws of solid mechanics [3],

$$\frac{d\Phi}{dz} = \frac{T}{GJ} \quad \text{or} \quad T = GJ\frac{d\Phi}{dz} \tag{5.39}$$

where G is the shear modulus, and J the torsional constant, defined as

$$J = \int_A r^2 dA \tag{5.40}$$

see Fig. 5.46b. For a circular section

$$J = \frac{\pi R^4}{2} = \frac{\pi D^4}{32} \tag{5.41}$$

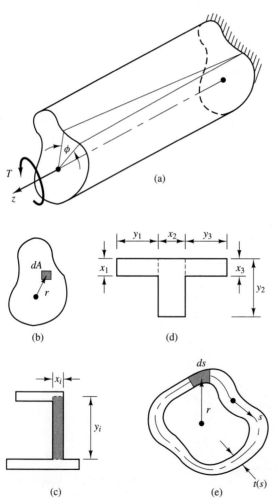

Figure 5.46 Torsion of thin-walled sections.

where R is the section radius and D the diameter. For a rectangular section, evaluation of Eq. 5.40 is more difficult. If x and y denote, respectively, the smaller and larger side, the solution can be expressed in series form

$$J = \frac{x^3 y}{3}\left(1 - .63\frac{x}{y} + .052\frac{x^5}{y^5} - \cdots\right)$$

For narrow rectangular sections with small x/y ratios, it follows that

$$J \approx \frac{x^3 y}{3}$$

and for a section that can be idealized as a series of narrow rectangles of thickness x_i and width y_i (Fig. 5.46c and d), we obtain

$$J \approx \sum_i \frac{x_i^3 y_i}{3} \tag{5.42}$$

The solution for closed sections is given by Bredt's formula (Fig. 5.46e):

$$J = \frac{4A^2}{\oint \frac{ds}{t(s)}} \tag{5.43}$$

where A is the area enclosed by the section wall's centerline, s is the circumferential coordinate, $t(s)$ the thickness as a function of s, and $\oint$ denotes a line integral.

EXAMPLE 5.12

Compare the torsional constants of the two thin-walled steel sections shown in Fig. 5.47. The only difference between them is that the first section has a small gap, whereas the second one is closed.

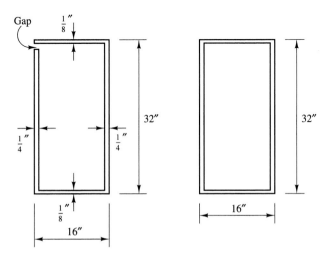

Figure 5.47   Example 5.12.

Solution   Since the first section is not closed, we use Eq. 5.42 to obtain

$$J_1 = (2)\left(\frac{1}{3}\right)\left[(32)(.25)^3 + (16)(.125)^3\right] = .354 \text{ in}^4$$

The second section is closed, that is, Eq. 5.43 applies, such that

$$J_2 = \frac{4(16 \times 32)^2}{\frac{(2)(32)}{.25} + \frac{(2)(16)}{.125}} = 2048 \text{ in}^4$$

Note that the closure of the gap increases the torsional constant for the section by the factor $2048/.354 = 5785$.   ∎

Shear stresses in a solid section due to torsional moments are computed in close analogy to bending stresses ($f_b = Mc/I$) as

$$v = \frac{Tr}{J} \tag{5.44}$$

For a thin-walled closed section, the shear stress is computed as

$$v = \frac{T}{2tA} \tag{5.45}$$

where A is again the area enclosed by the section centerline, and t is the thickness at the point where the shear stress is calculated. The *shear flow* is then defined as

$$q = vt = \frac{T}{2A} = \text{const} \tag{5.46}$$

that is, the shear flow q is the same at any point s in Fig. 5.46e; if the thickness becomes smaller, the shear stress v increases, and vice versa.

In a rectangular section of dimensions x and y, the maximum shear stress occurs at the middle of the longer side:

$$v = \frac{T}{\alpha x^2 y} \tag{5.47}$$

where α varies from .208 for a square to .333 for $y \gg x$. Thus, in analogy to Eq. 5.42, we can approximate the maximum shear stress in a section made up of a series of rectangles as

$$v = \frac{T}{\sum_i (x_i^2 y_i / 3)} \tag{5.48}$$

The completely different responses of closed and open sections to torsion can best be illustrated by showing the shear stresses v or shear flow vt needed to equilibrate an externally applied torque (Fig. 5.48). In solid sections, shear stresses vary from zero at the centroid to some maximum value. In a circular section, the average stress is thus only half the maximum. Moreover, the lever arm r of an infinitesimal force vdA varies from R (very effective) to zero (not effective at all); see Fig. 5.48b.

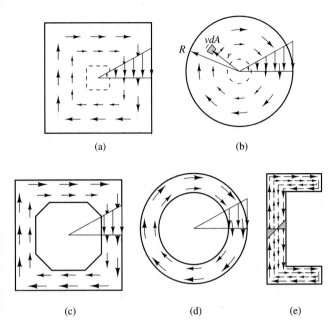

Figure 5.48 Torsion-induced shear stresses in solid, closed, and open sections.

(a)   (b)   (c)   (d)   (e)

In a tubular section, the material is stressed to the maximum almost everywhere, and the internal lever arm is large for all $v\,dA$ forces. Thus, in an efficient torsional member, the material is concentrated near the periphery, as in a tube or square box. To avoid stress concentrations near the interior corners of box sections, it is common practice to use fillets, as shown in Fig. 5.48c. The worst type of section is a thin-walled open section (Fig. 5.48e), in which shear stresses have large gradients and small internal lever arms.

It so happens that the differential equation that governs the torsion problem in linear elasticity is similar to the equation for a thin membrane subjected to normal pressure. This coincidence is the basis for the *soap film* or *membrane* analogy, which is extremely helpful in visualizing important aspects of torsion. Consider the cross sections of different torsional members cut out from the top side of a closed box and covered with thin soap films, which will deform if the box is pressurized (Fig. 5.49). Two facts constitute the analogy:

1. the volume under the deformed soap film is proportional to the torsional rigidity of the section;
2. the steepest slope of the membrane at a given point is proportional to the shear stress at that point.

Using this analogy, all of the previously made observations can be easily visualized:

1. open, thin-walled sections (Fig. 5.49a) allow volumes (torques) to build up only at the expense of steep slopes (high shear stresses);

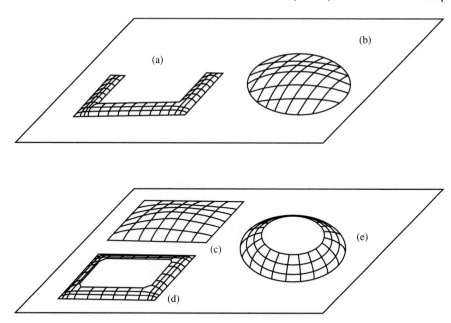

Figure 5.49 Soap film analogy.

2. solid sections (Fig. 5.49b and c) permit large volumes to build up, but near the section centroid slopes are very small, that is, the material is underutilized;

3. in closed, thin-walled sections (Fig. 5.49d and e), the entire interior areas A are displaced, creating large volumes with moderate slopes in the soap film. (Note how the fillets in the square box corners reduce stress concentrations.)

Although the foregoing discussion is strictly limited to elastic methods, it is made very clear that thin-walled open concrete sections are not very effective in resisting torsion. Solid sections, such as rectangles with side aspect ratios not too much different from unity, are suitable for torsion members, except that the concrete near the centroid is underutilized. By ignoring it, we assume indirectly that the torsion is carried entirely by an outer tube. This is in fact one of the two important assumptions (given below) underlying the provisions of the ACI Code for torsion design. (The background for these provisions can be found in [4], which also forms the basis for much of what follows.)

a) before cracking, a thin-walled outer tube is effective in resisting the torsion;

b) after cracking, the contribution of the concrete to resist torsion is negligible, and all torsion is assumed to be carried by an equivalent space truss consisting of closed transverse stirrups, longitudinal reinforcement, and diagonal concrete compression struts.

The thin-walled outer tube can be defined such that the stirrups are centered along its middle surface, Fig. 5.50. If the area enclosed by the stirrups is A_o and its perimeter

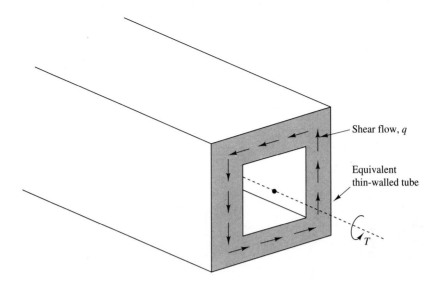

Figure 5.50 Equivalent thin-walled tube.

length p_h, then the equivalent tube thickness can be approximated as,

$$t = \frac{A_o}{p_h}$$

For example, for a section with square stirrups enclosing a 24-by-24-in area (576 in^2) and a perimeter of 96 in, the equivalent tube thickness is $t = 576/96 = 6$ in. For rectangular sections, t turns out to be approximately $\frac{1}{6}$ to $\frac{1}{4}$ of the minimum width of the section.

After cracking, the concrete is assumed to have no torsional shear strength left. This assumption is reminiscent of that for flexural design, where the concrete is assumed to have no tensile strength after cracking, but it is contrary to that for shear design, where the concrete is assumed to carry as much shear after cracking as before. (This inconsistency of the ACI Code gives cause to the belief that the shear design provisions will be changed in future editions.) Note that the logical follow-up assumption is that the capacity of concrete to carry shear, V_c, is not affected by torsion.

When the applied torsion exceeds the cracking strength of the thin-walled tube, the cracks tend to spiral around the member as shown in Fig. 5.51, forming approximately 45° angles with the edges of any side or wall of the tube. The torsion is resisted by an imaginary space truss, with the closed stirrups and longitudinal reinforcement in the corners acting as tension members and the diagonal concrete struts as compression members (Fig. 5.53a). This is equivalent to the simple plane truss analogy for shear design (Fig. 5.23). For a rectangular section, let us designate the truss width as x_o and the height as y_o. Consequently, the enclosed area is $A_o = x_o y_o$ and the perimeter $p_h = 2(x_o + y_o)$.

Before determining the torsional reinforcement required to carry a given factored torque T_u, it is necessary to distinguish between those cases where torsional moments are

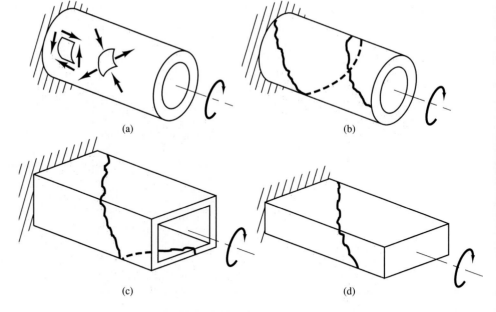

(a)   (b)

(c)   (d)

Figure 5.51   Torsional cracking.

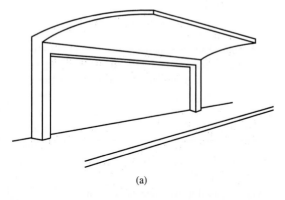

(a)

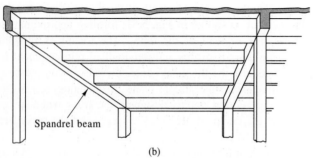

Spandrel beam

(b)

Figure 5.52   Equilibrium and compatibility torsion. (a) Equilibrium torsion: design torque may not be reduced; (b) compatibility torsion: design torque in spandrel beam may be reduced (courtesy of the American Concrete Institute).

required to maintain equilibrium (*equilibrium torsion*) and those where moments arise in members of statically indeterminate structures that undergo twisting to maintain compatibility of deformations (*compatibility torsion*). The structure of Fig. 5.52a illustrates the first case, in which the full torque is needed to maintain equilibrium. In the structure of Fig. 5.52b the torsional moments in the spandrel beam can be reduced by redistribution of internal forces after cracking. If the other structural members have sufficient strength to carry the applied loads, the torsional moment in the spandrel beam is not needed for equilibrium and may be reduced as described later.

Let us now derive an estimate of the torque at which the concrete cracks. In pure torsion, the principal tensile stress σ_1 is equal to the maximum shear stress v (see Fig. 5.8b). With Eq. 5.45, we obtain

$$\sigma_1 = v = \frac{T}{2tA} \tag{5.49}$$

Defining $A = A_{cp}$ as the area enclosed by the perimeter p_{cp} of the concrete section (i.e., A_o plus the concrete cover all around), and assuming that the tube has an equivalent thickness $t = \frac{3}{4}A_{cp}/p_{cp}$, and its centerline encloses an area $\frac{2}{3}A_{cp}$, Eq. 5.49 becomes

$$\sigma_1 = v = \frac{T p_{cp}}{A_{cp}^2} \tag{5.50}$$

By setting the tensile strength of concrete equal to $4\sqrt{f_c'}$, which is less than the regular $7.5\sqrt{f_c'}$ because the material is in biaxial tension-compression, Eq. 5.50 yields an estimate for the cracking torque,

$$T_{cr} = 4\sqrt{f_c'}\,\frac{A_{cp}^2}{p_{cp}} \tag{5.51}$$

The ACI Code allows that torsion be neglected if it does not exceed $T_{cr}/4$, that is, if

$$T_u \le \phi\sqrt{f_c'}\,\frac{A_{cp}^2}{p_{cp}} \tag{5.52}$$

where $\phi = 0.85$ is the strength reduction factor.

The required torsional reinforcement is readily determined from the forces that the various members in the space truss carry. According to Eq. 5.46, the shear flow in a tubular member is

$$q = vt = \frac{T}{2A} \tag{5.53}$$

Note that A is now the area enclosed by the centerline of the tube (or the closed stirrup) that we had designated as A_o. The total shear force acting on any particular wall is found by multiplying q with the wall width, for example, for the left vertical wall of the section in Fig. 5.53a,

$$V_2 = q y_o = \frac{T}{2A_o} y_o \tag{5.53}$$

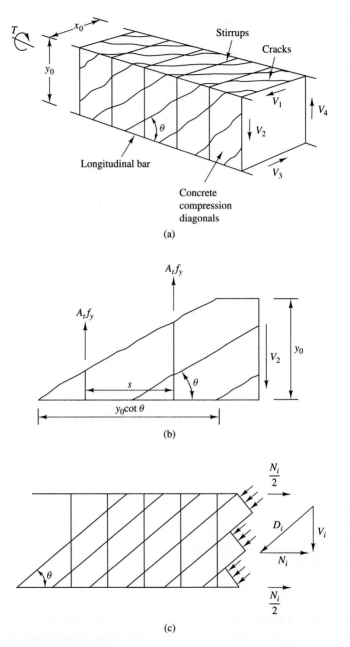

Figure 5.53 Thin-walled tube and space truss analogy: (a) equivalent space truss; (b) vertical equilibrium; (c) resolution of shear force into diagonal compression and axial tension (Courtesy of the American Concrete Institute).

As Fig. 5.53b illustrates, this force needs to be equilibrated by the vertical component of the concrete compression strut. Since a typical crack crosses $n = y_o \cot\theta / s$ stirrups, where s is the stirrup spacing, equilibrium requires that the combined forces of n stirrups at yield equal the vertical component of the compression strut, which is equal to V_2,

$$V_2 = \frac{A_t f_{yv} y_o}{s} \cot\theta = \frac{T y_o}{2 A_o}$$

where A_t is the cross-sectional area of the stirrup and f_{yv} its yield strength. The nominal torsional capacity thus becomes

$$T_n = \frac{2 A_o A_t f_{yv}}{s} \cot\theta \qquad (5.54)[11\text{-}22]$$

If, as usual, cracks are assumed to form the angle $\theta = 45°$ with the horizontal, then the stirrup size required to resist the factored torsion $T_u = \phi T_n$ follows from

$$A_{t,\text{req}} = \frac{T_u s}{2\phi A_o f_{yv}} \qquad (5.55)$$

This needs to be added to the transverse reinforcement required to carry shear, Eq. 5.24,

$$A_{v,\text{req}} = \frac{V_s s}{d f_y}$$

Recognizing that for shear reinforcement a closed stirrup provides two legs with combined area A_v, whereas for torsional reinforcement a similar stirrup provides only one leg of area A_t, the combined stirrup requirement becomes

$$A_{t,\text{req}} = \frac{T_u s}{2\phi A_o f_y} + \frac{V_s s}{2 d f_y} \qquad (5.56)$$

and for a preselected stirrup of size A_t, the required spacing becomes

$$s_{\text{req}} = \frac{2 A_t f_y}{\frac{T_u}{\phi A_o} + \frac{V_s}{d}} \qquad (5.57)$$

The force in the diagonal compression strut, $D_i = V_i / \sin\theta$ (Fig. 5.53c), has a horizontal component $N_i = V_i \cot\theta$. Since the shear flow q is constant over the width of the wall, N_i acts at the midpoint and can be replaced by two forces $N_i / 2$ at the top and bottom, or rather in the corners, provided reinforcing bars are placed there to carry these forces. Similar forces result from the loading on the other three sides, so that for a rectangular tube, the total longitudinal force is

$$N = 2(N_1 + N_2) = 2\cot\theta (V_1 + V_2)$$

and with Eq. 5.53,

$$N = \frac{T_n}{2 A_o} 2(x_o + y_o) \cot\theta = \frac{T_n p_h}{2 A_o} \cot\theta \qquad (5.58)$$

where p_h is the perimeter of area A_o. Longitudinal reinforcement A_l with yield strength f_{yl} must be provided to carry the entire force N, that is,

$$N = A_l f_{yl}$$

or

$$A_{l,\text{req}} = \frac{T_n p_h}{2 A_o f_{yl}} \cot\theta \qquad (5.59)$$

The steel requirement can be expressed in terms of the required stirrup area A_t by substituting Eq. 5.54 for T_n,

$$A_{l,\text{req}} = \left(\frac{A_t}{s}\right) p_h \left(\frac{f_{yv}}{f_{yl}}\right) \cot^2\theta \qquad (5.60)[11\text{-}23]$$

For $\theta = 45°$ and $f_{yv} = f_{yl}$, this simplifies to

$$A_{l,\text{req}} = \frac{A_t}{s} p_h$$

Note that this steel area should be distributed uniformly around the perimeter of the section. However, the designer is free to select the angle θ anywhere between $30°$ and $60°$. The result will be that the stirrup requirement is reduced and the longitudinal steel requirement increased, or vice versa.

When torsion is combined with flexure, the tensile flexural and torsional longitudinal reinforcement requirements are additive, while flexural compression reduces the longitudinal steel required for torsion.

When combining torsion with shear (Fig. 5.54a), we realize that on one side of a tube the shear stresses are additive and on the other side they cancel. Assuming that after torsional cracking, $A = 0.85 A_o$ and $t = A_o/p_h$, the shear stress due to torsion is

$$v = \frac{T_u}{1.7 A_o t} = \frac{T_u p_h}{1.7 A_o^2}$$

In one wall this adds to the shear stress $V_u/b_w d$. The combined shear stress should not exceed the maximum shear strength. According to Eq. 5.26, the maximum shear to be carried by stirrups is $8\sqrt{f_c'}$. Combined with the concrete shear strength, $v_c = V_c/b_w d$ (see Sec. 5.2), we find

$$\frac{V_u}{b_w d} + \frac{T_u p_h}{1.7 A_o^2} \le \phi \left(\frac{V_c}{b_w d} + 8\sqrt{f_c'}\right) \qquad (5.61)[11\text{-}20]$$

In a solid section, the shear stresses due to direct shear vary parabolically over the entire section depth, while the torsional stresses act only on the (imaginary) outer tube (Fig. 5.54b). A straight addition of these two stresses appears to be overconservative, and it is more reasonable to combine them as the square root of the sum of the squares,

$$\sqrt{\left(\frac{V_u}{b_w d}\right)^2 + \left(\frac{T_u p_h}{1.7 A_o^2}\right)^2} \le \phi \left(\frac{V_u}{b_w d} + 8\sqrt{f_c'}\right) \qquad (5.62)[11\text{-}19]$$

The upper limit in both cases assures that crack widths remain small, rather than to assure that the concrete diagonal struts do not crush in compression before the stirrups yield in tension. In fact, it can be shown [4] that combined shear and torsion will cause crushing of concrete before exceeding acceptable crack widths only if $f_c' < 1324$ psi, which is not the case in practical situations.

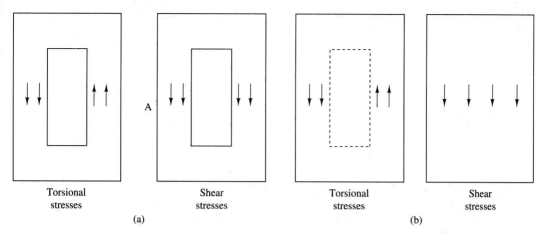

Torsional   Shear   Torsional   Shear
stresses   stresses   stresses   stresses
        (a)   (b)

Figure 5.54   Addition of torsional and shear stresses: (a) hollow section; (b) solid section (courtesy of the American Concrete Institute).

If the applied torque exceeds the cracking torsion, we would like to assure that a minimum amount of reinforcement is present to carry the torsion released by the concrete. Tests have shown that one percent of steel by volume is generally adequate, that is,

$$\frac{A_{l,\min}\, s + A_t p_h}{A_{cp} s} \geq 0.01$$

or

$$A_{l,\min} = 0.01 A_{cp} - \frac{A_t p_h}{s} \tag{5.63}$$

The ACI Code expresses this requirement in the form

$$A_{l,\min} = \frac{5\sqrt{f_c'}}{f_{yl}} A_{cp} - \frac{A_t p_h}{s}\frac{f_{yv}}{f_{yl}} \tag{5.64}[\mathit{11\text{-}25}]$$

which is somewhat less than that given by Eq. 5.63 for standard material properties.

When torsional reinforcement is required, the Code minimum area of transverse closed stirrups is given by

$$A_v + 2A_t \geq \frac{50 b_w s}{f_{yv}} \tag{5.65}[\mathit{11\text{-}24}]$$

Comparing this with the minimum requirement for shear reinforcement alone (Eq. 5.21), Eq. 5.65 recognizes that a closed stirrup provides only a single leg of area A_t to resist torsion.

Recalling the special case of compatibility torsion, we note that the ACI Code permits the maximum factored torsional moment T_u to be reduced to

$$T_u' = \phi 4\sqrt{f_c'}\left(\frac{A_{cp}^2}{p_{cp}}\right) \tag{5.66}$$

in which case the correspondingly redistributed bending moments and shears in the adjoining members are to be used to design those members. Note that T'_u is equal to the cracking moment T_{cr} of Eq. 5.51, except for the strength reduction factor. In other words, the applied torque may be reduced as low as the member's cracking moment.

We are now in a position to summarize the design for torsion in the following step-by-step procedure, assuming the section has already been dimensioned for flexure and shear.

Step 1.   Check whether torsion must be considered. Neglect torsion if

$$T_u \leq \phi\sqrt{f'_c}\,\frac{A^2_{cp}}{P_{cp}}$$

Step 2.   Check whether the section has the capacity to resist the applied torque. If for a hollow section,

$$\frac{V_u}{b_w d} + \frac{T_u p_h}{1.7 A^2_o} > \phi\left(\frac{V_c}{b_w d} + 8\sqrt{f'_c}\right)$$

or for a solid section,

$$\sqrt{\left(\frac{V_u}{b_w d}\right)^2 + \left(\frac{T_u p_h}{1.7 A^2_o}\right)^2} > \phi\left(\frac{V_u}{b_w d} + 8\sqrt{f'_c}\right)$$

even maximum reinforcement is insufficient, that is, the section size must be increased.

Step 3.   Compute the required spacing of closed stirrups with cross-sectional area A_t, to resist both shear and torsion,

$$s_{\text{req}} = \frac{2A_t f_y}{\dfrac{T_u}{\phi A_o} + \dfrac{V_s}{d}}$$

Step 4.   Check minimum transverse steel,

$$s_{\text{max}} = \frac{A_v f_{yv}}{50 b_w}$$

where A_v accounts for both shear and torsional reinforcement.

Step 5.   Determine the additional longitudinal reinforcement required for torsion,

$$A_l = \frac{T_n p_h}{2A_o f_{yl}} \geq \frac{5\sqrt{f'_c}}{f_{yl}} A_{cp} - \frac{A_t p_h}{s}\frac{f_{yv}}{f_{yl}}$$

EXAMPLE 5.13

Two simply supported parallel box beams are rigidly connected at midspan by a cross beam carrying a concentrated live load of 50 kips as shown in Fig. 5.55. A stiffness analysis determined that each of the two beams carries a concentrated 25-kip live load and a 3-kip dead load at midspan, plus a factored torsional moment of 20 ft-kips in addition to its own

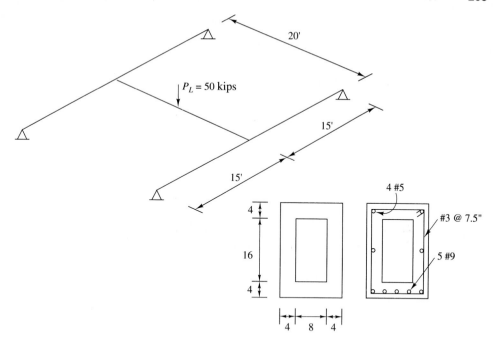

Figure 5.55   Example 5.13.

dead weight. Design the beams for flexure, shear, and torsion. Given: 150 pcf concrete; $f_c' = 4000$ psi; $f_y = 60,000$ psi.

a) *Flexural design*

Dead weight: $w_d = \dfrac{16 \times 24 - 8 \times 16}{144}(.15) = .267$ kips/ft

Factored moment:

$$M_u = 1.4 \left(\frac{.267 \times 30^2}{8} + \frac{3 \times 30}{4} \right) + 1.7 \left(\frac{25 \times 30}{4} \right) = 392 \text{ ft-kips}$$

Required flexural steel: $d = 22,\; a = 4,\; A_s = \dfrac{(392)(12)}{(.9)(60)(22 - 2)} = 4.36$ in^2.

Check $a = \dfrac{(4.36)(60)}{(.85)(4)(16)} = 4.8$, say OK. Use 5 #9, with 5.0 in^2.

b) *Shear design*

Factored shear: $V_u = 1.4 \left[(.267) \left(15 - \dfrac{22}{12} \right) + \dfrac{3}{2} \right] + 1.7 \left[\dfrac{25}{2} \right] = 28.3$ kips

Concrete shear strength: $V_c = 2\sqrt{4000}(8)(22) = 22.3$ kips

Use #3 stirrups with $s_{\max} = \min \left\{ \dfrac{22}{2} = 11;\; 24;\; \dfrac{(.22)(60,000)}{(50)(8)} = 33 \right\} = 11$ in

Check $V_s = \frac{V_u}{\phi} - V_c = \frac{28.3}{.85} - 22.3 = 11$ kips

$s_{req} = \frac{(.22)(60)(22)}{11} = 26.4$ in OK. Use #3 stirrups @ 11 in.

c) *Torsion design*

1. $A_{cp} = (16)(24) = 384$ in^2, $p_{cp} = 2(16 + 24) = 80$ in

$T_{min} = \phi\sqrt{f_c'}\frac{A_{cp}^2}{p_{cp}} = 0.85\sqrt{4000}\frac{384^2}{80} = 99,088$ in-lb $= 8.26$ ft-kips

$T_{min} < T_u = 20$ ft-kips. Torsion needs to be considered.

2. Section capacity: $\phi(10\sqrt{f_c'}) = (0.85)(10)\sqrt{4000} = 538$ psi

$A_o = (12)(20) = 240$ in^2, $p_h = (2)(12 + 20) = 64$ in

$\frac{V_u}{b_w d} + \frac{T_u p_h}{1.7A_o^2} = \frac{28,300}{(8)(22)} + \frac{(240,000)(64)}{(1.7)(240)^2} = 318 < 538$ OK.

3. Stirrup spacing for combined shear and torsion:

$s_{req} = \frac{2A_t f_y}{\frac{T_u}{\phi A_o} + \frac{V_s}{d}} = \frac{(2)(0.11)(60)}{\frac{(20)(12)}{(.85)(240)} + \frac{11}{22}} = 7.9$ in. Say 7.5 in.

Note that $A_t = \frac{T_u s}{2\phi A_o f_{yv}} = \frac{(240)(7.5)}{(2)(.85)(240)(60)} = .074$ in^2 @ 7.5 in

$A_v = \frac{V_s s}{d f_y} = \frac{(11)(7.5)}{(22)(60)} = .062$ in^2 @ 7.5 in

Combined $2A_t + A_v = (2)(.074) + .062 = .21$ in$^2 < .22$ in^2 provided by #3 stirrups spaced at 7.5 in.

4. Minimum transverse steel:

$\frac{50b_w s}{f_{yv}} = \frac{(50)(8)(7.5)}{60,000} = .05 < A_v + 2A_t = .21$ OK.

5. Determine longitudinal steel requirement:

$A_{l,min} = \frac{5\sqrt{f_c'}}{f_{yl}}A_{cp} - \frac{A_t}{s}p_h\frac{f_{yv}}{f_{yl}} = \frac{5\sqrt{4000}}{60,000}(384) - \frac{.074}{7.5}(64)(1) = 1.39$ in^2

$A_{l,req} = \frac{T_u p_h}{2\phi A_o f_{yl}} = \frac{(240)(64)}{(2)(.85)(240)(60)} = .62$ in$^2 < 1.39$ in^2

Divide by 6 bars, $\frac{1.39}{6} = .23$ in^2/bar

Add 4 #5 longitudinal bars, with 0.31 in^2 each (see Fig. 5.55).

The capacity of the 5 #9 bars provided for flexure with 5.0 in^2 exceeds the 4.36 in^2 required by more than $(2)(.23) = 0.46$ in^2. The solution shown in Fig. 5.55 satisfies all requirements.

Note that theoretically the top bars would not be needed because the torsional tension is overstressed by flexural compression. However, the torque is constant throughout the member length, whereas the bending moment decreases toward the supports. Also, the corner bars are needed to form the space truss and permit the stirrups to be tied into a reinforcing cage. Thus, the above solution is appropriate.

Since this structure is an example with compatibility torsion, the reduced torsional moment of Eq. 5.66 may be used, $T'_u = 4T_{min} = (4)(99,088) = 396,352$ in-lbs = 33.0 ft-kips $> T_u$, that is, since $T_u < \phi T_{cr}$, a reduction due to cracking is inappropriate. ■

SUMMARY

1. Diagonal tension (or principal) stresses in an uncracked homogeneous beam are computed by

$$\sigma = \frac{f}{2} \pm \sqrt{\frac{f^2}{4} + v^2}$$

where $f = (My/I)$ is the bending stress, and $v = (VQ/Ib)$ is the shear stress at y, with $Q = \int_{\eta=y}^{h} \eta \, dA$.

2. Shear reinforcement is generally more conservatively designed than flexural reinforcement to avoid sudden or brittle shear failures and to assure gradual or ductile flexural failure modes.

3. Web-shear cracks can be expected if, in a region of small moment and large shear, the shear force exceeds

$$V_c = 3.5\sqrt{f'_c} b_w d$$

4. Flexure-shear cracks can be expected at the much smaller shear force

$$V_c = 1.9\sqrt{f'_c} b_w d$$

in a region of large moment and large shear, because a larger diagonal tension is acting on a concrete area reduced by flexural cracks.

5. The various effects of shear reinforcement are:
   - web reinforcement carries shear directly, as demonstrated by the truss analogy;
   - by traversing diagonal cracks, web reinforcement makes aggregate interlock possible;
   - by keeping diagonal cracks from opening, web reinforcement inhibits the propagation of cracks into the concrete compression zone;
   - by tying longitudinal bars into the main bulk of the concrete, the stirrups increase the dowel shear effect;
   - hoop confinement increases the concrete strength and ductility.

6. The shear reinforcement design according to the ACI Code can be organized in the following step-by-step procedure:
   Step 1. Draw the nominal shear diagram for the factored beam loading, $V_n = (V_u/\phi)$.
   Step 2. Draw the moment diagram for the factored beam loading (only for use of the refined shear strength formula).

Step 3. On the shear diagram, superimpose the concrete shear capacity diagram, given by

$$V_c = \left(1.9\sqrt{f_c'} + 2500\rho_w \frac{V_u d}{M_m}\right) b_w d$$

$$\leq 3.5\sqrt{f_c'}b_w d\sqrt{1 + \frac{N_u}{500A_g}}$$

where $M_m = M_u - N_u(4h - d/8)$, and N_u is the factored axial force (positive for compression). Alternatively,

$$V_c = 2\left(1 + \frac{N_u}{2000A_g}\right)\sqrt{f_c'}b_w d$$

can be used, if N_u is positive (compression), or

$$V_c = 2\left(1 + \frac{N_u}{500A_g}\right)\sqrt{f_c'}b_w d$$

if N_u is negative (tension). If there is no axial force, substitute M_u for M_m and satisfy the condition $(V_u d/M_u) \leq 1$.

Step 4. Select a stirrup size, typically #3 or #4.

Step 5. Check if anywhere, $V_s > 8\sqrt{f_c'}b_w d$. If yes, increase the web width or reduce the load.

Step 6. Determine the stirrup spacing according to regions:

If $(V_u/\phi) \leq (V_c/2)$, no stirrups are needed.

If $(V_c/2) < (V_u/\phi) \leq V_c + V_{s,min}$, minimum shear reinforcement (or maximum stirrup spacing) is sufficient:

$$s_{max} = \min\left\{\frac{d}{2}; 24''; \frac{A_v f_y}{50b_w}\right\} \quad \text{if} \quad V_s \leq 4\sqrt{f_c'}b_w d$$

$$= \min\left\{\frac{d}{4}; 12''; \frac{A_v f_y}{50b_w}\right\} \quad \text{if} \quad V_s > 4\sqrt{f_c'}b_w d$$

If $(V_u/\phi) > V_c + V_{s,min}$, determine the required stirrup spacing according to

$$s = \frac{A_v f_y d}{V_s} \quad \text{where} \quad V_s = \frac{V_u}{\phi} - V_c$$

Step 7. Select a reasonable and practical spacing pattern enveloping the requirements of Step 6.

7. The minimum basic development length for bars is

$$\frac{l_d}{d_b} = \frac{3}{40}\frac{f_y}{\sqrt{f_c'}}\frac{\alpha\beta\gamma\lambda}{\left(\frac{c+k_{tr}}{d_b}\right)} \quad \text{with} \quad \left(\frac{c+k_{tr}}{d_b}\right) \leq 2.5$$

where
- for top reinforcement, bar location factor $\alpha = 1.3$
- coating factor $\beta = 1.2$ or 1.5 for epoxy-coated bars
- bar size factor $\gamma = 0.8$ for #6 and smaller bars
- lightweight aggregate concrete factor $\lambda = 1.3$ or $6.7\sqrt{f_c'}/f_{ct} \geq 1.0$
- transverse reinforcement index $k_{tr} = A_{tr}f_{yt}/1500sn$
- and a factor for excess reinforcement, $A_{s,\text{required}}/A_{s,\text{provided}}$.

8. Near simple supports and inflection points, the following requirement holds:

$$l_d \leq \frac{M_n}{V_u} + l_a$$

where $l_a = \max\{12d_b; d\}$. If the ends of the reinforcement are confined by a compressive reaction, M_n may be multiplied with 1.3.

9. The most efficient cross sections to resist torsion are of hollow circular or square shape.

10. The torsional constant is computed as

$$J = \frac{\pi R^4}{2} \quad \text{for circular sections}$$

$$J \approx \sum_i \frac{x_i^3 y_i}{3} \quad \text{for sections composed of narrow rectangular segments}$$

$$J = \frac{4A^2}{\oint \frac{ds}{t(s)}} \quad \text{for closed thin-walled sections}$$

11. The design procedure for torsion, after a section has been dimensioned and reinforced for flexure and shear, is as follows.

   Step 1. Check whether torsion must be considered. Neglect torsion if

$$T_u \leq \phi\sqrt{f_c'}\,\frac{A_{cp}^2}{P_{cp}}$$

   Step 2. Check whether the section has the capacity to resist the applied torque. If for a hollow section,

$$\frac{V_u}{b_w d} + \frac{T_u p_h}{1.7 A_o^2} > \phi\left(\frac{V_c}{b_w d} + 8\sqrt{f_c'}\right)$$

   or for a solid section,

$$\sqrt{\left(\frac{V_u}{b_w d}\right)^2 + \left(\frac{T_u p_h}{1.7 A_o^2}\right)^2} > \phi\left(\frac{V_u}{b_w d} + 8\sqrt{f_c'}\right)$$

   even maximum reinforcement is insufficient, that is, the section size must be increased.

   Step 3. Compute the required spacing of closed stirrups with cross-sectional area

A_t, to resist both shear and torsion,

$$s_{req} = \frac{2A_t f_y}{\frac{T_u}{\phi A_o} + \frac{V_s}{d}}$$

Step 4.   Check minimum transverse steel,

$$s_{max} = \frac{A_v f_{yv}}{50b_w}$$

where A_v accounts for both shear and torsional reinforcement.

Step 5.   Determine the additional longitudinal reinforcement required for torsion,

$$A_l = \frac{T_n p_h}{2A_o f_{yl}} \geq \frac{5\sqrt{f_c'}}{f_{yl}} A_{cp} - \frac{A_t p_h}{s} \frac{f_{yv}}{f_{yl}}$$

REFERENCES

1. ACI-ASCE Committee 326, "Shear and Diagonal Tension," *ACI Journal*, vol. 59, no. 2, Feb. 1962.
2. Collins, M. P., and Mitchell, D., *Prestressed Concrete Structures*, Prentice-Hall, Englewood Cliffs, NJ, 1991.
3. Popov, E. P., *Engineering Mechanics of Solids*, Prentice-Hall, Englewood Cliffs, NJ, 1990.
4. MacGregor, J. G., and Ghoneim, M. G., "Design for Torsion," *ACI Structural Journal*, vol. 92, no. 2, March–April 1995.

PROBLEMS

For all problems below, unless otherwise specified, assume normal-weight concrete weighing 150 pcf, with $f_c' = 4000$ psi and steel with $f_y = 60,000$ psi. Concrete cover = 2 in.

5.1 A simply supported beam is reinforced with 4 #8 bars and #3 stirrups @ 10-in spacing, as shown. Assuming a live load–dead load ratio of 2, determine the dead and live load that this beam is allowed to carry according to the ACI Code.

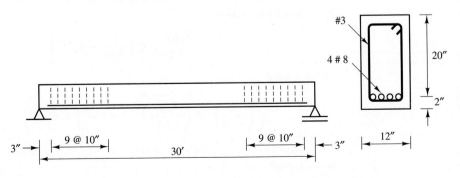

Figure P5.1

5.2 The T-beam shown carries two 50-kip live loads at its thirdpoints, in addition to its own weight. It has been determined that 5 #11 bars are required for flexural reinforcement. Determine the necessary shear reinforcement and show it in a neatly drawn sketch of the beam elevation.

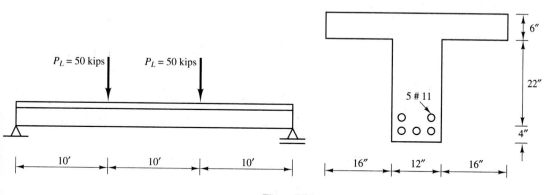

Figure P5.2

5.3 Use Eq. 5.10 (ACI Code formula [11-5]) to compute V_c for the previous problem and judge whether the added effort is justified.

5.4 A one-way joist floor system has a simple span of 25 ft and is reinforced with 2 #6 bars per joist, as shown.
  (a) Based on both flexural and shear capacity, how much live load (in psf) is this joist floor system permitted to carry in addition to its own weight?
  (b) Determine the amount of reinforcement needed if the live load were raised to 200 psf. Indicate the number and size of bars per joist.
  (c) Determine the amount of shear reinforcement (stirrup size and spacing at the supports only) needed for a live load of 200 psf.

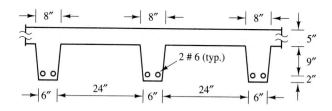

Figure P5.4

5.5 The slab-and-beam floor system shown is reinforced with 4 #8 bars per beam and spans a simple span of 36 ft.
  (a) What is the live load (in psf) that this floor system is allowed to carry in addition to its own weight, on the basis of its flexural capacity?
  (b) Determine the amount of shear reinforcement for an allowable live load at least equal to that of problem **(a)**.

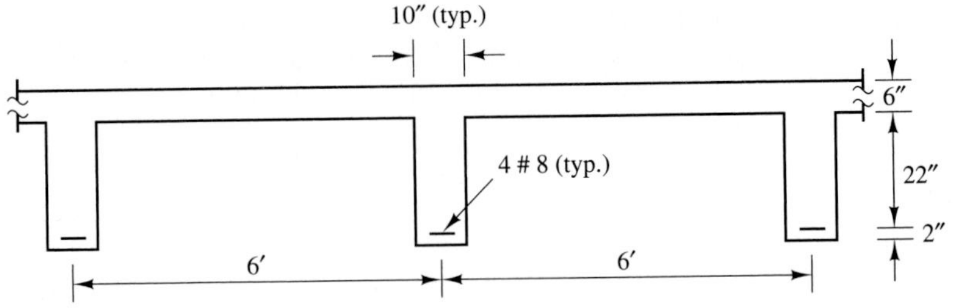

10″ (typ.)

4 # 8 (typ.)

6″

22″

2″

6′   6′

Figure P5.5

5.6 A 9-in thick one-way slab is reinforced with #4 bars spaced at $7\frac{1}{2}$ inches and carries a live load of 100 psf in addition to its own weight. (Use $d = 7.5$ inch).
   (a) Determine the longest permissible simple span, based on the slab's flexural capacity.
   (b) Determine the longest permissible simple span, based on the slab's shear strength.
   (c) Determine the span range (if any) for which the slab's shear strength is critical.

5.7 A 6-in roof slab is supported by parallel frames, each consisting of a 16-ft column and a 50-ft-long beam, spaced 15 feet apart as shown. The beam cross section is 15×36 in, the column cross section is 15×15 in. The controlling design load case is dead weight plus 300 psf snow load (live load).
   (a) Draw the shear, moment, and axial force diagrams due to factored loads for a typical frame.
   (b) Qualitatively but neatly draw the deflected shape of the frame and highlight any inflection points.
   (c) Determine the positive and negative reinforcement required for the beam and check it against the Code's minimum and maximum requirements.
   (d) Determine the cutoff points of the negative reinforcement.
   (e) Design the beam for shear.

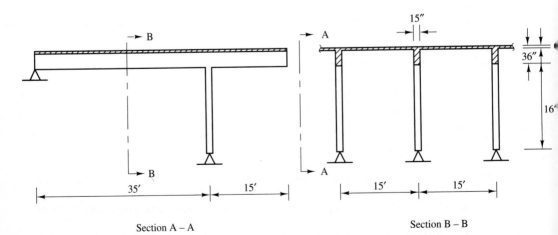

Section A – A   Section B – B

Figure P5.7

5.8 A simply supported beam is to be designed to carry a dead load of 400 lb/ft plus a live load with triangular distribution as shown, with a maximum intensity of 1000 lb/ft at midspan.
   (a) Determine the required amount of flexural reinforcement and show the bars in a neatly drawn sketch of the beam section.
   (b) Check if shear reinforcement is necessary. If so, indicate the stirrup size and spacing in a sketch of the beam elevation.

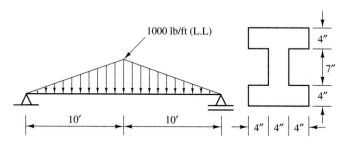

Figure P5.8

5.9 A simply supported beam 24 ft in length carries a factored uniform load of $w_u = 1.5$ kips/ft, plus a load of $P_u = 12$ kips applied at midspan with a 30-inch eccentricity as shown. Determine the required beam cross sectional dimensions and the necessary reinforcement.

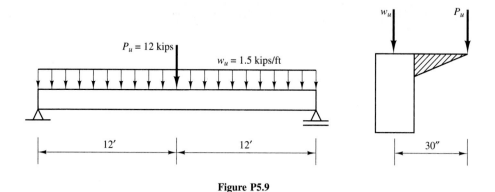

Figure P5.9

5.10 The one-way floor slab system shown carries a uniform live load of 125 psf in addition to its own weight. The spandrel beams have gross cross sectional dimensions of $h = 20$ in and $b = 12$ in. Determine the moments in a typical 1-foot-wide slab strip and the maximum torque and uniform load, for which the spandrel beams must be designed if

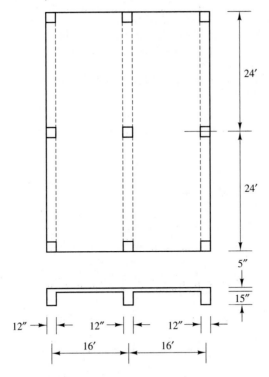

24'

24'

5"

15"

12" →| |← 12" →| |← 12" →| |←

16'   16'

Figure P5.10

(a) the spandrel beam restraint moment according to ACI Sec. 8.3.3 is used;

(b) maximum moment redistribution for compatibility torsion is used.

5.11 Write a subroutine in Fortran or any other programming language to determine the necessary spacing of stirrups in a beam. Input: beam dimensions b_w and d; beam span L; material properties f'_c and f_y; flexural steel ratio ρ_w; bending moment M_u and shear force V_u at $L/20$ intervals; axial force N_u. Output: stirrup size and required stirrup spacing s at $L/20$ intervals.

Continuous Beams and Frames

6.1 GENERAL

Continuity plays an important role in structural engineering. It has a number of important advantages, but it also has a few drawbacks that could motivate the structural engineer in certain cases to forsake it. First the advantages: increases in *strength* and *stiffness*, and *redundancy*.

Consider the simple beam of Fig. 6.1a with a concentrated load at midspan, which produces the maximum moment and deflection as indicated. If the beam were rigidly connected to an adjacent beam, this neighbor beam would share the load by reducing the directly loaded span's end rotation (Fig. 6.1b). As a result, the statical moment of $PL/4$ would be split between a midspan moment of $(13/64)PL$ (or 81% of $PL/4$) and a negative moment provided by the neighbor beam, which makes up the remaining 19% of the total. In other words, the continuous beam requires only 81% of the strength of an equivalent simply supported beam to carry the same load P. The beam's stiffness is affected even more, with the simple beam deflection reduced by 64%. This increase in stiffness is significant, as it permits the use of members with increased slenderness. Applied to bridge structures, this advantage translates into longer spans. Also, the slope discontinuities over the supports of a series of simply supported spans for a long highway or railroad bridge can impair riding comfort, especially at higher speeds. These are eliminated in a continuous bridge.

The stiffness and strength increases are even more impressive in a uniformly loaded interior span of a beam that is continuous over a large number of spans (Fig. 6.2).

275

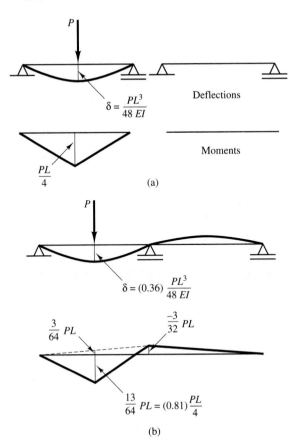

Figure 6.1 Effect of continuity on deflections and moments: (a) simply supported beam; (b) continuous beam.

Because of symmetry, the slopes over the supports can be assumed to be zero, which is equivalent to stipulating that the typical interior span is fixed at both ends. Strengthwise, the (negative) end moments account for two-thirds of the total statical moment of $wL^2/8$, while the (positive) field or midspan moment carries the remaining one-third share. This means that a continuous beam requires only two-thirds the strength of a simply supported beam to carry the same uniform load. In addition, the midspan deflection is $wL^4/384EI$, that is, merely one-fifth of that of an equivalent simply supported span!

As will be shown later, a plastic analysis of a *prismatic* continuous beam (i.e., one that has the same cross-sectional properties at all sections, such as a rolled-steel beam) indicates that, at collapse, the total statical moment is shared equally by the end and midspan moments. Denoting the plastic moment capacity by M_p, it follows that $M_{\text{end}} = M_{\text{mid}} = wL^2/16 = M_p$ (Fig. 6.3). Compared with the results of an elastic analysis, this result points to a 33% capacity increase, and compared with a simply supported beam, the capacity is now increased by 100%. In other words, to carry the same load, a beam with half of a simply supported beam's capacity is sufficient.

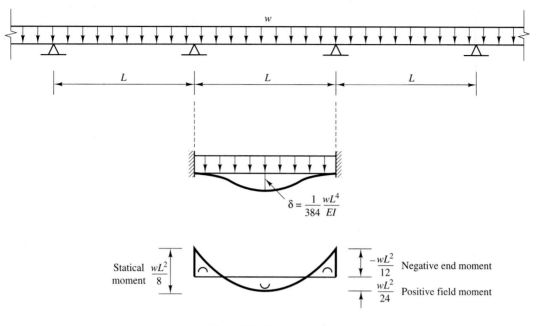

Figure 6.2   Beam fixed at both ends.

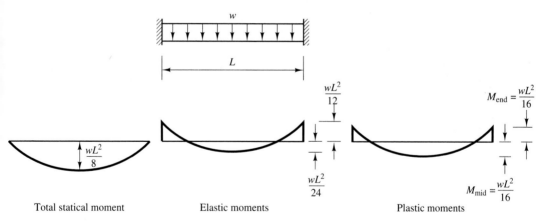

Figure 6.3   Moments from elastic and plastic analysis.

Perhaps the most essential advantage of continuity is redundancy. In fact, in structural parlance, continuity and redundancy are often used as synonymous terms. A simply supported beam forms a collapse mechanism as soon as its capacity is reached, that is, as soon as a single plastic hinge forms. The same is true for any statically determinate structure. A continuous or statically indeterminate structure, on the other hand, does not necessarily collapse when the first plastic hinge forms, because other, less stressed portions of the structure may carry the excess load. As moments are redistributed through rotations in the plastic hinges, strength reserves are activated, which statically determinate structures do not have. For each degree of redundancy, one additional plastic hinge is needed to form a collapse mechanism.

As a fundamental concept of structural safety, redundancy is an essential element of modern structural design philosophy. Should some nonredundant load-bearing member lose its capacity to carry load, for example, as a result of an accident, the consequences for the remaining structure are likely to be catastrophic. However, for a structure with redundancy, there may very well be a chance that hidden strength reserves are activated, thereby preventing catastrophic collapse. For this reason, some current bridge design codes offer premiums for redundancy. In Sec. 6.4, we shall further explore the advantages of redundant or continuous structures using plastic analysis.

Members of steel and timber structures are typically articulated, and special design measures must be taken to achieve continuity, unless the members are long enough to span several fields. For cast-in-place concrete structures, continuity is an inherent property of the material, but reinforcing bars still must be positioned appropriately to ensure continuity (Fig. 6.4). The reinforcement placed to resist negative moments is

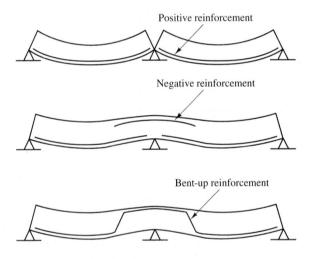

Figure 6.4 Effect of negative reinforcement.

therefore referred to as *negative reinforcement*. It is also possible to bend the bars at the inflection points in such a way that they resist both positive and negative moments. If a concrete beam cannot be cast as one monolithic unit in a single pour, the construction joints are still crossed by reinforcement as needed. The joints themselves should be cleaned and roughed for better bond between the old and new concrete.

Let us now address the disadvantages or adverse effects of continuous or statically indeterminate structures.

1. It lies in the nature of determinate structures that temperature changes, creep and shrinkage deformations, and support settlements cause no internal forces or stresses

(Fig. 6.5). This is not the case with indeterminate structures. The so-called *secondary* effects may not be secondary in magnitude at all and can considerably complicate a design. There are two reasons, though, for not being overly alarmist about this. First, truly unusual moments due to the above-mentioned causes can be reduced through plastic moment redistribution. Secondly, the creep behavior of concrete tends to relieve these secondary stresses if sufficient time is available. Therefore, daily temperature cycles are of much more concern than annual or seasonal cycles.

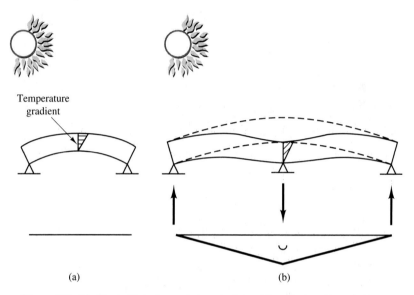

Figure 6.5 Combined effect of temperature gradient and continuity: (a) no moments; (b) temperature-induced moments.

2. Movable live loads are likely to cause moment reversals in continuous structures (Fig. 6.6), so a given section may have to be designed both for a negative and a positive moment. Since members are often provided with nominal reinforcement on the compression side anyway, this drawback may not be very significant.

3. Near interior supports of a continuous beam, large moments occur simultaneously with large shear forces. This coincidence has an adverse effect on the concrete's capacity to carry shear, and somewhat more shear reinforcement may be needed.

4. If a structure is to be assembled out of precast elements, the achievement of continuity in the field is not straightforward, and well-engineered connection details are needed.

5. A significant adverse aspect of continuous structures is the fact that they are more difficult to analyze and their behavior more difficult to understand. Until the 1930s, engineers were not overly comfortable with the indeterminate structural analysis tools available. This is one reason why they were reluctant to design continuous

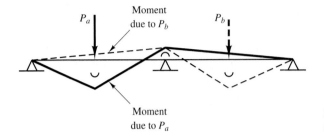

Moment due to P_a

Figure 6.6   Moment reversal due to movable live load.

structures. Much of the credit for changing this situation is due to Hardy Cross, whose introduction of moment distribution was instrumental in helping engineers analyze continuous structures and understand their behavior so that they could feel more confident designing them. Thus, continuous structures have been designed and built on a wide scale for only half a century.

The key to the successful practical analysis and design of continuous beams and frames is a thorough understanding of their behavior. Much of this understanding is the result of practical experience. A statement like this is not overly encouraging to young engineers who are just getting started. In Sec. 6.5, an attempt will be made to assist them in their efforts to accelerate this learning process. Basically, it rests on the ability to draw the deflected shapes of loaded structures, complete with inflection points, and to correlate these deflected shapes with qualitatively drawn moment diagrams. Experienced designers know how to sketch the deflected shapes of structures with sufficient accuracy. By identifying the approximate locations of inflection points (i.e., points of zero moment), a structure is rendered statically determinate for analysis purposes, so the determination of the moment diagram is reduced to an exercise in statics. It is not uncommon that internal forces and moments are computed with no more than 10 or 20% error, strictly on the basis of intuition and a thorough knowledge of statics. Such accuracy is generally sufficient for preliminary design purposes. Approximate analysis methods are particularly useful if applied to highly indeterminate multistory building frames.

A final comment on the complication that continuity introduces into the design process is in order. The reader will recall the fact that statically determinate structures can be analyzed using only equations of statics. The determination of deformations, which requires knowledge of section and material properties, may be desirable but is not necessary to compute internal forces and moments. The analysis of statically indeterminate structures, on the other hand, is possible only with the knowledge of section and material properties, since the conditions of compatibility can be satisfied only by computing deflections. This requirement introduces one of the basic dilemmas of the structural engineer: in order to analyze a structure, we need to know the section properties (i.e., cross-sectional dimensions) of all structural elements, but in order to determine the member sizes, we must analyze the structure for the internal forces and moments for which they are to be designed. In other words, we cannot analyze a structure before we have designed it, and we cannot design it before we have analyzed it. The common escape route from this vicious cycle is illustrated in the flow diagram of Fig. 6.7, and can be described as follows:

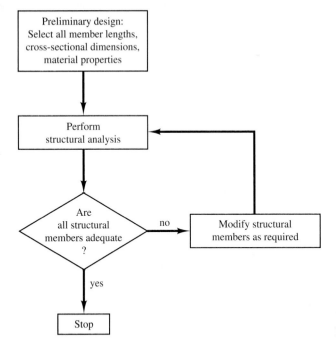

Figure 6.7   Design-analysis cycle.

1. The first and often most important step is the *preliminary design.* Here, the designer determines the layout of the structure, complete with span lengths, bay widths, column heights, etc., (if not predetermined by the architect) as well as a complete selection of cross-sectional dimensions of every structural element. The basic guideline is the designer's golden rule: *assume what is not yet specified or known.* (See Sec. 1.6.)

2. The preliminary design allows us to determine all loads, including the dead load, and to perform an initial indeterminate structural analysis, which furnishes all internal forces, moments, and deflections.

3. With the forces and moments known from step 2, it is possible to correct the sizes of structural members if necessary, to select reinforcement, etc.

4. If the member dimensions determined in step 3 are significantly different from those selected in step 1, the structure will have to be reanalyzed.

Convergence of this design/analysis cycle is normally rapid, and an experienced engineer will, as a rule, come up with a preliminary design that requires at most one reanalysis.

Strictly speaking, only the relative (as opposed to the absolute) stiffnesses of the members meeting at a common joint are needed for the indeterminate analysis. However, structural analyses are typically performed using computer programs, and these require section data as input. Still, it is useful to remember that only the relative stiffnesses determine the outcome of the analysis. For this reason, it is standard practice to input the properties of uncracked gross sections. This assumption implies that the ratios of cracked to uncracked moments of inertia are approximately equal for all members framing into a common joint. For columns, which generally have higher steel percentages and (because of axial compression) are likely to be cracked less than girders, it is advisable to use transformed section properties when computing relative stiffnesses. In the case of T-beams, it is necessary to account for the stiffness contribution of the flange. For consistency, the same effective flange width can be used as in the design for strength. In negative moment regions, the top flange is in tension and assumed cracked, so it can be ignored in relative stiffness computations.

6.2 METHODS OF ANALYSIS

The analysis of statically indeterminate structures is generally taught in separate courses on structural analysis, and numerous textbooks have been written on the subject [1, 2, 3]. Here, we will assume that the reader is somewhat familiar with these methods of analysis. The short summary on the following pages shall serve as a review of the basic underlying concepts.

First, however, it is appropriate to recall the place of structural analysis within the overall design process. Isolated from its ultimate purpose, structural analysis can be viewed as a branch of applied mathematics, with a claim to exactness within certain limits. For example, if a reaction necessary for equilibrium is 40 kips, this would be the exact or right answer, whereas 42 kips would be wrong, and even 40.1 kips simply inaccurate. However, in the real world of structural engineering, we must recognize the various sources of uncertainty, such as span lengths, cross-sectional dimension tolerances, and especially material properties (which are significant in the case of indeterminate structures) and loads. Given the large variability of live loads alone, it becomes obvious that it really does not matter whether the reaction is 40 kips, 40.1 kips, or even 42 kips. As for the necessary precision of calculations, the old 5% rule dating from the days of the slide rule appears to be a reasonable guideline.

The analysis of indeterminate structures involves three different sets of equations:

1. equations of statics, which assure equilibrium;
2. equations of compatibility, to assure that members that fit together before load application still do so after load application;
3. equations that relate member forces to member deformations.

The classical methods of structural analysis, generally known as the force (or flexibility) method and the displacement (or stiffness) method, date back to the previous century.

In the force method, the structural system is simplified by making it statically determinate in such a way that equilibrium and the force-displacement relationships are satisfied; but there are violations of compatibility (errors in geometry). To get the correct answers, appropriate equations of compatibility must be solved for the unknown redundant forces that need to be removed to render the system statically determinate.

In the displacement method, the structure is simplified in such a way that the compatibility conditions and force-displacement relationships are satisfied; but there are violations of equilibrium (errors in statics). To get the correct answers, appropriate equations of equilibrium must be solved for the unknown kinematic displacements.

For practical analysis, a variation of the displacement method, known as the slope-deflection method, has enjoyed wide use. However, the numerical effort required to solve the equations had placed severe limitations on the practical size of problems. It was Hardy Cross' contribution in the 1930s to relax these limitations to such an extent that considerably larger indeterminate systems could be analyzed. In his *moment distribution* method, the slope-deflection equations are written in a different form and solved by an iterative scheme with clear physical interpretations. The usefulness of this procedure for hand calculations was so obvious that it became the standard analysis method for over thirty years. During the last few decades, the moment distribution method has been widely replaced by computer-based matrix analysis methods in engineering practice. These are almost exclusively based on the displacement method and involve the solution of large sets of simultaneous equations. Moment distribution still serves well for checking purposes and quick hand calculations, as well as in situations where the use of computers is not really practical or justified.

All analysis methods are derived from the same classical principles of structural theory, and common to all of them is the concept of *stiffness*. Physically, stiffness is an easily understood property. The stiffness of a simple spring is identical with the spring constant and equal to the force required to cause a unit elongation (Fig. 6.8a):

$$F = kr$$

where F is the force acting on the spring, r is the spring elongation, and k is the spring constant. For a bar element with area A, elastic modulus E, and length L (Fig. 6.8b), the force-displacement relationship reads

$$F = \frac{EA}{L}r$$

that is, the bar stiffness is defined as $k = EA/L$. If we recognize that a bar element has, in general, two kinematic degrees of freedom, r_1 and r_2 (Fig. 6.8c), we must redefine the stiffness. Instead of a simple scalar, we are now dealing with a matrix of *stiffness coefficients*. A stiffness coefficient is defined as the force required to cause a unit displacement, while all other end displacements are kept zero. The *stiffness matrix* for the bar element of Fig. 6.8c can thus be written as

$$\left\{ \begin{array}{c} F_1 \\ F_2 \end{array} \right\} = \left[\begin{array}{cc} EA/L & -EA/L \\ -EA/L & EA/L \end{array} \right] \left\{ \begin{array}{c} r_1 \\ r_2 \end{array} \right\} \tag{6.1}$$

The beam element of Fig. 6.8d has four degrees of freedom, that is, a rotation and a transverse displacement at each end. By a displacement we now imply a *generalized displacement*, which can be either a displacement or a rotation. Similarly, a force is now implied to be a generalized force, which may be either a force or a moment. To

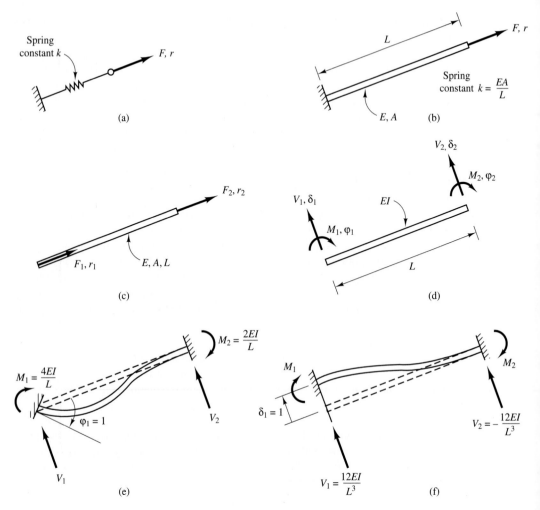

Figure 6.8 Spring constant and stiffness coefficients: (a) spring; (b) bar element (1 DOF); (c) bar element (2 DOFs); (d) beam element; (e) load case $\varphi_1 = 1$, (f) load case $\delta_1 = 1$.

determine the individual coefficients of the four-by-four stiffness matrix for the beam element of Fig. 6.8d, we introduce one unit displacement at a time, keeping the other three zero. Starting with the application of a unit rotation, $\phi_1 = 1$ (Fig. 6.8e), a moment $M_1 = 4EI/L$ is required. The *carry-over* moment needed to keep the other end from rotating is $M_2 = 2EI/L$. The vertical reactions required to prevent the beam ends from displacing are equal and opposite, $V_2 = -V_1 = 6EI/L^2$. Similarly, we can compute all four reactions necessary to hold the beam in the displaced configuration of Fig. 6.6f: $M_1 = M_2 = -6EI/L^2$ and $V_1 = -V_2 = 12EI/L^3$. Computing the corresponding reactions for either a unit rotation or displacement at the other end is straightforward because we can make use of symmetry, but we must pay careful attention to the sign convention of Fig. 6.8d. We summarize the results for all four load cases as the four columns of the beam stiffness matrix:

$$\begin{Bmatrix} M_1 \\ V_1 \\ M_2 \\ V_2 \end{Bmatrix} = \begin{bmatrix} 4EI/L & -6EI/L^2 & 2EI/L & 6EI/L^2 \\ -6EI/L^2 & 12EI/L^3 & -6EI/L^2 & -12EI/L^3 \\ 2EI/L & -6EI/L^2 & 4EI/L & 6EI/L^2 \\ 6EI/L^2 & -12EI/L^3 & 6EI/L^2 & 12EI/L^3 \end{bmatrix} \begin{Bmatrix} \phi_1 \\ \delta_1 \\ \phi_2 \\ \delta_2 \end{Bmatrix} \tag{6.2}$$

If the right end of the beam has a pinned support, it follows that $M_2 = 0$. Together with the assumption $\delta_2 = 0$, it can be shown that Eq. 6.2 reduces in this case to

$$\begin{Bmatrix} M_1 \\ V_1 \end{Bmatrix} = \begin{bmatrix} 3EI/L & -3EI/L^2 \\ -3EI/L^2 & 3EI/L^3 \end{bmatrix} \begin{Bmatrix} \phi_1 \\ \delta_1 \end{Bmatrix} \tag{6.3}$$

In other words, the stiffness (the moment needed to produce a unit rotation) of a beam with the far end pinned, is three-fourths of that of a beam whose far end is fixed. In general, then, a stiffness coefficient k_{ij} is defined to be the force (or moment) required to hold the ith displacement (or rotation) in place, when the jth degree of freedom undergoes a unit displacement (or rotation), while all others are locked in place.

Determining the stiffness matrices for all members of a structure constitutes the first phase of structural analysis. The second phase requires that the member stiffness matrices be assembled such that both equilibrium and compatibility conditions are satisfied at each joint.

Consider, for example, the continuous beam of Fig. 6.9, consisting of two beam elements. Let

$$M_1 = (4EI/L)_1 \phi_1 \tag{6.4a}$$

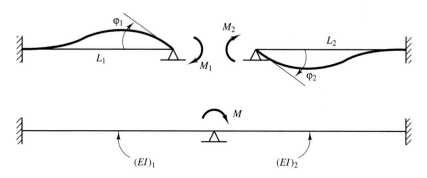

Figure 6.9   Two-span continuous beam.

be the moment required to rotate the right end of the left beam element about the angle ϕ_1, and

$$M_2 = (4EI/L)_2 \phi_2 \tag{6.4b}$$

the moment required to rotate the left end of the right beam element about the angle ϕ_2, where, for compatibility's sake,

$$\phi_1 = \phi_2 = \phi \tag{6.5}$$

The moment needed to impose the angle ϕ on both beam elements simultaneously must satisfy the following equilibrium condition.

$$M = M_1 + M_2 = [(4EI/L)_1 + (4EI/L)_2]\phi \qquad (6.6)$$

As we have now made use of all three sets of equations, namely, force-displacement relationships (Eqs. 6.4), compatibility or continuity (Eq. 6.5), and equilibrium (Eq. 6.6), the problem can be considered solved. Note that in Eq. 6.6 the stiffness coefficients are additive.

EXAMPLE 6.1

Determine the bending moments in the continuous beam of Fig. 6.10.

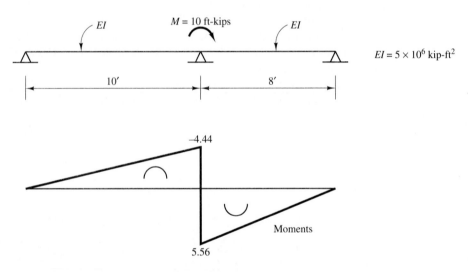

Figure 6.10   Example 6.1.

Solution   In contrast to the case of Fig. 6.9, the beams' far ends are pinned, so the rotational stiffness coefficients, according to Eq. 6.3, are $k = 3EI/L$. Modifying Eq. 6.6 accordingly, the joint rotation over the interior support follows as

$$\phi = \frac{M}{k_1 + k_2} = \frac{M}{(3EI/L)_1 + (3EI/L)_2}$$

$$= \frac{10}{(\frac{1}{10} + \frac{1}{8})(3)(5 \times 10^6)} = 2.963 \times 10^{-6} \text{ Rad}$$

Moment in left beam:

$$M_1 = \left(\frac{3EI}{L}\right)_1 \phi = \frac{(3)(5 \times 10^6)}{10}(2.963 \times 10^{-6}) = 4.44 \text{ ft-kips}$$

Moment in right beam:

$$M_2 = \left(\frac{3EI}{L}\right)_2 \phi = \frac{(3)(5 \times 10^6)}{8}(2.963 \times 10^{-6}) = 5.56 \text{ ft-kips.}$$

Note that the moments in the two beams are proportional to their stiffnesses, in this case, inversely proportional to their span lengths. Stiff members attract more moment than flexible ones. ∎

EXAMPLE 6.2

Determine the bending moments in the continuous beam of Fig. 6.11.

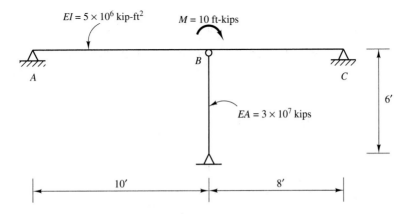

Figure 6.11   Example 6.2.

Solution Note that the only difference between this and the previous example is the addition of the pin-connected column, which has the effect of making the vertical displacement of support B nonzero. Thus, we need to introduce this vertical displacement δ as a second degree of freedom. For the stiffness of the right beam, we use Eq. 6.3 directly. For the left beam, we can use it too, except the signs of the off-diagonal coupling terms change (see Fig. 6.8d for sign convention). For the pin-connected column, the contributing stiffness coefficient is EA/L. Adding the various stiffness coefficients leads to the two equilibrium equations for joint B:

$$\left\{\begin{array}{c} M \\ P \end{array}\right\} = \begin{bmatrix} \left(\dfrac{3EI}{L}\right)_1 + \left(\dfrac{3EI}{L}\right)_2 & \left(\dfrac{3EI}{L^2}\right)_1 - \left(\dfrac{3EI}{L^2}\right)_2 \\[2ex] \left(\dfrac{3EI}{L^2}\right)_1 - \left(\dfrac{3EI}{L^2}\right)_2 & \left(\dfrac{3EI}{L^3}\right)_1 + \left(\dfrac{3EI}{L^3}\right)_2 + \left(\dfrac{EA}{L}\right)_{col} \end{bmatrix} \left\{\begin{array}{c} \phi \\ \delta \end{array}\right\}$$

Substituting numerical values,

$$\left\{\begin{array}{c} 10 \\ 0 \end{array}\right\} = 5 \times 10^6 \begin{bmatrix} (0.3 + 0.375) & (0.03 - .046875) \\[2ex] (0.03 - .046875) & \left(.003 + .005859 + \dfrac{1}{6} \times \dfrac{30}{5}\right) \end{bmatrix} \left\{\begin{array}{c} \phi \\ \delta \end{array}\right\}$$

from which $\phi = 2.964 \times 10^{-6}$ and $\delta = .0496 \times 10^{-6}$. The bending moments follow as

$$M_1 = (3EI/L)_1\phi + (3EI/L^2)_1\delta = 4.45$$

$$M_2 = (3EI/L)_2\phi - (3EI/L^2)_2\delta = 5.55$$

As we can see, the axial deformability of the column has a minimal effect on the beam moments, because the column axial stiffness is large compared with the flexural stiffnesses of the beam elements. This is true in general. A member is usually much more effective in axial or direct action than in bending. ■

EXAMPLE 6.3

Compute the bending moments for the continuous beam shown in Fig. 6.12a using the method of moment distribution.

Solution   We will illustrate this solution method by following a step-by-step procedure and referring to Fig. 6.12b.

1. Compute the relative stiffness factors (also known as distribution factors) of all beam elements framing into a common joint. These factors are needed because moments applied to a joint are resisted by the beam elements in proportion to their relative stiffnesses. At joint B, the left beam element AB has the stiffness $4EI/L = 0.125(4EI)$, while for the right beam element BC, $4EI/L = 0.1(4EI)$. Together, both elements supply a stiffness of $(0.125+0.1)(4EI) = 0.225(4EI)$, to which beam AB contributes $0.125/0.225(100) = 55.6\%$ and beam BC contributes $0.1/0.225(100) = 44.4\%$. These are the distribution factors, entered in line 2 of Fig. 6.12b, and any applied moments must be shared by the two beams according to these relative stiffnesses. For joint C, the relative stiffnesses are .5455 for beam element CB and .4545 for beam element CD. Since beam element A is the only one contributing stiffness to joint A, its relative stiffness is 1.0.

2. Compute fixed-end moments (FEM), that is, lock all joints against rotation and compute the moments due to any external loads applied to them. For span AB, FEM $= \pm(PL/8) = \pm[(5)(8)/8] = \pm5$ ft-kips. For span BC, FEM $= \pm(wL^2/12) = \pm[(2)(12)^2/12] = \pm16.67$ ft-kips. These are recorded in line 3. They are positive if they rotate a joint clockwise.

3. Release joint A. To restore equilibrium, a moment equal and opposite to the now unbalanced fixed-end moment of 5 ft-kips must be applied. As joint A is allowed to rotate clockwise, a carry-over moment half the size of the applied moment develops at the far end, signified by the right-pointing arrow in line 4.

4. The total out-of-balance moment at joint B is now $16.67 - 5 - 2.5 = 9.17$ ft-kips. To restore equilibrium, we release joint B and apply a moment of -9.17 ft-kips, of which $(.556)(-9.17) = -5.09$ ft-kips is supplied by beam AB, and the remaining -4.08 ft-kips by beam BC. Similarly, equilibrium is restored at joint C to balance the fixed-end moment of -16.67 ft-kips. Releasing joints B and C results in four carry-over moments, signified by four arrows. When joint A is released a second time to balance the -2.54 ft-kips moment, it causes its own carry-over moment at B. Note that the moment at joint D is not out of balance, because this joint is fixed.

5. Joints B and C are again out of balance, but the errors in equilibrium are considerably smaller than those of the first cycle. Four further cycles of relaxation reduce the

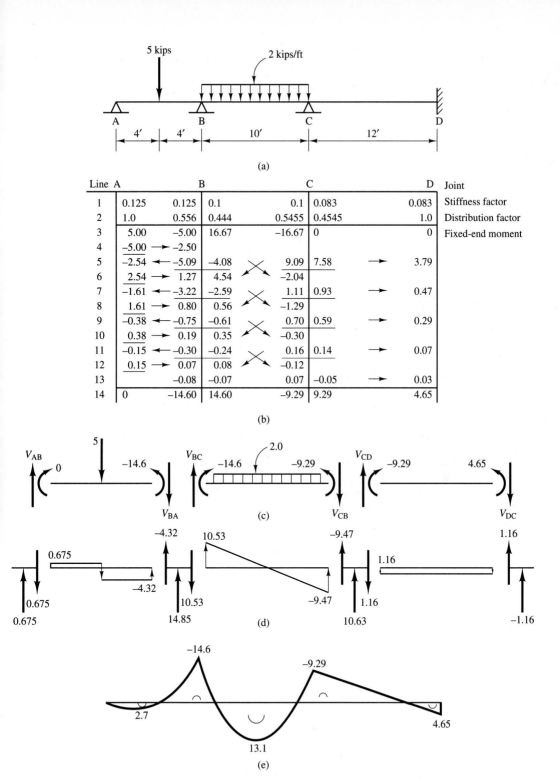

Figure 6.12 Example 6.3: Moment distribution: (a) continuous beam; (b) moment distribution; (c) internal forces and moments; (d) shear forces and reactions; (e) moment diagram.

unbalanced moments to 0.15 and −0.12 ft-kips, respectively (line 12). After this redistribution, the carry-over moments are now negligible.

6. Final moments are obtained by summing up all moments of lines 3 through 13 in line 14.

7. To determine "beam" bending moments, the "joint" moments of line 14 must be converted to the common sign convention (positive moments produce tension at the bottom, see Fig. 6.12c). Shear forces are readily computed from statics (Fig. 6.12d) by

$$V_{AB} = \frac{-14.6 + (5)(4)}{8} = 0.675 \text{ kips}$$

$$V_{BA} = 0.675 - 5 = -4.325 \text{ kips}$$

$$V_{BC} = \frac{14.6 - 9.29 + (2)(10)(5)}{10} = 10.53 \text{ kips}$$

$$V_{CB} = -(2)(10) + 10.53 = -9.47 \text{ kips}$$

$$V_{CD} = -V_{DC} = \frac{9.29 + 4.65}{12} = 1.16 \text{ kips}$$

8. The reactions follow from the shear forces (Fig. 6.12d).

9. The complete beam bending moment diagram (Fig 6.12e) is obtained by superimposing the simple-span moment diagrams of the individual beams, due to external loads, and the moments of Fig. 6.12c. ∎

EXAMPLE 6.4

Repeat the previous problem, taking advantage of the fact that joint A is pin-supported.

Solution We shall follow the same step-by-step procedure as before (Fig. 6.13), this time pointing out the differences.

Line	A	B		C		D	Joint
1	0	$(\frac{3}{4})(0.125)$ 0.1		0.1	0.083	0.083	Stiffness factor
2		0.484	0.516	0.5455	0.4545	1.0	Distribution factor
3		−7.50	16.67	−16.67	0	0	Fixed-end moment
4		−4.44	−4.73	9.09	7.58	3.79	
5			4.54	−2.36			
6		−2.20	−2.34	1.29	1.07	0.54	
7			0.65	−1.17			
8		−0.31	−0.34	0.64	0.53	0.26	
9			0.32	−0.17			
10		−0.15	−0.17	0.09	0.08	0.04	
11			0.05	−0.09			
12		−0.02	−0.03	0.05	0.04	0.02	
13	0	−14.62	14.62	−9.30	9.30	4.65	

Figure 6.13 Example 6.4: Moment distribution with modified stiffness coefficient.

1. By recognizing that the "far" end is pin-supported, the stiffness of beam AB at joint B is now $3EI/L$ instead of $4EI/L$, that is, $k = \frac{3}{4}(0.125)(4EI) = 0.09375(4EI)$. This change affects the distribution factors at joint B to $[0.09375/(0.09375 + 0.1)](100) = 48.4\%$ and $100 - 48.4 = 51.6\%$. At joint C, relative stiffness factors are the same as before.

2. The fixed-end moment for a beam with a concentrated midspan load and fixed at only one side is $-(3/16)PL = -[(3)(5)(8)/16] = -7.5$ ft-kips. The other fixed-end moments remain unchanged.

3. When releasing joint B, no carry-over moment reaches joint A because it is already permanently released. Aside from this change, the successive relaxation of moments proceeds as before, except that we notice a slightly faster convergence. Final results obviously must agree with the previous results, except for minor differences due to roundoff errors.   ■

EXAMPLE 6.5

Determine the bending moments in the frame problem of Fig. 6.14 by moment distribution.

Solution   Because of symmetry, no horizontal displacements can take place; therefore, only the members' contributions to the rotational stiffnesses of the joints are of concern. That is, the moments may be distributed as if the three members were collinear, as in a continuous beam. The necessary individual solution steps are straightforward, as illustrated in Fig. 6.14b.

The symmetry of the frame is clearly reflected in these solution steps and begs for a shortcut. In fact, such a shortcut exists. Let us reconsider the beam element of Fig. 6.8d. If each rotation ϕ_1 at the left end is accompanied by a symmetric rotation $\phi_2 = -\phi_1$ at the right end, it can easily be shown from Eq. 6.2 that the bending moments necessary to deform the beam element into such a symmetric configuration are $M_1 = -M_2 = 2EI/L$; the effective stiffness for the symmetric deformation pattern is reduced by a factor of 2. If this reduction factor is accounted for, we can indeed distribute moments in only half the frame (Fig. 6.14c). Since no carry-over moments are generated, a single moment relaxation yields the final answers. Now that's some shortcut! A moment distribution analysis that can give us the correct answers in a single step is a powerful tool indeed.   ■

EXAMPLE 6.6

Solve the problem of Fig. 6.15 using the direct stiffness method of matrix analysis (see also Example 6.3).

Solution   At the three degrees of freedom (the three joint rotations illustrated in Fig. 6.15), the fixed-end moments due to external loads must be equilibrated by the moments caused by the "near" and "far" joint rotations. For joint A, the moment equilibrium equation is readily extracted from Eq. 6.2:

$$\left(\frac{4EI}{L}\right)_{AB} \phi_A + \left(\frac{2EI}{L}\right)_{AB} \phi_B = \frac{PL_{AB}}{8} \tag{6.7}$$

For joint B, two beam elements must be considered for the moment equilibrium equation:

$$\left(\frac{2EI}{L}\right)_{AB} \phi_A + \left(\frac{4EI}{L}\right)_{AB} \phi_B + \left(\frac{4EI}{L}\right)_{BC} \phi_B + \left(\frac{2EI}{L}\right)_{BC} \phi_C = \frac{wL_{BC}^2}{12} - \frac{PL_{AB}}{8} \tag{6.8}$$

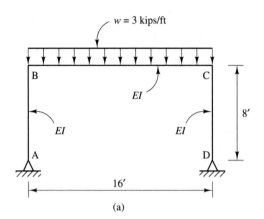

(a)

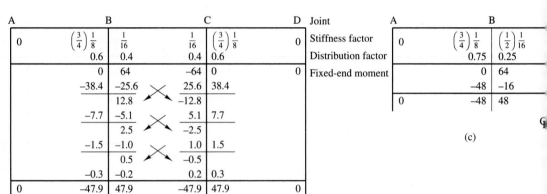

Table (b):

A	B		C		D
0	$\left(\frac{3}{4}\right)\frac{1}{8}$	$\frac{1}{16}$	$\frac{1}{16}$	$\left(\frac{3}{4}\right)\frac{1}{8}$	0
	0.6	0.4	0.4	0.6	
	0	64	-64	0	0
	-38.4	-25.6	25.6	38.4	
		12.8	-12.8		
	-7.7	-5.1	5.1	7.7	
		2.5	-2.5		
	-1.5	-1.0	1.0	1.5	
		0.5	-0.5		
	-0.3	-0.2	0.2	0.3	
0	-47.9	47.9	-47.9	47.9	0

(b)

Table (c):

	A	B	
Joint			
Stiffness factor	0	$\left(\frac{3}{4}\right)\frac{1}{8}$	$\left(\frac{1}{2}\right)\frac{1}{16}$
Distribution factor		0.75	0.25
Fixed-end moment		0	64
		-48	-16
	0	-48	48

(c)

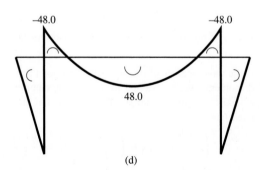

-48.0   -48.0

48.0

(d)

Figure 6.14 Example 6.5: Moment distribution of portal frame. (a) Portal frame; (b) moment distribution—full frame; (c) moment distribution—half frame.

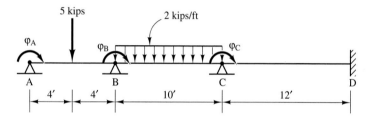

Figure 6.15 Example 6.6: Three rotational degrees of freedom.

Likewise, for joint C,

$$\left(\frac{2EI}{L}\right)_{BC}\phi_B + \left(\frac{4EI}{L}\right)_{BC}\phi_C + \left(\frac{4EI}{L}\right)_{CD}\phi_C = -\frac{wL_{BC}^2}{12} \tag{6.9}$$

Equations 6.7 through 6.9 assure equilibrium at all joints that are free to rotate while guaranteeing that beam elements at a common joint rotate the same amount, that is, compatibility is maintained. We can rewrite these three equations in matrix form, as follows

$$\begin{bmatrix} \left(\dfrac{4EI}{L}\right)_{AB} & \left(\dfrac{2EI}{L}\right)_{AB} & 0 \\[2ex] \left(\dfrac{2EI}{L}\right)_{AB} & \left(\dfrac{4EI}{L}\right)_{AB}+\left(\dfrac{4EI}{L}\right)_{BC} & \left(\dfrac{2EI}{L}\right)_{BC} \\[2ex] 0 & \left(\dfrac{2EI}{L}\right)_{BC} & \left(\dfrac{4EI}{L}\right)_{BC}+\left(\dfrac{4EI}{L}\right)_{CD} \end{bmatrix} \begin{Bmatrix} \phi_A \\ \phi_B \\ \phi_C \end{Bmatrix} = \begin{Bmatrix} \dfrac{PL_{AB}}{8} \\[2ex] \dfrac{wL_{BC}^2}{12} - \dfrac{PL_{AB}}{8} \\[2ex] -\dfrac{wL_{BC}^2}{12} \end{Bmatrix}$$

Substituting numerical values

$$EI\begin{bmatrix} 0.5 & 0.25 & 0 \\ 0.25 & 0.9 & 0.2 \\ 0 & 0.2 & 0.733 \end{bmatrix} \begin{Bmatrix} \phi_A \\ \phi_B \\ \phi_C \end{Bmatrix} = \begin{Bmatrix} 5.0 \\ 11.67 \\ -16.67 \end{Bmatrix}$$

and solving for the joint rotations, we find

$$\phi_A = .479/EI \qquad \phi_B = 19.04/EI \qquad \phi_C = -27.94/EI$$

so the end moments in beam AB become

$$M_A = \left(\frac{4EI}{L}\right)_{AB}\phi_A + \left(\frac{2EI}{L}\right)_{AB}\phi_B - 5.0 = (0.5)(.479) + (.25)(19.04) - 5.0 = 0$$

$$M_B = \left(\frac{2EI}{L}\right)_{AB}\phi_A + \left(\frac{4EI}{L}\right)_{AB}\phi_B + 5.0 = (.25)(.479) + (0.5)(19.04) + 5.0 = 14.64$$

Similarly, for beam BC we get

$$M_B = \left(\frac{4EI}{L}\right)_{BC}\phi_B + \left(\frac{2EI}{L}\right)_{BC}\phi_C - 16.67 = (0.4)(19.04) + (0.2)(-27.94) - 16.67 = -14.64$$

$$M_C = \left(\frac{2EI}{L}\right)_{BC} \phi_B + \left(\frac{4EI}{L}\right)_{BC} \phi_C + 16.67 = (0.2)(19.04) + (0.4)(-27.94) + 16.67 = 9.30$$

And with $\phi_D = 0$ and no fixed-end moments present, we get for beam CD

$$M_C = \left(\frac{4EI}{L}\right)_{CD} \phi_C = (.333)(-27.94) = -9.30$$

$$M_D = \left(\frac{2EI}{L}\right)_{CD} \phi_C = (.167)(-27.94) = -4.65$$

These results agree, of course, with those obtained earlier by moment distribution. This time we have solved the three equations simultaneously instead of iteratively; but note that we had to first solve for the joint rotations, which are usually of little interest. ∎

In many cases of practical interest, the spans of a continuous beam are approximately equal and uniformly loaded. The resulting moment diagrams for some cases are shown in Fig. 6.16. As the number of spans increases, the bending moments for the interior spans approach those of a beam fixed at both ends.

As can be seen in Fig. 6.16b for the three-span case, the positive moments for the center and outer spans are disparate. Since three-span bridges are very common (for example, for highway overcrossings), it seems logical and esthetically pleasing to shorten the side spans relative to the center span. If the side spans are approximately 80% of the center span, all three maximum positive moments for uniform load are equal. For esthetic reasons, the side spans often are chosen to be 70% of the center span or less. For long-span bridges, it is both economical and pleasing to the eye to make the beam nonprismatic by varying its depth over the entire length or only near the interior supports, using either straight or curved haunches (Fig. 6.17).

For regular continuous beams, the ACI Code permits the use of moment coefficients [Sec. 8.3] provided that: a) the spans do not vary by more than 20%; b) the loads are uniformly distributed; c) the unit live load does not exceed three times the unit dead load; and d) members are prismatic. The Code's moment coefficients are depicted in Fig. 6.18, expressed in terms of uniformly distributed load w and clear span length l_n. Note that the negative moments are taken at the face of the supports because the high-moment peaks are of only theoretical significance, as real members and supports have finite cross-sectional dimensions. A comparison between Figs. 6.18 and 6.16 shows that the Code coefficients are conservative. Therefore, a more refined analysis can sometimes be justified. As computers are generally available, such analyses pose no problems. However, if computers and easy-to-use programs are not readily on hand, and the moment coefficients cannot be used because any one of the conditions for their use is violated, then a moment distribution analysis can often provide quick answers with adequate accuracy. If the loading results in sidesway deformations, the moment distribution analysis becomes very involved and is less likely to be used in today's design practice.

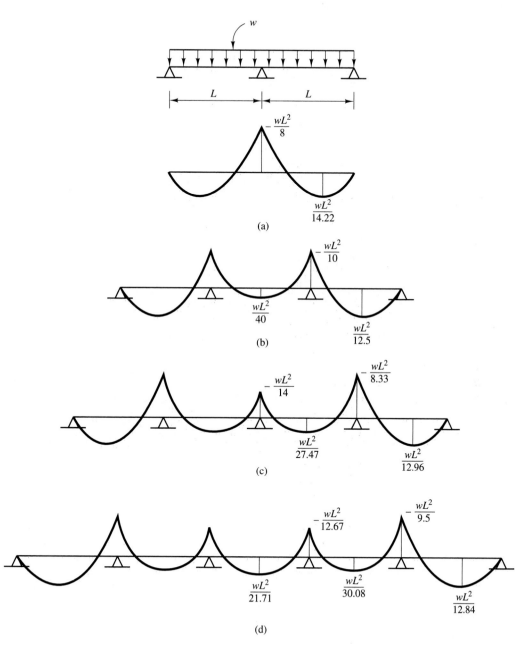

Figure 6.16   Moments in uniformly loaded continuous beams with equal spans: (a) two equal spans; (b) three equal spans; (c) four equal spans; (d) five equal spans.

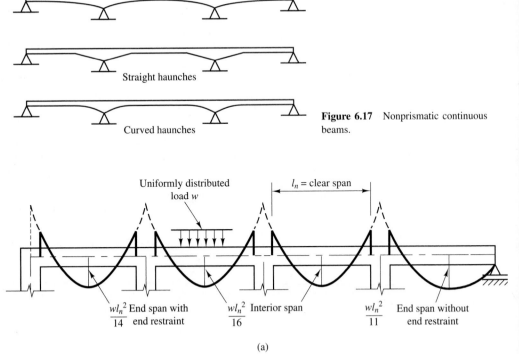

Figure 6.17   Nonprismatic continuous beams.

Straight haunches

Curved haunches

Figure 6.18   ACI-Code moment coefficients: (a) positive moments; (b) negative moments.

6.3 INFLUENCE LINES AND MOMENT ENVELOPES

Beams and girders should be designed to withstand the worst combination of loads to which they may be subjected during their useful lifetime. These load combinations include dead load, live load, wind, etc. as discussed earlier. While the magnitude and distribution of dead loads can be determined fairly accurately, the randomness of live loads requires special consideration with regard to both magnitude and distribution. The magnitude or intensity of live loads is generally prescribed by codes. It is the responsibility of the designer, however, to determine the worst possible arrangement of live loads to which the beam or girder may be subjected.

Consider, for example, the highway bridge shown in Fig. 6.19. Safety requires that the bridge possess sufficient moment capacity to resist the worst possible effect of the truck loading shown—*at each section*. A truck will always find the "critical" position as it crosses a bridge, even if the designer cannot. To determine the live load *moment envelope*, the following step-by-step procedure is used.

1. Select a number of sections along the entire bridge. Do not use too many (because of the large amount of work involved), but select enough to construct reliable moment envelopes. Each span is typically divided into four to ten equal segments (Fig. 6.19a).
2. For any section x selected in step 1, find the position in which the truck produces the largest positive moment, M_{max}, at x (Fig. 6.19b).
3. For the same section x, find the position in which the truck produces the largest negative moment, M_{min}.
4. Repeat steps 2 and 3 for all other sections x, and draw the maximum and minimum moment envelope curves (Fig. 6.19b) by connecting, respectively, all maximum and minimum moment values computed for the various sections x.
5. Add the moments due to dead weight, wind, etc. (Fig. 6.20) to the live load moment envelope curves of step 4.
6. Design the bridge to resist the combined maximum and combined minimum moment at every section x.

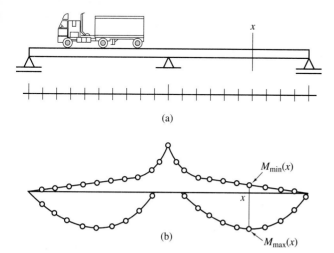

(a)

(b)

$M_{min}(x)$

x

$M_{max}(x)$

Figure 6.19 Moment envelope for highway bridge live load: (a) highway bridge loading; (b) typical moment envelope.

Shear envelope curves are constructed analogously to moment envelopes, except that absolute values can be used, since signs in shear design are generally unimportant.

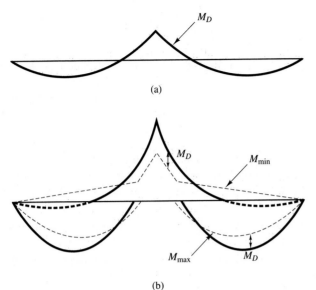

(a)

(b)

Figure 6.20 Combined dead-load and live-load moment envelope: (a) dead-load moments; (b) combined moments.

For buildings, a similar procedure can be followed. However, instead of coping with a moving live load such as a truck or railroad locomotive, we may opt to either fully load an entire span (or sometimes a partial span) or not load it at all, unless extremely heavy concentrated loads are involved (Fig. 6.21). For the design of a typical continuous building girder, the worst of all possible load combinations will need to be determined. In moment-resistant building frames, the live load applied to one floor generally has little effect on the bending moments of other floors. Therefore, the ACI Code permits us to assume that the live load is applied only to the floor or roof under consideration, and that the far ends of columns built integrally with the structure are fixed.

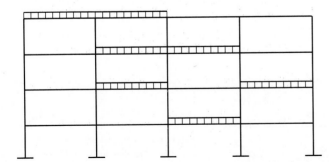

Figure 6.21 Building with variable live load.

The most difficult step in the construction of moment envelopes is the determination of the critical truck position or live-load pattern. This is greatly simplified by the use of *influence lines*.

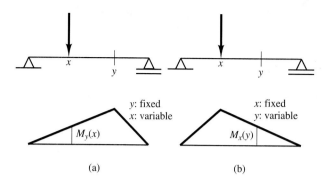

Figure 6.22   (a) Moment influence line; (b) moment diagram.

An influence line indicates the value for some quantity (e.g., a bending moment) at a fixed point y if a unit load is acting at an arbitrary point x. That is, y is fixed, whereas x is variable or "moving" (Fig. 6.22a). This should not be confused with the traditional moment diagram, which shows the bending moment at any point y for a load at a fixed point x, that is, where y is variable and x fixed (Fig. 6.22b). For notation, we can designate the bending moment diagram by $M_x(y)$ (moment at y due to a load fixed at x) and the moment influence line by $M_y(x)$ (moment at fixed y due to a load at x). To construct an influence line, we make use of the *Müller-Breslau Principle* [1], which shall be stated here without proof:

> If an internal moment, shear, or reaction component is considered to act through some small rotation or displacement, thereby displacing the structure, the curve of the displaced structure will be, to some scale, the influence line for the moment, shear, or reaction component.

To find the influence line for the bending moment M at some section y, we insert a hinge at y and introduce a unit rotation. The resulting deflected shape of the beam is the influence line for M (Fig. 6.23a). To find the influence line for the shear force at y, we insert a sliding hinge at y (permitting a relative displacement of the two faces, but no relative rotation) and introduce a unit relative displacement, which deflects the beam into the influence line for V (Fig. 6.23b). Similarly, to find the influence line for the right or left support reaction, we remove the respective support and introduce a unit displacement (Fig. 6.23c and d).

When dealing with continuous beams and frames, the Müller-Breslau Principle remains just as useful, because with some practice it is still possible to sketch the deflected shapes qualitatively (Fig. 6.24). Such sketches of influence lines enable us to determine the most critical positions for live loads. For example, after drawing the influence line for the bending moment over support B of the four-span beam of Fig. 6.24d, it is obvious that trucks should be positioned only in the first, second, and fourth span to produce the largest negative moment at B. A truck in the third span would cause a positive moment at B and thus reduce the effect of the other three trucks. Similarly, sketches of influence lines for building frames reveal that a maximum positive moment at midspan of a beam is caused by a checkerboard loading (Fig. 6.24e). For maximum moments at member

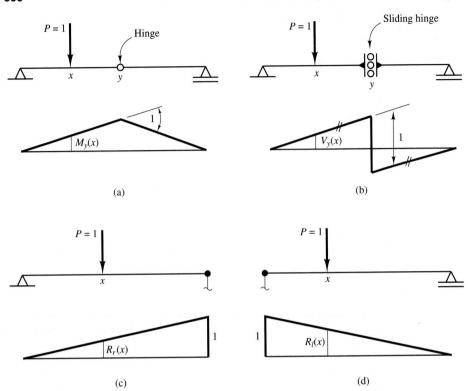

Figure 6.23 Influence lines for simply supported beam: (a) moment at y; (b) shear at y; (c) right reaction; (d) left reaction.

ends, the influence lines (deflected shapes) depend on the relative stiffnesses of beams and columns. Therefore, a conclusive if only qualitative determination of the critical loading is not possible without a moment distribution or similar analysis. Figure 6.24f depicts one possible outcome for the moment at B. However, as mentioned earlier, the ACI Code permits us to consider only one floor at a time.

EXAMPLE 6.7

For the three-span continuous girder of an office building (Fig. 6.25a), construct a qualitative moment envelope for unit live load applied to any one span, independently of the other two spans.

Solution   For the sake of simplicity, we construct moment envelope curves for the maximum and minimum moments at the five points numbered in Fig. 6.25a, while making use of symmetry.

    1. For the critical moments at point 1, we require the influence line for this moment, M_1 (Fig. 6.25b). To produce a maximum positive moment, spans 1 and 3 need to be loaded, while loading only span 2 will cause the largest negative moment at

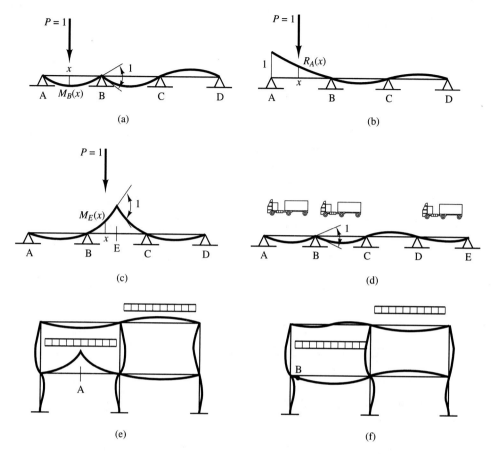

Figure 6.24 Influence lines and critical loadings for indeterminate structures: (a) moment at B; (b) reaction at A; (c) moment at E; (d) critical loading for moment at B; (e) critical loading for moment at A; (f) critical loading for moment at B.

point 1. The corresponding moment curves are shown in Fig. 6.25c and d. Because of symmetry, these moment curves are valid also for $M_{5,\text{max}}$ and $M_{5,\text{min}}$.

2. For the extreme moments at point 2, we require the influence line for M_2 (Fig. 6.25e). For the largest negative effect, spans 1 and 2 need to be loaded, while the maximum positive moment is caused by loading on the third span only. The resulting moment curves are shown in Fig. 6.25f and g. Because of symmetry, these moment curves are mirror images of those for point 4.

3. For the extreme moments at point 3, we construct the influence line for M_3 (Fig. 6.25h). Loading only span 2 maximizes M_3, while loading spans 1 and 3 causes the largest negative moment at point 3. These two loadings are identical to those of step 1 (Fig. 6.25d and c).

4. We superimpose all moment curves obtained so far and identify the moment envelopes (Fig. 6.25i). Note that for live load alone, each section of the beam may be subjected

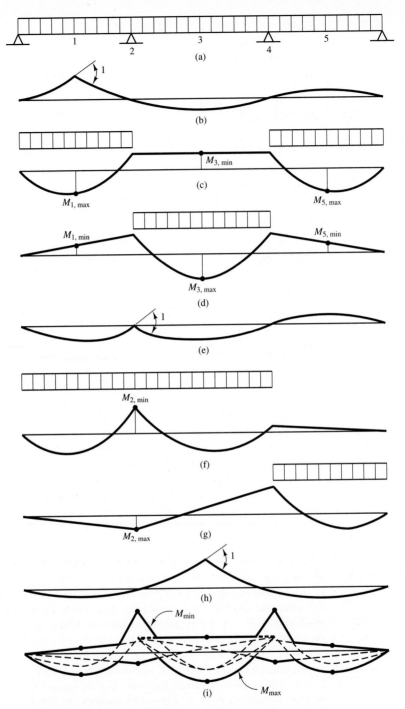

Figure 6.25 Example 6.7: Construction of moment envelopes for dead plus live load: (a) continuous beam; (b) influence line for M_1; (c) loading for $M_{1,\max}$ and resulting moment diagram; (d) loading for $M_{1,\min}$ and resulting moment diagram; (e) influence line for M_2; (f) loading for $M_{2,\min}$ and resulting moment diagram; (g) loading for $M_{2,\max}$ and resulting moment diagram; (h) influence line for M_3; (i) moment envelopes.

302

to a positive or negative moment, depending on the live load arrangement. If dead-load effects are small compared to live-load effects, much of the beam will have to be designed for moment reversals.   ∎

6.4 PLASTIC ANALYSIS AND MOMENT REDISTRIBUTION

At the end of Chapter 3, the concept of a plastic hinge was introduced, together with ductility, which indicates the amount of inelastic rotation the hinge may undergo before the concrete crushes in compression, followed by total collapse of the beam. In the context of a simply supported beam, the significance of either the plastic hinge or the rotational ductility are not obvious. As a matter of fact, there is no significance; once such a beam is overloaded, it simply forms a plastic hinge at midspan and collapses.

If the beam is statically indeterminate, the situation is completely different. Consider, for example, the uniformly loaded beam of Fig. 6.26a, fixed at both ends. A linear elastic structural analysis of this beam leads to the bending moment diagram of Fig. 6.26b, with negative moments of $-wL^2/12$ at the fixed ends and a positive midspan moment of $wL^2/24$. Note that the algebraic sum of fixed-end and midspan moment equals $wL^2/8$, the *total statical moment*, which is the midspan moment in an equivalent beam with simple end supports. In order to develop the negative support moments, the beam obviously must be reinforced with negative steel, as shown in Fig. 6.26c. Let us now suppose that equal amounts of reinforcement are provided at the supports and at midspan, which supply these sections with a nominal flexural capacity M_p. In other words, if the bending moment at any one of these sections reaches M_p, a plastic hinge will develop there. According to the *elastic* moment diagram of Fig. 6.26b, the bending moments at the supports are always twice the moment at midspan. As the uniform load is increased up to some level w_1 at which the end moments produce plastic hinges, the midspan moment $w_1L^2/24$ has reached only half of the beam's capacity M_p (Fig. 6.26d). It now becomes apparent why the behavior of a statically indeterminate beam is fundamentally different from that of a simply supported or statically determinate one: if we choose to increase the load beyond w_1, the beam will not collapse, because the midspan section still has a flexural reserve capacity of $M_p - (M_p/2) = M_p/2$. For any additional load w_2, the beam will behave as if it were simply supported, except that it has *plastic* instead of *real* hinges at the ends (Fig. 6.26e). The bending moment diagram for such a beam is familiar (Fig. 6.26f). Since the midspan reserve flexural capacity is $M_p/2$, it is straightforward to determine the load w_2, which will cause a third hinge to form, this one at midspan (i.e., $w_2 = 4M_p/L^2$). The third hinge causes a *collapse mechanism*, and the total load that leads to failure is

$$w_{\text{tot}} = w_1 + w_2 = \frac{12M_p}{L^2} + \frac{4M_p}{L^2} = \frac{16M_p}{L^2} \tag{6.10}$$

Compared to a simply supported beam with a load capacity of $8M_p/L^2$, the fixed boundary conditions and *full moment redistribution* activate a 100% increase in load-carrying

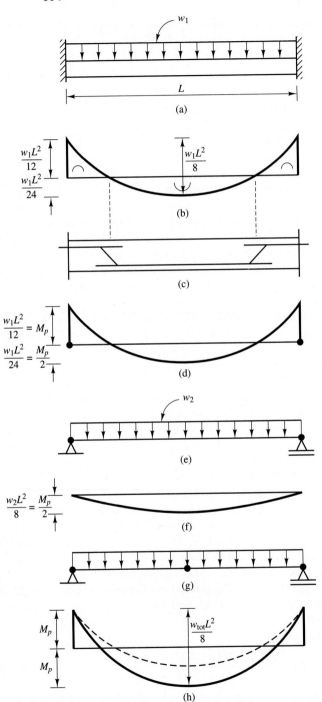

Figure 6.26 Moment redistribution in beam fixed at both ends: (a) fixed-end beam; (b) elastic moment diagram; (c) reinforcement; (d) plastic hinges at fixed ends; (e) "simply supported" beam with two plastic hinges; (f) moment diagram for simply supported beam; (g) three plastic hinges form collapse mechanism; (h) elastic and (final) plastic moment diagram.

capacity. Figure 6.26h illustrates the moment redistribution: as the load is increased from w_1 to $w_1 + w_2$, the moment diagram changes from the dotted to the solid line, while the inflection points move toward the supports. Whether or not the beam ends have sufficient rotational ductility to permit the complete redistribution of moments without prior concrete crushing is a matter that we shall explore later. Also, provisions must be made to ensure that the beam not fail prematurely in shear or bond.

The result of Eq. 6.10 could have been obtained in a single step, using statics alone, as illustrated in Fig. 6.26h. At failure, the total static moment due to load, $w_{tot}L^2/8$, is equal to the sum of the moment capacities at the ends and midspan, that is, $2M_p$, from which it follows directly that $w_{tot} = 16M_p/L^2$.

EXAMPLE 6.8

A 16-ft-long beam is fixed at both ends and reinforced with 2 #8 bars, which are bent up at the inflection points to serve as positive reinforcement at midspan and as negative reinforcement at the supports (Fig. 6.27). Determine the uniform load under which the first two plastic hinges form, and the load under which the beam collapses, provided no premature shear or bond failure occurs and the sections have sufficient rotational capacity to permit full plastic moment redistribution. Given: $f_y = 60$ ksi; $f_c' = 3000$ psi.

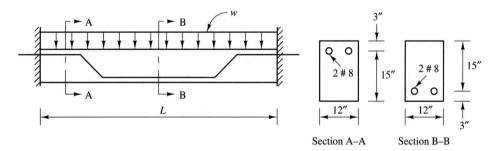

Figure 6.27   Example 6.8.

Solution

Steel capacity: $T = C = A_s f_y = (2)(.79)(60) = 94.8$ kips

Depth of compression block: $a = \dfrac{A_s f_y}{.85 f_c' b} = \dfrac{94.8}{(.85)(3)(12)} = 3.1$ in

Nominal moment capacity: $M_p = T\left(d - \dfrac{a}{2}\right) = \dfrac{(94.8)(15 - 3.1/2)}{12} = 106.3$ ft-kips

Note that the actual or nominal moment capacity counts, not the ultimate capacity, which includes the strength reduction factor.

Load at which the first two hinges form: $w_1 = \dfrac{12M_p}{L^2} = \dfrac{(12)(106.3)}{16^2} = 4.98$ kips/ft

Additional load that will cause formation of third plastic hinge:

$$w_2 = \frac{4M_p}{L^2} = \frac{(4)(106.3)}{16^2} = 1.66 \text{ kips/ft}$$

Collapse load: $w_{\text{tot}} = w_1 + w_2 = 4.98 + 1.66 = \underline{6.64 \text{ kips/ft}}$ ■

The ACI Code [Sec. 8.4] recognizes the capacity of concrete members to redistribute moments. Provided the reinforcing limit of Eq. 3.51 is satisfied, the negative moments calculated by elastic theory for the end supports may be increased or decreased by $20[1 - ((\rho - \rho')/\rho_b)]$ percent. When computing midspan or positive moments, this modification of negative moments shall be taken into account such that statics remains satisfied.

EXAMPLE 6.9

Making use of the Code's allowance for plastic moment redistribution, determine the flexural reinforcement for a beam with the same dimensions as in Example 6.8 to support a factored uniform load of 5.0 kips/ft.

Solution

Elastic support moment: $M_u = \dfrac{w_u L^2}{12} = \dfrac{(5)(16)^2}{12} = 106.7$ ft-kips

Elastic midspan moment: $M_u = \dfrac{w_u L^2}{24} = \dfrac{(5)(16)^2}{24} = 53.3$ ft-kips

Balanced steel ratio: $\rho_b = .85\beta_1 \dfrac{f_c'}{f_y} \dfrac{87}{87 + f_y} = (.85)^2 \dfrac{3}{60} \dfrac{87}{87 + 60} = .0214$

Required negative steel:

With $a = 3$ in, $A_s = \dfrac{M_u}{\phi f_y (d - a/2)} = \dfrac{(106.7)(12)}{(.9)(60)(15 - 1.5)} = 1.76$ in^2;

check: $a = \dfrac{A_s f_y}{.85 f_c' b} = \dfrac{(1.76)(60)}{(.85)(3)(12)} = 3.4.$ OK

Steel ratios: $\rho = \dfrac{1.76}{(12)(15)} = .0098, \rho' = 0$

Maximum redistribution allowance: $20\left(1 - \dfrac{.0098 - 0}{.0214}\right) = 10.8\%$

Adjusted negative moment: $M_u = (1 - .108)(106.7) = 95.2$ ft-kips

Adjusted positive moment: $M_u = 53.3 + (.108)(106.7) = 64.8$ ft-kips

Note Fig. 6.28 and that the sum of negative plus positive moments, 160 ft-kips, remains unchanged.

Required negative steel: $a = 3$ in, $A_s = \dfrac{(95.2)(12)}{(.9)(60)(15 - 1.5)} = 1.57$ in^2;

Use 2 #8 bars, with 1.58 in^2.

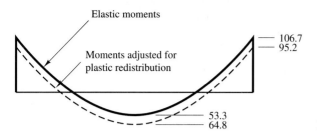

Figure 6.28   Example 6.9.

Required positive steel: $a = 2$ in, $A_s = \dfrac{(64.8)(12)}{(.9)(60)(15-1)} = 1.03$ in^2;

Use 2 #7 bars, with 1.2 in^2.   ∎

In structural steel design, we normally deal with rolled sections. Their plastic moment capacities are constant along an entire member unless reinforcing plates are attached locally. In reinforced concrete design, it is extremely simple to vary the flexural moment capacities of members by varying the amount of flexural reinforcement. This has an effect, of course, on the beam's load-carrying capacity. Suppose a beam is fixed at both ends and has the capacities M_1 at the left support, M_2 at the right support, and M_3 at midspan (Fig. 6.29). In this case, failure will occur if

$$\frac{wL^2}{8} \geq \frac{M_1 + M_2}{2} + M_3 \tag{6.11a}$$

or if

$$w \geq \frac{8}{L^2}\left(\frac{M_1 + M_2}{2} + M_3\right) \tag{6.11b}$$

irrespective of the moments computed by a detailed statically indeterminate elastic analysis. Strictly speaking, Eq. 6.11 is not exact, because for $M_1 \neq M_2$, the maximum positive moment does not occur at midspan. However, if M_1 and M_2 are not too different, then the solution of Eq. 6.11 is usually sufficiently accurate for practical purposes.

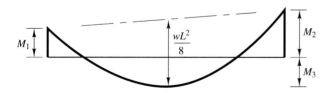

Figure 6.29   Ultimate moments in beam with variable capacities.

EXAMPLE 6.10

A beam with a 20-ft span is fixed at both ends and reinforced such that the left support capacity is 150 ft-kips, the right support capacity is 200 ft-kips, and the midspan capacity is 180 ft-kips (Fig. 6.30a). Determine the uniform load levels at which the first, second, and third hinges form.

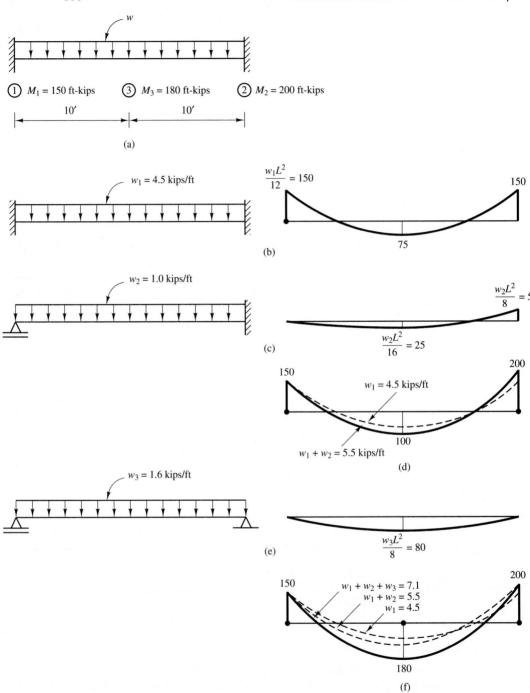

Figure 6.30 Example 6.10: (b) beam fixed at both ends; (c) beam with one hinge; (d) moments due to $w_1 + w_2$; (e) beam with two hinges; (f) moments due to $w_1 + w_2 + w_3$.

Solution The first hinge occurs at the left support, when the elastic moment, $w_1 L^2/12$, reaches the beam's capacity of 150 ft-kips (Fig. 6.30b), that is,

$$w_1 = \frac{12M_1}{L^2} = \frac{(12)(150)}{(20)^2} = 4.5 \text{ kips/ft}$$

Upon further load increase, the moment at section 1 cannot increase. The plastic hinge there has the effect that the beam reacts to additional load w_2 as if it were "hinged" at 1 and fixed at 2, for which the right support moment is now $w_2 L^2/8$, or twice the midspan moment (Fig. 6.30c). The reserve capacity at section 2 is $200 - 150 = 50$ ft-kips, and for the midspan section 3, it is $180 - 75 = 105$ ft-kips. As the load increases, a second hinge will occur at the right support. This happens when the moment $w_2 L^2/8$ due to the added load w_2 exhausts the reserve capacity of 50 ft-kips, that is, when

$$w_2 = \frac{8M}{L^2} = \frac{(8)(50)}{(20)^2} = 1.0 \text{ kips/ft}$$

The total load is now $4.5 + 1.0 = 5.5$ kips/ft, and the moments in the beam are those of Fig. 6.30d. At this point, only the midspan section has any reserve capacity left, namely, $180 - 75 - 25 = 80$ ft-kips. For any additional load w_3, the beam behaves as if it were simply supported, with "hinges" at both ends (Fig. 6.30e), and a midspan moment $w_3 L^2/8$, so that the remaining 80 ft-kips capacity is good for another

$$w_3 = \frac{8M}{L^2} = \frac{(8)(80)}{(20)^2} = 1.6 \text{ kips/ft}$$

The third hinge forms and collapse occurs at a load $w_{\text{tot}} = 5.5 + 1.6 = 7.1$ kips/ft. The moment diagram for this case is shown in Fig. 6.30f, in which both previous moment curves are indicated by dotted lines. This result could have been obtained in a single step using statics, that is, Eq. 6.11b:

$$w = \left(\frac{150 + 200}{2} + 180 \right) \frac{8}{20^2} = 7.1 \text{ kips/ft} \qquad \blacksquare$$

EXAMPLE 6.11

A three-span continuous beam has a uniform nominal flexural capacity of 150 ft-kips throughout (Fig. 6.31a). Determine the theoretical failure load based on elastic moments and the actual failure load based on plastic moments.

Solution According to elastic theory (Fig. 6.16b), the critical moments occur at the interior supports, $M = w_1 L^2/10$, and the associated load level is

$$w_1 = \frac{10M}{L^2} = \frac{(10)(150)}{(20)^2} = 3.75 \text{ kips/ft}$$

Note that the maximum positive moment in the side spans, $w_1 L^2/12.5 = 120$ ft-kips, does not quite occur at midspan. At midspan, the moment is $w_1 L^2/8 - \frac{1}{2} w_1 L^2/10 = 112.5$ ft-kips. Ignoring this 7% error, we assume the reserve capacity at the center of the side spans to be $150 - 112.5 = 37.5$ ft-kips, and for the center span, $150 - 37.5 =$

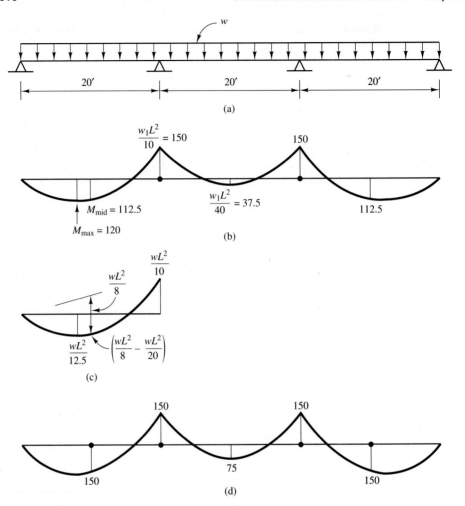

Figure 6.31 Example 6.11: (a) three-span beam; (b) moments for $w_1 = 3.75$ kips/ft (two plastic hinges); (c) statics of side span; (d) moments for $w = 4.5$ kips/ft (four plastic hinges).

112.5 ft-kips. The side spans develop midspan hinges before the center span does, and these lead to collapse when the static moment $w_2 L^2/8$ equals the midspan reserve capacity of 37.5, that is, failure occurs at

$$w_{tot} = 3.75 + \frac{(37.5)(8)}{(20)^2} = 4.5 \text{ kips/ft}$$

Theoretically, the center span collapses only when the static moment due to any additional load, $w_3 L^2/8$, reaches this span's reserve capacity of 112.5 ft-kips, that is, at a total load of $w = 3.75 + (112.5)(8)/20^2 = 6.0$ kips/ft. However, once the side spans have collapsed, they cease to restrain the center span, so the actual failure load would indeed be 4.5 kips/ft.   ■

The previous examples illustrate that the ultimate loads and collapse mechanisms are governed by the *section capacities* and not by the elastic moments, which are based on *relative member stiffnesses*. For the design of concrete structures this is an important distinction because, as a rule, we follow the ultimate strength design philosophy. For working stress design, elastic moments would be more appropriate than plastic moments after moment redistribution. However, for ultimate strength design, it is consistent to design sections for demands based on true failure moments, that is, on final moments, after redistribution has taken place.

In making use of plastic moment redistribution, the designer has an effective tool with which to control the collapse mechanism by appropriately allocating moment capacities and, at the same time, to minimize the total material requirements or cost.

EXAMPLE 6.12

The three-span continuous beam of Fig. 6.32a supports a dead load of 1.0 kips/ft and a live load of 2.0 kips/ft. Determine the required steel reinforcement areas at sections 1, 2, and 3

a) based on elastic moments;

b) by allocating uniform strength capacities;

c) by assuring that all three spans collapse under the same load;

d) compare the results quantitatively.

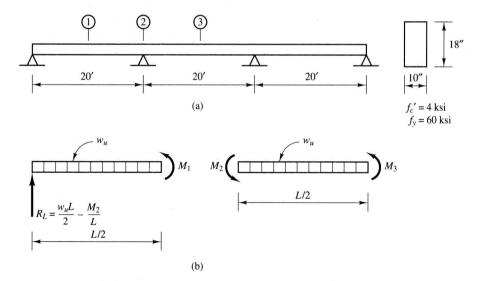

Figure 6.32   Example 6.12: (a) three-span beam; (b) free body diagrams of half spans.

Solution

a) With $w_u = 1.4w_D + 1.7w_L = (1.4)(1.0) + (1.7)(2.0) = 4.8$ kips/ft, the elastic moments (see Fig. 6.16b) follow as

$$M_1 = \frac{w_u L^2}{12.5} = 153.6 \text{ ft-kips}$$

$$M_2 = \frac{w_u L^2}{10} = 192.0 \text{ ft-kips}$$

$$M_3 = \frac{w_u L^2}{40} = 48.0 \text{ ft-kips}$$

Using the design method of Chapter 4, we find the following reinforcing areas:

$$A_{s1} = 2.47 \text{ in}^2 \qquad A_{s2} = 3.25 \text{ in}^2 \qquad A_{s3} = 0.69 \text{ in}^2$$

b) To provide uniform strength, we know (from Example 6.10) that the collapse load for the critical side spans is

$$w_u = \frac{12M_u}{L^2} \quad \text{or} \quad M_{u,\text{req}} = \frac{w_u L^2}{12} = \frac{(4.8)(20)^2}{12} = 160 \text{ ft-kips}$$

and $A_{s,\text{req}} = 2.59 \text{ in}^2$, for all three sections.

c) To determine the required capacities for which the three spans collapse at the same load, we need to derive the necessary equilibrium equations for the collapse state. From the free body diagrams for the side and center span segments (Fig. 6.32b), we establish the following respective relationships:

$$M_1 + \frac{M_2}{2} = \frac{w_u L^2}{8} \tag{6.12a}$$

$$M_2 + M_3 = \frac{w_u L^2}{8} \tag{6.12b}$$

This problem, with two equations and three unknowns, is indeterminate. In other words, there are infinitely many solutions, and we may want to choose the one that requires the least amount of steel. *Optimization problems* of this kind are encountered frequently in structural engineering. A solution can be found analytically. Alternatively, we may select various values for M_2 (which makes the problem determinate again), compute the total steel required for each choice of M_2, and select the case that requires the least total amount. For example, we may select $M_2 = 160$ ft-kips, assuming that the 16.7% reduction from the elastic moment of 192 ft-kips is possible through plastic moment redistribution. From Eq. 6.12, it follows that

$$M_3 = \frac{w L^2}{8} - M_2 = 240 - 160 = 80 \text{ ft-kips}$$

$$M_1 = \frac{w L^2}{8} - \frac{M_2}{2} = 240 - 80 = 160 \text{ ft-kips}$$

The required steel areas follow as $A_{s1} = A_{s2} = 2.54 \text{ in}^2$; $A_{s3} = 1.19 \text{ in}^2$.

d) To compare total steel requirements, we should multiply the various bar areas with their respective lengths and add the results. For simplicity, let us assume that all bars are approximately of the same length, so that the total weight of reinforcing steel is proportional to $A_{s1} + A_{s2} + A_{s3}$. Thus, for the three cases we have

a) $A_{s,\text{tot}} = 2.47 + 3.25 + 0.69 = 6.41$ in^2 (elastic moments)
b) $A_{s,\text{tot}} = 2.59 + 2.59 + 2.59 = 7.77$ in^2 (uniform strength)
c) $A_{s,\text{tot}} = 2.54 + 2.54 + 1.19 = 6.27$ in^2 (near-optimum solution).

The uniform-strength solution is clearly uneconomical, because the center span has positive reinforcement well in excess of what is needed. The near-optimum solution requires slightly less steel than the elastic moment solution, but the difference as well as any further savings that may be realized by an exact solution of the optimization problem is negligible in view of the available sizes of rebars. However, savings can be realized when, at certain sections, the steel requirement based on strength falls well below the Code's minimum steel requirement.   ∎

All examples of plastic moment redistribution discussed so far have made two implicit assumptions. First, it was assumed that all members were designed such that no premature shear failure would occur and that all bars have sufficient development lengths. Second, it was assumed that each section with a plastic hinge had sufficient plastic rotation capacity or ductility to permit full plastic moment redistribution. Such ductility is not assured automatically. Rather we must check whether in fact the available plastic rotation capacity θ_{cap}, exceeds θ_{req}, the rotation required for full moment redistribution. For this purpose, we must compute both θ_{cap} and θ_{req}.

Consider the moment-curvature curve of Fig. 6.33. By definition, the tension steel starts to yield at the yield moment M_y and yield curvature ϕ_y. Upon further loading, a small moment increase from M_y to M_p is accompanied by rapidly increasing curvature. For analysis purposes, it can be assumed that a full plastic hinge develops at curvature ϕ_y and failure occurs at curvature ϕ_u when the concrete reaches its crushing strain (see Chapter 3). The plastic rotation associated with the change in curvature, $\phi_u - \phi_y$, is not actually concentrated in a plastic hinge, but rather distributed over a plastic region of length l_p. The plastic rotation capacity θ_{cap} can therefore be expressed as

$$\theta_{\text{cap}} = (\phi_u - \phi_y)l_p \tag{6.13}$$

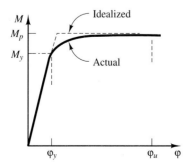

Figure 6.33   Idealized moment-curvature curve.

Fig. 6.34 shows the strain distributions both at first yield and at failure. At incipient yield, the maximum concrete strain is

$$\epsilon_{ce} = \phi_y kd = \frac{kd}{d - kd}\epsilon_y \tag{6.14}$$

The yield curvature follows as

$$\phi_y = \frac{\epsilon_{ce}}{kd} = \frac{\epsilon_y}{d - kd} \tag{6.15}$$

whereas the ultimate curvature is given by

$$\phi_u = \frac{\epsilon_{cu}}{c} = \frac{.003}{c} \tag{6.16}$$

Substituting Eqs. (6.15) and (6.16) into Eq. (6.13), we find

$$\theta_{cap} = \left(\frac{\epsilon_{cu}}{c} - \frac{\epsilon_{ce}}{kd}\right)l_p \tag{6.17}$$

The equivalent length of the plastic hinge l_p is difficult to determine. One of the empirical formulas that have been proposed is

$$l_p = 0.08l + 0.15d_b f_y \tag{6.18}$$

where l is the beam span length and d_b the bar diameter [4].

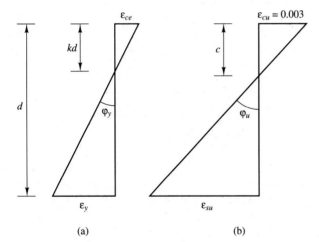

Figure 6.34   Strains at (a) first yield; and (b) failure.

EXAMPLE 6.13

Check whether the beam of Fig. 6.35 has sufficient plastic rotation capacity to permit full moment redistribution. How much moment redistribution would be permitted by the ACI Code? Given: $f'_c = 4$ ksi; $f_y = 40$ ksi; $n = 8$; $E_c = 3600$ ksi.

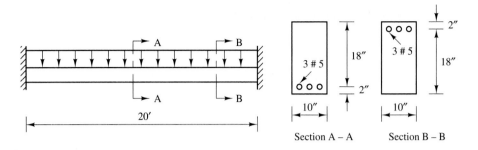

Section A – A   Section B – B

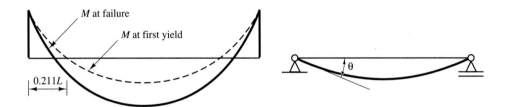

Figure 6.35   Example 6.13.

Solution

Nominal moment capacity: $a = \dfrac{A_s f_y}{.85 f_c' b} = \dfrac{(3)(.31)(40)}{(.85)(4)(10)} = 1.09$;

$$M_n = A_s f_y \left(d - \frac{a}{2}\right) = (.93)(40)\frac{18 - .54}{12} = 54.1 \text{ ft-kips.}$$

The first two hinges appear at: $w_1 = \dfrac{12 M_n}{L^2} = \dfrac{(12)(54.1)}{20^2} = 1.62 \text{ kips/ft.}$

Collapse load after full moment redistribution:

$$w_2 = \frac{16 M_n}{L^2} = \frac{(16)(54.1)}{20^2} = 2.16 \text{ kips/ft.}$$

While the load increases from w_1 to w_2, the beam behaves as if it were simply supported, with end rotations in the plastic hinges equal to

$$\theta_{\text{req}} = \frac{(w_2 - w_1)L^3}{24EI} = \frac{(2.16 - 1.62)(20)^3}{24EI} = \frac{180}{EI}$$

The length of the plastic hinge region is estimated to be

$$l_p = .08l + .15 d_b f_y = (.08)(240) + (.15)(.625)(40) = 22.95 \text{ in}$$

Next, the concrete strain reserve capacity needs to be evaluated.

Neutral axis position at incipient yield: With $n\rho = (8)\dfrac{.93}{(10)(18)} = .0413$,

$$k = -.0413 + \sqrt{(.0413)^2 + (2)(.0413)} = .249, \text{ and } kd = (.249)(18) = 4.48 \text{ in}$$

Concrete strain when steel starts to yield:

$$\epsilon_{ce} = \epsilon_y \frac{kd}{d - kd} = \frac{f_y}{E_s}\frac{kd}{d - kd} = \frac{40}{29,000}\frac{4.48}{18 - 4.48} = .00046$$

Depth of concrete compression block: $a = 1.09$ in, $c = \dfrac{ua}{\beta_1} = \dfrac{1.09}{.85} = 1.28$ in

Plastic rotation capacity: $\theta_{cap} = \left(\dfrac{.003}{1.28} - \dfrac{.00046}{4.48}\right)(22.95) = .0514$ Rad

Cracked moment of inertia:

$$I_{cr} = \frac{b}{3}(kd)^3 + nA_s(d - kd)^2 = \frac{(10)(4.48)^3}{3} + (8)(.93)(13.52)^2 = 1660 \text{ in}^4$$

Required rotation capacity: $\theta_{req} = \dfrac{(180)(12)^2}{(3600)(1660)} = .00434$ Rad

As $\theta_{req} \ll \theta_{cap}$, the beam has sufficient ductility to undergo full plastic moment redistribution, that is, to redistribute $\left(\dfrac{wL^2/12 - wL^2/16}{wL^2/12}\right) 100 = 25\%$.

Maximum moment redistribution permitted by the ACI Code:

With balanced steel ratio, $\rho_b = (.85)^2\dfrac{4}{40}\left(\dfrac{87}{87 + 40}\right) = .0495$,

and $\rho = \dfrac{.93}{(10)(18)} = .00516$, maximum allowable $= 20\left(1 - \dfrac{.00516 - 0}{.0495}\right) = 17.9\%$.

∎

Note that the maximum amount of moment redistribution permitted by the ACI Code, $20[1 - ((\rho - \rho')/\rho_b)]$, has an upper bound of 20%. This limit is rather conservative for members with high ductility. On the other hand, a design based on plastic analysis results that differ appreciably from elastic analysis results may experience moment redistribution even under service loads, with the danger of unsightly cracking and the potential of subsequent steel corrosion in the plastic hinge region. For this reason, it is good design practice to limit the amount of moment redistribution.

The rotational ductility of a flexural member is very much a function of the amount and kind of lateral reinforcement, because such reinforcement creates a state of triaxial compression under which the concrete strength, and in particular its failure strain, increases tremendously (see Chapter 2). This effect is utilized in the design of concrete structures to resist strong earthquake ground motions [4]. In response to such loads, concrete frames may undergo large lateral displacements, which are possible only after the formation of plastic hinges. If these plastic regions are well confined with lateral reinforcement, they are capable of withstanding several cycles of inelastic load reversals without failure. Moreover, the large amount of energy dissipated in the hinges is largely responsible for the survival of such structures during destructive earthquakes and is the basis for ductile moment-resisting frame construction.

6.5 APPROXIMATE ANALYSIS AND DESIGN

Before we learn how to design continuous structures, we must understand how such structures behave under load. We will then be in a position to control structural behavior. The capability of "telling the structure how to behave" is an exciting aspect of the art of structural design, but it can be learned only over years of design practice—building on one's own experience as well as that of seasoned designers. Therefore, the discussion and examples of this section can be no more than an introduction for young engineers. Subsequent steps will have to be taken by themselves.

We start by drawing the deflected shapes and corresponding moment diagrams of statically determinate beams and frames (Fig. 6.36)—a straightforward exercise. For more complicated cases, however, a step-by-step approach may be necessary. Consider, for example, the frame of Fig. 6.37a, for which the deflected shape is not so obvious. Let us first imagine the beam to be very stiff, which is equivalent to locking the left frame corner against rotation. In this case, only the left column can flex under the horizontal load (Fig. 6.37b). When the upper left corner is released and allowed to rotate clockwise, as the moment in the left column dictates, the right corner will likewise rotate, but counterclockwise. Moreover, because of the carry-over effect, it will rotate only half the amount of the left corner. In the absence of a horizontal reaction at the roller support, the right column must remain straight and moment-free when the upper right corner rotates (Fig. 6.37c).

When sketching the deflected shapes of indeterminate structures, it is useful to sketch them together with the corresponding, if only approximate, moment diagrams, because the inflection points coincide with points of zero moment. If the positions of all inflection points are known or assumed, the determination of the moment diagram is reduced to an exercise in statics. In other words, the structure is effectively made "statically determinate." In addition, it helps to establish upper and lower bounds for stiffness ratios and to use common sense when interpolating between the two extremes. For example, consider the frame of Fig. 6.38a, which is statically indeterminate to the first degree. Let us start by assuming the columns are very stiff compared with the beam, so that the beam is effectively fixed at both ends (Fig. 6.38b). We have encountered this case before and know how the beam moment diagram looks (Fig. 6.38c). We even know (or can easily figure out) that the inflection points are located a distance $0.21L$ from the corners. Because of equilibrium, the column moments must equal the beam's fixed-end moments. Assuming very stiff columns is equivalent to assuming the corners are locked against rotation. If we relax this assumption and let the corners rotate in the direction of the fixed-end moments, we get the deflected shape of Fig. 6.38d, which indicates that the inflection points are moving toward the corners during the relaxation process. At the same time, the bending moment curve for the beam is dropping a certain amount, because the columns, not as stiff as originally assumed, cannot take the full fixed-end moments.

We can approach the problem from the other extreme—the lower bound—by assuming that the columns have no stiffness compared with the beam. In this case, the beam is effectively simply supported, with the resulting moment diagram of Fig. 6.38g

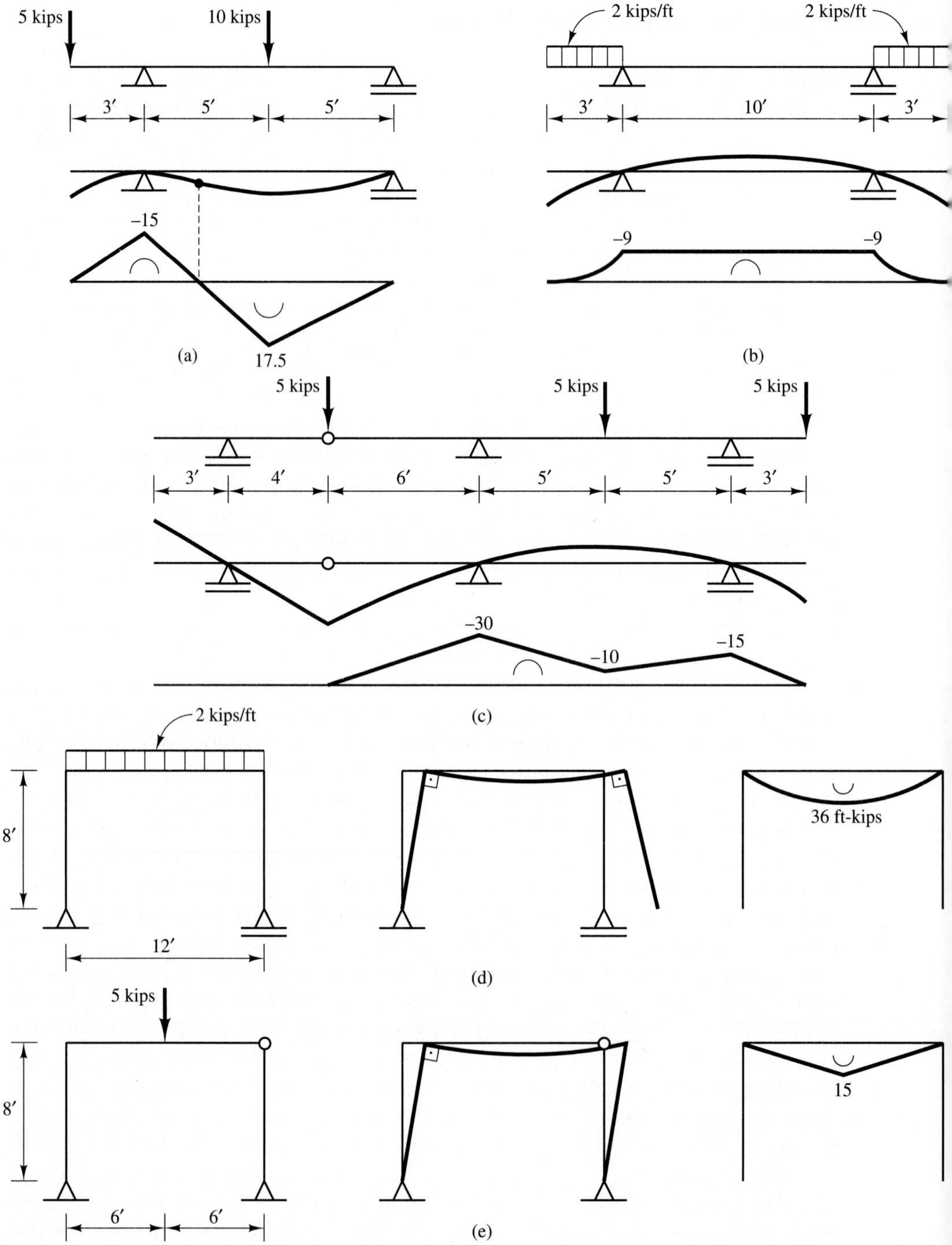

Figure 6.36 Deflected shapes and bending moments of statically determinate beams and frames.

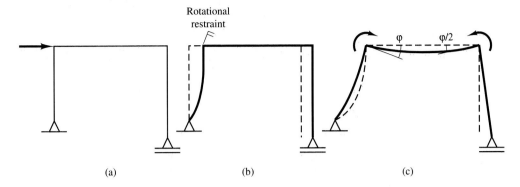

Figure 6.37 Constructing deflected shape of frame.

and the deflected shape of Fig. 6.38f. For compatibility the columns must bend with the beams, even though their stiffnesses are so small that their moments remain effectively zero. Next, we must correct our initial assumption and recognize that the columns do in fact have stiffness. Therefore, the left column tends to rotate the left corner counterclockwise, while the right column rotates the right corner clockwise. Both rotations introduce countercurvature into the beam, with the inflection points moving from the corners inwards (Fig. 6.38h). The moment diagram for the beam is "raised" in this process to the position of Fig. 6.38i, which, of course, should coincide with that of Fig. 6.38e.

   As another example, let us use the continuous beam of Fig. 6.39a to graphically illustrate moment distribution. Starting with the two interior beam joints fixed against rotation, we obtain the deflected shape of Fig. 6.39b and the moment diagram of Fig. 6.39c. The relaxation of the second support rotation results in the deflected shape of Fig. 6.39d and the moment diagram of Fig. 6.39e. Note, during this relaxation step, how the inflection point in the first span moves to the right as the negative moment over the first interior support decreases. The relaxation of the rotational restraint over the third support results in the deflected shape of Fig. 6.39f and the moment diagram of Fig. 6.39g. Upon renewed release of the second interior joint to restore equilibrium, the bending moment is reduced further, with another slight shift of the inflection point to the right.

   Figure 6.40 shows the deflected shapes of some additional continuous beams and frames, drawn qualitatively together with their corresponding approximate moment diagrams for the loadings shown. If the deflected shape is known, we can qualitatively sketch the moment diagram and vice versa. Students are encouraged to thoroughly practice sketching such diagrams and to solve problems to exercise their engineering common sense. Problem sets 1 and 2 at the end of this chapter contain a good number of examples for such practice.

   Approximate analysis results allow a quick determination of steel reinforcement requirements, which is often sufficiently accurate for preliminary design purposes. Consider, for example, the portal frame of Fig. 6.41a, subjected to a single horizontal load P. If the "exact" deflected shape were known, complete with the position of the

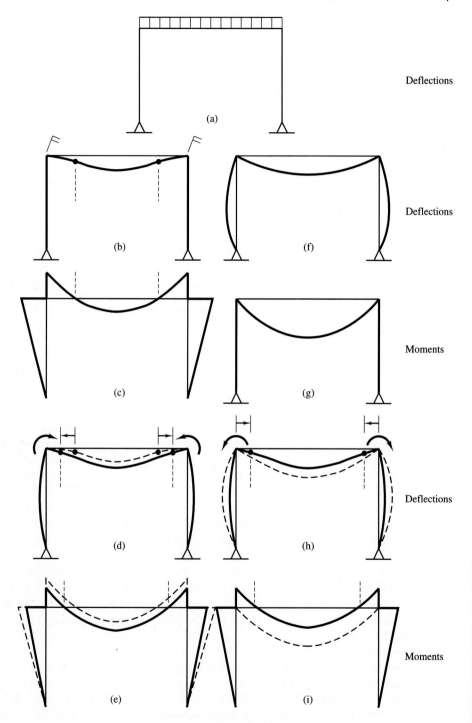

Figure 6.38 Deflected shape of portal frame.

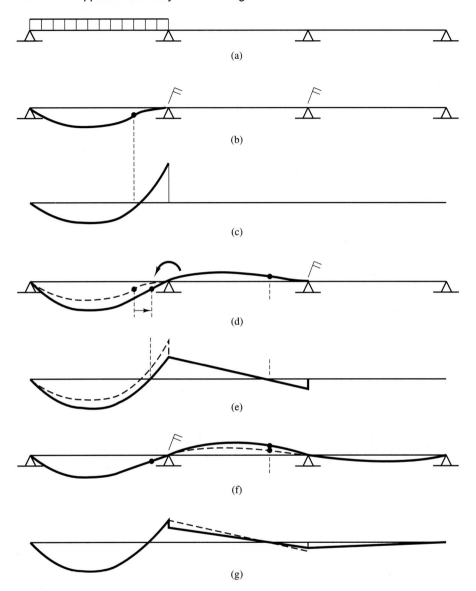

Figure 6.39 Successive relaxation of joints in three-span continuous beam.

inflection points, the determination of all bending moments would be straightforward, as it requires only the use of statics: The total applied moment Ph is resisted equally by the two columns for symmetry reasons, that is, each column must carry a total moment $Ph/2$. The only remaining uncertainty is the position of the inflection points in the columns, which depends on the stiffness of the beam relative to that of the columns.

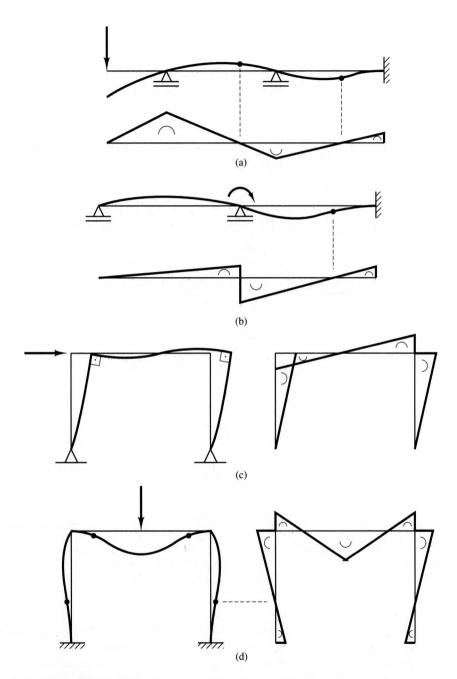

Figure 6.40 Deflected shapes and corresponding moment diagrams for statically indeterminate beams and frames.

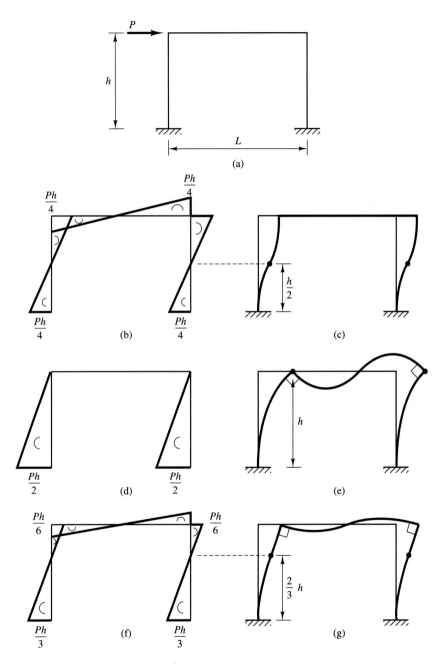

Figure 6.41   Horizontally loaded portal frame.

If the beam were infinitely stiff, the inflection points must, again because of symmetry, be located at midheight, that is, $h/2$ (Fig. 6.41b and c). If the beam were infinitely flexible, the inflection points (or rather the points of zero moment) would be located at the beam-column joints, that is, at h (Fig. 6.41d and e). Since the real beam stiffness is somewhere between zero and infinity, we know that the inflection points must be located somewhere between h and $h/2$. Assuming they are located $2/3h$ above ground (Figs. 6.41f and g) leads to an approximate moment diagram, with $M = Ph/3$ at the column bases and $M = Ph/6$ at the column tops and beam ends. We can now say with confidence that these results do not greatly differ from the correct results. Moreover, whatever errors do exist can be disregarded, as long as the frame has sufficient ductility and the columns have enough flexural strength to cover the sum of the moments at top and bottom, namely, $Ph/2$.

EXAMPLE 6.14

Design the portal frame columns of Fig. 6.42a for exact and approximate moments due to the horizontal force shown and compare the results.

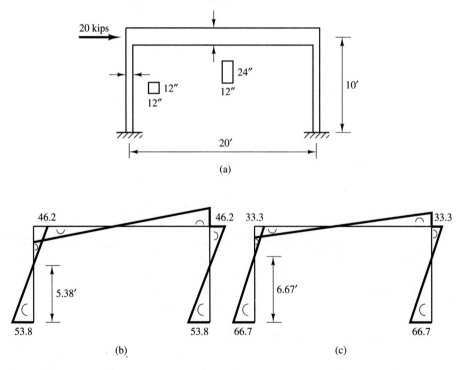

Figure 6.42 Example 6.14: (a) frame; (b) "exact" moments; (c) "approximate" moments.

Solution   The exact moments are computed by any structural analysis method and are given in Fig. 6.42b. Using the design method of Chapter 4, we determine the steel required for the base moments of 53.8 ft-kips as $A_s = 1.25$ in^2 (use 3 #6 bars), and for the moments at the top of the columns, $M = 46.2$ ft-kips, which requires $A_s = 1.04$ in^2 (use 2 #7 bars). The inflection points in this case are located $(53.8/100)(10) = 5.38$ ft above ground.

Had we estimated the inflection points to be $2h/3 = 6.67$ ft above ground, the column moments at the top and bottom would be 33.3 ft-kips and 66.7 ft-kips, respectively (Fig. 6.42c). Note that the sum is the same as before, that is, $33.3 + 66.7 = 46.2 + 53.8 = 100$ ft-kips $= Ph/2$, because the equations of statics still must be satisfied in both cases. The steel areas required to cover the approximate moments are now $A_s = 0.72$ in^2 (2 #6 bars) and $A_s = 1.62$ in^2 (3 #7 bars) for the top and bottom, respectively. We note that there exists a small but insignificant difference between the total steel requirements for the two cases (2.29 in^2 vs. 2.34 in^2). This difference is magnified by the discreteness of available bar sizes (3 #6 and 2 #7 vs. 2 #6 and 3 #7). Significant, however, is the fact that the structure will try to correct the designer's "mistake," or rather respond to the designer's true intentions: as the horizontal load is applied gradually, the exact elastic moment at the column tops will at one point reach the design capacity 33.3 ft-kips. As a result, the steel will start to yield, forming a plastic hinge at the top. The missing moment capacity, $46.2 - 33.3 = 12.9$ ft-kips, will have to be supplied by the base of the column, where in fact it is available since the design capacity exceeds the required design moment by the same amount, namely, $66.7 - 53.8 = 12.9$ ft-kips. Provided the ductility of the plastic hinge at the column top is sufficient to permit plastic moment redistribution, the structure will eventually make up the strength deficit at the top with the strength reserve at the bottom. In effect, the structure can forgive the designer's laziness, inability, or unwillingness to compute exact elastic moments. In other words, the designer can make the structure behave in such a way as he or she wishes it to behave, not as elastic theory tells it to behave.

■

When reinforcing a frame such as the one in Example 6.14 for the horizontal force to the right, we must be careful to always place the steel on the tension side. To identify the tension sides, it is a good practice to always indicate the sense of bending ($\smile$ or $\frown$) in the moment diagrams. Plus or minus prefixes to numerical values make sense for beams with a "top" and "bottom" only if the common sign convention is used that positive moments produce tension at the bottom. For vertical or inclined members such a sign convention is arbitrary. The frame of Fig. 6.42 would thus be reinforced as shown in Fig. 6.43. Near the inflection points, reinforcing bars can be bent at 45° angles to change sides such that they always end up on the tension side. If not bent, their cutoff points must satisfy the development length requirements, as discussed in Chapter 5. Of course, if the horizontal load is reversible (e.g., if it were due to wind), the frame must be reinforced symmetrically.

An experienced engineer will note that the reinforcement in the left frame corner (Detail A) will tend to pull out, as it is anchored by nothing more than a thin concrete cover inadequate for this task. In fact, it is not straightforward to detail the reinforcement

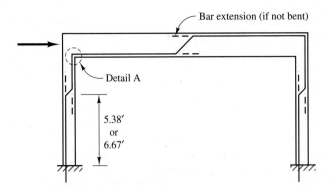

Figure 6.43 Reinforcement of frame for horizontal load.

of frame corners subject to moments that open up the corner. Some reinforcing details are suggested in Fig. 6.44 to anchor the flexural reinforcement. The purpose of the diagonal bars is to control the cracks [5].

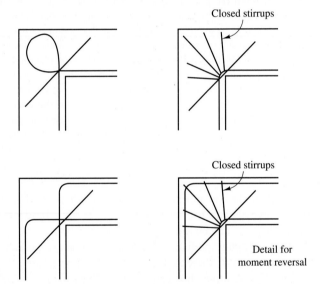

Figure 6.44 Reinforcing details for frame corners.

EXAMPLE 6.15

Design the interior span of a three-span continuous beam (Fig. 6.45a) for exact and approximate moments due to a uniform factored load of $w_u = 4$ kips/ft. Qualitatively indicate the position of the reinforcement. Given: 8-by-16-in cross section, $f_c' = 4$ ksi; $f_y = 60$ ksi.

Solution We start by computing the exact moments with the help of Fig. 6.16b. These moments are shown in Fig. 6.45b. The steel requirements are determined in the usual manner, $A_s^- = 1.64$ in^2 (2 #7 bars and 1 #6 bar), and $A_s^+ = 0.37$ in^2 (controlled by $\rho_{min} = .0033$) or 2 #4 bars. The sum of negative and positive steel provided is therefore $(2)(0.2) + (1)(0.44) + (2)(0.6) = 2.04$ in^2.

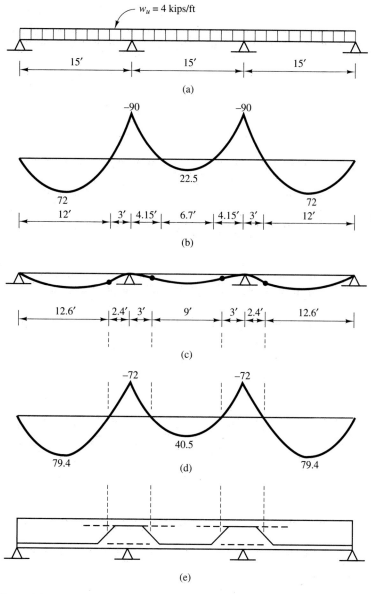

Figure 6.45 Example 6.15: (a) continuous beam; (b) "exact" moments; (c) approximate deflected shape; (d) approximate moments; (e) qualitative reinforcement.

Suppose that for some reason we are content with finding only approximate moments. For example, we draw the deflected shape of the beam and estimate the position of the inflection points to be 3 ft away from the supports (Fig. 6.45c) instead of the correct 4.15 ft. The problem is now "statically determinate." The midspan moment (with $L^* = 9$ ft between inflection points) is $w(L^*)^2/8 = (4)(9)^2/8 = 40.5$ ft-kips, and the negative moments are $(4)(15)^2/8 - 40.5 = 72$ ft-kips. The maximum positive moments in the side spans can also be determined easily, using statics alone:

Left reaction: $R_L = \left[\dfrac{wL^2}{2} - M_{\text{sup}}\right]/L = [(4)(15)^2/2 - 72]/15 = 25.2$ kips

Point of zero shear: $x = \dfrac{R_L}{w} = \dfrac{25.2}{4} = 6.3$ ft

Inflection point: $2x = (2)(6.3) = 12.6$ ft

Maximum positive moment: $M = \dfrac{w(L^*)^2}{8} = \dfrac{(4)(12.6)^2}{8} = 79.4$ ft-kips

The steel requirements are now, $A_s^- = 1.27$ in^2 (or 3 #6 bars) and $A_s^+ = 0.67$ in^2 (or 2 #6 bars), amounting to $(5)(.44) = 2.20$ in^2.

For comparison, we may want to design the center span as if it were simply supported. The gross statical moment is $(4)(15)^2/8 = 112.5$ ft-kips and requires a steel area of $A_s = 2.15$ in^2 (or 2 #8 bars and 1 #7 bar with 2.18 in^2). We see, in all three cases, an approximately equal total amount of steel is needed.

     A cautionary note: we are free to choose the positions of only two inflection points, which render the problem statically determinate. The other two inflection points are then fixed in the locations determined above. Arbitrarily locating all four inflection points is bound to cause errors in statics. ∎

     In conclusion, we have learned some of the most important lessons that a designer of concrete structures needs to know:

1. *The total statical moment must be correct.* This moment is $wL^2/8$ for uniformly distributed load w on a continuous beam of span length L between continuous or simple supports.

2. *Provide sufficient steel reinforcement to cover the total statical moment.* Since steel area is not strictly proportional to moment, this rule is true only in an approximate sense.

3. *Divide the total steel between positive and negative steel areas.* Using the elastic moment diagram as a guide is one option, but not the only one. Approximate distribution between negative and positive steel is permissible, within limits.

4. *Place the steel always on the tension side.* If the steel is on the wrong side of the member, it is not going to do much good, no matter how much is provided. In fact, it would hardly be better than having no steel at all.

5. *Provide for sufficient ductility to let the structure adjust the moments through plastic moment redistribution.* Ductility is achieved by keeping the reinforcing percentage low and by confining potential plastic hinge regions with closely spaced lateral reinforcement.

6. *Design the structure conservatively against shear and bond.* This way, the structure behaves exactly the way the designer wishes it to behave—in a ductile manner, without premature failure in either shear or bond.

As a final comment, we should stress that this entire mechanism by which the designer controls the behavior of a structure applies only to continuous or statically indeterminate structures with redundancy, which permits the internal redistribution of moments.

SUMMARY

1. The advantages of continuity (statical indeterminacy, redundancy) are as follows:
   a) an increase in strength;
   b) an appreciable increase in stiffness, which results in much smaller displacements, or, for the same target deflections, allows more slender members;
   c) alternative load paths, which add a new dimension to the structural safety against accidental loads;
   d) redundancy, together with ductility, allows plastic moment redistribution, thereby providing the designer with a tool to control the structure's failure behavior.
2. The disadvantages of continuity are as follows:
   a) secondary stresses of significant magnitude are possible, and are caused by temperature, creep and shrinkage, support settlements, etc.;
   b) potential moment reversals caused by movable live loads;
   c) highly stressed regions near interior supports, caused by the coincidence of high flexure and shear stresses, possibly exacerbated by high vertical stresses (due to columns) and torsional stresses (due to spandrel beams);
   d) difficulty in achieving continuity with precast concrete elements;
   e) indeterminate structures are more difficult to analyze.
3. Elastic analysis of indeterminate structures is based on relative stiffnesses of members (such as $4EI/L$ for a member fixed at the far end). Plastic analysis is based on the relative capacities of members (such as their nominal moment capacities M_n).
4. The method of moment distribution offers rapid solutions for continuous beams and frames, especially for checking purposes, as long as no sidesway deformations are present. The following steps are involved:
   Step 1. Compute rotational member stiffnesses: $k_i = 4EI/L$ for beams with two rotational degrees of freedom, $3EI/L$ for a beam with one end hinged, $2EI/L$ for a beam undergoing symmetric deformations.
   Step 2. Compute the distribution factor $k_i / \sum_j k_j$ for each joint with a rotational degree of freedom.
   Step 3. Compute fixed-end moments. For uniformly distributed load, $M_f = \pm wL^2/12$ (if one end is pinned, $wL^2/8$). For a concentrated midspan load, $M_f = \pm PL/8$ (if one end is pinned, $(3/16)PL$).

Step 4. Distribute moments according to the distribution factors, by relaxing one joint at a time such that moment equilibrium is restored.

Step 5. Continue moment distribution until out-of-balance moments are negligibly small.

Step 6. Use equations of statics to compute moments and shears for each member separately, taking into account locally applied loads.

5. A bending moment diagram is a function that shows the bending moment at an arbitrary point y as a result of a load at a fixed position x. A moment influence line is a function that shows the bending moment at a fixed point y for a load at an arbitrary position x.

6. Influence lines are constructed by use of the Müller-Breslau Principle: If an internal moment, shear, or reaction component is considered to act through some small rotation or displacement, thereby displacing the structure, the curve of the displaced structure will be, to some scale, the influence line for the moment, shear, or reaction component.

7. Moment envelopes for continuous beams subjected to movable live load (such as highway bridges) are constructed using the following step-by-step procedure:

Step 1. Select a number of sections along the entire bridge.

Step 2. For any section x, find the position in which a truck or series of trucks produces the largest positive moment, M_{max}, at x.

Step 3. For the same section x, find the position in which the truck or series of trucks produces the largest negative moment, M_{min}.

Step 4. Repeat Steps 2 and 3 for all other sections x and draw the moment envelope curves.

Step 5. Add the moments due to dead load and other loads to the minimum and to the maximum moment envelope curves.

Shear envelope curves are constructed in an exactly analogous way.

8. Provided the reinforcing ratio ρ is less than $\frac{1}{2}\rho_b$, the ACI Code allows plastic moment redistribution by reducing the negative moments up to $20[1-((\rho-\rho')/\rho_b)]$ percent, provided the positive moment is increased by such an amount that the total statical moment remains the same.

9. The overall design of continuous reinforced concrete frame structures allows the engineer considerable freedom in controlling the structural behavior as well as some latitude in the choice of an approximate analysis method, provided the following conditions are met:

   a) The total statical moment must be correct.

   b) Sufficient steel reinforcement needs to be provided to cover the total statical moment.

   c) The total steel area may be divided between positive and negative steel areas in an approximate way, within limits.

   d) The steel always must be placed on the tension side.

  e) The structural members need to have sufficient ductility to permit plastic moment redistribution.

  f) The structure needs to be designed conservatively against shear and bond to assure that it behaves as the designer intends.

REFERENCES

1. Norris, C. H., Wilbur, J. B., and Utku, S., *Elementary Structural Analysis,* 3d ed., McGraw-Hill Book Co., New York, 1976.

2. Leet, K. M., *Fundamentals of Structural Analysis,* Macmillan Publishing Co., New York, 1988.

3. West, H. H., *Analysis of Structures,* 2d ed., John Wiley and Sons, New York, 1989.

4. Paulay, T., and Priestley, M. J. N., *Seismic Design of Reinforced Concrete and Masonry Buildings,* John Wiley and Sons, New York, 1992.

5. Skettrup, E., Strabo, J., Andersen, N. H., and Brondum-Nielsen, T., "Concrete Frame Corners," *ACI Journal,* Nov.–Dec., 1984.

PROBLEMS

For all problems below, unless otherwise specified, assume normal-weight concrete weighing 150 pcf, with $f'_c = 4000$ psi; and steel with $f_y = 60,000$ psi. Concrete cover $= 2$ in.

  6.1 For each of the statically determinate structures shown in Fig. P6.1 and the loadings indicated,
  (a) draw the moment diagram;
  (b) qualitatively draw the deflected shape;
  (c) redraw the undeformed structure with finite member depths and qualitatively indicate the position of reinforcement (positive or negative) required for the specified loadings.

  6.2 For each of the statically indeterminate structures shown in Fig. P6.2 and the loadings indicated,
  (a) qualitatively draw the deflected shape and indicate the positions of inflection points;
  (b) draw the moment diagram corresponding to (a);
  (c) redraw the undeformed structure with finite member depths and qualitatively indicate the position of reinforcement (positive or negative) required for the specified loadings.

  6.3 For each of the structures in Fig. P6.3 with the indicated loadings,
  (a) qualitatively draw the deflected shape and indicate the positions of inflection points; use upper and lower bound assumptions for beam or column stiffnesses to arrive at an approximate answer;
  (b) draw the moment diagram corresponding to the deflected shape of (a);
  (c) redraw the undeformed structure with finite member depths and qualitatively indicate the position of reinforcement (positive or negative) required for the specified loadings;
  (d) compute the moment diagram using moment distribution and compare the results with the approximate results of (b).

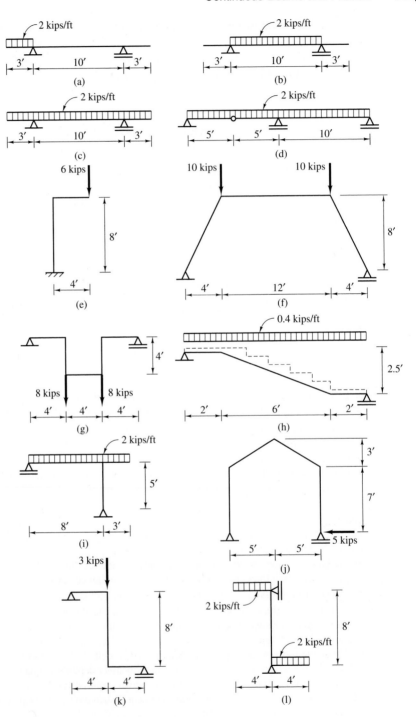

Figure P6.1

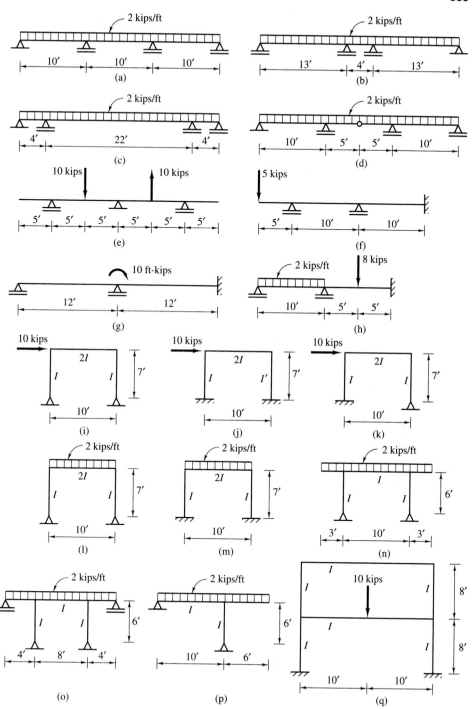

Figure P6.2

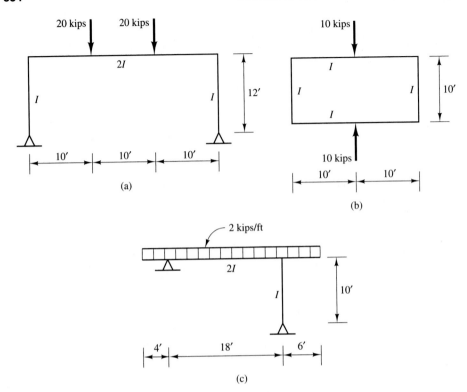

Figure P6.3

6.4 The box girder bridge shown in Fig. P6.4 is continuous over three spans and carries a live load of 500 psf in addition to its own weight.
    (a) Draw the moment and shear diagrams for a unit load $w = 1.0$ kips/ft, using either an "exact" or reasonably approximate method.
    (b) Determine the maximum negative and positive moments for the specified design loads.
    (c) Determine the amount of flexural reinforcement required in the center and side spans, both at the supports and at midspan. Select bars and show them in neatly drawn sketches of the cross section and girder elevation.
    (d) Check the flexural steel areas against the ACI Code's upper and lower limits.
    (e) Compute the shear reinforcement required for the center span.

6.5 A portal frame is fixed at both supports and carries a factored load of 2 kips/ft (Fig. P6.5).
    (a) Compute and draw the bending moment diagram.
    (b) Neatly draw the qualitative deflected shape of the frame.
    (c) Compute reinforcing steel requirements for both beam and columns (ignoring axial forces). Check the steel requirements for the Code minimum and show the steel bars in a sketch of the frame.

6.6 The portal frame shown in Fig. P6.6 shall be designed for a uniform dead load of 1500 lb/ft and a concentrated live load of 15 kips, both acting on the beam.
    (a) Compute and draw the bending moment diagram (suggested method: moment distribution).

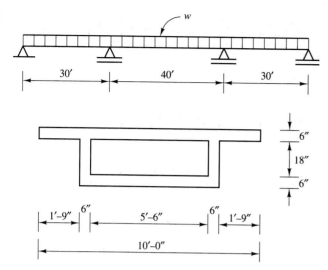

Figure P6.4

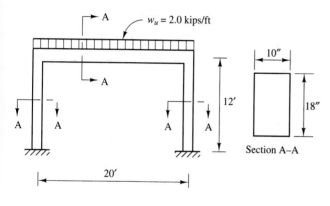

Figure P6.5

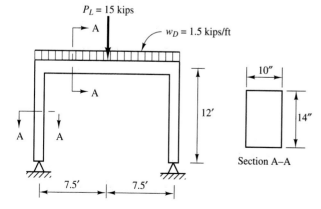

Figure P6.6

(b) Neatly draw the qualitative deflected shape of the frame.

(c) Determine the required reinforcement for both the beam and columns (ignoring axial forces) and show the selected bars in a neatly drawn sketch of the frame.

6.7 A three-span floor system (Fig. P6.7) is to be constructed of precast girders placed seven feet on center, which carry an eight inch cast-in-place slab on top as shown. The girders consist of three segments, including a "drop-in" segment, the connections of which can be considered hinges, thus rendering the system statically determinate.

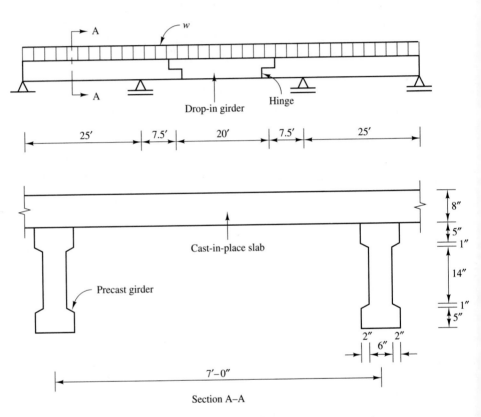

Section A–A

Figure P6.7

(a) Compute and draw the shear and moment diagrams for the three-span girder, for a uniformly distributed load $w = 1.0$ kips/ft.

(b) Determine the amount of flexural reinforcement for the drop-in girder segment to carry its own weight, plus its share of the weight of the (wet) concrete slab, plus a construction load of 10 psf (use the same overload factor as for live load), if no internal shoring of the girder is used during concreting. Satisfy the Code's minimum reinforcement requirement.

(c) Assuming live load is applied only after the concrete deck has hardened and can serve as the compression flange for the girder, determine how much live load (in psf) the same drop-in girder can carry.

(d) Describe in specific detail but without actual numerical calculations how you would determine the required size and spacing of dowels that tie the girders and the slab together.

6.8 A single-lane two-span highway ramp structure is to be designed for dead weight and standard AASHTO HS-20 truck loading (Fig. P6.8).

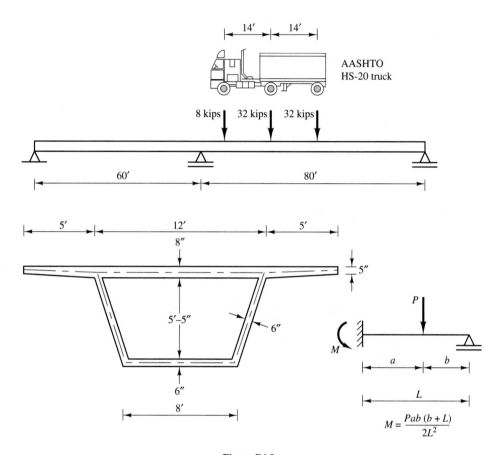

Figure P6.8

(a) Determine the dead weight of the single-cell box girder used for the bridge in kips/ft.
(b) Determine and draw the moment diagram for dead weight only. (Suggested method: moment distribution.)
(c) Draw the qualitative influence lines for the bending moment at the interior support and for the bending moment at midspan of the right span.
(d) Determine the approximate positions of one or two trucks such as to produce the maximum negative moment at the interior support and the maximum positive moment at midspan of the right span. Indicate the position of the truck(s) relative to the supports.

(e) For the truck loading of problem (d) which maximizes the positive midspan moment in the right span, compute and draw the bending moment diagram. Use moment distribution. The fixed-end moment due to a concentrated load in an arbitrary position is given in Fig. P6.8.

(f) Design the longitudinal flexural reinforcement for the midspan section of the right span for dead weight plus truck loading [from problem (e)] and impact [$I = 50/(L + 125)$].

6.9 The beam shown in Fig. P6.9 has been designed to carry a dead load of 400 lb/ft and a live load of 600 lb/ft. Check the adequacy of the design and propose any corrections that you deem necessary.

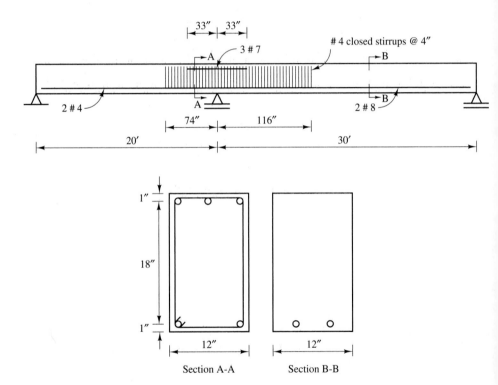

Figure P6.9

6.10 A highway overpass bridge is to be constructed without interrupting traffic (Fig. P6.10). Precast beams are positioned onto the supports to carry a cast-in-place slab plus a construction load of 20 psf, in addition to their own weight, as simply supported beams, until the slab has hardened.

(a) Determine the amount of steel for a typical precast girder required to carry its own weight, the weight of its share of the concrete slab, and the 20 psf construction live load.

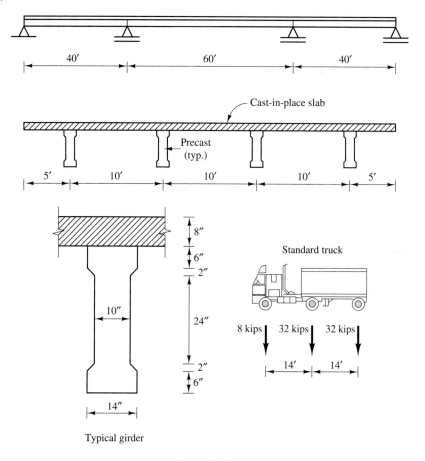

Figure P6.10

(b) After the deck has hardened, it acts compositely with the precast girders in carrying traffic load. Qualitatively draw the influence line for the continuous beam moment at the left interior support. Place one or more standard trucks into the positions where they produce approximately the largest negative moment over the interior support.

(c) For the truck loading of problem (b) compute the negative moment over the interior support. Assume that one girder will carry one full traffic lane. You may use an approximate method of analysis, but the results must satisfy statics.

(d) Determine the amount of negative reinforcement necessary to carry the live load moment of problem (c).

6.11 A 6-in-thick roof slab is supported by frames spaced 15 ft apart. Each frame consists of a 36-by-15-in beam with cantilever overhangs, and two 15-by-15-in columns, as shown in Fig. P6.11. The controlling design load shall be dead weight plus 300 psf snow load (live load).

(a) Draw the shear, moment, and axial force diagrams for the frame due to factored loads. You may use an approximate method of analysis, but the results must satisfy statics.

(b) Neatly draw the qualitative deflected shape of the frame and indicate the positions of inflection points.

(c) Determine the flexural reinforcement required for the beam and satisfy the ACI Code's upper and lower limitations on reinforcing steel.

(d) Determine the cutoff points of the negative reinforcement.

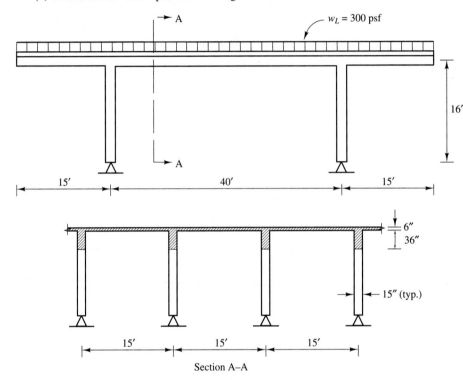

Section A–A

Figure P6.11

6.12 A 5-in floor slab is supported by integral beams spaced 6 ft apart, which are continuous over two spans and fixed at one end (Fig. P6.12). In addition to its own weight, the floor system must support a live load of 50 psf and a transverse wall at midspan of the left span. The loads due to the wall are: 1200 lb/ft dead load and 800 lb/ft live load.

(a) Indicate in a sketch the factored distributed load in kips/ft and the factored concentrated load in kips to be carried by a typical beam.

(b) What are the combined factored static moments for the two spans?

(c) Compute the bending moments for the loads of problem (a). Draw the moment diagram, including numerical values for maximum positive and negative moments. (Suggested method: moment distribution.)

(d) Determine the required amount of flexural reinforcement for the floor system.

(e) Show the reinforcement in a sketch of the floor beam, including approximate cutoff points.

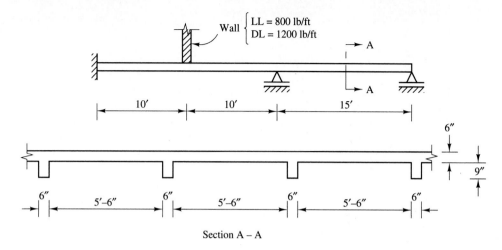

Section A – A

Figure P6.12

Design of Columns

7.1 INTRODUCTION

Columns are structural members designed to carry axial loads. In practice, a column is hardly ever subjected to only an axial load, because such a load, say P, is most likely applied with some eccentricity e with regard to the column centroid, which causes a bending moment, $M = Pe$ (Fig. 7.1). Columns are not the only members subjected to axial compression. Also, beams may carry axial forces in addition to bending moments. Certain truss members, arch ribs, etc., are further examples of the more general category of *compression members*. For simplicity (and reasons of convention) we shall refer

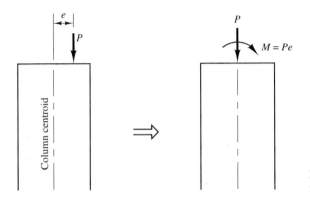

Figure 7.1 Column subjected to axial force and bending.

to these as columns or *beam-columns*. In general, then, a beam-column is a member subjected to both an axial load P and a moment M, or an axial force P applied at an eccentricity $e = M/P$.

If the eccentricity e were exactly zero, the member would experience no bending, let alone tensile stresses, so in theory no reinforcing steel would be needed. In practice, however, we always reinforce columns with a minimum amount of longitudinal steel, for a number of reasons. First, it is a precaution against accidental eccentricities or unforeseen horizontal loads, which cause bending. Next, steel reinforcement gives the column some ductility. Finally, because concrete creeps under load, steel reinforcement, which does not, attracts a disproportionate amount of that load, thereby effectively reducing the amount of creep deformations. Sometimes, in particular for highly stressed members, the steel reinforcement is provided in the form of structural shapes (Fig. 7.2). Such columns are known as *composite* compression members.

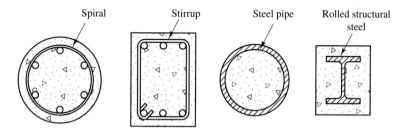

Figure 7.2 Typical column sections.

When subjected to high axial loads, reinforcing bars tend to buckle. This can be avoided by restraining them by lateral ties spaced at a certain distance or by a continuous spiral (Fig. 7.3). One of the reasons why the concrete in a *spiral reinforced* column is much better confined than in a *tied column* is the pitch of the spiral, which is usually smaller than the spacing of the individual ties. The result is an enhancement of the concrete's apparent compressive strength and a significant increase in its ductility. This can be explained as follows.

Consider a concrete column encased in a steel pipe (Fig. 7.4a). Due to Poisson's ratio, the concrete tends to expand laterally if subjected to a vertical compression load. Because of the lateral restraint offered by the steel pipe against such lateral expansion, the concrete experiences a state of triaxial compression, under which its strength increases considerably, and its deformability (i.e., ductility) is increased by an order of magnitude, as we have seen in Chapter 2. In fact, such concrete-filled tubes are enormously strong, ductile, and tough. Punching holes into the pipe will weaken the pipe itself, but will not have much effect on the properties of the confined concrete unless the holes are very large. If a continuous spiral replaces the tube, much of the confinement effect is retained because the concrete will form little arches to bridge the space between adjacent steel hoops, which effectively maintain the state of triaxial compression (Fig. 7.4b). A column thus confined will typically fail only after the spiral yields in tension, if the

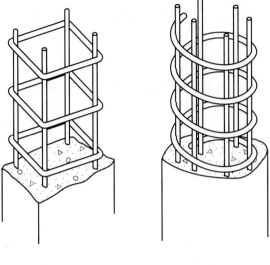

Figure 7.3   Tied and spiral reinforced column.

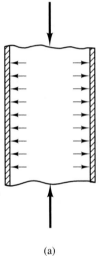

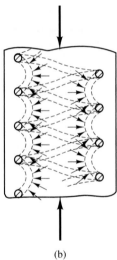

(a)   (b)

Figure 7.4   Pipe and spiral confined concrete.

lateral pressure is high enough. Once the spiral yields, and possibly before, the concrete not confined by the spiral, that is, the outer cylindrical shell, will spall off, as it is not tied into the main body of the column.

A tied column fails when both the concrete and the longitudinal reinforcement reach their capacities, that is,

$$P_n = 0.85 f'_c A_c + A_{st} f_y \qquad (7.1)$$

where A_c is the net concrete area and A_{st} is the total longitudinal steel area. At or near that load, the reinforcing bars usually buckle between ties, while the concrete is crushed in compression (Fig. 7.5a). This type of failure is rather sudden. A spiral reinforced column fails much more gradually. Near the load at which a tied column would fail, a spiral reinforced column only loses the outside shell. After further load increase, the spiral will yield, and at some point all longitudinal bars will buckle laterally as a unit (Fig. 7.5b). The failure load is not necessarily much higher than that for a tied column (Fig. 7.5c), but the large strains associated with the yielding of the spiral provide the ductility and toughness that are so important in modern design philosophy.

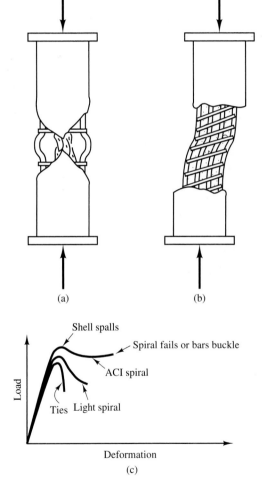

(a)   (b)

Shell spalls

Spiral fails or bars buckle

ACI spiral

Load

Ties   Light spiral

Deformation

(c)

Figure 7.5   Failure modes of tied and spiral reinforced columns.

The failure load and ductility are strongly dependent on the amount of steel in the spiral. A column with a very light spiral is, in essence, not much different than a tied column. On the other hand, an overly heavy spiral offers little added benefit and is thus a waste of steel. Not only is a column's usefulness greatly impaired after the loss of the outside shell (requiring costly repairs), but the column is likely to buckle in the mode shown in Fig. 7.5b, before the spiral is fully stressed. Therefore, the goal is to provide just the right amount of steel in the spiral to assure a ductile failure mode. Less steel than that amount implies a lower ductility. Any excess of steel beyond that amount barely increases the column strength and is therefore wasted.

In order to determine the proper amount of steel, we shall stipulate that the strength of the column after the loss of the outer shell but with the activated spiral shall not be less than the strength of the column with an intact outer shell. This original strength was given in Eq. 7.1, rewritten as

$$P_1 = 0.85 f_c'(A_g - A_{st}) + A_{st} f_y \tag{7.2}$$

where A_g is the gross column area. The strength of the column after loss of the outer shell can be written as

$$P_2 = 0.85 f_c^*(A_c - A_{st}) + A_{st} f_y \tag{7.3}$$

where A_c is the concrete core area, measured to the outside diameter of the spiral, and f_c^* is the strength of the confined concrete, which benefits from the triaxial compression generated by the spiral. Equating Eqs. 7.2 and 7.3 and recognizing that the steel area is small compared with the concrete area, we get

$$f_c^* = f_c' \frac{A_g}{A_c} \tag{7.4}$$

that is, the spiral should be heavy enough to increase the concrete's strength by the factor A_g/A_c to make up for the loss of the outer shell.

Whereas the conventional definition of ρ is a ratio of steel to concrete area, the spiral steel is specified as a volumetric ratio ρ_s, that is, as the volume of the spiral itself divided by the enclosed concrete core volume (Fig. 7.6),

$$\rho_s = \frac{V_{\text{spiral}}}{V_{\text{core}}} = \frac{A_{sp}\pi D_c}{s \frac{\pi D_c^2}{4}} = \frac{4 A_{sp}}{s D_c} \tag{7.5}$$

where A_{sp} is the cross-sectional area of the spiral, s the spiral pitch, and D_c the concrete core diameter, measured to the outside diameter of the spiral. Horizontal equilibrium on a slice of the concrete core of height s requires

$$2 A_{sp} f_y = f_h D_c s \tag{7.6}$$

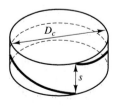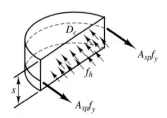

Figure 7.6 Horizontal equilibrium of column slice.

Solving for f_h, the horizontal pressure, and substituting the definition of ρ_s (Eq. 7.5), gives

$$f_h = \frac{2A_{sp}f_y}{sD_c} = \frac{\rho_s f_y}{2} \tag{7.7}$$

In Chapter 2 we saw that the strength enhancement of concrete under triaxial compression can be expressed approximately as

$$f_c^* = f_c' + 4.1 f_h \tag{7.8}$$

Substituting Eq. 7.7 into Eq. 7.8 and equating it to Eq. 7.4, we can solve the resulting expression for the required amount of spiral steel,

$$\rho_s = \frac{1}{2.05}\left(\frac{A_g}{A_c} - 1\right)\frac{f_c'}{f_y} \tag{7.9}$$

In the ACI Code, this requirement is rounded to give

$$\rho_s = 0.45\left(\frac{A_g}{A_c} - 1\right)\frac{f_c'}{f_y} \tag{7.10}$$

Spiral reinforced columns are somewhat more difficult to build and contain more steel per unit volume than tied columns. Their slightly higher cost should be weighed against the preferable behavior under overload and at failure, which is recognized explicitly by the ACI Code, as we shall see later. In design practice, however, architectural considerations often take priority.

In the next few sections we will consider *short* columns. These are members that fail in any one of the compression modes just mentioned, rather than buckle, as would be expected of *long* or *slender* compression members. The additional aspects to be considered for the design of such slender columns will conclude this chapter.

7.2 SERVICE LOAD BEHAVIOR

A column or compression member subjected to only a moderate bending moment is not likely to be cracked. To show this, we compute the concrete stress at a point some distance y away from the neutral axis by simply superimposing the direct stress P/A and the bending stress due to the moment $M = Pe$:

$$f_c = \frac{P}{A} \pm \frac{My}{I} = \frac{P}{A}\left(1 \pm \frac{eAy}{I}\right) \tag{7.11}$$

where the cross-sectional area A and moment of inertia I, to be "exact," should be computed for the transformed section. Steel stresses at the point y follow from strain compatibility as

$$f_s = nf_c \tag{7.12}$$

where $n = E_s/E_c$ is the modular ratio. It is an easy matter to determine the maximum eccentricity e (or moment M) at which a load P causes only compression stresses on a section. We simply set the stress on the tension side equal to zero, that is, we select the minus sign in Eq. 7.11 and set $f_c = 0$:

$$f_c = \frac{P}{A}\left(1 - \frac{eAy}{I}\right) = 0$$

Since the maximum stress occurs at the extreme fiber, that is, at $y = c$, the maximum eccentricity becomes

$$e = \frac{I}{Ac} \equiv k \qquad (7.13)$$

This is the definition of the *kern eccentricity* or *kern distance* (Fig. 7.7). As long as $e \leq k$, the entire section is free of tensile stresses and therefore uncracked. If the axial force

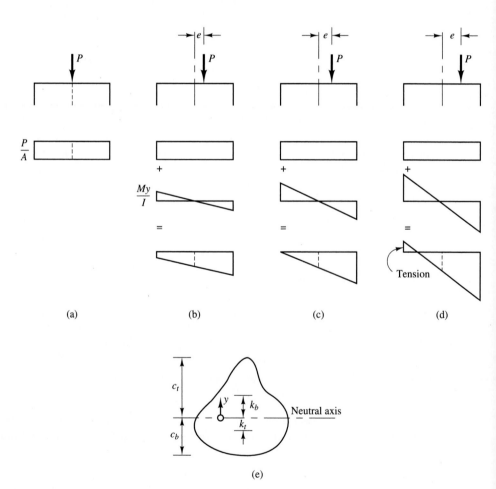

Figure 7.7 Definition of kern distances: (a) $e = 0$; (b) $e < k$; (c) $e = k$; (d) $e > k$; (e) kern distances.

is applied outside the kern area, that is, $e > k$, tensile stresses appear. These will crack the concrete if they exceed the modulus of rupture f_r.

In a nonsymmetrical section, the top and bottom extreme fibers are of unequal distance from the neutral axis—c_t and c_b, respectively—and therefore such a section has two different kern distances (Fig. 7.7e):

$$k_t = \frac{I}{Ac_b} \qquad k_b = \frac{I}{Ac_t} \tag{7.14}$$

For a rectangular section, with $A = bh$, $I = bh^3/12$, and $c_t = c_b = h/2$, the kern distances are $k_t = k_b = h/6$, that is, the kern area has a total depth of $h/3$. If the same section were subjected to bending about the vertical axis, the kern distance would be $k = b/6$. As long as stresses due to biaxial bending follow linear superposition, the kern area boundaries must be straight lines, that is, the kern area has the diamond shape shown in Fig. 7.8a. The kern areas of some other sections are also shown in Fig. 7.8. To restate: an axial load causes only compression stresses in the column, as long as it is applied inside the (shaded) kern area.

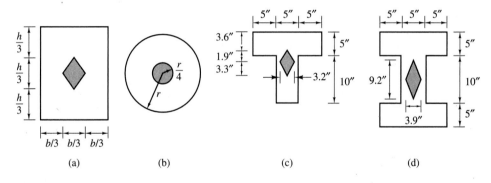

Figure 7.8 Typical kern areas.

EXAMPLE 7.1

For a column with the section shown in Fig. 7.9, determine whether a 100-kip force and a 500–in-kips moment about the y-axis will cause the concrete to crack. Given: $n = 8$; $f_r = 300$ psi.

Solution

Transformed section area:

$$A_t = A_g + (n - 1)A_{st} = (12)(20) + (8 - 1)(4)(1.0) = 268 \text{ in}^2$$

Transformed moment of inertia of the uncracked section:

$$I_t = \frac{bh^3}{12} + (n - 1)A_{st}y_s^2 = \frac{(12)(20)^3}{12} + (7)(4)(7.5)^2 = 9575 \text{ in}^4$$

Load eccentricity: $e = \dfrac{M}{P} = \dfrac{500}{100} = 5 \text{ in}$

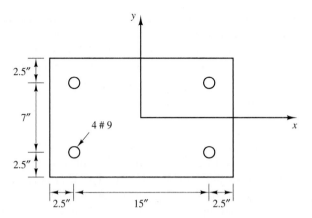

Figure 7.9 Example 7.1: Column section.

Maximum tensile stress:

$$f_c = \frac{P}{A_t}\left(1 - \frac{eA_tc}{I_t}\right) = -\frac{100}{268}\left(1 - \frac{(5)(268)(10)}{9575}\right) = .149 \text{ ksi} < 0.3 \text{ ksi}$$

Thus, the section is not likely to be cracked.

Alternatively, we can determine the kern distance,

$$k = \frac{I_t}{A_t c} = \frac{9575}{(268)(10)} = 3.572 \text{ in} < e = 5$$

that is, the load causes a maximum tensile stress of

$$f_c = \frac{P}{A_t}\left(1 - \frac{e}{k}\right) = -\frac{100}{268}\left(1 - \frac{5}{3.572}\right) = .149 \text{ ksi} \qquad \blacksquare$$

When the tensile stresses exceed the concrete modulus of rupture, the determination of steel and concrete stresses is more complicated because properties of the cracked section must be computed, similar to what was done for beams subjected to service load bending moments. Figure 7.10 shows a column section with steel area A_s on the tension side and steel area A'_s on the compression side, subjected to a load P with an eccentricity e sufficiently large to crack the section. Also shown are the strain and stress variations, as well as the internal force resultants. To locate the neutral axis, we take moments about the point of external load application:

$$bkd\frac{f_c}{2}\left(e - c_t + \frac{kd}{3}\right) + A'_s f'_s(e - c_t + d') - A_s f_s(e - c_t + d) = 0 \qquad (7.15)$$

This equation contains four unknowns, namely, the neutral axis position kd, the maximum concrete stress f_c, and the steel stresses f_s and f'_s. From similar strain (or stress) triangles, we can express the steel stresses in terms of the concrete stress by

$$f_s = \frac{d - kd}{kd}nf_c \qquad (7.16)$$

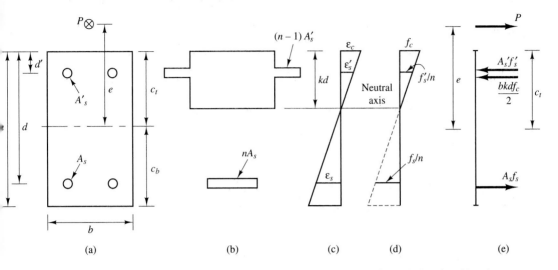

Figure 7.10 Cracked column section under service load: (a) section; (b) cracked section; (c) strains; (d) stresses; (e) internal forces.

and

$$f_s' = \frac{kd - d'}{kd} n f_c \tag{7.17}$$

Substitution of these expressions into Eq. 7.15 leads to an equation with only one remaining unknown, kd:

$$(kd)^3 \frac{b}{6} + (kd)^2 \frac{b}{2}(e - c_t) + kdn[(A_s + A_s')(e - c_t) + d'A_s' + dA_s]$$

$$- n(A_s d + A_s' d')(e - c_t) - nA_s' d'^2 - nA_s d^2 = 0 \tag{7.18}$$

Since this is a cubic equation for kd, some numerical method should be used to solve it.

To compute stresses, we utilize the requirement of vertical equilibrium of internal forces, that is,

$$P - bkd \frac{f_c}{2} - A_s' f_s' + A_s f_s = 0 \tag{7.19}$$

Substitution of Eqs. 7.16 and 7.17 into Eq. 7.19 yields the maximum concrete stress

$$f_c = \frac{P}{nA_s' \left(1 - \frac{d'}{kd}\right) - nA_s \left(\frac{d}{kd} - 1\right) + \frac{bkd}{2}} \tag{7.20}$$

and with f_c known, the steel stresses follow from Eqs. 7.16 and 7.17.

EXAMPLE 7.2

Determine the stresses in the section of Example 7.1, if the 100-kip load is applied with an eccentricity of 8 in.

Solution   Substituting $h = 20$, $b = 12$, $e = 8$, $c_t = 10$, $n = 8$, $A_s = A_s' = 2.0$, $d' = 2.5$, and $d = 17.5$, into Eq. 7.18 leads to

$$2(kd)^3 - 12(kd)^2 + 256kd - 4360 = 0$$

Rewritten in the form

$$kd = \left[\frac{4360 + 12(kd)^2 - 256kd}{2}\right]^{1/3} = [2180 + 6(kd)^2 - 128kd]^{1/3}$$

this equation is readily solved by iteration. Starting with $kd = 8$, the solution converges after two iterations to $kd = 11.45$. Substituting this value into Eqs. 7.20, 7.16, and 7.17 for the required stresses yields

$$f_c = \frac{100}{(8)(2)\left[2 - \frac{2.5}{11.45} - \frac{17.5}{11.45}\right] + \frac{(12)(11.45)}{2}} = 1.375 \text{ ksi}$$

$$f_s = \frac{17.5 - 11.45}{11.45}(8)(1.375) = 5.8 \text{ ksi}$$

$$f_s' = \frac{11.45 - 2.5}{11.45}(8)(1.375) = 8.6 \text{ ksi}$$

The stress in the concrete surrounding the tensile steel is $5.8/n = 5.8/8 = .725$ ksi. As this is well in excess of the assumed modulus of rupture, $f_r = 0.3$ ksi, the original assumption of a cracked section was correct.   ∎

This example illustrates that the determination of elastic column stresses is far from simple, as it involves the lengthy solution of a cubic equation. Fortunately, service load behavior is not usually a design concern. The overriding issue is a column's ultimate strength and safety against failure, so the advantage of ultimate over allowable stress design is even more obvious here than in the case of beams.

7.3 ULTIMATE STRENGTH AND INTERACTION DIAGRAM

For a column subjected to a perfectly concentric load, the failure load was given by Eq. 7.1 as

$$P_n = 0.85 f_c' A_c + A_{st} f_y$$

If a load is applied with an eccentricity e, the column must fail at a lower load, because the compressive stress due to the added bending moment causes the concrete to crush sooner. We shall now determine the failure load (or nominal load capacity) P_n for a given section if the load is applied at a given eccentricity e. This problem can also be restated: for a given axial load P_n, find the eccentricity e (or moment $P_n e$) that will cause the section to fail.

The section is shown in Fig. 7.11, together with strains, stresses, and resultant internal forces at the time of failure. We shall assume that the steel on the tension side is stressed below the yield point—an assumption easily verified or corrected if found to be wrong. For the steel on the compression side, on the other hand, it is safe to assume

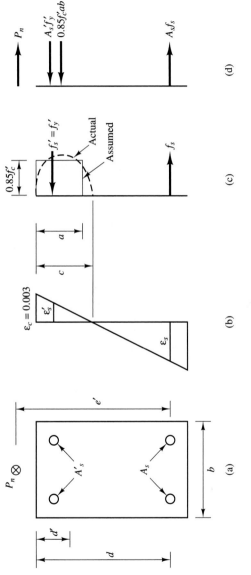

Figure 7.11 Column section at failure: (a) section; (b) strains; (c) stresses; (d) internal forces.

that it yields at failure. Even if a strain compatibility check of the kind we made for compression steel in doubly reinforced beams (Sec. 4.4) should indicate a stress below yield, concrete creep due to dead load alone tends to increase the stress in the steel with time, so it is likely to be considerably higher than indicated by linear elastic theory. If in doubt, we can always check the stress in the compression steel and correct the equations below as needed.

Vertical equilibrium of all forces in Fig. 7.11d requires

$$P_n = 0.85 f'_c ab + A'_s f'_y - A_s f_s \tag{7.21}$$

while moment equilibrium about the tension steel leads to

$$P_n e' = 0.85 f'_c ab \left(d - \frac{a}{2} \right) + A'_s f'_y (d - d') \tag{7.22}$$

where e' is the load eccentricity with respect to the tension steel, that is, $e' = e + (d - d')/2$. Equations 7.21 and 7.22 are two equations with three unknowns: failure load P_n, depth of concrete compression block a, and tension steel stress f_s. Compatibility of strains will have to furnish the remaining equation. Similar strain triangles in Fig. 7.11b place the neutral axis at

$$c = \frac{.003}{.003 + \epsilon_s} d = \frac{.003}{.003 + \frac{f_s}{E_s}} d = \frac{87,000}{87,000 + f_s} d$$

and the depth of the concrete compression block follows as

$$a = \beta_1 c = \frac{87,000}{87,000 + f_s} \beta_1 d \tag{7.23}$$

where $\beta_1 = .85$ for $f'_c \le 4000$ psi (see Eq. 3.41). Equation 7.23 can be solved for f_s:

$$f_s = \frac{\beta_1 d - a}{a} 87,000 \tag{7.24}$$

Of course, this value cannot exceed f_y. Substitution of Eq. 7.24 into Eqs. 7.21 and 7.22 results in two equations for the two unknowns, P_n and a. Elimination of P_n leads to a single equation for a—again a cubic equation. Since iteration is needed to solve such an equation anyway, we might as well rearrange the entire solution process in the following order:

Step 1. Assume a value for a. This choice should be inversely proportional to the eccentricity ratio e/h. For starters, $a = h/2$ may be a good value.

Step 2. Compute the corresponding stress in the tension steel using Eq. 7.24:

$$f_s = \frac{\beta_1 d - a}{a} 87,000 \le f_y$$

Step 3. Compute the nominal load capacity from Eq. 7.22:

$$P_n = \left[0.85 f'_c ab \left(d - \frac{a}{2} \right) + A'_s f'_y (d - d') \right] / e'$$

Step 4.   Use Eq. 7.21 to establish a better estimate for *a*:

$$a = \frac{P_n - A'_s f'_y + A_s f_s}{.85 f'_c b}$$

Step 5.   Return to Step 2 with the improved value for *a* and continue the iteration until convergence.

EXAMPLE 7.3

Compute the nominal load capacity of the column of Example 7.1 for an eccentricity of 6 in. Given: $f_y = f'_y = 60,000$ psi; $f'_c = 4000$ psi.

Solution

Step 1.   Assume $a = 8$ in.

Step 2.   Tension steel stress:

$$f_s = \frac{(.85)(17.5) - 8}{8}(87) = 74.8 \text{ ksi} > f_y, \text{ that is, } f_s = 60 \text{ ksi.}$$

Step 3.   Nominal load capacity, with $e' = 6 + \left(\dfrac{17.5 - 2.5}{2}\right) = 13.5$ in:

$$P_n = \left[(.85)(4)(8)(12)\left(17.5 - \frac{8}{2}\right) + (2)(60)(17.5 - 2.5)\right]/13.5 = 459.7 \text{ kips.}$$

Step 4.   Improved concrete compression block depth:

$$a = \frac{459.7 - (2)(60) + (2)(60)}{(.85)(4)(12)} = 11.27 \text{ in.}$$

Step 5.   Second iteration: $f_s = 27.8$ ksi, $P_n = 537.5$ kips, $a = 11.59$ in. Third iteration: $f_s = 24.6$ ksi, $P_n = 543.4$ kips, $a = 11.58 \approx 11.59$ in.

Answer: $\underline{P_n = 543 \text{ kips.}}$ ■

Before determining the failure load on the premise that the tensile steel is stressed to the yield point, let us consider the case where the tensile steel reaches the yield point at exactly the instant when the extreme concrete fiber reaches the crushing strain. As in beam design, this case is referred to as the *balanced condition*. With $f_s = f_y$, Eq. 7.23 becomes

$$a = \frac{87,000}{87,000 + f_y}\beta_1 d \tag{7.25}$$

Of practical importance is the symmetrically reinforced column, with $A'_s = A_s$ and $f'_y = f_y$. The nominal load capacity at balanced conditions, referred to as *balanced load* P_b, is then found by substituting Eq. 7.25 into 7.21:

$$P_b = .85 f'_c ab = .85 \beta_1 bd f'_c \frac{87,000}{87,000 + f_y} \tag{7.26}$$

The eccentricity associated with P_b follows from Eq. 7.22 as

$$e_b = \frac{1}{P_b}\left[.85 f'_c ab \left(d - \frac{a}{2}\right) + A'_s f_y(d - d')\right] - \frac{d - d'}{2} \tag{7.27}$$

EXAMPLE 7.4

For the column of Example 7.1, compute the load P_b and its eccentricity e_b at balanced strain conditions.

Solution

Depth of concrete compression block: $a = \dfrac{87}{87 + 60}(.85)(17.5) = 8.80$ in

Balanced load: $P_b = (.85)(4)(8.80)(12) = \underline{359 \text{ kips}}$

Balanced eccentricity:

$$e_b = \frac{1}{359}\left[(.85)(4)(8.80)(12)\left(17.5 - \frac{8.80}{2}\right) + (2)(60)(17.5 - 2.5)\right] - 7.5 = \underline{10.6 \text{ in}}$$

∎

The discussion up to this point clearly demonstrates that the nominal column capacity cannot be characterized by a single value, as opposed to the nominal moment capacity of a beam. The axial force capacity P_n is dependent on the bending moment M_n or the eccentricity e, and conversely, the moment capacity M_n is a function of P_n. This relationship is best illustrated with an *interaction diagram* or *failure curve* (Fig. 7.12), in which the ordinate represents the nominal load capacity P_n and the abscissa the nominal moment capacity M_n. For a given load P_n, a typical point S on this curve indicates the moment M_n at which the column will fail, or the force P_n that will cause the column to fail if applied together with the moment M_n, or eccentricity $e = M_n/P_n$. Note that the slope of the radial line from the origin to point S is given by $1/e = P_n/M_n$.

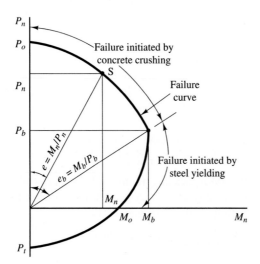

Figure 7.12 Column interaction diagram.

Four characteristic points in the interaction diagram require further comment. For $M_n = 0$, the curve intersects the P_n-axis at the value $P_n = P_o$ for compression and

$P_n = P_t$ in tension. These points represent the ideal cases of a column loaded exactly at its centroid, with the failure in compression load given by Eq. 7.1, and in tension by $P_n = A_{st} f_y$. At the other extreme, we have $P_n = 0$ and $M_n = M_o$, the case of a column loaded only with a bending moment. We call this a beam (rather than a column) problem and have discussed it in Sec. 4.4 on doubly-reinforced beams, where the nominal moment capacity of such a beam has been found to be

$$M_o = \left(A_s f_y - A_s' f_s'\right)\left(d - \frac{a}{2}\right) + A_s' f_s'(d - d')$$ (7.28)

The fourth characteristic point in Fig. 7.12 is associated with balanced strain conditions and clearly separates the failure curve into two distinct branches. For $P_n > P_b$, failure of the column is initiated by crushing of the concrete. In this domain, the nominal capacity P_n decreases with increasing moment, because concrete will reach the crushing strain sooner. If $P_n < P_b$, failure is initiated by the yielding of the tension steel. In this case, the moment capacity increases with an increase of the axial load, because the compression stresses due to the axial load reduce the tensile stresses in the steel due to bending. By delaying the point of yielding and failure, they create additional bending capacity. This beneficial effect of the axial load must be considered in design: When determining the most critical live load case for a building, it is possible that partial live load, with its smaller axial loads combined with the associated bending moments, is more critical than full live load axial forces with the corresponding bending moments.

Returning to the special case $e = 0$, we should note that at ultimate conditions, e should not be measured from the usual elastic centroid of the section, but rather from the *plastic centroid*. By definition, the plastic centroid is positioned such that at failure all stresses or internal forces acting on a section be in moment equilibrium and all concrete stresses be uniform (Fig. 7.13):

$$\bar{y}_p = \frac{\sum_i A_i f_i y_i}{\sum_i A_i f_i}$$ (7.29)

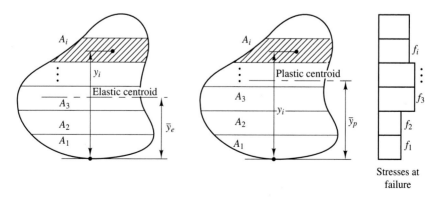

Figure 7.13 Elastic and plastic centroid.

where A_i is the ith partial area, f_i its uniform stress at failure, and y_i the distance of its centroid from a common reference point. Note that for a homogeneous section, f_i is the same for each partial area and therefore cancels out in Eq. 7.29, so that in this case the plastic centroid coincides with the elastic centroid:

$$\bar{y}_e = \frac{\sum_i A_i y_i}{\sum_i A_i} \tag{7.30}$$

EXAMPLE 7.5

For the section shown in Fig. 7.14, compute the position at which an axial force P_n causes no bending. Given: $f_y = 60$ ksi; $f'_c = 3000$ psi; $n = 9$.

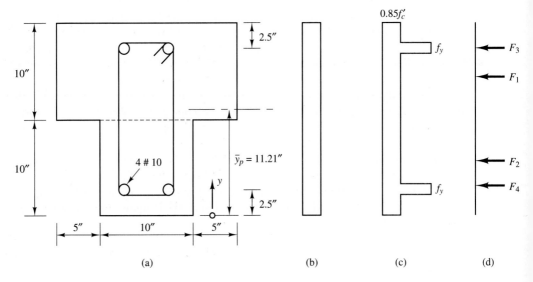

Figure 7.14 Plastic centroid for Example 7.5: (a) section; (b) strains at failure; (c) stresses at failure; (d) internal forces.

Solution Dividing the concrete section into the two partial areas shown in Fig. 7.14a, we compute four internal forces:

$$F_1 = .85 f'_c A_{c1} = (.85)(3)(10)(20) = 510 \text{ kips}$$

$$F_2 = .85 f'_c A_{c2} = (.85)(3)(10)(10) = 255 \text{ kips}$$

$$F_3 = F_4 = A_s \left(f_y - .85 f'_c \right) = (2)(1.27)(60 - .85 \times 3) = 146 \text{ kips}$$

To locate the plastic centroid according to Eq. 7.29, we take moments about the bottom edge of the section:

$$\bar{y}_p = \frac{(510)(15) + (255)(5) + (146)(2.5 + 17.5)}{510 + 255 + (2)(146)} = \underline{11.21 \text{ in}}$$

When designing for ultimate loads, eccentricities of axial loads should be measured with regard to this plastic centroid and not the elastic centroid, which is located for the transformed section at

$$\bar{y}_e = \frac{(10)(20)(15) + (10)(10)(5) + (8)(4)(1.27)(10)}{200 + 100 + (8)(4)(1.27)} = \underline{11.47 \text{ in}}$$ ∎

As this example shows, the distance between the elastic and plastic centroid is often so small that it can be ignored. For symmetrical sections it is always zero.

7.4 EFFECT OF HIGH-STRENGTH MATERIALS AND INTERMEDIATE BARS

A glance at the interaction diagram of Fig. 7.12 shows that large or small moments and axial forces must be understood in relative terms. A large moment implies either that the axial force is relatively small or that the eccentricity is large relative to the column size.

For columns subjected to small bending moments, the position of reinforcing bars has little effect on the ultimate capacity. In such cases, it is appropriate to arrange the bars uniformly along the column perimeter. For larger moments it is more advantageous to place the bars for maximum internal lever arm near the faces parallel to the axis of bending (Fig. 7.15). If, in the latter case, intermediate bars are used in between, they are not likely to yield at failure because of their closeness to the neutral axis. For the same reason, it would not be economical to employ high-strength steel for columns subjected to large bending moments. For small moments the opposite is true. For example, in tall buildings, axial loads increase almost linearly, from roof to basement, while column bending moments increase very little. Therefore, the relative moments are small in the lower stories of such buildings, making the use of high-strength materials more economical.

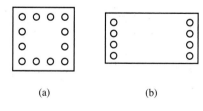

(a)   (b)

Figure 7.15   Reinforcing bar arrangements: (a) relatively small moment; (b) relatively large moment.

EXAMPLE 7.6

The column shown in Fig. 7.16a is loaded with an eccentricity of $e = 9$ in. Determine its nominal load and moment capacity and the effectiveness of the intermediate bars in positions 2 and 3. Given: $f'_c = 6$ ksi; $f_y = 75$ ksi.

Solution   We shall generalize the step-by-step procedure outlined in Sec. 7.3 by including the intermediate bars.

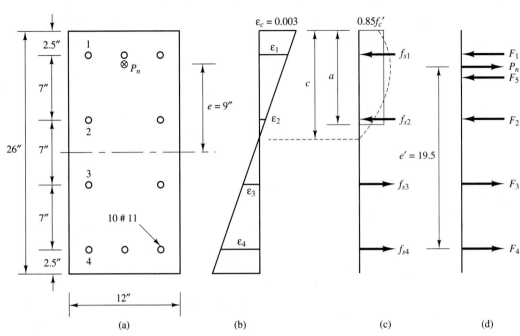

Figure 7.16 Example 7.6: (a) section; (b) strains; (c) stresses; (d) internal forces.

Step 1. Assume a value for a, say $a = 12$ in.

Step 2. With the neutral axis position, $c = a/\beta_1 = 12/.75 = 16$ in, similar strain triangles of Fig. 7.16b lead to the internal forces

$$\epsilon_1 = \frac{c - d'}{c}(.003) = \frac{16 - 2.5}{16}(.003) = .00253$$

$$f_{s1} = \epsilon_1 E = (.00253)(29{,}000) = 73.4 \text{ ksi}$$

$$F_1 = f_{s1} A_{s1} = (73.4)(3)(1.56) = 343.5 \text{ kips}$$

and, similarly,

$$f_{s2} = \frac{16 - 9.5}{16}(87) = 35.3, \quad F_2 = (35.3)(2)(1.56) = 110.3 \text{ kips}$$

$$f_{s3} = \frac{16 - 16.5}{16}(87) = -2.7, \quad F_3 = (-2.7)(2)(1.56) = -8.5 \text{ kips}$$

$$f_{s4} = \frac{16 - 23.5}{16}(87) = -40.8, \quad F_4 = (-40.8)(3)(1.56) = -190.9 \text{ kips}$$

$$F_5 = .85 f'_c ab = (.85)(6)(12)(12) = 734.4 \text{ kips}$$

Step 3. Determine the nominal failure load from the moment equilibrium condition about F_4, that is,

$$P_n(9 + 10.5) = F_1(21) + F_2(14) + F_3(7) + F_5\left(23.5 - \frac{a}{2}\right)$$

from which

$$P_n = [(343.5)(21) + (110.3)(14) - (8.5)(7) + (734.4)(23.5 - 6)]/19.5 = 1105 \text{ kips}$$

Step 4. Compute a new value for a from the condition of vertical equilibrium:
$P_n = F_1 + F_2 + F_3 + F_4 + .85 f'_c ab$ that is,

$$a = \frac{1105 - 343.5 - 110.3 + 8.5 + 190.9}{(.85)(6)(12)} = 13.90 \text{ in.}$$

Step 5. Second iteration, with $c = 13.90/.75 = 18.53$: $F_1 = 351.0$, $F_2 = 132.3$, $F_3 = 29.7$, $F_4 = -109.2$, $F_5 = 850.7$, $P_n = 1206$, $a = 13.10$.

Third iteration, with $c = 17.47$: $F_1 = 348.9$, $F_2 = 123.8$, $F_3 = 15.1$, $F_4 = -140.5$, $F_5 = 801.7$, $P_n = 1167$, $a = 13.39 \approx 13.10$ in.

Answer: $P_n = \underline{1167 \text{ kips.}}$

Repeating the iteration without the intermediate bars in positions 2 and 3, we get:

$a = 12$, $c = 16$, $f_{s1} = 73.4$, $F_1 = 343.5$, $F_4 = -190.9$, $F_5 = 734.4$, as before, but now,

$P_n = [(343.5)(21) + (734.4)(23.5 - 6)]/19.5 = 1029$,

$$a = \frac{1029 - 343.5 + 190.9}{(.85)(6)(12)} = 14.32.$$

Try $a = 14$, $c = 18.67$, $f_{s1} = 75.0$, $F_1 = 351.0$, $F_4 = -105.4$, $F_5 = 856.8$, $P_n = 1103$, $a = 14.01 \approx 14.0$.

Answer: $P_n = \underline{1103 \text{ kips.}}$

Thus, the elimination of 4 out of 10 bars (or 40%) reduces the column's load-carrying capacity by just 5%. According to our definition, the moment in this case would be relatively large. ∎

7.5 BIAXIAL BENDING

Columns are often subjected to bending about two orthogonal axes, in addition to an axial force. This is the case whenever the moments of beams framing into a column in two orthogonal directions are not balanced by moments on the opposite sides. This situation is most obvious in corner columns. The determination of the ultimate strength of such columns is much more complicated because the second bending moment adds a third dimension to the problem. The corresponding interaction diagram is represented by a failure surface in three-dimensional space with the three coordinates P_n, M_{nx}, and M_{ny} (Fig. 7.17). Any point S on this surface corresponds to failure for a given set of P_n, M_{nx}, and M_{ny} or P_n, e_x, and e_y. If either one of the two moments is zero, the surface degenerates to the interaction curve discussed before.

The accurate determination of failure loads requires lengthy calculations and is seldom justified. For practical purposes, it is adequate to approximate the actual interaction

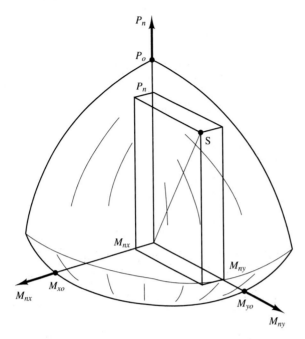

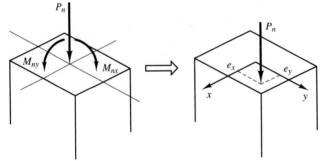

Figure 7.17 Failure surface for column subjected to biaxial bending.

surface by

$$\frac{P_n}{P_{xo}} + \frac{P_n}{P_{yo}} - \frac{P_n}{P_o} = 1 \tag{7.31}$$

where P_n is the nominal failure load applied with eccentricities e_x and e_y; P_{xo} is the failure load with eccentricity e_x only; P_{yo} is the failure load with eccentricity e_y only; and P_o is the failure load for a concentrically loaded column.

In order to interpret Eq. 7.31, consider the case $e_y = 0$, which means $P_{yo} = P_o$. Substituting this into Eq. 7.31 leads to $P_n = P_{xo}$, which is in agreement with the definition of P_{xo}. The presence of a nonzero eccentricity e_y has the effect of reducing P_n. Note that Eq. 7.31 can also be written as

$$\frac{1}{P_n} = \frac{1}{P_{xo}} + \frac{1}{P_{yo}} - \frac{1}{P_o} \tag{7.32}$$

EXAMPLE 7.7

The column of Fig. 7.18 is loaded with eccentricities $e_x = 6$ in and $e_y = 3$ in. What is the approximate nominal failure load P_n? Given: $f_y = 60$ ksi; $f'_c = 4$ ksi.

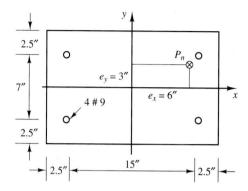

Figure 7.18   Example 7.7.

Solution   The failure load applied with no eccentricities is given by Eq. 7.1:

$$P_o = .85 f'_c (bh - A_{st}) + A_{st} f_y = (.85)(4)(12 \times 20 - 4) + (4)(60) = 1042 \text{ kips}$$

For eccentricity $e_x = 6$ in, we have already determined the failure load in Example 7.3 to be $P_n = P_{xo} = 543$ kips. For $e_y = 3$ in, we similarly obtain $P_{yo} = 568$ kips. Substituting these values into Eq. 7.32 leads to

$$\frac{1}{P_n} = \frac{1}{543} + \frac{1}{568} - \frac{1}{1042} \quad \text{from which} \quad \underline{P_n = 378 \text{ kips}}$$

■

7.6 STRENGTH REDUCTION FACTOR

The design of columns is controlled by strength requirements. Service load criteria, such as deflection and crack limitations, are of less importance for columns than for beams. Only the relative horizontal displacements of adjacent floors in multistory buildings, called *interstory drifts*, must be controlled to remain within prescribed limits. However, the overriding concern for the design of columns is a sufficient margin of safety against failure, in other words, ultimate strength. For this reason, columns are almost always designed using the USD method. The overload factors are the same as those that control the design of beams, for example, 1.4 for dead load and 1.7 for live load. However, the strength reduction factor ϕ is different, namely, more conservative than for beams, for the following reasons:

1. It is more difficult to pour the concrete into column forms (especially for long columns), avoid air pockets, and obtain a quality comparable to that of beams.

2. While the strength of beams is controlled mostly by the steel reinforcement, the strength of columns depends to a much larger extent on the concrete, whose quality cannot be controlled as well as that of manufactured steel.

3. On construction sites with poor supervision, column cross-sectional reductions for electrical conduits, pipes, etc., can often be encountered in violation of design specifications.

4. Finally, the consequences of failure of a column are more catastrophic than when the strength of a beam is exceeded.

For these reasons, the ACI Code prescribes a strength reduction factor of $\phi = 0.7$ for tied columns and $\phi = 0.75$ for spiral reinforced columns. The larger value for the latter reflects the higher strength and better ductility of such columns. Applying the strength reduction factor to the failure curve of a column has the effect of "shrinking" it (Fig. 7.19a). In addition, two special cases require further comment. For very small moments, earlier editions of the Code prescribed a minimum eccentricity (or moment) associated with the failure load. To guard against accidental eccentricities or alignment tolerances, the current edition requires a 20% reduction of the nominal axial load strength for tied columns and 15% for spiral reinforced columns, Thus, the ultimate or reduced axial load capacity of a spiral reinforced column is

$$P_u = .85\phi P_o = .85\phi[.85 f'_c(A_g - A_{st}) + f_y A_{st}] \qquad (7.33)[10\text{-}1]$$

with $\phi = .75$, and that of a tied column

$$P_u = .80\phi P_o = .80\phi[.85 f'_c(A_g - A_{st}) + f_y A_{st}] \qquad (7.34)[10\text{-}2]$$

where $\phi = 0.7$, and A_{st} is the total area of longitudinal reinforcement (bars or steel shapes).

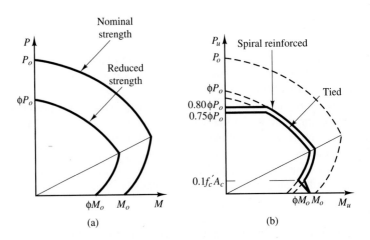

Figure 7.19 Column design curves.

At the other extreme, a column loaded with a very small axial load or no axial load at all may be treated like a beam, for which the strength reduction factor is $\phi = 0.9$. In order to eliminate this apparent inconsistency, the ACI Code permits the determination of ϕ using linear interpolation between the case, $P_u = 0$ (with $\phi = 0.9$), and the case of the small axial load, $P_u = 0.1 f'_c A_g$ (with $\phi = 0.7$ or 0.75), provided $f_y \leq 60$ ksi and $(d - d')/h \geq 0.70$. In the rare case that $\phi P_b < 0.1 f'_c A_g$, the interpolation is performed between $P_u = \phi P_b$ and $P_u = 0$.

These two extreme case modifications result in the column design curves of Fig. 7.19b. In Appendix B, a number of design graphs are reproduced from the ACI Design Handbook [1], all of which have the characteristic shape of Fig. 7.19b.

EXAMPLE 7.8

Is the column shown in Fig. 7.20 strong enough to support an axial load due to dead weight of 120 kips and live load of 200 kips, applied at zero eccentricity? Given: $f_y = 60$ ksi; $f'_c = 4$ ksi.

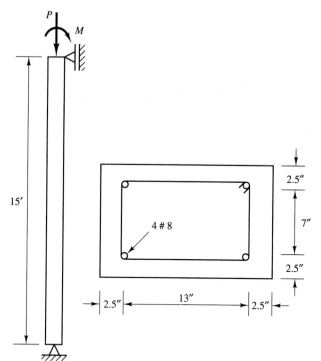

Figure 7.20   Example 7.8.

Solution

Factored load: $P_u = (1.4)(120) + (1.7)(200) = 508$ kips

Nominal strength: $P_o = (.85)(4)(18 \times 12 - 3.14) + (3.14)(60) = 912$ kips

Ultimate strength: $P_u = .80\phi P_o = (.80)(.70)(912) = 511$ kips > 508 kips, <u>OK</u>   ■

EXAMPLE 7.9

Suppose the column of the previous example is subjected to a factored axial load of only 20 kips, in addition to a factored moment of 50 ft-kips. Determine the applicable strength reduction factor.

Solution   Since $f_y \leq 60$ ksi, and $(d - d')/h = 13/18 = .72 > .70$, linear interpolation for ϕ applies.

Balanced load: $\phi P_b = (.70)(.85)^2(12)(15.5)(4)\dfrac{87}{87 + 60} = 223$ kips

Other cutoff load: $0.1 f_c' A_g = (0.1)(4)(12)(18) = 86.4 < 223$

For $P_u = 0$, $\phi = 0.9$

For $P_u = 86$, $\phi = 0.7$

For $P_u = 20$, $\phi = 0.9 - \dfrac{(0.2)(20)}{86} = \underline{0.85}$ ∎

7.7 REINFORCING DETAILS

Column reinforcement is subject to a number of detailing requirements to assure satisfactory column behavior:

1. The minimum amount of longitudinal reinforcement shall be 1% of the gross area of the column section. This amount of steel provides a minimum of ductility and resistance to bending, even if computations show that no bending exists. Also, longitudinal reinforcement is very effective in reducing long-term creep and shrinkage deformations. Under the sustained application of compression loads, stress is transferred gradually from the (creeping) concrete to the (noncreeping) steel. This stress transfer can become significant for small reinforcing ratios and lead to the yielding of the reinforcing bars, with attendant side effects. The minimum amount of steel assures satisfactory behavior.

2. The maximum amount of longitudinal reinforcement shall be 8% of the gross area of the column section. Because it is difficult to place the concrete and avoid congestion, reinforcing ratios in excess of 4 or 5% are seldom used in practice. If congestion is a problem, it may be advisable to bundle two, three, or four bars together (Fig. 7.21a). If reinforcing bars must be lap spliced, the reinforcing percentage should not exceed 4%.

3. The minimum number of longitudinal reinforcing bars shall be six in a circular arrangement, and four in a rectangular arrangement (Fig. 7.21b). For other shapes, one bar should be provided at each apex or corner. For example, a triangular column should contain at least three bars, a hexagonal section six bars, etc.

4. The clear distance between longitudinal bars shall neither be less than $1\frac{1}{2}$ bar diameters nor $1\frac{1}{2}$ in. This provision is intended to assure proper placement of

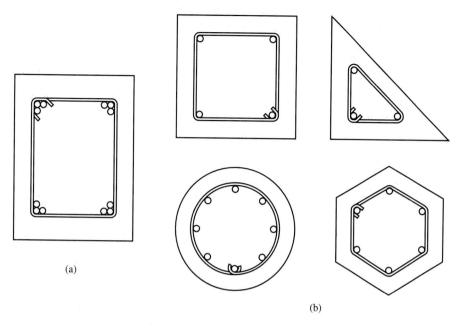

Figure 7.21   Column reinforcement details: (a) bundled bars; (b) minimum number of bars.

concrete without honeycombs and voids and is also related to the requirement that the maximum size of coarse aggregate be no larger than $\frac{3}{4}$ of the clear spacing between bars.

Longitudinal bars must be held securely in place during placement and vibration of the concrete, and they must be prevented from buckling. This is achieved by lateral reinforcement, that is, a continuous spiral in spiral reinforced columns, and closed ties in tied columns. These are subject to a number of detail requirements. A spiral must satisfy, among others, the following criteria:

1. the size shall be not less than $\frac{3}{8}$ in in diameter;
2. the clear spacing between spirals (the pitch) shall not exceed 3 in, nor be less than 1 in;
3. the minimum amount of spiral reinforcement was given by Eq. 7.10

$$\rho = 0.45 \left(\frac{A_g}{A_c} - 1 \right) \frac{f'_c}{f_y}$$

Standard spiral sizes are #3, #4, and #5 of hot-rolled or cold-drawn material, smooth or deformed. ACI Code Sec. 7.10.4 contains some additional spiral detail requirements.

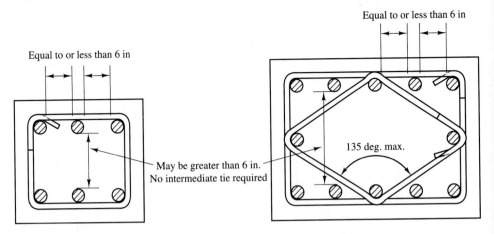

Figure 7.22 ACI Code requirements for lateral support of column reinforcement.

If ties are used for lateral reinforcement, the following restrictions must be observed, among others (see ACI Code Sec. 7.10.5):

1. for longitudinal bars #10 or smaller, ties should be of size #3 or larger;
2. for longitudinal bars #11 or larger and bundled longitudinal bars, ties should be of size #4 or larger;
3. the vertical spacing of ties shall not exceed 16 longitudinal bar diameters, 48 tie bar diameters, or the smallest dimension of the compression member;
4. every corner and alternate longitudinal bar shall have lateral support provided by the corner of a tie with an included angle of not more than 135° (Fig. 7.22);
5. no bars shall be farther than 6 in clear on each side along the tie from a laterally supported bar (Fig. 7.22).

7.8 DESIGN OF SHORT COMPRESSION MEMBERS

The equations that control the design of rectangular columns, derived in Sec. 7.3, are briefly restated here for recollection (see Fig. 7.11). Vertical equilibrium results in the equation

$$P_n = .85 f'_c ab + A'_s f'_s - A_s f_s$$

Moment equilibrium taken about the tension steel requires

$$P_n e' = .85 f'_c ab \left(d - \frac{a}{2} \right) + A'_s f'_s (d - d')$$

Stresses in the tension and compression steel, respectively, relate to a, the depth of the concrete compression block, by

$$f_s = \frac{\beta_1 d - a}{a} 87{,}000 \le f_y$$

$$f_s' = \frac{a - \beta_1 d'}{a} 87{,}000 \le f_y$$

as a result of strain compatibility conditions. Similar equations can be derived for circular and other cross sections. Before using these equations in a practical step-by-step design procedure, a few general remarks about the column design problem are in order. This problem can be stated in at least two different forms:

1. Given a section, what is the ultimate capacity P_u and M_u?
2. Find a section that can carry the given factored loads P_u and M_u.

The eccentricity $e = M_u/P_u$ is usually known as the result of a structural analysis. Since the design problem in its second form above is not solvable in a straightforward way, we normally proceed by assuming all cross-sectional dimensions and then determining the amount of reinforcing steel required.

The yield strengths f_y and f_y' of the longitudinal reinforcing bars are normally 60 ksi. For highly stressed columns (such as those in the lower stories of tall buildings), characterized by small M_u/P_u ratios, higher yield strengths are economical. For large M_u/P_u ratios, this is not the case to the same extent, as was shown earlier, especially for intermediate bars.

High concrete strength f_c' is helpful in reducing the size of the cross section. This is true mostly for the highly stressed columns in tall buildings. Real estate developers like to remind us that they cannot collect rent on space occupied by columns, and therefore they place a premium on the smallest possible column cross sections. An enormous increase in the strength of commercially available concrete has taken place in recent years. Concretes with over 10,000 psi strengths are now readily available in many parts of the country and have been partially responsible for the appearance of increasingly taller concrete buildings in such cities as Chicago, Houston, and New York. At the time of this writing, the world's tallest concrete building, at 1228 ft, is the 78-story Central Plaza Building in Hong Kong [2]. Concrete with strengths up to 8700 psi was used in its lower columns.

The concrete protection requirements for reinforcement are the same for columns as for beams, that is, typically $1\frac{1}{2}$ in, unless the concrete is exposed to weather or in contact with the ground, in which case the requirement is 2 in or more.

We shall now present a long-hand, step-by-step procedure for the design of short columns with symmetrically reinforced rectangular cross sections, intended to resist a given factored load P_u and moment M_u. A short-hand procedure using design aids will follow later.

Step 1. Compute the required nominal column strength from the factored applied load and moment:

$$P_n = \frac{P_u}{\phi} \qquad M_n = \frac{M_u}{\phi} = \frac{P_u e}{\phi} \tag{7.35}$$

Step 2. Assume a section with dimensions b, h, d, and d'. A cross section with trial area $A_g = 2P_n/f'_c$ implies that approximately half the section is in compression at failure. Depending on the e-value, an economical section will be either more rectangular or square in shape.

Step 3. Assume a value for the depth of the concrete compression block a, which will be a function of the e/h ratio. $a = h/4$ may be a reasonable starting value.

Step 4. Compute stresses in the tension and compression steel:

$$f_s = \frac{\beta_1 d - a}{a} 87{,}000 \le f_y$$
$$f'_s = \frac{a - \beta_1 d'}{a} 87{,}000 \le f_y \tag{7.36}$$

Step 5. Compute the required steel areas:

$$A_s = A'_s = \frac{P_n e' - .85 f'_c a b (d - \frac{a}{2})}{f'_s (d - d')} \tag{7.37}$$

Step 6. Compute a new concrete compression block depth:

$$a = \frac{P_n - A_s(f'_s - f_s)}{.85 f'_c b} \tag{7.38}$$

Step 7. Repeat Steps 4, 5, and 6 until convergence is achieved. When continuing with Step 4, convergence can be accelerated by averaging the new and old values for a.

Step 8. If A_s is too large, the section depth must be increased. Remember that the total steel area $A_s + A'_s$ should generally not exceed 4%, and never 8%, of the gross column area. If A_s is too small, that is, if $A_s + A'_s$ is less than 1% of the gross area, the section size may be decreased.

EXAMPLE 7.10

Design a tied column that is required to carry a factored load of $P_u = 320$ kips and a moment of $M_u = 100$ ft-kips. Given: $f'_c = 3000$ psi; $f_y = 60$ ksi.

Solution

Step 1. With $\phi = 0.7$, the required nominal strength is $P_n = \dfrac{320}{0.7} = 457$ kips;

$$M_n = \frac{(100)(12)}{0.7} = 1714 \text{ in-kips.}$$

Step 2. For the relatively small eccentricity of $e = 1200/320 = 3.75$ in, we can expect most of the section to be in compression at failure, that is, $A_{req} = \dfrac{P_n}{.85 f'_c} =$

$\dfrac{457}{(.85)(3)} = 179$ in^2. Try a 15-by-12-in section, with $h = 15$, $b = 12$, $d = 12$,

$d' = 3$, and $e' = 3.75 + \dfrac{12-3}{2} = 8.25$ in.

Steps 3, 4, 5. The remaining calculations shall be organized in a tabular format, making use of the following expressions:

$$f_s = \frac{\beta_1 d - a}{a} 87 = \frac{(.85)(12) - a}{a} 87 = \frac{10.2 - a}{a} 87 \le 60 \qquad \text{(a)}$$

$$f_s' = \frac{a - \beta_1 d'}{a} 87 = \frac{a - (.85)(3)}{a} 87 = \frac{a - 2.55}{a} 87 \le 60 \qquad \text{(b)}$$

$$A_s = \frac{(457)(8.25) - (.85)(3)a(12)(12 - a/2)}{f_s'(12 - 3)}$$

$$= \frac{3770 - 30.6a(12 - a/2)}{9 f_s'} \qquad \text{(c)}$$

$$a = \frac{457 - A_s(f_s' - f_s)}{(.85)(3)(12)} = \frac{457 - A_s(f_s' - f_s)}{30.6} \qquad \text{(d)}$$

Trial a	f_s	f_s'	A_s	a_{new}	Next trial a
	Eq. (a)	Eq. (b)	Eq. (c)	Eq. (d)	$\dfrac{a_{new} + a}{2}$
10.00	1.74	60.0	3.01	9.20	9.60
9.60	5.44	60.0	3.06	9.48	9.54
9.54	6.02	60.0	3.07	9.52	(Close enough)

The required steel area can be supplied by 2 #11 bars on each side. The four bars supply 6.25 in^2 or $\rho_g = 6.25/180 = 0.035$, a reasonable reinforcing ratio. ∎

In principle, the same iterative procedure can be applied to columns with different cross-sectional shapes and reinforcing patterns. For example, for a circular column (Fig. 7.23a), the procedure might be modified as follows.

Step 1. Compute the nominal column strength:

$$P_n = \frac{P_u}{\phi} \qquad M_n = \frac{M_u}{\phi} = \frac{P_u e}{\phi}$$

Note that circular columns are normally spiral reinforced, so that $\phi = 0.75$.

Step 2. Assume a section with diameter h and the diameter h' of the circle prescribed by the steel bars. The section size depends on the load eccentricity, but an area $A_g = 2 P_n / f_c'$ serves as a reasonable starting point.

Step 3. Assume a value for the depth of the concrete compression block a.

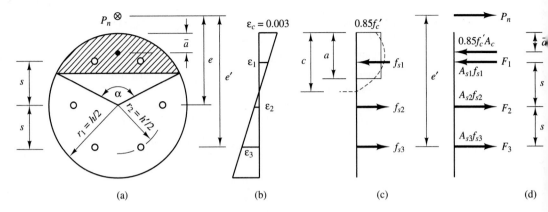

(a)   (b)   (c)   (d)

Figure 7.23 Circular column section at failure: (a) section; (b) strains; (c) stresses; (d) internal forces.

Step 4. Using strain compatibility conditions, compute stresses in all reinforcing bars. Thus, for the six-bar arrangement of Fig. 7.23, the stresses are given by

$$f_{s1} = \frac{a - \beta_1(r_1 - s)}{a} 87{,}000 \le f_y$$

$$f_{s2} = \frac{\beta_1 r_1 - a}{a} 87{,}000 \le f_y \tag{7.39}$$

$$f_{s3} = \frac{\beta_1(r_1 + s) - a}{a} 87{,}000 \le f_y$$

Step 5. Compute the concrete area under compression:

$$A_c = \frac{r_1^2}{2}\left(\frac{\pi}{180}\alpha - \sin\alpha\right) \tag{7.40}$$

where

$$\alpha = 2 \text{ arc cos}\left(1 - \frac{a}{r_1}\right) \tag{7.41}$$

Step 6. Use moment equilibrium to determine the required steel area for each pair of bars:

$$A_s = \frac{P_n e' - .85 f_c' A_c(s + r_1 - \bar{a})}{2sf_{s1} - sf_{s2}} \tag{7.42}$$

where

$$\bar{a} = r_1 - \frac{(2r_1 \sin\frac{\alpha}{2})^3}{12A_c} \tag{7.43}$$

is the position of the centroid of the concrete compression block (Fig. 7.23a). Note that Eq. 7.43 is valid only for the specific bar arrangement of Fig. 7.23a.

Step 7. Use vertical force equilibrium to update the concrete area under compression:

$$A_c = \frac{P_n - A_s f_{s1} + A_s f_{s2} + A_s f_{s3}}{.85 f_c'} \tag{7.44}$$

Step 8. Since it is difficult, for a given A_c, to solve Eq. 7.40 for α and then for a from Eq. 7.41, it is more advantageous to assume a new value either for α or a, and to continue the iteration by repeating Steps 5, 6, and 7 until sufficient convergence has been achieved.

EXAMPLE 7.11

Design a short circular column to carry $P_u = 500$ kips and $M_u = 127$ ft-kips. The longitudinal steel is to be contained within a #3 spiral. Given: $f_c' = 4000$ psi; $f_y = 60$ ksi.

Solution

Step 1. $P_n = \dfrac{500}{.75} = 667$ kips, $M_n = \dfrac{(127)(12)}{.75} = 2032$ in-kips, $e = \dfrac{2032}{667} = 3.05$ in.

Step 2. Assume an average stress at failure of 2.5 ksi, then, $A_g = 667/2.5 = 266.8$ in^2. Try an 18 in column, with $A_g = 254$ in^2.

The remaining calculations are again organized in tabular format, based on the following numerical expressions. With $r_1 = 9$, $r_2 = 6.5$, $s = r_2 \sin 60 = 5.63$, $e' = e + s = 8.68$, and A_s denoting the area required by a pair of bars:

$$\alpha = 2 \text{ arc cos}\left(1 - \frac{a}{r_1}\right) = 2 \text{ arc cos}\left(1 - \frac{a}{9}\right) \tag{a}$$

$$A_c = \frac{r_1^2}{2}\left(\frac{\pi}{180}\alpha - \sin\alpha\right) = (40.5)\left(\frac{\pi}{180}\alpha - \sin\alpha\right) \tag{b}$$

$$\bar{a} = r_1 - \frac{(2r_1 \sin\frac{\alpha}{2})^3}{12 A_c} = 9 - (486)\frac{\sin^3\frac{\alpha}{2}}{A_c} \tag{c}$$

$$f_{s1} = \frac{a - \beta_1(r_1 - s)}{a}87 = \frac{a - 2.86}{a}87 \le 60 \tag{d}$$

$$f_{s2} = \frac{\beta_1 r_1 - a}{a}87 = \frac{7.65 - a}{a}87 \le 60 \tag{e}$$

$$f_{s3} = \frac{\beta_1(r_1 + s) - a}{a}87 = \frac{12.44 - a}{a}87 \le 60 \tag{f}$$

$$A_s = \frac{P_n e' - .85 f_c' A_c(s + r_1 - \bar{a})}{2s f_{s1} - s f_{s2}} = \frac{5790 - 3.4 A_c(14.63 - \bar{a})}{11.26 f_{s1} - 5.63 f_{s2}} \tag{g}$$

$$A_c = \frac{P_n + A_s(f_{s3} + f_{s2} - f_{s1})}{.85 f_c'} = \frac{667 + A_s(f_{s3} + f_{s2} - f_{s1})}{3.4} \tag{h}$$

Continue the iteration until the two values for A_c are sufficiently close.

Trial a	α	A_c	$\bar{a}$	f_{s1}	f_{s2}	f_{s3}	A_s	A_c
	Eq. (a)	Eq. (b)	Eq. (c)	Eq. (d)	Eq. (e)	Eq. (f)	Eq. (g)	Eq. (h)
9.0	180.0	127.2	5.179	59.4	−13.05	33.25	2.294	169.7
12.0	218.9	180.2	6.739	60.0	−31.54	3.19	1.120	167.1
11.0	205.7	163.0	6.237	60.0	−26.50	11.39	1.380	165.7
11.2	208.3	166.4	6.337	60.0	−27.58	9.63	1.322	165.9

The new and previous values for A_c agree sufficiently. Use 6 #8 bars with a combined area $A_{st} = 4.71$ in^2.

According to Eq. 7.10, the spiral requires the following amount of steel,

$$\rho_s = 0.45 \left(\frac{A_g}{A_c} - 1 \right) \frac{f'_c}{f_y} = (.45) \left(\frac{254}{\pi 7^2} - 1 \right) \frac{4}{60} = .0195$$

For a #3 spiral, $A_{sp} = .11$ in^2, and from Eq. 7.5, the required pitch can be determined,

$$s = \frac{4 A_{sp}}{\rho_s D_c} = \frac{(4)(.11)}{(.0195)(14)} = 1.61 \text{ in. Use } s = 1.5 \text{ in.}$$

This pitch satisfies the lower limit of 1 in and upper limit of 3 in. ∎

As the previous examples illustrate, the determination of steel requirements is not straightforward, even if all other design parameters have been preselected. Convergence of the iteration can be slow, or if the initial choice of a is poor, the iteration can diverge. In this case, different values may have to be tried.

In engineering practice, the column design task is made considerably easier by the availability of design aids developed by ACI [1] and CRSI [3]. For a number of rectangular and circular cross sections with regular reinforcing patterns, design interaction curves are available in nondimensional form that allow easy determination of steel requirements. Some of these graphs are reproduced in Appendix B. The graph labels start with a letter, which indicates how the reinforcement is placed in a rectangular section: E for maximum lever arm, L for normal to the axis of bending, and R and C for placement around the perimeter of a rectangular and circular section, respectively. The remainder of the graph labels contains values of f'_c, f_y, and the geometric quantity γ, which is illustrated in Fig. 7.24.

With the help of such graphic aids, the design of a column is reduced to the following steps.

Step 1. Assume a cross section with dimensions b, h, d, and d' (for a rectangular section) or h and h' for a circular section (Fig. 7.24). As a starting point, $A_g = 2P_n / f'_c$ may be an appropriate choice.

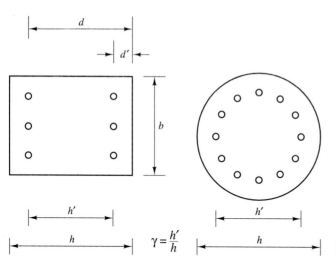

$$\gamma = \frac{h'}{h}$$

Figure 7.24   Cross-sectional dimensions.

Step 2.   Determine the γ-value: $\gamma_{\text{rect}} = \dfrac{d - d'}{h}$ or $\gamma_{\text{circ}} = \dfrac{h'}{h}$

Step 3.   Compute coordinates for entering the design charts:

$$\frac{\phi P_n}{A_g} = \frac{P_u}{A_g} \quad \text{and} \quad \frac{\phi P_n}{A_g} \frac{e}{h} = \frac{M_u}{A_g h}$$

Step 4.   Select the appropriate design chart for the γ-value computed in Step 2. Enter with the coordinate values of Step 3 and graphically determine the required reinforcing ratio ρ_g. The steel area then follows as $A_{st} = \rho_g A_g$. If no chart is available for the computed γ-value, linear interpolation will be necessary. For other cross-sectional shapes and nonstandard reinforcing patterns, for example, nonsymmetrical bar arrangements, resort to an iterative procedure, as outlined earlier. To ease the designer's burden, a number of computer programs that automatically solve the design problem for many practical cases are available.*

EXAMPLE 7.12

Solve the problem of Example 7.10 using design charts.

Solution

Step 1.   The design graphs do not spare us the effort to estimate a trial section. We shall use the same 15-by-12-in section of Example 7.10.

Step 2.   Let $d = 12$ and $d' = 3$, then $\gamma = 9/15 = 0.6$.

---

*For example, PCACOL, a computer program for the design and investigation of reinforced concrete column sections, and IRRCOL, a computer program for columns with irregular shapes, were developed by and are available from the Portland Cement Association, Skokie, IL.

Step 3. Compute coordinates for the design chart:

$$\frac{\phi P_n}{A_g} = \frac{P_u}{A_g} = \frac{320}{180} = 1.78 \text{ ksi}; \quad \frac{\phi P_n}{A_g}\frac{e}{h} = \frac{M_u}{A_g h} = \frac{(100)(12)}{(180)(15)} = .444.$$

Step 4. From Graph E3-60.60 in Fig. B.2 of Appendix B, we find $\rho_g = .034$, so that $A_{st} = \rho_g A_g = (.034)(180) = 6.12 \text{ in}^2$. Use 4 #11 bars with $A_{st} = 6.25 \text{ in}^2$. (Note that the design charts satisfy the reinforcing limits, that is, $.01 < \rho_g < .08$.) ∎

EXAMPLE 7.13

Repeat the problem of Example 7.11 using design charts.

Solution

Step 1. As in the long-hand solution, it is necessary to choose a trial area. We shall again try an 18-in column, with $A_g = 254 \text{ in}^2$.

Step 2. Let $h' = 13$, so that $\gamma = \dfrac{13}{18} = .722$.

Step 3. Compute coordinates for the design chart:

$$\frac{P_u}{A_g} = \frac{500}{254} = 1.97 \text{ ksi}; \quad \frac{M_u}{A_g h} = \frac{(127)(12)}{(254)(18)} = .333.$$

Step 4. Since $\gamma = .722 \approx .75$, enter Graph C4-60.75 in Fig. B.16 of Appendix B to find $\rho_g = .017$, so that $A_{st} = (.017)(254) = 4.32 \text{ in}^2$. Use 6 #8 bars with $A_{st} = 4.71 \text{ in}^2$. ∎

In both design problems, the design charts led to the same reinforcement as the long-hand iterative procedures did. This is not quite an accident, since the discreteness of available steel bars tends to neutralize numerical roundoff errors. The long-hand solution in Example 7.11 led to a steel requirement of 3.94 in², whereas the design chart gave 4.32 in². This discrepancy is well within the accuracy with which the design charts can be read.

7.9 BUCKLING OF SLENDER COLUMNS

We have thus far assumed that columns are sufficiently short to fail in compression or flexure rather than in buckling. Columns that do not satisfy this condition are classified as *long* or *slender* columns. Strictly speaking, the theory outlined below will show that there is no clean-cut dividing line between short and long columns. (For practical situations, the ACI Code provides a guideline, which will be presented later.)

Consider a perfectly straight column, free to rotate at both ends, and subject to an axial force only. To determine the critical load at which the column buckles, we need to solve its differential equation. Equilibrium of a free-body segment of the column in its buckled state (Fig. 7.25) requires

$$Py = -EIy'' \tag{7.45}$$

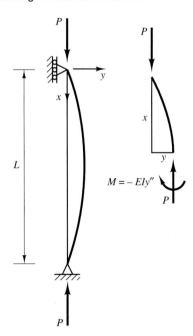

Figure 7.25 Column in buckled configuration.

where y'' is the column curvature at point x. Strictly speaking, the curvature is given by $1/\rho = y''/(1 - y'^2)^{3/2}$, but for small displacements, that is, when buckling starts, y' is small and the linearization of the buckling problem $(1/\rho \approx y'')$ is acceptable. Introducing the parameter

$$\lambda^2 = \frac{P}{EI} \tag{7.46}$$

into Eq. 7.45 leads to the following simple ordinary differential equation of second order

$$y'' + \lambda^2 y = 0 \tag{7.47}$$

with the general solution

$$y = A \sin \lambda x + B \cos \lambda x \tag{7.48}$$

The boundary condition $y(0) = 0$ requires that $B = 0$, and from the other boundary condition, $y(L) = 0$, it follows that

$$A \sin \lambda L = 0$$

where either $A = 0$ or $\sin \lambda L = 0$. Since $A = 0$ leads to the trivial solution $y \equiv 0$, we are more interested in the other case, which applies if

$$\lambda L = n\pi \qquad n = 1, 2, \ldots \tag{7.49}$$

and, with the definition of Eq. 7.46,

$$P = \left(\frac{n\pi}{L}\right)^2 EI \qquad n = 1, 2, \ldots \tag{7.50}$$

that is, the differential equation has an infinite number of solutions. However, we are interested most in the lowest solution, which means that $n = 1$. This leads to the critical load

$$P_c = \frac{\pi^2 E I}{L^2} \tag{7.51}$$

or the *Euler buckling load*, named after Leonhard Euler (1707–83), who derived this solution in 1759. Interestingly, the moment of inertia I had not yet been defined, and Euler introduced in its place a geometric quantity that he called *stiffness moment*, which he speculated to be proportional to either h^2 or h^3.

Introducing the *radius of gyration*, that is,

$$r^2 = \frac{I}{A} \tag{7.52}$$

and the *effective length factor k*, Eq. 7.51 can also be written as

$$P_c = \frac{\pi^2 E A}{(kL/r)^2} \tag{7.53}$$

where kL/r is the *slenderness ratio* and the most important parameter in buckling problems. Note that the critical load is inversely proportional to the square of the slenderness ratio. In other words, if the length of a column is doubled, the buckling load is reduced by factor 4.

The effective length factor k is a function of the column boundary conditions. Consider, for example, a column fixed at one end and free to move at the other end (Fig. 7.26b). The buckled shape is the same as that of a column twice the length, pinned at both ends. Therefore, the critical load for this column is the same as if the column were twice as long and pinned at both ends, that is, $k = 2$. For reasons of symmetry, the column of Fig. 7.26c (fixed at both ends) has inflection points at the quarter points. Here, the half-sine wave typical of the Euler column is of length $L/2$, that is, the critical load is the same as that of an Euler column of half the length: $k = 0.5$. In the case of Fig. 7.26d, both ends are fixed but the top is not restrained against lateral movement, resulting in an inflection point at midheight. The two quarter-sine waves add up to an equivalent half-sine wave of length L, so that $k = 1$. In the case of Fig. 7.26e, $k = 0.7$, and the buckled shape of Fig. 7.26f is a quarter-sine wave, that is, $k = 2$.

Columns of actual concrete structures are rarely completely fixed or perfectly pinned. They are usually restrained by slabs, girders, or footings with some finite stiffness. Therefore, the effective lengths of such columns (shown in Fig. 7.27) are somewhere in between those of the idealized cases of Fig. 7.26. For example, the case of a column with partial end constraints and no possibility of sidesway is somewhere between the fixed and pinned case, that is, $1/2 \le k \le 1$. If sidesway is not prevented, the effective length factor is at least equal to 1 for infinitely rigid girders. The more flexible the girders, the less restraint they can offer to the column. In the extreme case of zero girder stiffness, the column is unstable and the critical load is zero as $k \to \infty$. Consider the example of a simple portal frame (Fig. 7.28). If it is braced against sidesway, the effective length of the columns is somewhat less than h. With sidesway, the effective

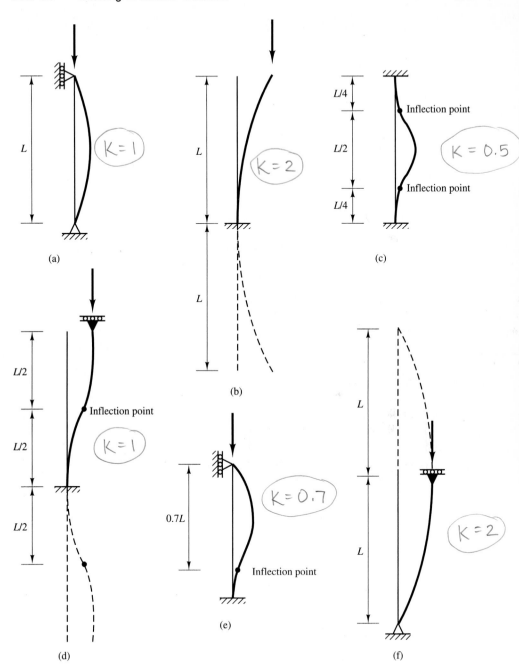

Figure 7.26   Effective length factors for basic boundary conditions: (a) $k = 1$; (b) $k = 2$; (c) $k = \frac{1}{2}$; (d) $k = 1$; (e) $k = 0.7$; (f) $k = 2$.

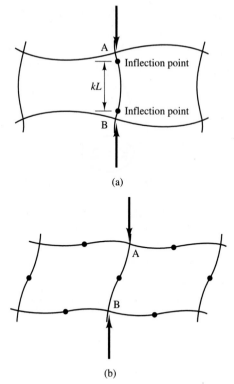

(a)

(b)

Figure 7.27 Buckling of partially restrained columns: (a) sidesway prevented ($\frac{1}{2} \leq k \leq 1$); (b) sidesway uninhibited ($1 \leq k < \infty$).

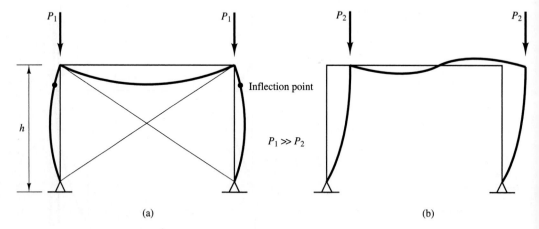

(a)   (b)

Figure 7.28 Buckling of braced and unbraced portal frame: (a) braced frame ($0.7 \leq k \leq 1$); (b) unbraced frame ($k \geq 2$).

length is at least $2h$. In the case of equal beam and column stiffnesses, $k = 0.875$ and 2.33, respectively, that is, lateral bracing increases the buckling strength of the frame by a factor $(2.33/0.875)^2 = 7$. Therefore, if buckling must be considered, structural means (e.g., shear walls) should be devised to prevent sidesway, whereupon the carrying capacities of the columns are increased considerably.

The effective length factor is a function of the effective restraint the column gets from the beams, or more specifically, of the relative stiffnesses of beams and columns at the respective column end, that is,

$$\psi_A = \frac{\sum (EI/L)_{column}}{\sum (EI/L)_{beam}} \qquad \text{(at end A)}$$

$$\psi_B = \frac{\sum (EI/L)_{column}}{\sum (EI/L)_{beam}} \qquad \text{(at end B)}$$

(7.54)

The determination of the actual effective length factor k requires the solution of an equation equivalent to Eq. 7.49, but that equation is now transcendental. For practical purposes, nomograms, called the Jackson and Moreland Alignment Charts, have been developed (Fig. 7.29) that permit a rapid graphic determination of k. These alignment charts are reproduced from the commentary of the ACI Code.

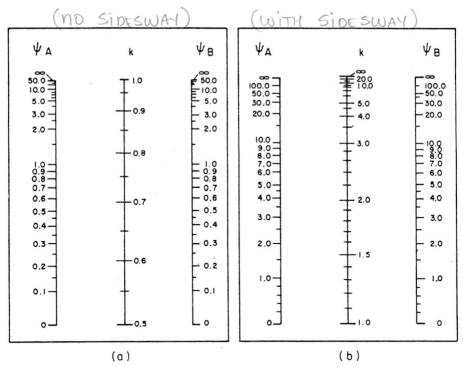

(a)   (b)

ψ = Ratio of $\sum(EI/\ell_c)$ of compression members to $\sum(EI/\ell)$ of flexural members in a plane at one end of a compression member

k = Effective length factor

Figure 7.29 Alignment charts for effective length factors: (a) braced frames; (b) unbraced frames. (Authorized reprint from the ACI Code commentary.)

The beam stiffnesses in Eq. 7.54 should be modified with the following multipliers:

	Sidesway	No Sidesway
Far beam end hinged	0.5	1.5
Far beam end fixed	0.667	2.0

For a column fixed at end A, this would be equivalent to assuming the presence of an infinitely stiff beam at A, so $\psi_A = 0$. In practice, we can substitute $\psi_A = 0.1$ or even 1.0 for unbraced frames, because columns are rarely completely fixed. Likewise, a pinned boundary condition implies the absence of any beam restraint, so $\psi_A = \infty$. As there is usually some restraint, it is more appropriate to use $\psi_A = 100$ or even 10 for braced frames. The rigidity EI of the beams can be calculated on the basis of the moments of inertia of the cracked transformed sections and the rigidity of the compression members using uncracked transformed section properties. Alternatively, for columns with slenderness ratios below 60, it may be permissible to use half of the beam's gross moment of inertia to account for the cracking, and for the compression member, the gross section moment of inertia.

EXAMPLE 7.14

Determine the effective length factors for the five columns in the frame of Fig. 7.30.

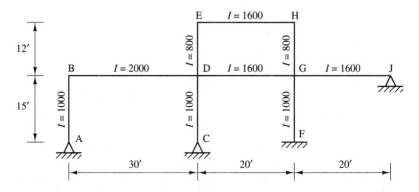

Figure 7.30   Example 7.14.

Solution   Since Young's modulus is the same for all members, it cancels out in Eq. 7.54 and only the I/L ratios are of concern.

Column AB: $\psi_A = 10$ for pinned end (instead of ∞); $\psi_B = \dfrac{1000/15}{2000/30} = 1.0$.

From Fig. 7.29a (no sidesway), find $k = 0.86$.

Column CD: $\psi_C = 10$; $\psi_D = \dfrac{800/12 + 1000/15}{2000/30 + 1600/20} = .909.$

From Fig. 7.29a (no sidesway), find $k = 0.85$.

Column FG: $\psi_F = 0.1$ for fixed end (instead of 0);

$$\psi_G = \dfrac{800/12 + 1000/15}{1600/20 + 1600/20 \times 1.5} = .667 \text{ (far end of beam GJ pinned).}$$

From Fig. 7.29a (no sidesway), find $k = 0.63$.

Column DE: $\psi_D = .909$; $\psi_E = \dfrac{800/12}{1600/20} = .833.$

From Fig. 7.29b (with sidesway), find $k = 1.27$.

Column GH: $\psi_G = \dfrac{800/12 + 1000/15}{1600/20 + 1600/20 \times 0.5} = 1.111 \text{ (far end of beam GJ pinned);}$

$$\psi_H = \dfrac{800/12}{1600/20} = .833.$$

From Fig. 7.29b (with sidesway), find $k = 1.30$.   ∎

The buckling problem discussed so far is primarily of theoretical interest because it implies that the axial load is applied exactly at the column centroid. In reality, this is seldom the case, and a moment is almost always present, due either to a small eccentricity or the *initial crookedness* of the column. Suppose the initial column shape before application of load can be approximated by the sine function

$$y_o = \delta_o \sin \frac{\pi x}{L} \tag{7.55}$$

where δ_o is the maximum deviation from straightness at midheight (Fig. 7.31). The equation of moment equilibrium of a free body at point x (Eq. 7.45) is now replaced by

$$Py = -EI(y'' - y_o'') \tag{7.56}$$

from which, again with Eq. 7.46,

$$y'' + \lambda^2 y = -\delta_o \frac{\pi^2}{L^2} \sin \frac{\pi x}{L} \tag{7.57}$$

This is now a nonhomogeneous differential equation, with the general solution

$$y = A \sin \lambda x + B \cos \lambda x + \frac{\delta_o}{1 - \frac{\lambda^2 L^2}{\pi^2}} \sin \frac{\pi x}{L} \tag{7.58}$$

The boundary conditions $y(0) = 0$ and $y(L) = 0$ require, respectively, that $A = 0$ and $B = 0$, so that only the particular integral remains:

$$y = \frac{\delta_o}{1 - \frac{\lambda^2 L^2}{\pi^2}} \sin \frac{\pi x}{L} = \frac{\delta_o}{1 - \frac{P}{P_c}} \sin \frac{\pi x}{L} \tag{7.59}$$

where $P_c = \pi^2 EI/L^2$ is the Euler load (Eq. 7.51). The maximum displacement at midheight, $x = L/2$, is thus

$$y_{max} = \frac{\delta_o}{1 - \frac{P}{P_c}} \tag{7.60}$$

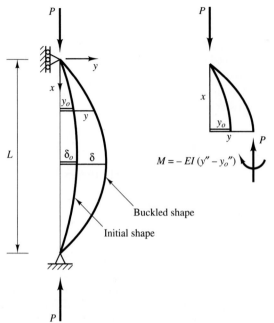

$$M = -EI\,(y'' - y_o'')$$

Buckled shape

Initial shape

Figure 7.31 Buckling of initially crooked column.

and the maximum moment follows as

$$M_o = \frac{1}{1 - \frac{P}{P_c}} M_o \qquad (7.61)$$

where $M_o = Pe$ is the moment acting on an initially straight column, or $M_o = Py_o$ for the initially crooked column. The factor $1/[1 - (P/P_c)]$ is thus the *moment magnification factor*.

The relationship between axial load and lateral displacement is shown qualitatively in Fig. 7.32 for three different values of initial crookedness. As can be seen, displacements can become large well before the critical load is reached. However, in the theoretical case of zero initial crookedness, the column will not displace at all until the critical load is reached, and at $P = P_c$, sudden instability occurs. This point on the P-δ diagram is called the *bifurcation point*.

For columns with some initial crookedness (which is always present) or accidental eccentricity, no matter how small, the axial force is always accompanied by a bending moment, and therefore bending deformations. Consider the column of Fig. 7.33. As the load P is applied with the eccentricity e, the column is subjected to a constant primary bending moment $M_o = Pe$, which causes bending deformations δ with a maximum at midheight. These deformations give rise to additional bending moments $M = P\delta$, which are usually small and referred to as secondary moments. In small-displacement analysis, they are neglected. However, the total moment is $Pe + P\delta = P(e + \delta)$, and the secondary moment obviously cannot be neglected if δ is of the same order of magnitude as e. Likewise, for axial loads close to the theoretical buckling load, a small eccentricity or initial crookedness can have a large effect, as was observed in Fig. 7.32. This is why

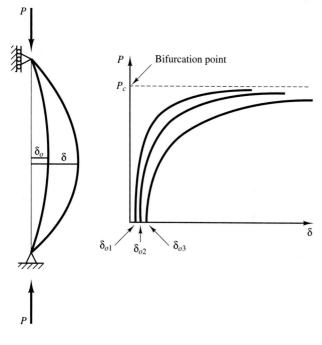

Figure 7.32 Load-deflection curves for initially crooked columns.

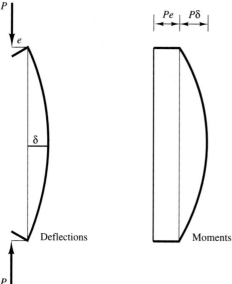

Figure 7.33 Bending moments in beam-column.

the *Pδ-effect* cannot be ignored in buckling problems. It indicates the amount by which the axial force magnifies the primary bending moment M_o. The variation of the bending moment as a function of the slenderness ratio is shown in Fig. 7.34a. For a constant P, M increases asymptotically as the critical slenderness ratio is approached. Similarly, the bending moment grows unbounded, as the axial load reaches the critical load, for a given slenderness ratio (Fig. 7.34b). The member will fail when M and P reach the values M_n and P_n, respectively.

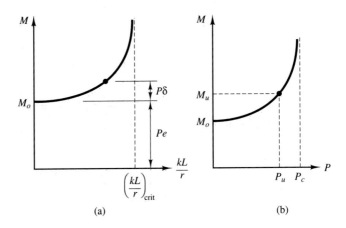

Figure 7.34   Effect of slenderness ratio and axial load on column bending moment.

The column of Fig. 7.33, even if perfectly straight, is subjected to an axial force and a bending moment. Such a member is called a *beam-column*, and strictly speaking all compression members are beam-columns, because small eccentricities or initial crookednesses cannot be avoided. Figure 7.35 illustrates the more general case of a beam-column subjected to an axial force P and two arbitrary end moments M_1 and M_2, where by definition $|M_2| > |M_1|$, and the moment sign convention is such that the M_1/M_2 ratio is positive if the moments bend the column in single curvature (Fig. 7.35c). Conversely, the M_1/M_2 ratio is assumed negative if the moments bend the column in double curvature (Fig. 7.35d). The equation of equilibrium for this case reads

$$Py + M_1 + \frac{M_2 - M_1}{L}\, x = -EIy'' \tag{7.62}$$

or

$$y'' + \lambda^2 y = \lambda^2 \left(\frac{M_1 - M_2}{L}\right)\frac{x}{L} - \frac{\lambda^2}{P}M_1 \tag{7.63}$$

and has the general solution

$$y = A\sin\lambda x + B\cos\lambda x + \frac{M_1 - M_2}{P}\frac{x}{L} - \frac{M_1}{P} \tag{7.64}$$

where the constants of integration need to be determined from the boundary conditions. Rather than finding the closed form solution for the moment magnification factor in this more general case, we present only the approximate solution, which is commonly used for design purposes,

$$\delta_b = \frac{C_m}{1 - \dfrac{P}{P_c}} \tag{7.65}$$

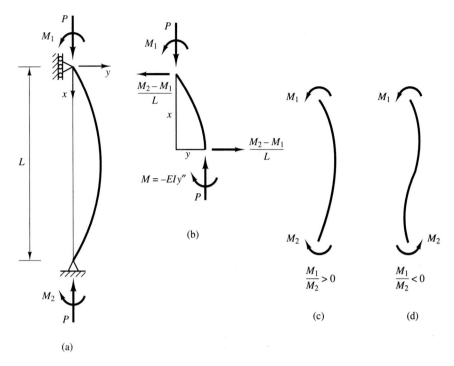

Figure 7.35   Beam-column with arbitrary end moments.

where

$$C_m = 0.6 + 0.4 \frac{M_1}{M_2} \tag{7.66}$$

Further details can be found in the specialized literature on stability of structures [4, 5, 6].

7.10 DESIGN OF SLENDER COLUMNS

In the general case it is necessary to perform a second-order analysis of a structure for factored forces and moments. In other words, a linear elastic analysis based on small-displacement theory is not adequate. Moreover, factors such as material nonlinearities, concrete cracking, shrinkage, and creep need to be considered as well. Only under certain conditions spelled out in the ACI Code (including, for example, that $kl/r < 100$) is it permissible to use the moment magnifier method, that is, to multiply the moments determined from an ordinary first-order analysis with the moment magnification factors as described in the previous section. In this case it is essential to distinguish between frames with and without sidesway. Whether a frame is braced or not is determined by comparing the lateral stiffness of the bracing members with that of the columns.

Alternatively, this determination can be made by determining

a) whether the increase in column end moments due to second-order effects does not exceed five percent of the first-order end moments, or

b) whether

$$Q = \frac{\sum P_u \Delta_o}{V_u l_c} \leq 0.05 \tag{7.67}$$

where $\sum P_u$ and V_u are the total factored vertical load and story shear, respectively, in the story in question; Δ_o is the first-order relative deflection between the top and bottom of that story due to V_u; and l_c is the column length (ACI Code Sec. 10.11.4).

Aside from their largely different effective length factors, there is a fundamental difference between frames with and without sidesway. In a braced frame, each column in a given story is "on its own," and its safety against buckling can be assessed by comparing the factored axial load and end moments with its buckling strength. In a building not restrained against sidesway, a column is unlikely to buckle without all the other columns in the same story buckling at the same time. In this case, safety against buckling should be assessed by comparing the sum of all factored axial loads with the combined buckling strengths of all columns in a given floor.

Turning first to compression members in *non-sway frames* (ACI Code Sec. 10.12), the Code permits us to ignore slenderness effects if

$$\frac{k l_u}{r} \leq 34 - \frac{12 M_1}{M_2} \tag{7.68}$$

with $M_1/M_2 \geq -0.5$. The design moment is then determined by

$$M_c = \delta_{ns} M_2 \tag{7.69}$$

where

$$\delta_{ns} = \frac{C_m}{1 - \dfrac{P_u}{0.75 P_c}} \geq 1.0 \tag{7.70}$$

with

$$C_m = 0.6 + 0.4 \frac{M_1}{M_2} \geq 0.4 \tag{7.71}$$

if no transverse loads between supports are present (otherwise, $C_m = 1.0$). M_1 and M_2 are, respectively, the absolute smaller and larger end moments. Since a column is much stiffer in the buckling mode involving counter-curvature (Fig. 7.35d), its buckling strength is considerably larger than in the case where the bending moments produce single-curvature bending (Fig. 7.35c). For pinned columns, use $C_m = 1.0$. If computations show that there are no moments or very small moments at both column ends, a minimum eccentricity of $e = 0.6 + 0.03h$ in should be used for the axial load P_u, that is,

$$M_{2,\min} = P_u(0.6 + 0.03h) \tag{7.72}$$

Note that the Euler buckling load appearing in Eq. 7.70,

$$P_c = \frac{\pi^2 EI}{(kl_u)^2} \tag{7.73}$$

is proportional to the effective column stiffness EI. This may be taken as

$$EI = \frac{0.2E_c I_g + E_s I_{se}}{1 + \beta_d} \tag{7.74}$$

or, more conservatively, as

$$EI = \frac{0.40E_c I_g}{1 + \beta_d} \tag{7.75}$$

where E_c and E_s are, respectively, the Young's moduli for concrete and steel; I_g is the moment of inertia of the gross concrete section; I_{se} is the moment of inertia of the reinforcement about the centroidal axis of the member cross section; and β_d is the ratio of maximum factored dead or sustained load to maximum total factored load. The β_d factor accounts for the fact that, due to creep deformations under a sustained load (usually dead load), the steel assumes a proportionally larger stress than indicated by linear strain compatibility. Thus, the concrete's remaining strain capacity is reduced. Moreover, the steel is stressed closer to yield and is therefore less effective in resisting buckling than in a column where sustained load constitutes a smaller fraction of the total column load.

l_u is the unsupported length of the column and k the effective length factor, which we know depends on the column's end boundary conditions. These in turn depend on the relative stiffnesses of the attached frame members.

Finally, the radius of gyration of a section, that is,

$$r = \sqrt{\frac{I}{A}} \tag{7.76}$$

may be approximated by $0.3h$ for a rectangular section of depth h, or by $0.25D$ for a circular section of diameter D.

EXAMPLE 7.15

A column with the loads indicated in Fig. 7.36 is part of a building frame that is restrained against sidesway. Determine the required reinforcement for the case (a) in which the moments produce counter-curvature, and (b) in which the moments produce single-curvature bending. Given: $f_c' = 3$ ksi; $f_y = 60$ ksi; $EI = 28{,}000$ kip-ft^2; $kl_u = 15$ ft.

Solution
a) *Counter-Curvature Bending*
Check if slenderness effect needs to be considered:

$$\frac{kl_u}{r} = \frac{(15)(12)}{(0.3)(16)} = 37.5$$

$$\frac{M_1}{M_2} = -\frac{80}{100} = -0.8; \text{ use } -0.5$$

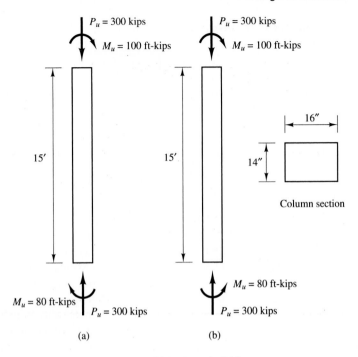

Figure 7.36 Example 7.15

$34 - (12)(-0.5) = 40 > 37.3$. Column can be treated as short, but we shall determine the moment magnification factor anyway.

Euler buckling load: $P_c = \dfrac{\pi^2 EI}{(kl_u)^2} = \dfrac{\pi^2(28{,}000)}{15^2} = 1228$ kips;

$C_m = 0.6 - 0.4\dfrac{80}{100} = 0.28;$ use 0.4

Check minimum moment:

$P_u(0.6 + .03h) = (300)(0.6 + .03 \times 16)/12 = 27$ ft-kips < 100 ft-kips   **OK**

Moment magnification factor: $\delta_{ns} = \dfrac{C_m}{1 - \dfrac{P_u}{0.75 P_c}} = \dfrac{0.4}{1 - \dfrac{300}{(0.75)(1228)}} = 0.59;$ use 1.0

Design loads: $M_c = 100$ ft-kips; $P_u = 300$ kips

Design chart coordinates:

$\dfrac{P_u}{A_g} = \dfrac{300}{224} = 1.34;$ $\dfrac{M_c}{A_g h} = \dfrac{(100)(12)}{(224)(16)} = 0.33;$ $\gamma = \dfrac{16 - 5}{16} = .69$

From Fig. B.2 for $\gamma = 0.60$, find $\rho_g = 0.015$

From Fig. B.3 for $\gamma = 0.75$, find $\rho_g = 0.012$

For $\gamma = 0.69$, linear interpolation gives $\rho = .012 + \dfrac{.75 - .69}{.75 - .60}(.015 - .012) = .013$

and $A_{st} = (.013)(224) = 2.91$ in^2.   Use 4 #8 bars.

b) *Single-Curvature Bending*

Check if slenderness effect needs to be considered:

$$\frac{kl_u}{r} = 37.5 > 34 - (12)(0.5) = 28. \text{ Column is slender.}$$

$$C_m = 0.6 + 0.4\frac{80}{100} = 0.92$$

Moment magnification factor: $\delta_{ns} = \dfrac{0.92}{1 - \dfrac{300}{(0.75)(1228)}} = 1.36$

Design loads: $M_c = (1.36)(1200) = 1637$ in-kips; $P_u = 300$ kips

Design chart coordinates: $\dfrac{P_u}{A_g} = 1.34$; $\dfrac{M_c}{A_g h} = \dfrac{1637}{(224)(16)} = 0.46$; $\gamma = .69$

From Fig. B.2 for $\gamma = 0.60$, find $\rho_g = 0.030$

From Fig. B.3 for $\gamma = 0.75$, find $\rho_g = 0.024$

For $\gamma = 0.69$, find $\rho = .024 + \dfrac{.06}{.15}(.030 - .024) = .026$ and $A_{st} = (.026)(224) = 5.82$ in^2.

Use 6 #9 bars.   ∎

When dealing with unbraced or *sway frames*, we must distinguish, for each separate load combination, between those loads that cause appreciable sidesway (i.e., lateral loads) and those that do not (generally gravity loads). The factored end moments due to loads that cause sidesway are designated as M_{1s} and M_{2s}. These need to be magnified because of the $P\delta$ effect. The end moments due to loads that do not cause sidesway, M_{1ns} and M_{2ns}, are not magnified and can be determined by a first-order analysis. The magnified sway moments $\delta_s M_s$ can be computed using one of the following three methods:

a) a second-order elastic frame analysis, using reduced moment of inertia values (e.g., $0.35I_g$ for beams and $0.70I_g$ for columns) in lieu of the expressions of Eq. 7.74 or 7.75 for braced frames, to reflect the actual stiffness of frame members immediately prior to failure. Also, the β_d term is zero, except in the unusual case that the lateral loads that cause the sidesway are permanent;

b) using the expression

$$\delta_s M_s = \frac{M_s}{1 - Q} \geq M_s \tag{7.77}$$

provided $\delta_s \leq 1.5$, where Q was given by Eq. 7.67;

c) using the moment magnifier method,

$$\delta_s M_s = \frac{M_s}{1 - \dfrac{\sum P_u}{0.75\sum P_c}} \geq M_s \tag{7.78}$$

where $\sum P_u$ is the sum of all factored vertical loads in a story, and $\sum P_c$ is the sum of the corresponding Euler buckling strengths. The 0.75 factor signifies the strength reduction factor ϕ (note also Eq. 7.70). Equation 7.78 guards against the case that all columns in a story buckle at the same time.

The design moments for the compression member then become

$$M_1 = M_{1ns} + \delta_s M_{1s} \tag{7.79}$$

$$M_2 = M_{2ns} + \delta_s M_{2s} \tag{7.80}$$

Slenderness effects can be ignored if

$$\frac{kl_u}{r} < 22 \tag{7.81}$$

But for columns with a high slenderness ratio or heavy loads, specifically if

$$\frac{l_u}{r} > \frac{35}{\sqrt{\dfrac{P_u}{f_c' A_g}}} \tag{7.82}$$

it is possible that the bending moment between ends exceeds the larger of the two end moments (Fig. 7.37). In this case the design moment should be computed as

$$M_c = \delta_{ns}(M_{2ns} + \delta_s M_{2s}) \tag{7.83}$$

Even if there are no sidesway-producing lateral loads present, a sway frame can become unstable under gravity loads alone. To guard against this failure mode, the ACI Code prescribes a check that depends on the method used to compute the sway moments:

a) If the sway moments are determined by a second-order analysis, the ratio of lateral deflections determined in such a second-order analysis to the first-order lateral deflections caused by 1.4 times dead load, 1.7 times live load, plus lateral load shall not exceed 2.5. This check should be performed for the structure as a whole as well as for any individual story that is significantly more flexible than the others.

b) If the sway moments are determined using Eq. 7.77, the value for Q computed using $\sum P_u$ for 1.4 times dead load plus 1.7 times live load shall not exceed 0.60, which corresponds to a value $\delta_s = 2.5$.

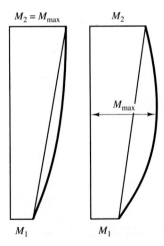

$M_2 = M_{max}$ M_2

M_{max}

M_1 M_1

Figure 7.37   Column with maximum moment between ends

c) If the moment magnifier method (Eq. 7.78) is used to compute sway moments, then δ_s shall be less than 2.5 for $\sum P_u$ and $\sum P_c$, corresponding to the factored dead and live load.

EXAMPLE 7.16

The portal frame of Fig. 7.38 is fixed at the foundations and carries a uniform dead load of $w_D = 1.0$ kips/ft, a uniform live load of $w_L = 1.2$ kips/ft, and a horizontal force due to wind, $H_W = 8$ kips. Determine the required column reinforcement. Given: $f_c' = 3$ ksi; $f_y = 60$ ksi; $EI_{col} = 20,000$ kips-ft^2; $EI_{beam} = 145,000$ kips-ft^2; $kl_u = 14$ ft; 2 in concrete cover.

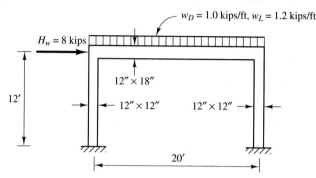

Figure 7.38   Example 7.16

Solution

a) *Moment distribution for unit loads:*

Beam stiffness factor (symmetric): $\dfrac{2EI}{L} = \dfrac{(2)(145,000)}{20} = 14,500$

Column stiffness factor: $\dfrac{4EI}{L} = \dfrac{(4)(20,000)}{12} = 6667$

Distribution factors: $DF_{beam} = \dfrac{14.5}{14.5 + 6.7} = 0.68;$ $DF_{col} = 0.32$

Fixed-end moment: $M_F = \dfrac{wL^2}{12} = \dfrac{(1)(20)^2}{12} = 33.3$ ft-kips

Column top moment after distribution: $(0.32)(33.3) = 10.7$ ft-kips

Column bottom moment: $(0.5)(10.7) = 5.3$ ft-kips

For horizontal load, the beam can be assumed to be very stiff, so that the column inflection points appear at midheight, and the column moments at top and bottom due to $H = 1$ kip are equal to $\dfrac{(1)(12)}{4} = 3$ ft-kips.

b) *Design the column first for load case "dead plus live load," and later check the design for load cases involving wind:*

$$w_u = (1.4)(1) + (1.7)(1.2) = 3.44 \text{ kips/ft}$$

Factored moments:

$$M_{2ns} = (10.7)(3.44) = 36.8 \text{ ft-kips}; \quad M_{1ns} = -(5.3)(3.44) = -18.4 \text{ ft-kips}$$

Factored axial force: $P_u = (3.44)(10) = 34.4$ kips

Euler buckling load: $P_c = \dfrac{\pi^2(20,000)}{14^2} = 1007$ kips

Moment magnification factor:

$$C_m = 0.6 - 0.4\frac{18.4}{36.8} = 0.4; \; \delta_b = \frac{0.4}{1 - \frac{34.4}{(.75)(1007)}} = .42; \quad \text{use } 1.0$$

Design loads: $P_u = 34.4$ kips; $M_c = 36.8$ ft-kips; $\gamma = \dfrac{7}{12} = .583 \approx .60$

Design chart coordinates: $\dfrac{P_u}{A_g} = \dfrac{34.4}{144} = .24; \; \dfrac{M_c}{A_g h} = \dfrac{(36.8)(12)}{(144)(12)} = .26$

From Fig. B.2 find $\rho_g = .011$.

c) *Check design for load case "dead load plus wind":*
Column top moments: $M_{2ns} = (1.4)(1)(10.7) = 15.0; \; M_{2s} = (1.7)(8)(3) = 40.8$
Column bottom moments: $M_{1ns} = (1.4)(1)(5.3) = 7.5; \; M_{1s} = (1.7)(8)(3) = 40.8$
Factored axial force: $P_u = (1.4)(1)(10) = 14$ kips
Euler buckling load: $P_c = 1007$ kips (as before)

Moment magnification factor: $\delta_s = \dfrac{1}{1 - \frac{(2)(14)}{(.75)(2)(1007)}} = 1.02$

Design loads: $P_u = 14$ kips; $M_c = 15.0 + (1.02)(40.8) = 56.6$ ft-kips

Design chart coordinates: $\dfrac{P_u}{A_g} = \dfrac{14}{144} = 0.1; \; \dfrac{M_c}{A_g h} = \dfrac{(56.6)(12)}{(144)(12)} = .39$

From Fig. B.2 find $\rho_g = .022$ (more critical than load case "dead plus live").

d) *Check design for load case "dead plus live plus wind":*
(note the reduction factor of 0.75 for this load combination):
Column top moments: $M_{2ns} = (.75)(3.44)(10.7) = 27.6; \; M_{2s} = (.75)(1.7)(8)(3) = 30.6$

Column bottom moments: $M_{1ns} = \dfrac{27.6}{2} = 13.8; \; M_{1s} = 30.6$

Factored axial force: $P_u = (.75)(3.44)(10) = 25.8$ kips

Moment magnification factor: $\delta_s = \dfrac{1}{1 - \frac{(2)(25.8)}{(.75)(2)(1007)}} = 1.04$

Design loads: $P_u = 25.8$ kips; $M_c = 27.6 + (1.04)(30.6) = 59.4$ ft-kips

Design chart coordinates: $\dfrac{P_u}{A_g} = \dfrac{25.8}{144} = 0.18; \; \dfrac{M_c}{A_g h} = \dfrac{(59.4)(12)}{(144)(12)} = .41$

From Fig. B.2, find $\rho_g = .023$ (most critical load case).
Required steel: $A_{st} = (.023)(144) = 3.31$ in^2. Use 4 #9 bars.

e) *Check for individual column buckling case:*

$$\frac{l_u}{r} = \frac{(12)(12)}{(0.3)(12)} = 40 \; < \; 35/\sqrt{\frac{P_u}{f_c' A_g}} = 35/\sqrt{\frac{34.4}{(3)(144)}} = 124 \; \underline{\text{OK}} \quad \blacksquare$$

EXAMPLE 7.17

Design the first-story columns of a three-story, two-bay office building (Fig. 7.39). Given: $f_c' = 3000$ psi; $f_y = 60$ ksi; $E_c = 3122$ ksi; $E_s = 29,000$ ksi. In addition, a conventional structural analysis has determined the following forces and moments:

Dead load: $P_D^{\text{ext}} = 89$ kips; $P_D^{\text{int}} = 152$ kips; $M_D^{\text{ext}} = 6.3$ ft-kips; $M_D^{\text{int}} = 0$

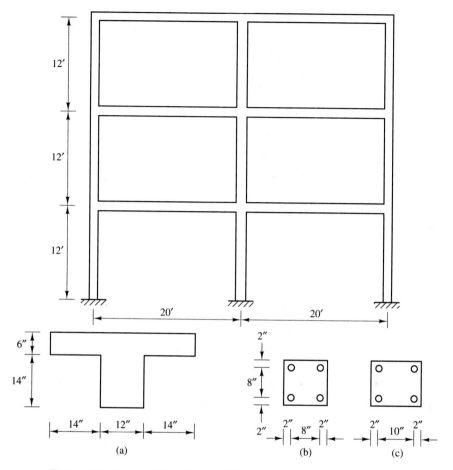

Figure 7.39 Example 7.17: (a) typical beam section; (b) exterior column section; (c) interior column section.

Live load: P_L^{ext} = 48 kips; P_L^{int} = 96 kips; M_L^{ext} = 6.6 ft-kips; M_L^{int} = 0
Partial live load (single half floor loaded):

$\quad\quad P_{L'}^{ext}$ = 20 kips; $P_{L'}^{int}$ = 20 kips; $M_{L'}^{ext}$ = 6.6 ft-kips; $M_{L'}^{int}$ = 6 ft-kips
Earthquake: P_E^{ext} = 5 kips; P_E^{int} = 0; M_E^{ext} = 25 ft-kips; M_E^{int} = 50 ft-kips.

Solution We shall design the columns as short columns first, then check their adequacy in the case that the slenderness effect needs to be considered.

a) *Design for the load case "dead plus live load plus earthquake":*
Interior column:

$\quad\quad P_u^{int} = (.75)[(1.4)(152) + (1.7)(96) + 0] = 282$ kips
$\quad\quad M_u^{int} = (.75)(1.7)(1.1)(50) = 70$ ft-kips

$\quad\quad$(Note: if earthquake is substituted for wind, add load factor 1.1.)

Design chart coordinates:

$$\frac{P_u}{A_g} = \frac{282}{168} = 1.68; \quad \frac{M_u}{A_g h} = \frac{(70)(12)}{(168)(12)} = .42; \quad \gamma = \frac{10}{14} = .71$$

From the design graphs, find $\rho_g = .030$ or $A_{st} = (.03)(168) = 5.0 \text{ in}^2$.

Use 6 #9 bars with 6.0 in².

Exterior column:

$$P_u^{\text{ext}} = (.75)[(1.4)(89) + (1.7)(48) + (1.7)(1.1)(5)] = 162 \text{ kips}$$
$$M_u^{\text{ext}} = (.75)[(1.4)(6.3) + (1.7)(6.6) + (1.7)(1.1)(25)] = 50 \text{ ft-kips}$$

Design chart coordinates:

$$\frac{P_u}{A_g} = \frac{162}{144} = 1.125; \quad \frac{M_u}{A_g h} = \frac{(50)(12)}{(144)(12)} = .35; \quad \gamma = \frac{8}{12} = .67$$

From the design graphs, find $\rho_g = .013$ or $A_{st} = (.013)(144) = 1.87 \text{ in}^2$. Use 4 #7 bars.

b) *Check design for load case "dead plus live load":*

Interior column:

$$P_u^{\text{int}} = (1.4)(152) + (1.7)(96) = 376 \text{ kips}; \quad M_u^{\text{int}} = 0$$

Design chart coordinates:

$$\frac{P_u}{A_g} = \frac{376}{168} = 2.24; \quad \frac{M_u}{A_g h} = 0; \text{ find } \rho_g = .025 < .031 \text{ OK}$$

Exterior column:

$$P_u^{\text{ext}} = (1.4)(89) + (1.7)(48) = 206 \text{ kips}$$
$$M_u^{\text{ext}} = (1.4)(6.3) + (1.7)(6.6) = 20 \text{ ft-kips}$$

Design chart coordinates:

$$\frac{P_u}{A_g} = \frac{206}{144} = 1.43; \quad \frac{M_u}{A_g h} = \frac{(20)(12)}{(144)(12)} = .14; \text{ find } \rho_g < \rho_{\text{min}} \text{ OK}.$$

c) *Check design for load case "dead plus partial live load":*

Interior column:

$$P_u^{\text{int}} = (1.4)(152) + (1.7)(20) = 247 \text{ kips}$$
$$M_u^{\text{int}} = (1.4)(0) + (1.7)(6) = 10.2 \text{ ft-kips}$$

Design chart coordinates:

$$\frac{P_u}{A_g} = \frac{247}{168} = 1.47; \quad \frac{M_u}{A_g h} = \frac{(10.2)(12)}{(168)(12)} = .06; \text{ find } \rho_g < \rho_{\text{min}} \text{ OK}$$

Exterior column:

$$P_u^{\text{ext}} = (1.4)(89) + (1.7)(20) = 159 \text{ kips}$$
$$M_u^{\text{ext}} = (1.4)(6.3) + (1.7)(6.6) = 20 \text{ ft-kips}$$

Design chart coordinates:

$$\frac{P_u}{A_g} = \frac{159}{144} = 1.10; \quad \frac{M_u}{A_g h} = \frac{(20)(12)}{(144)(12)} = .14; \text{ find } \rho_g < \rho_{\text{min}} \text{ OK}.$$

d) *Check Slenderness Effect:*

Moments of inertia (gross sections):

$$I_b = 13,346 \text{ in}^4; \quad I_c^{\text{ext}} = 1728 \text{ in}^4; \quad I_c^{\text{int}} = 2744 \text{ in}^4;$$

Interior column:

$$\psi_{top} = \frac{\sum(EI/L)_{column}}{\sum(EI/L)_{beam}} = \frac{(2)(2744)/12}{(2)(13,346)/20} = .34$$

$\psi_{bot} = 0.1$ (rather than 0 for fixed end)

From nomogram find effective length factor, $k = 1.08$

Radius of gyration: $r = (0.3)(14) = 4.2$ in

Slenderness ratio: $\dfrac{kl_u}{r} = \dfrac{(1.08)(12)(12)}{4.2} = 37 > 22.$

Consider slenderness in critical load case "dead plus live load plus earthquake."

Factored loads:

$$P_u = 282 \text{ kips}; \; M_{2ns} = 0; \; M_{2s} = 70 \text{ ft-kips}; \; P_D = (.75)(1.4)(152) = 160 \text{ kips}$$

$$E_c = 3122; \; E_s = 29,000; \; I_g = 2744; \; I_{se} = (6.0)(5)^2 = 150; \; \beta_d = \frac{160}{282} = .567$$

Column rigidity: $EI = \dfrac{(0.2)(3122)(2744) + (29,000)(150)}{1 + .567} = 3.87 \times 10^6$ kips-in^2

Euler buckling load: $P_c = \dfrac{\pi^2 3.87 \times 10^6}{(1.08 \times 12 \times 12)^2} = 1579$ kips.

Exterior column:

$$\psi_{top} = \frac{(2)(1728)/12}{(13,346)/20} = .43; \; \psi_{bot} = 0.1, \text{ and } k = 1.1$$

Radius of gyration: $r = (0.3)(12) = 3.6$ in

Slenderness ratio: $\dfrac{kl_u}{r} = \dfrac{(1.1)(12)(12)}{3.6} = 44$

Factored loads: $P_u = 162$ kips; $M_{2ns} = 17$ ft-kips; $M_{2s} = 33$ ft-kips

$P_D = (.75)(1.4)(89) = 93$ kips

$$E_c = 3122; \; E_s = 29,000; \; I_g = 1728; \; I_{se} = (2.41)(4)^2 = 38.6; \; \beta_d = \frac{93}{162} = .574$$

Column rigidity: $EI = \dfrac{(0.2)(3122)(1728) + (29,000)(38.6)}{1 + .574} = 1.40 \times 10^6$ kips-in^2

Euler buckling load: $P_c = \dfrac{\pi^2 1.40 \times 10^6}{(1.1 \times 12 \times 12)^2} = 549$ kips.

Interior column:

Moment magnification factor: $\delta_s = \dfrac{1}{1 - \frac{162+282+162}{(.75)(549+1579+549)}} = 1.48$

Design moment: $M_c = M_{2ns} + \delta_s M_{2s} = 0 + (1.48)(70) = 103$ ft-kips

Check minimum moment:

$$P_u(0.6 + .03h) = (282)(0.6 + .03 \times 14)/12 = 24 \text{ ft-kips} < 103 \text{ ft-kips} \underline{\text{OK}}$$

Design graph coordinates: $\dfrac{P_u}{A_g} = 1.68$ (as before); $\dfrac{M_c}{A_g h} = \dfrac{(103)(12)}{(168)(12)} = .62; \; \gamma = .71$

Find $\rho_g = .045$ and $A_{st} = (.045)(168) = 7.56 > 6.0$ in^2

The 6 #9 bars are not adequate. <u>Use 6 #10</u>.

Exterior column:

Design moment: $M_c = 17 + (1.48)(33) = 66$ ft-kips

Check minimum moment:

$$P_u(0.6 + .03h) = (162)(0.6 + .03 \times 12)/12 = 13 \text{ ft-kips} < 67 \text{ ft-kips} \underline{\text{OK}}$$

Design graph coordinates:

$$\frac{P_u}{A_g} = 1.125 \text{ (as before)}; \quad \frac{M_c}{A_g h} = \frac{(66)(12)}{(144)(12)} = .46; \gamma = .67$$

Find $\rho_g = .023$ and $A_{st} = (.023)(144) = 3.31 > 2.41 \text{ in}^2$

The 4 #7 bars are not adequate. Use 4 #9.

Check for individual buckling of interior column:

$$\frac{l_u}{r} = \frac{144}{4.2} = 34 \text{ vs } 35/\sqrt{\frac{P_u}{f_c' A_g}} = 35/\sqrt{\frac{376}{(3)(144)}} = 37 \underline{\text{OK}}$$

If the check were negative, we would need to compute the non-sway moment magnification factor. For interior columns, moments are zero. For exterior columns:

$$M_1 = -0.5M_2, C_m = 0.4, \text{ and } \delta_{ns} = \frac{0.4}{1 - \frac{206}{(.75)(549)}} = 0.8. \text{ Use 1.0, that is, design}$$

moments would not change in our case.   ∎

SUMMARY

1. The reinforcing ratio for the ACI spiral is given by

$$\rho_s = 0.45 \left(\frac{A_g}{A_c} - 1 \right) \frac{f_c'}{f_y}$$

Less steel reduces the column's ductility. More steel increases the strength and ductility only marginally.

2. The kern distances for a section are defined by

$$k_t = \frac{I}{A c_b} \quad \text{and} \quad k_b = \frac{I}{A c_t}$$

and determine the region within which an axial load can be applied without causing tension.

3. Column stresses under service loads are computed from the following equations.

$$bkd \frac{f_c}{2} \left(e - c_t + \frac{kd}{3} \right) + A_s' f_s'(e - c_t + d') - A_s f_s(e - c_t + d) = 0 \quad \text{(a)}$$

$$f_s = \frac{d - kd}{kd} n f_c \quad \text{(b)}$$

$$f_s' = \frac{kd - d'}{kd} n f_c \quad \text{(c)}$$

$$P - bkd \frac{f_c}{2} - A_s' f_s' + A_s f_s = 0 \quad \text{(d)}$$

$$f_c = \frac{P}{n A_s'(1 - \frac{d'}{kd}) - n A_s(\frac{d}{kd} - 1) + \frac{bkd}{2}} \quad \text{(e)}$$

To determine the neutral axis, substitute Eqs. b and c into a and solve the resulting cubic equation for kd. The stresses follow from Eqs. b, c, and e.

4. The nominal column capacity is controlled by the two equilibrium equations

$$P_n = 0.85 f_c' ab + A_s' f_y' - A_s f_s$$

$$P_n e' = 0.85 f_c' ab \left(d - \frac{a}{2} \right) + A_s' f_y' (d - d')$$

To determine the load capacity for a given eccentricity e', proceed as follows:

Step 1. Assume a value for a.

Step 2. Compute the corresponding tension steel stress,

$$f_s = \frac{\beta_1 d - a}{a} 87,000 \le f_y.$$

Step 3. Compute the nominal load capacity:

$$P_n = \left[0.85 f_c' ab \left(d - \frac{a}{2} \right) + A_s' f_y' (d - d') \right] / e'$$

Step 4. Use the vertical equilibrium equation to improve a:

$$a = \frac{P_n - A_s' f_y' + A_s f_s}{.85 f_c' b}$$

Step 5. Return to Step 2 with the improved value for a and continue the iteration until convergence.

5. Balanced load conditions are characterized by

$$P_b = .85 f_c' ab = .85 \beta_1 bd f_c' \frac{87,000}{87,000 + f_y}$$

$$e_b = \frac{1}{P_b} \left[.85 f_c' ab \left(d - \frac{a}{2} \right) + A_s' f_y (d - d') \right] - \frac{d - d'}{2}$$

The balanced point separates the two branches of the column interaction diagram or failure curve. For $P_u < P_b$, failure is initiated by steel yielding; for $P_u > P_b$, failure is initiated by crushing of concrete.

6. At failure, the eccentricity of the axial load, $e = e' - [(d - d')/2]$, should be measured from the plastic centroid, defined by

$$\bar{y}_p = \frac{\sum_i A_i f_i y_i}{\sum_i A_i f_i}$$

7. Intermediate bars and high-strength steels are usually economical if the moment is relatively small compared with the axial load, or the axial load is high, such as in the lower stories of tall buildings.

8. The failure load for a column subjected to biaxial bending can be approximated by the formula

$$\frac{1}{P_n} = \frac{1}{P_{xo}} + \frac{1}{P_{yo}} - \frac{1}{P_o}$$

9. The strength reduction factor for a tied column is $\phi = 0.7$; for a spiral-reinforced column, $\phi = 0.75$. For small eccentricities, an additional reduction factor of 0.8 (for tied columns) or 0.85 (for spiral-reinforced columns) is applied, such that for a tied column, $P_u = .80\phi[.85 f_c'(A_g - A_{st}) + f_y A_{st}]$, with $\phi = 0.7$. For small axial forces, ϕ can be interpolated linearly between 0.9 (for $P_u = 0$) and 0.7 (or 0.75) (for $P_u = 0.1 f_c' A_g$).

10. General long-hand design procedure for rectangular columns:

Step 1. Compute the required nominal column strength:

$$P_n = \frac{P_u}{\phi} \quad \text{and} \quad M_n = \frac{M_u}{\phi} = \frac{P_u e}{\phi}$$

Step 2. Assume a section with dimensions b, h, d and, d'; for example, start with $A_g = 2 P_n / f_c'$.

Step 3. Assume a value for a; for example, $a = h/4$.

Step 4. Compute steel stresses:

$$f_s = \frac{\beta_1 d - a}{a} 87{,}000 \le f_y \quad \text{and} \quad f_s' = \frac{a - \beta_1 d'}{a} 87{,}000 \le f_y$$

Step 5. Compute the required steel areas:

$$A_s = A_s' = \frac{P_n e' - .85 f_c' a b (d - \frac{a}{2})}{f_s' (d - d')}$$

Step 6. Compute a new concrete compression block depth:

$$a = \frac{P_n - A_s (f_s' - f_s)}{.85 f_c' b}$$

Step 7. Repeat Steps 4, 5, and 6 until convergence is achieved. Using $a_{i+1} = (a_i + a_{i-1})/2$ may speed up convergence.

Step 8. If A_s is too large ($\rho > .08$ or .04), the section depth must be increased. If A_s is too small ($\rho < .01$), the section size may be reduced.

11. Column design procedure using design graphs:

Step 1. Assume a cross section with dimensions b, h, d, d' and $\gamma = (d - d')/h$ (for a rectangular section) or h, h', and $\gamma = h'/h$ (for a circular section). For example, start with $A_g = 2 P_n / f_c'$.

Step 2. Compute coordinates for design graphs:

$$\frac{\phi P_n}{A_g} = \frac{P_u}{A_g} \quad \text{and} \quad \frac{\phi P_n}{A_g} \frac{e}{h} = \frac{M_u}{A_g h}$$

Step 3. Select the appropriate design graph to find ρ_g, and then $A_{st} = \rho_g A_g$.

12. The Euler buckling load is given by

$$P_c = \frac{\pi^2 E A}{(k l_u / r)^2} = \frac{\pi^2 E I}{(k l_u)^2}$$

where kl_u/r is the slenderness ratio and k the effective length factor (1.0 for pinned, 2.0 for fixed-free, 0.5 for fixed-fixed, and 0.7 for fixed-pinned end conditions).

13. Sidesway reduces the load capacity of a compression member considerably and should be prevented wherever possible.

14. The slenderness effect is to be considered if $kl_u/r > 34 - 12M_1/M_2$ (with sidesway prevented), or if $kl_u/r > 22$ (with sidesway not prevented).

    Note that $r = \sqrt{I/A}$ can be approximated by $0.3h$ or $0.3b$ for rectangular sections, and by $0.25D$ for circular sections. Also note the definitions $|M_2| > |M_1|$ and $M_1/M_2 > 0$ for single-curvature bending, and $M_1/M_2 < 0$ for counter-curvature bending.

15. A column in a non-sway frame is to be proportioned for the design moment

$$M_2 = \delta_{ns} M_2 = \frac{C_m}{1 - \frac{P_u}{.75 P_c}} M_2 \geq M_2$$

    with $C_m = 0.6 + 0.4(M_1/M_2) \geq 0.4$. To compute the Euler buckling load P_c, the column stiffness may be computed as

$$EI = \frac{0.2E_c I_g + E_s I_{se}}{1 + \beta_d} \quad \text{or} \quad EI = \frac{0.4E_c I_g}{1 + \beta_d}$$

16. Columns in sway frames are to be proportioned for the design moments

$$M_1 = M_{1ns} + \delta_s M_{1s} \quad \text{and} \quad M_2 = M_{2ns} + \delta_s M_{2s}$$

    where

$$\delta_s = \frac{1}{1 - \frac{\sum P_u}{0.75 \sum P_c}} \geq 1.0$$

    and moments with subscript ns result from loads not producing sidesway, while subscript s denotes moments resulting from loads that do cause sidesway.

17. For columns with

$$\frac{l_u}{r} > \frac{35}{\sqrt{\frac{P_u}{f'_c A_g}}}$$

    the maximum moment may occur between ends, and the design moment is

$$M_2 = \delta_{ns}(M_{2ns} + \delta_s M_{2s})$$

REFERENCES

1. ACI, *Design Handbook in Accordance with the Strength Design Method of ACI 318-89*, Vol. 2, *Columns*, American Concrete Institute, Detroit, MI, Special Publication SP-17A, 1990.

2. Kelsey, J., "World's Tallest RC Building Completed in Record Time," *Concrete International*, Dec. 1993.

3. CRSI, *CRSI Handbook*, 7th ed., Concrete Reinforcing Steel Institute, Schaumburg, IL, 1992.

4. Timoshenko, S. P., and Gere, J. M., *Theory of Elastic Stability*, 2d ed., McGraw-Hill Book Co., New York, 1961.

5. Chen, W. F., and Atsuta, T., *Theory of Beam-Columns*, 2 vols., McGraw-Hill Book Co., New York, 1977.

6. Bazant, Z. P. and Cedolin, L., *Stability of Structures: Elastic, Inelastic, Fracture, and Damage Theories*, Oxford University Press, New York, 1991.

PROBLEMS

For all problems below, unless otherwise specified, assume normal-weight concrete weighing 150 pcf, with $f'_c = 4000$ psi; steel with $f_y = 60,000$ psi; $\beta_d = 0.5$; concrete cover = 2 in; and columns braced normal to the plane of the paper.

7.1 Determine the kern areas of the column sections shown in Fig. P7.1 and indicate them in sketches drawn to scale. Given: $n = 8$.

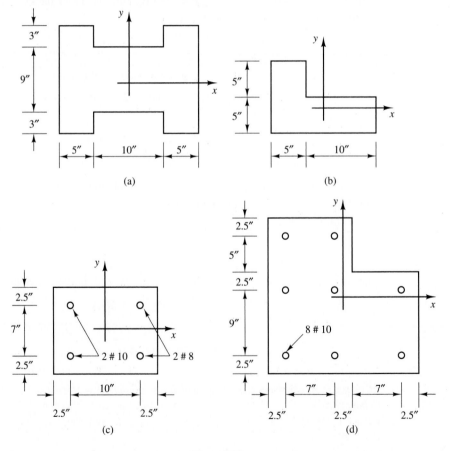

Figure P7.1

7.2 The two column sections shown (Fig. P7.2) are subjected to the service loads $P = 360$ kips and $M = 2520$ in-kips. Determine the maximum steel and concrete stresses. Given: $n = 8$.

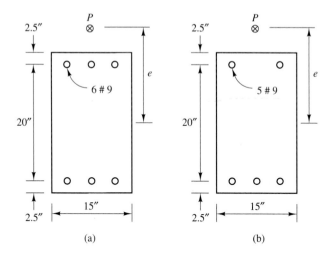

(a)   (b)   **Figure P7.2**

7.3 For the two column sections of Problem 7.2, determine the balanced loads and corresponding eccentricities.

7.4 For the column section shown in Fig. P7.4, locate both the elastic and plastic centroids, using transformed section properties. Given: $n = 8$.

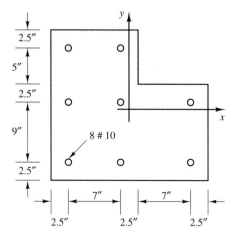

Figure P7.4

7.5 A structural wall with the section shown (Fig. P7.5) and $f'_c = 5$ ksi is loaded with an eccentricity of $e = 24$ in.
(a) Determine the nominal strength of the wall, P_n.
(b) Determine the effectiveness of the 8 #9 intermediate bars, that is, their contribution to the wall's load-carrying capacity in relation to their contribution to the total amount of steel.
(c) What is the increase in load capacity if $f_y = 75$ ksi is used for all bars instead of 60 ksi?

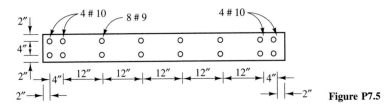

Figure P7.5

7.6 For a column with the section shown in Fig. 7.18 (Example 7.7) determine the load capacity if the load is applied with an eccentricity $e_y = 3$ in only, that is, verify the claim made in Example 7.7 that $P_{yo} = 568$ kips.

7.7 Using design charts, design a short column of rectangular section to carry a 30-kip dead load and a 40-kip live load, in addition to a 42-ft-kip dead-load moment and a 36 ft-kip live-load moment. After having determined the required reinforcement, verify the adequacy of your design using the long-hand iterative procedure.

7.8 The statically indeterminate frame shown in Fig. P7.8 supports a factored uniform load of 21.1 kips/ft. Design the column. Given: $I_b = 130{,}000 \text{ in}^4$.

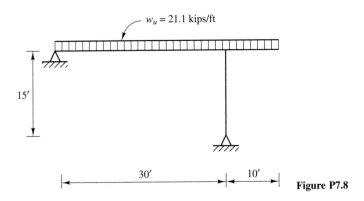

Figure P7.8

7.9 The statically indeterminate frame shown in Fig. P7.9 supports a factored uniform load of 6.125 kips/ft. Design the columns and check them for the slenderness effect. Given: Beam web width = 12 in; $I_b = 100{,}000 \text{ in}^4$.

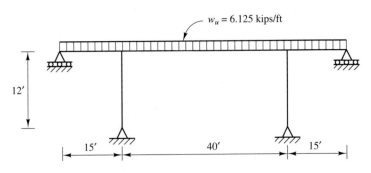

Figure P7.9

7.10 The portal frame shown in Fig. P7.10 is fixed at both supports and carries a factored load of 2.0 kips/ft.
   (a) Determine the required column reinforcement (top and bottom), neglecting the axial force.
   (b) Will the design be satisfactory, if the axial force is included?
   (c) Check the column design, including the slenderness effect.

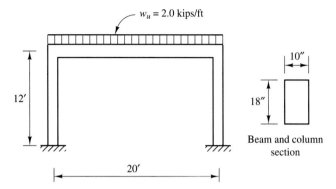

$w_u = 2.0$ kips/ft

12′

20′

10″

18″

Beam and column section

Figure P7.10

7.11 A smokestack with the cross section shown (Fig. P7.11) is to be designed to resist its own weight plus a horizontal wind load of 100 lb/ft.
   (a) Determine the required longitudinal reinforcement at the base. Use load factor 1.7 for wind and the actual concrete area for all calculations involving A_g.
   (b) Check the design with the slenderness effect and modify, if necessary.

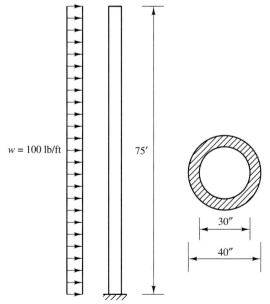

$w = 100$ lb/ft

75′

30″

40″

Figure P7.11

7.12 For each of the three column sections shown in Fig. P7.12, determine the axial load capacity P_u if the load is applied at an eccentricity of $e_x = 8$ in.

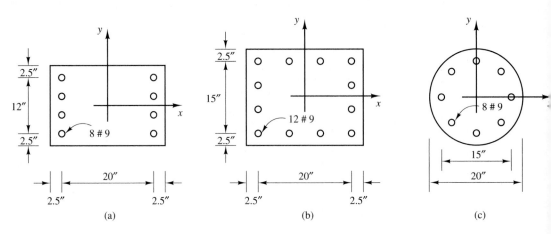

Figure P7.12

7.13 Determine for each of the three column sections of Problem 7.12 the failure loads P_u if they are applied with an $e_y = 5$ in eccentricity in addition to the eccentricity $e_x = 8$ in.

7.14 The columns of the two-bay frame shown in Fig. P7.14 are to be designed for
 (a) uniform dead load of 1.0 kips/ft;
 (b) uniform live load of 1.2 kips/ft;
 (c) horizontal wind load of 5 kips.
 Given: Beam section 20×12 in; column trial section 14×12 in.

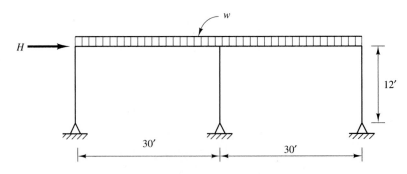

Figure P7.14

7.15 A column with a 15-by-20-in section, reinforced with 4 #10 bars as shown (Fig. P7.15), and pin-connected at both ends, carries a factored load of $P_u = 300$ kips at an eccentricity of 8 in. Determine the maximum unbraced length permitted by the ACI Code. Given: $EI = 80,000$ kips-ft^2.

7.16 Write a subroutine in Fortran or any other programming language to generate a typical point of a column interaction diagram. Assume the following data are provided as input: P_n, ϕ,

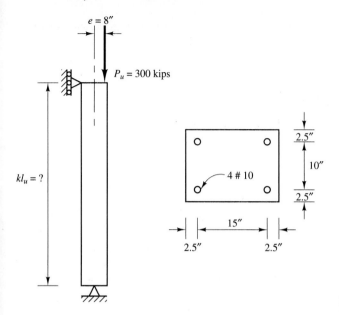

Figure P7.15

A_g, h, γ, ρ_g, f_c', and f_y. Using an iterative procedure such as given in Sec. 7.8, compute $M_u = \phi P_n e$. Disregard the special case $P_n < 0.1 f_c' A_g$.

7.17 Design a tied column, which is part of a braced frame, for the factored loads shown in Fig. P7.17. The unbraced length in the plane of bending is 14.4 ft. In the plane normal to the plane of bending, the unbraced length is 12.8 ft, and the bending moments are zero. Given: $\beta_d = 0.6$.

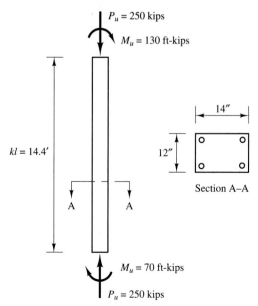

Figure P7.17

7.18 The two-story, one-bay frame of Fig. P7.18 is braced normal to its plane but free to sway in its own plane. Design the lower floor columns to resist the factored loads, $P_u = 190$ ft-kips, $M_u = M_{2s} = 60$ ft-kips. Given: $\beta_d = 0.6$.

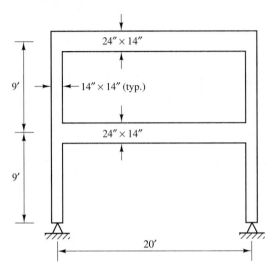

Figure P7.18

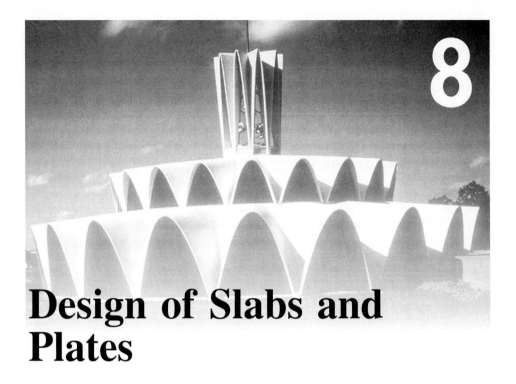

8

Design of Slabs and Plates

8.1 INTRODUCTORY REMARKS

The terms *plates* and *slabs* are often used interchangeably. Plates (or slabs) can be idealized as two-dimensional structural elements, whose middle surfaces are flat or plane and whose thicknesses are small compared with the other two dimensions. Loads are applied normal to their planes and are carried to the supports by bending. The analysis of plates is much more difficult than that of frame structures, because they are highly indeterminate internally, and the accurate determination of even their elastic behavior requires the solution of a difficult problem of elasticity theory. Theoretical or closed-form solutions are possible only for a limited number of cases. For all other cases, including those of practical significance, recourse must be taken to numerical methods or highly approximate analysis techniques.

In order to design concrete plates or slab structures properly, it is not really essential to be fully knowledgeable of classical plate theory, but some knowledge of it is definitely advantageous because it helps us interpret basic plate behavior and understand some of the design requirements of the ACI Code. Moreover, the simplified design methods laid out in the Code are based on a number of quite restrictive assumptions. When these are not applicable, recourse to more advanced design methods, based on plate theory, must be taken.

For these reasons, this chapter starts with a brief introduction to the theory of plates. For a more detailed treatment of the subject see, for example, the standard text by Timoshenko [1] or the classic paper by Westergaard [2]. Some basic knowledge of mechanics of solids, theory of elasticity, and partial differential equations is helpful. However, the reader can skip the next section without the danger of missing essential information necessary to comprehend the remainder of this chapter.

8.2 THEORY OF PLATES

8.2.1 Assumptions

The classical theory of plates is normally based upon the following assumptions:

1. the plate thickness is small compared with the other two plate dimensions;
2. loads are acting normal to the middle surface of the plate;
3. the plate thickness is constant;
4. plate deformations are described accurately enough by the plate deflections $w = w(x, y)$, which are assumed to be small compared with the thickness;
5. a straight line normal to the middle surface before load application shall remain straight and normal to the middle surface after load application;
6. the material is isotropic and homogeneous and has a linear-elastic stress-strain law.

The first two assumptions are part of the definition of a plate. A structural element or component, such as a beam or column, of which only one dimension is large compared with the other two, is normally idealized as a *one-dimensional* element. On the other hand, structural elements of which no single dimension is clearly large or small compared with the other two, such as massive dams, must be analyzed as three-dimensional solids.

The second assumption appears to be restrictive, but it really is not. If a plate is subjected to loads with components within the plane of the plate, then these load components can be dealt with separately as a *plane stress problem*, as long as displacements remain small, so that no coupling between bending and membrane actions takes place. Plate theory concerns itself only with the load components normal to the plate.

The remaining assumptions apply only to the classical linear, small-displacement, thin-plate theory to be discussed here. The fifth assumption, in particular, is also referred to as the *Kirchhoff* assumption and therefore the key to *Kirchhoff plate theory*. It is equivalent to the familiar assumption of plane sections in ordinary beam theory and the disregard of shear deformations. More advanced theories, such as *Reissner* or *Mindlin* plate theories, relax some of the above restrictions by, for example, including shear deformations. Other higher order theories consider large displacements and nonlinear stress-strain laws.

8.2.2 Stresses, Moments, and Shears

Before deriving the plate equation, it is necessary to define stresses, moments, and shears that are present in a plate responding to load. Figure 8.1 shows a segment cut from a plate, such that all components of stress relevant to the problem are set free. There

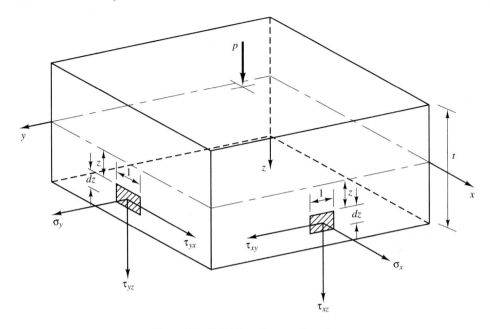

Figure 8.1 Definition of stresses in a plate.

are two normal stress components, σ_x and σ_y, acting along the x- and y-directions, respectively. The σ_z-stress acting in the z-direction is normally neglected in plate theory. The other remaining stress components are $\tau_{xy} = \tau_{yx}$, τ_{yz}, and τ_{xz}.

For design purposes, it is customary to deal with stress resultants rather than with stresses. For the discussion that follows, we therefore define the following bending moments

$$M_x = \int_{-t/2}^{t/2} \sigma_x z\, dz \qquad M_y = \int_{-t/2}^{t/2} \sigma_y z\, dz \qquad M_{xy} = M_{yx} = \int_{-t/2}^{t/2} \tau_{xy} z\, dz \qquad (8.1)$$

and shear forces

$$Q_x = \int_{-t/2}^{t/2} \tau_{xz}\, dz \qquad Q_y = \int_{-t/2}^{t/2} \tau_{yz}\, dz \qquad (8.2)$$

These are illustrated in Fig. 8.2 and shown as positive. For example, to establish the proper sign convention for M_x, we look at the "positive" face, that is, the section facing into the positive x-direction. We consider a positive σ_x-stress at a positive distance z away from the middle surface (Fig. 8.1) and determine the sense of the moment it contributes to M_x, that is, $dM_x = (\sigma_x \cdot dz \cdot 1) \cdot z$. The moment definitions of Eq. 8.1 are quite standard yet give cause for confusion, because the subscript indicates not the axis about which the moment acts but rather the stress component that gives rise to the moment.

All stress resultants defined by Eqs. 8.1 and 8.2 are referring to unit widths. While a beam shear V is a force with units of lb or kN, the plate shears Q_x and Q_y are now forces

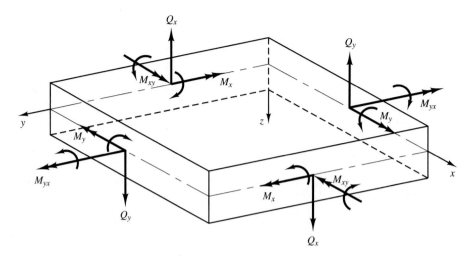

Figure 8.2 Moments and shears in a plate.

per unit length, such as kips/ft or kN/m. Likewise, bending moments are now given in units such as ft-kips/ft or kNm/m. In a very long and narrow plate, one which we would be inclined to call a beam, we can make the assumptions $\sigma_y = \tau_{xy} = \tau_{yx} = \tau_{yz} = 0$. The only surviving stress resultants are M_x and Q_x, the familiar beam shear force and bending moment, respectively.

It is also worthy of note that a general linear stress distribution over the thickness of a plate, such as the one shown in Fig. 8.3, gives rise not only to a bending moment M_x, but also to an in-plane stress resultant

$$N_x = \int_{-t/2}^{t/2} \sigma_x \, dz$$

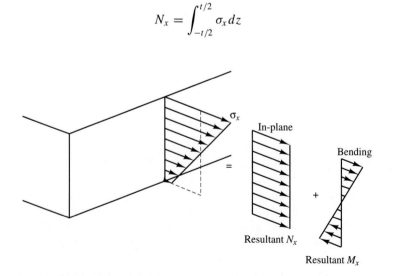

Figure 8.3 Bending moment and in-plane stress resultant.

and similarly

$$N_y = \int_{-t/2}^{t/2} \sigma_y \, dz \quad \text{and} \quad N_{xy} = \int_{-t/2}^{t/2} \tau_{xy} \, dz$$

These three stress resultants can only be nonzero if the applied loads have components within the plane of the plate, which we stipulated in the second assumption not to be the case.

The bending moment $M_{xy} = M_{yx}$ is called the *twisting moment*. It tends to curl the plate into a warped surface and is comparable to the torque in an ordinary beam.

8.2.3 The Plate Equation

In order to derive the plate equation, we start with the investigation of equilibrium of an infinitesimal element. Figure 8.4 illustrates all forces and moments that act on such an element. For example, let the M_x-moment at some section x be "M_x," expressed in ft-lb/ft. To obtain a moment with units ft-lb, we must multiply M_x with the length dy, over which it is acting. If the rate of change of M_x in the x-direction is $\partial M_x / \partial x$, then the total change in moment between the faces at x and $x + dx$, is $\partial M_x / \partial x \cdot dx$, so the moment per unit width at the face $x + dx$ becomes $M_x + (\partial M_x / \partial x) dx$. This must be multiplied again by the width dy, over which it is acting, in order to become a moment with units ft-lb.

Summing all vertical forces shown in Fig. 8.4 leads to

$$p\,dx\,dy + \left(Q_x + \frac{\partial Q_x}{\partial x} dx \right) dy - Q_x dy + \left(Q_y + \frac{\partial Q_y}{\partial y} dy \right) dx - Q_y dx = 0$$

which simplifies to

$$p + \frac{\partial Q_x}{\partial x} + \frac{\partial Q_y}{\partial y} = 0 \tag{8.3}$$

where p is the externally applied load on the plate, such as a uniform pressure given in lb/ft^2. Note that in a beam, in which Q_y would be insignificant, Eq. 8.3 reduces to

$$p = -\frac{dQ_x}{dx}$$

the well-known relationship between shear force and distributed load.

Taking moments about the x-axis at the face at y, we obtain

$$M_y dx - \left(M_y + \frac{\partial M_y}{\partial y} dy \right) dx + \left(Q_y + \frac{\partial Q_y}{\partial y} dy \right) dx\,dy + M_{xy} dy - \left(M_{xy} + \frac{\partial M_{xy}}{\partial x} dx \right) dy$$

$$+ \left(Q_x + \frac{\partial Q_x}{\partial x} dx \right) dy \frac{dy}{2} - Q_x dy \frac{dy}{2} + p\,dx\,dy \frac{dy}{2} = 0$$

Several terms in this equation cancel; others can be neglected because they contain three differential length multipliers, that is, they are of higher order. Thus this equation

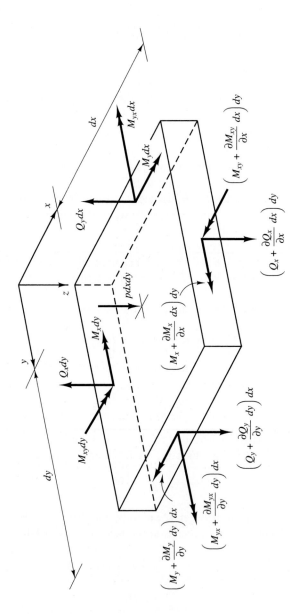

Figure 8.4 Forces and moments acting on infinitesimal plate element.

reduces to

$$\frac{\partial M_{xy}}{\partial x} + \frac{\partial M_y}{\partial y} - Q_y = 0 \qquad (8.4)$$

The last equilibrium equation follows from taking moments about the y-axis at the face at x:

$$\left(M_x + \frac{\partial M_x}{\partial x}dx\right)dy - M_x dy - \left(Q_x + \frac{\partial Q_x}{\partial x}dx\right)dydx + \left(M_{yx} + \frac{\partial M_{yx}}{\partial y}dy\right)dx - M_{yx}dx$$

$$+ Q_y dx\frac{dx}{2} - \left(Q_y + \frac{\partial Q_y}{\partial y}dy\right)dx\frac{dx}{2} - pdxdy\frac{dx}{2} = 0$$

from which

$$\frac{\partial M_x}{\partial x} + \frac{\partial M_{yx}}{\partial y} - Q_x = 0 \qquad (8.5)$$

In a beam, with $M_{yx} = 0$, Eq. 8.5 reduces to

$$Q_x = \frac{dM_x}{dx}$$

the well-known relationship between shear force and bending moment.

The three equilibrium equations (8.3, 8.4, and 8.5) can be combined by differentiating Eq. 8.4 with respect to y and Eq. 8.5 with respect to x, and substituting the results into Eq. 8.3. The result is

$$\frac{\partial^2 M_x}{\partial x^2} + 2\frac{\partial^2 M_{xy}}{\partial x \partial y} + \frac{\partial^2 M_y}{\partial y^2} = -p \qquad (8.6)$$

This is one equation with three unknowns, M_x, M_y, and M_{xy}. Similarly, Eqs. 8.3, 8.4, and 8.5 are three equations with five unknowns. In either case, we are short two equations to be able to derive a unique solution. The problem is statically indeterminate, so recourse must be taken to the deformational characteristics of the plate.

We shall start by recalling the fourth assumption made earlier, namely, that all plate deformations can be related to the plate deflection $w(x, y)$ or its derivatives. Consider some point A, at a distance z below the middle surface (Fig. 8.5). The horizontal displacements of point A in the x- and y-directions are, respectively,

$$u = -z\frac{\partial w}{\partial x}$$

$$v = -z\frac{\partial w}{\partial y} \qquad (8.7)$$

so that the strains in the x- and y-directions follow as

$$\epsilon_x = \frac{\partial u}{\partial x} = -z\frac{\partial^2 w}{\partial x^2} \qquad (8.8)$$

$$\epsilon_y = \frac{\partial v}{\partial y} = -z\frac{\partial^2 w}{\partial y^2} \qquad (8.9)$$

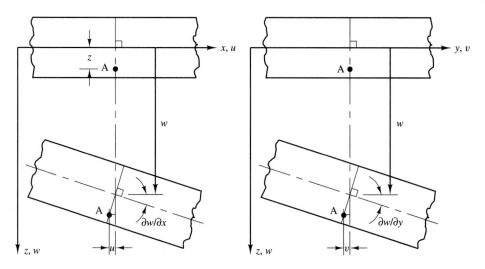

Figure 8.5 u- and v-displacements of a point.

These are the strain-displacement equations. Although they supply two more equations, they also introduce three new unknowns, namely, ϵ_x, ϵ_y, and w. The missing equations must be furnished by the stress-strain relationships, or Hooke's Law. In our case, these are

$$\sigma_x = \frac{E}{1-v^2}(\epsilon_x + v\epsilon_y) = -\frac{Ez}{1-v^2}\left(\frac{\partial^2 w}{\partial x^2} + v\frac{\partial^2 w}{\partial y^2}\right) \tag{8.10}$$

$$\sigma_y = \frac{E}{1-v^2}(\epsilon_y + v\epsilon_x) = -\frac{Ez}{1-v^2}\left(\frac{\partial^2 w}{\partial y^2} + v\frac{\partial^2 w}{\partial x^2}\right) \tag{8.11}$$

$$\tau_{xy} = G\left(\frac{\partial u}{\partial y} + \frac{\partial v}{\partial x}\right) = -2Gz\frac{\partial^2 w}{\partial x \partial y} \tag{8.12}$$

where the strain-displacement equations (8.8 and 8.9) have already been utilized. Now we have a sufficient number of equations to solve for the unknowns. We shall eliminate the three stress components as unknowns by substituting Eqs. 8.10, 8.11, and 8.12 into Eqs. 8.1 and 8.2, for example,

$$M_x = -\frac{E}{1-v^2}\left(\frac{\partial^2 w}{\partial x^2} + v\frac{\partial^2 w}{\partial y^2}\right)\int_{-t/2}^{t/2} z^2 dz$$

$$= -\frac{Et^3}{12(1-v^2)}\left(\frac{\partial^2 w}{\partial x^2} + v\frac{\partial^2 w}{\partial y^2}\right)$$

Introducing the abbreviation

$$D = \frac{Et^3}{12(1-v^2)} \tag{8.13}$$

we get

$$M_x = -D\left(\frac{\partial^2 w}{\partial x^2} + v\frac{\partial^2 w}{\partial y^2}\right) \tag{8.14}$$

and similarly

$$M_y = -D\left(\frac{\partial^2 w}{\partial y^2} + v\frac{\partial^2 w}{\partial x^2}\right) \tag{8.15}$$

$$M_{xy} = M_{yx} = -(1-v)D\frac{\partial^2 w}{\partial x \partial y} \tag{8.16}$$

$$Q_x = -D\frac{\partial}{\partial x}\left(\frac{\partial^2 w}{\partial x^2} + \frac{\partial^2 w}{\partial y^2}\right) \tag{8.17}$$

$$Q_y = -D\frac{\partial}{\partial y}\left(\frac{\partial^2 w}{\partial x^2} + \frac{\partial^2 w}{\partial y^2}\right) \tag{8.18}$$

Substitution of Eqs. 8.14, 8.15, and 8.16 into Eq. 8.6 results in

$$\frac{\partial^4 w}{\partial x^4} + 2\frac{\partial^4 w}{\partial x^2 \partial y^2} + \frac{\partial^4 w}{\partial y^4} \equiv \Delta\Delta w = \frac{p}{D} \tag{8.19}$$

This is the *plate equation*, a fourth-order partial differential equation for the plate deflection w. Once this equation has been solved for a given loading p and a given set of boundary conditions, the bending moments follow from Eqs. 8.14, 8.15, and 8.16, and the shear forces from Eqs. 8.17 and 8.18.

The quantity D (Eq. 8.13) is the *plate stiffness* and is equivalent to the beam stiffness EI. In fact, for a plate strip of unit width, the moment of inertia is $I = t^3/12$, and the elastic modulus is replaced by the effective elastic modulus $E' = E/(1 - v^2)$. If a regular beam is subjected to bending, it tends to expand laterally (due to Poisson's ratio) in the compression zone, and to contract in the tension zone (Fig. 8.6). In a typical plate strip, both lateral deformations are prevented by the neighboring plate strips. We can compute this restraint effect by considering the stress-strain equations

$$\epsilon_x = \frac{1}{E}(\sigma_x - v\sigma_y) \tag{8.20}$$

$$\epsilon_y = \frac{1}{E}(\sigma_y - v\sigma_x) \tag{8.21}$$

Compatibility between deformations of adjacent plate strips requires that $\epsilon_y = 0$, so that Eq. 8.21 becomes

$$\sigma_y = v\sigma_x \tag{8.22}$$

Substituting this back into Eq. 8.20 leads to

$$\epsilon_x = \frac{\sigma_x}{E'} \tag{8.23}$$

where

$$E' = \frac{E}{1 - v^2} \tag{8.24}$$

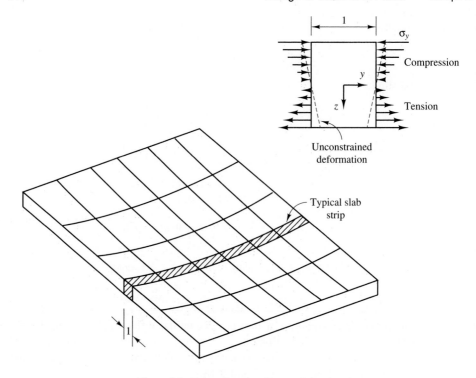

Figure 8.6   Poisson's ratio effect on slab strip.

is the effective elastic modulus in a *plane strain* problem. Figure 8.6 illustrates that the maintenance of compatibility produces σ_y-stresses, given by Eq. 8.22, which in turn produce an M_y-moment. This result is important for reinforced concrete plates, because even if they are subjected to bending in only one direction, they develop moments in the orthogonal direction as well, against which they must be reinforced.

In a beam, the deflection w is independent of y, so that in this case Eq. 8.19 reduces to

$$\frac{d^4 w}{dx^4} = \frac{p}{EI}$$

the well-known differential equation for an ordinary beam, with EI substituted for D.

8.2.4 Boundary Conditions

Any unique solution of a differential equation requires a set of boundary conditions. The plate equation (8.19) is no exception. The complete problem definition must include the support conditions along all plate boundaries. For the sake of simplicity, we shall restrict the following discussion to simple rectangular plates and illustrate it with the plate shown in Fig. 8.7. As in ordinary beam problems, there are three basic types of boundary conditions: fixed, simply supported, and free.

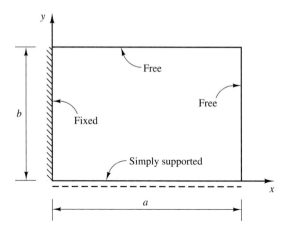

Figure 8.7 Plate boundary conditions.

1. *The fixed edge.* For the fixed edge of the plate shown by cross hatches in Fig. 8.7, the following conditions apply.

$$w = 0 \quad \text{and} \quad \frac{\partial w}{\partial x} = 0 \tag{8.25}$$

However, the support conditions require the slope and curvature to be zero along the support in the y-direction as well, that is,

$$\frac{\partial^2 w}{\partial x \partial y} = 0 \quad \text{and} \quad \frac{\partial^2 w}{\partial y^2} = 0 \tag{8.26}$$

Substituting the conditions of Eqs. 8.25 and 8.26 into Eqs. 8.14, 8.15, and 8.16 gives the additional information

$$M_x = -D \frac{\partial^2 w}{\partial x^2}$$

$$M_y = \nu M_x$$

$$M_{xy} = 0$$

Thus, along a fixed edge, only the twisting moment is zero.

2. *The simply supported edge.* At a simply supported edge, such as the one marked by a dashed line along the x-axis (Fig. 8.7), the following conditions must be satisfied:

$$w = 0 \quad \text{and} \quad M_y = 0 \tag{8.27}$$

In addition, neither slope nor curvature along the support in the x-direction is possible, so,

$$\frac{\partial w}{\partial x} = 0 \quad \text{and} \quad \frac{\partial^2 w}{\partial x^2} = 0 \tag{8.28}$$

If each of the second conditions in Eqs. 8.27 and 8.28 is substituted into Eqs. 8.14 and 8.15, the following further boundary conditions result:

$$\frac{\partial^2 w}{\partial y^2} = 0 \quad \text{and} \quad M_x = 0 \tag{8.29}$$

The boundary conditions for a simply supported edge, also known as *Navier's boundary conditions*, can therefore be written in the form

$$w = 0 \quad \text{and} \quad \frac{M_x + M_y}{1 + \nu} = 0 \tag{8.30}$$

or

$$w = 0 \quad \text{and} \quad \Delta w = \frac{\partial^2 w}{\partial x^2} + \frac{\partial^2 w}{\partial y^2} = 0 \tag{8.31}$$

Thus, along a simply supported edge, both bending moments vanish. The twisting moment M_{xy}, however, does not (in general) vanish. In fact, it has an effect on the reaction, which needs to be investigated further. In Fig. 8.8, the moment $M_{xy}dx$ is

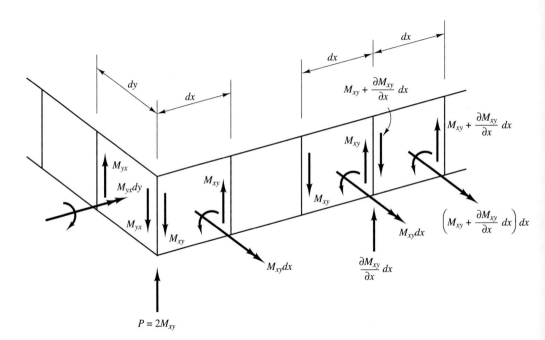

Figure 8.8   Action of twisting moment along boundary.

shown acting along a small length dx of the boundary. This can be interpreted as the moment produced by two "forces" M_{xy}, acting with the lever arm dx. In an adjacent segment, two forces $M_{xy}+(\partial M_{xy}/\partial x)dx$ produce a twisting moment of slightly different magnitude. At the boundary common to both plate segments the two forces cancel, except for the difference $(\partial M_{xy}/\partial x)dx$, which thus gives rise to a reaction force in addition to the regular shear force Q_y. The total or *Kirchhoff shear* now becomes

$$V_y = Q_y + \frac{\partial M_{xy}}{\partial x} \tag{8.32}$$

Similarly, along a simply supported edge parallel to the y-axis,

$$V_x = Q_x + \frac{\partial M_{xy}}{\partial y} \tag{8.33}$$

In a simply supported plate, the reaction must be equal to this effective shear force and not equal to Q_x or Q_y. In a plate corner, the two "forces" M_{xy} do not cancel, but rather are additive, causing a concentrated force $P = 2M_{xy}$. As a result, uniformly loaded, simply supported plates tend to lift off their supports in the corners. If they are tied down by anchors or walls situated on top along the edges, large diagonal moments can arise, against which the plates must be reinforced in order to prevent cracking in the corners (Fig. 8.9).

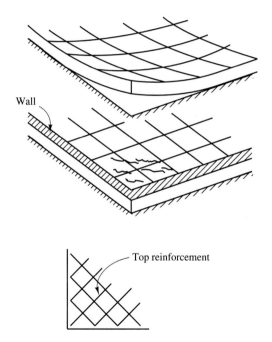

Top reinforcement

Figure 8.9   Reinforcement of slab corner.

3. *The free edge.*   As in beam theory, a free edge such as the one at $y = b$ (Fig. 8.7) is characterized by zero shear and zero moment. However, it is not the regular shear Q_y but rather the Kirchhoff shear V_y that vanishes along this edge. Thus, the boundary conditions for the free edge are

$$M_y = 0 \quad \text{and} \quad V_y = Q_y + \frac{\partial M_{xy}}{\partial x} = 0 \tag{8.34}$$

Using the definitions in Eqs. 8.15, 8.16, and 8.18, these boundary conditions read in terms of the derivatives of the plate deflection w:

$$\frac{\partial^2 w}{\partial y^2} + v \frac{\partial^2 w}{\partial x^2} = 0$$

$$\frac{\partial^3 w}{\partial y^3} + (2 - v) \frac{\partial^3 w}{\partial x^2 \partial y} = 0$$

(8.35)

8.2.5 Navier Solution of Simply Supported Plate

General solutions of the plate equation (8.19), subject to arbitrary boundary conditions, are very difficult to obtain. Closed-form solutions are available only for a limited number of cases and may be found in the literature [1]. As an instructive illustration, let us consider one particular case of practical significance: a rectangular, uniformly loaded plate of side lengths a and b and simply supported along all four sides (Fig. 8.10). The solution is attributed to Navier (1785–1836).

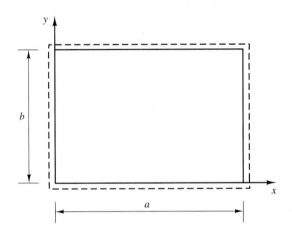

Figure 8.10 Simply supported rectangular plate.

The boundary conditions for this case are given by Eq. 8.31 as

$$w = 0 \quad \text{and} \quad \Delta w = 0$$

along $x = 0$ and a, and along $y = 0$ and b. The trial solution function will be of the form

$$w_{nm}(x, y) = C_{nm} \sin \frac{n\pi x}{a} \sin \frac{m\pi y}{b} \qquad n, m = 1, 2, \ldots$$

which satisfies all boundary conditions along the four edges, with C_{nm} being a free constant yet to be determined. The complete solution will then be the sum of trial solutions for all possible n and m values, that is,

$$w(x, y) = \sum_{n=1}^{\infty} \sum_{m=1}^{\infty} C_{nm} \sin \frac{n\pi x}{a} \sin \frac{m\pi y}{b}$$

(8.36)

provided the given loading $p(x, y)$ can be developed into a Fourier series, that is, the following series of sine functions

$$p(x, y) = \sum_{n=1}^{\infty} \sum_{m=1}^{\infty} p_{nm} \sin \frac{n\pi x}{a} \sin \frac{m\pi y}{b} \tag{8.37}$$

For most practical loadings, this is indeed the case, that is, the Fourier coefficients p_{nm} can be determined from the expression

$$p_{nm} = \frac{4}{ab} \int_{x=0}^{a} \int_{y=0}^{b} p(x, y) \sin \frac{n\pi x}{a} \sin \frac{m\pi y}{b} dx dy \tag{8.38}$$

For example, for uniform load, $p(x, y) = p_o = $ const, the coefficients p_{nm} are readily computed to be

$$p_{nm} = \frac{16}{nm\pi^2} p_o \tag{8.39}$$

By assuming that the given loading can be expressed in the form of the double series (Eq. 8.37), the only unknowns that remain to be determined in the solution of Eq. 8.36 are the coefficients C_{nm}. These will have to satisfy both the plate equation (8.19), and the boundary conditions. Performing the various differentiations on $w(x, y)$ of Eq. 8.36, as prescribed by the plate equation, and substituting Eqs. 8.36 and 8.37 into Eq. 8.19, leads to

$$D\Delta\Delta w = \sum_{n=1}^{\infty} \sum_{m=1}^{\infty} D \left(\frac{n^2}{a^2} + \frac{m^2}{b^2} \right)^2 \pi^4 C_{nm} \sin \frac{n\pi x}{a} \sin \frac{m\pi y}{b}$$

$$= \sum_{n=1}^{\infty} \sum_{m=1}^{\infty} p_{nm} \sin \frac{n\pi x}{a} \sin \frac{m\pi y}{b}$$

Equating each term of these double series separately, the unknown coefficients are determined to be

$$C_{nm} = \frac{p_{nm}}{D \left(\frac{n^2}{a^2} + \frac{m^2}{b^2} \right)^2 \pi^4} \tag{8.40}$$

The final solution thus becomes

$$w(x, y) = \sum_{n=1}^{\infty} \sum_{m=1}^{\infty} \frac{p_{nm}}{D \left(\frac{n^2}{a^2} + \frac{m^2}{b^2} \right)^2 \pi^4} \sin \frac{n\pi x}{a} \sin \frac{m\pi y}{b} \tag{8.41}$$

With the deflection known throughout the plate, moments and shears can be determined according to Eqs. 8.14 through 8.18. The number of *harmonics*, that is, the number of terms in the solution of Eq. 8.41 needed for accurate results, depends on the type of loading, on the response quantity sought (deflection, moment, shear), and on the acceptable error. For a uniformly distributed load, a few terms in each direction (say, 3 or 4) are often sufficient. For concentrated loads, convergence is much slower, but such loads rarely occur in concrete design practice. Even realistic wheel loads of trucks are distributed over an area of at least one square foot. If the load is indeed highly concentrated, 50 to 100 nonzero terms may be required in each direction. Deflections converge fastest. As Eq. 8.41 indicates, the solution convergence is of the order $1/n^4$. Bending moments require two differentiations of Eq. 8.41, thus, convergence is only of the order $1/n^2$. Shear forces, which according to Eqs. 8.17 and 8.18 are proportional to third derivatives of w, converge only with $1/n$, that is, they require the largest number of terms for an accurate determination.

In any case, the amount of numerical work required even for simple problems calls for the use of computers. Moreover, problems that arise in the design of realistic concrete structures, such as slabs that are continuous over several fields, with or without integral beams, walls, or column supports, or with cutouts or nonrectangular plan shapes, cannot be solved at all in closed form. Numerical methods such as finite element or grid analyses incorporated in general computer programs, must be used to obtain useful solutions to such practical problems. Such programs are now generally available. Their use can turn out to be both costly and time consuming if the preparation of input data and proper verification and interpretation of output results are considered.

The practice of engineering without computers is no longer imaginable, and engineering offices are now certain to have access to some computing facilities. Young designers may feel happy about this situation, believing that it eliminates the need to familiarize themselves with the awfully imposing plate theory outlined in this section. However, as was mentioned earlier, such complacency is ill advised. Computer programs do make mistakes, no matter how thoroughly checked and verified they have been through years of use by hundreds or thousands of engineers. The careless user who enters incorrect input data will receive only incorrect results in return, and it takes not only experience but courage to question the validity of such results and to verify them by some approximate alternative method. Designers who ultimately sign the drawings and construction documents bear the final reponsibility. Whether they performed the computer analysis themselves or delegated the task to someone else, they can in no instance blame the computer for a faulty design. In the event of a dispute, the computer is a poor defendant in a court of law.

Whether or not the use of a general computer program is contemplated, designers are well advised, for checking purposes, to perform approximate analyses that do not require a major computational effort. If the computer is forsaken altogether and the design based on only a highly simplified analysis, then the design will have to be accordingly conservative. This is the approach suggested by the ACI Code, which will be discussed in some detail later. However, even if the design is based on the results of a refined computer analysis, the designer should be able to check the results by some approximate method.

8.3 YIELD LINE THEORY

The classical plate theory introduced in the previous section is usually restricted to linear elastic plate behavior. If nonlinear behavior is to be considered, as is the case at ultimate conditions near failure, closed-form solutions become so difficult as to be almost insignificant for practical purposes. Even if numerical methods such as the finite element method are used, the effort required to solve realistic problems is usually prohibitive for most practical problems. There exists an analysis method, however, which with relatively modest effort permits the computation of failure loads and collapse mechanisms of plates. This is called the *yield line theory*, originally proposed by Ingerslev, but expanded and widely propagated by Johansen [3].

When designing a frame using the ASD method, the appropriate method of analysis is linear elastic analysis because, under service loads, structural behavior remains essentially linear elastic. When the USD method is used to design the same frame structure, plastic analysis is the more appropriate technique, because it accounts for the moment redistribution prior to failure in a direct way. If a frame is analyzed by a linear elastic method and then designed for ultimate strength, inconsistencies arise, as we saw in Chapter 6, which the ACI Code attempts to address with an allowance for some limited amount of plastic moment redistribution.

An exactly analogous situation exists in slab design. When a slab is designed for allowable stresses, classical plate theory is the appropriate tool for analysis. However, when the same slab is designed for ultimate strength, we should use an analysis method that establishes consistent relationships between ultimate moment capacities of the plate and the collapse load, just as plastic analysis does for frames. Since current design practice clearly favors the USD method, it is only logical that a plastic analysis method applicable to plate structures be used. Yield line theory is this tool and should be the method of choice.

As in plastic frame analysis, yield line theory is applicable only if all failure modes other than bending are eliminated (i.e., shear and bond failures). Also, a minimum ductility capacity must be assured, that is, reinforcing ratios should be small, so that failure is always initiated by yielding of the steel, followed by large deflections and finally crushing of the concrete in compression. As a *yield line* is a simple extension of a plastic hinge into two dimensions, it plays the same key role in a plate structure as a plastic hinge does in a frame structure for plastic moment redistribution. Most commonly encountered slabs have the low reinforcing ratios necessary to make yield line theory applicable.

Aside from being compatible with USD philosophy, yield line theory is of striking simplicity, which derives from the fact that it is based on the same two simple concepts as plastic frame analysis:

1. equations of static equilibrium at failure, which can be substituted by the principle of virtual work; and
2. the axiom that a slab will always fail in such a way that the load required to collapse the slab is a minimum.

The second axiom is generally known as the *lower bound theorem* of plastic analysis [4] and sometimes referred to as the axiom of *nature's smartness*. We shall now illustrate the basic principles of yield line theory in a series of examples of increasing complexity.

EXAMPLE 8.1

Determine the uniform load w_u (in lb/ft^2), which will cause a rectangular one-way slab, simply supported at two opposite edges and having the ultimate moment capacity M_u (in ft-lb/ft), to fail (Fig. 8.11).

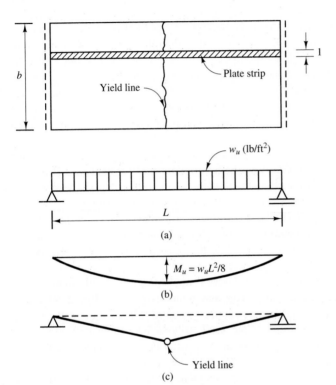

Figure 8.11 Example 8.1: Simply supported one-way slab. (a) One-way slab; (b) moment at failure; (c) deflections at failure.

Solution The applied load w_u causes a moment $w_u L^2/8$ in a slab strip of unit width, which at failure must be equal to the slab's ultimate moment capacity M_u. Once the steel starts to yield, a plastic hinge or yield line forms at midspan. The deflections associated with the concentrated angle change in the yield line are assumed to be large compared with the elastic plate deflections prior to formation of the yield line (Fig. 8.11c). When $w_u L^2/8 = M_u$, the slab cannot redistribute any additional load, so the single yield line triggers a failure mechanism, that is, failure occurs at the load:

$$w_u = \frac{8M_u}{L^2}$$

The problem statement can be reversed: Determine the midspan moment capacity needed to carry the factored load w_u. The answer, of course, would be

$$M_u = \frac{w_u L^2}{8}$$

■

This example is particularly simple, because the slab behaves essentially as a simply supported beam of great width. Therefore such a slab is readily designed by using the methods described in Chapter 4.

EXAMPLE 8.2

The one-way slab of Fig. 8.12 is fixed at both ends and reinforced with #6 bars @ 6 in on center, both at the supports and at midspan. What load w_u will cause the slab to fail? Given: $f'_c = 4$ ksi, $f_y = 60$ ksi.

Solution   The nominal flexural capacity of a 12-in-wide slab strip is found as readily as that of a beam:

Depth of compression block: $a = \dfrac{A_s f_y}{.85 f'_c b} = \dfrac{(2)(.44)(60)}{(.85)(4)(12)} = 1.29$ in

Nominal moment capacity:

$$M_n = A_s f_y \left(d - \frac{a}{2} \right) = (.88)(60) \left(5 - \frac{1.29}{2} \right) = 230 \text{ in-kips/ft.}$$

The load that causes the steel at the supports to yield is found from

$$\frac{w_1 L^2}{12} = M_n, \quad \text{that is,} \quad w_1 = \frac{12 M_n}{L^2} = \frac{(12)(230)}{(12)(10)^2} = 2.3 \text{ kips/ft}^2$$

At this load the slab does not fail, but yield lines form at the supports (like plastic hinges in a fixed beam), and at midspan the moment of $w_1 L^2/24$ is only half of the slab's capacity. For any additional load w_2, the slab will behave as if it were simply supported, with yield lines at the supports, until the midspan reserve capacity of $M_{res} = 230 - 115 = 115$ in-kips/ft is exhausted, that is, until the simple-span moment $w_2 L^2/8$ equals the reserve capacity M_{res}, so that

$$w_2 = \frac{8 M_{res}}{L^2} = \frac{(8)(115)}{(12)(10)^2} = 0.77 \text{ kips/ft}^2$$

At this load, a third yield line forms at midspan, which triggers a failure mechanism. The failure load is thus

$$w_u = w_1 + w_2 = 2.3 + 0.77 = \underline{3.07 \text{ kips/ft}^2} \qquad\blacksquare$$

The reader may recall that we solved a similar problem in Chapter 6 in conjunction with plastic analysis of beams (see Example 6.8). It is worthwhile, however, to reinforce the notion that yield line theory is in fact plastic frame analysis extended to two dimensions. The reader should also be reminded that the strength reserve extracted by this kind of plastic moment redistribution cannot be predicted by an elastic analysis.

EXAMPLE 8.3

What load will cause the slab of the previous example to fail if the #6 bars at the right and left supports are replaced by #7 and #8 bars, respectively?

Solution   In this case, the exact position of the third yield line is unknown a priori, because the problem is no longer symmetric. Calling this unknown location x, and denoting the three

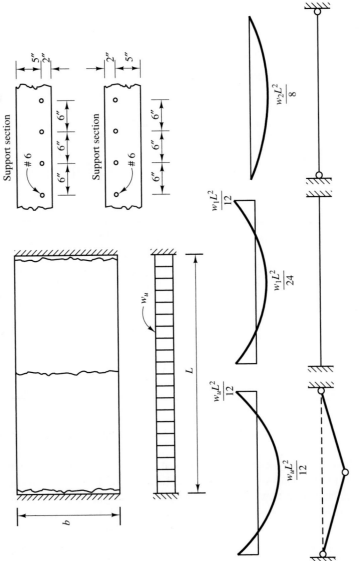

Figure 8.12 Example 8.2: One-way slab fixed at both ends.

nominal moment capacities by M_a, M_b, and M_c, we draw the deflected shape of the slab at failure (Fig. 8.13a). Taking moments on the free body of the left slab segment (Fig. 8.13b) furnishes one equilibrium equation,

$$M_a + M_b = \frac{w_u x^2}{2}$$

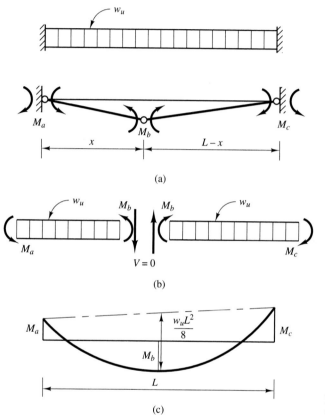

(a)

(b)

(c)

Figure 8.13 Example 8.3: One-way slab with unequal moment capacities.

while the right slab segment satisfies a second equilibrium equation,

$$M_b + M_c = \frac{w_u (L - x)^2}{2}$$

Note that the shear force at x is zero because the center yield line forms at the point of maximum moment, where the shear is zero. Since the three moment capacities are given (or can be computed), there remain two unknowns, x and w_u. Solving these two equations simultaneously, we find

$$x = \frac{L\sqrt{M_a + M_b}}{\sqrt{M_a + M_b} + \sqrt{M_b + M_c}} \tag{8.42}$$

and

$$w_u = \frac{2}{L^2} \left(\sqrt{M_a + M_b} + \sqrt{M_b + M_c} \right)^2 \tag{8.43}$$

From Example 8.2 we know that $M_b = 230$ in-kips/ft, while at the left support

$$a = \frac{(2)(.79)(60)}{(.85)(4)(12)} = 2.32 \quad \text{and} \quad M_a = (1.58)(60)\left(5 - \frac{2.32}{2}\right) = 364 \text{ in-kips/ft}$$

and at the right support

$$a = \frac{(2)(.60)(60)}{(.85)(4)(12)} = 1.76 \quad \text{and} \quad M_c = (1.2)(60)\left(5 - \frac{1.76}{2}\right) = 297 \text{ in-kips/ft}$$

Substituting these capacities into Eq. 8.43 gives the failure load

$$w_u = \frac{2}{(12)(10^2)} \left(\sqrt{364 + 230} + \sqrt{230 + 297} \right)^2 = \underline{3.73 \text{ kips/ft}^2}$$

while Eq. 8.42 gives the position of the center yield line

$$x = \frac{10\sqrt{364 + 230}}{\sqrt{364 + 230} + \sqrt{230 + 297}} = 5.15 \text{ ft}$$

Thus, the center yield line is slightly closer to the right support, that is, further away from the stronger left support. If the two support moment capacities are not too different, the center yield line can be assumed to occur at midspan. In this case, it is an easy matter to establish static moment equilibrium (Fig. 8.13c) as

$$\frac{w_u L^2}{8} = M_b + \frac{M_a + M_c}{2} = 230 + \frac{364 + 297}{2} = 560.5 \text{ in-kips/ft}$$

from which the collapse load follows to be $w_u = \frac{(8)(560.5)}{(12)(10)^2} = 3.74 \text{ kips/ft}^2$. This is obviously close enough to the "exact" failure load of 3.73 kips/ft^2 computed earlier. And again, in Example 6.10, we solved a similar problem with a beam instead of a slab. ∎

EXAMPLE 8.4

Determine the moment capacity of the simply supported square plate of Fig. 8.14 required to support a uniform load of 2.0 kips/ft^2, if the plate is to be equally reinforced in the x- and y-directions.

Solution   To start, we use intuition or engineering common sense to establish that the two diagonal yield lines of Fig. 8.14a form the failure mechanism. At failure, each one of the four plate triangles must be in equilibrium. However, before deriving the necessary equilibrium equation, we need to study the equilibrium of an infinitesimal element (Fig. 8.14c), the sides of which are of length dx, dy, ds, and parallel, respectively, to the x- and y-axis and some yield line oriented at an angle α. If the reinforcements in the x- and y-directions are known, the corresponding flexural capacities M_{ux} and M_{uy} can be determined. With the help of Fig. 8.14c, the bending moment M_u and twisting moment T_u along the inclined yield line can be expressed in terms of M_{ux} and M_{uy}:

$$M_u ds = M_{uy} dx \cos \alpha + M_{ux} dy \sin \alpha$$

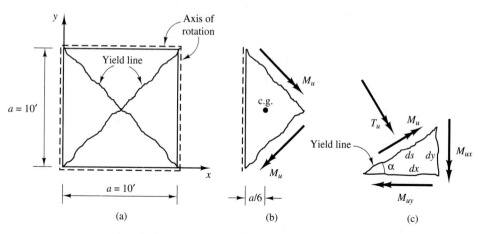

Figure 8.14   Example 8.4: Simply supported square plate.

or

$$M_u = M_{uy} \cos^2 \alpha + M_{ux} \sin^2 \alpha \qquad (8.44)$$

and

$$T_u ds = M_{uy} dx \sin \alpha - M_{ux} dy \cos \alpha$$

or

$$T_u = (M_{uy} - M_{ux}) \sin \alpha \cos \alpha \qquad (8.45)$$

A plate with equal reinforcement in the x- and y-directions is said to have *isotropic reinforcement*, for which $M_{ux} = M_{uy}$, and the above two equations reduce to $M_u = M_{ux} = M_{uy}$ and $T_u = 0$. Note that, in practice, the distance d from the extreme compression fiber to the steel centroid is different for the two directions. In order to make $M_{ux} = M_{uy}$, the bar spacings must be adjusted accordingly.

Returning to a quarter plate (Fig. 8.14b), we know now that along the diagonal yield line the moment at failure is equal to M_u and no twisting moment is present. The total load on the plate segment is $w_u a^2/4$, and its moment arm about the simply supported edge is $a/6$. The projection of the internal moment M_u along the simple support is $M_u(a/\sin 45°) \sin 45° = M_u a$, so moment equilibrium about the support edge requires

$$M_u a = \frac{w_u a^2}{4} \frac{a}{6} \quad \text{or} \quad M_u = \frac{w_u a^2}{24}$$

The required moment capacity in our example:

$$M_u = \frac{(2)(10)^2}{24} = 8.33 \text{ ft-kips/ft} = \underline{100 \text{ in-kips/ft}} \qquad \blacksquare$$

EXAMPLE 8.5

Determine the failure load for the simply supported rectangular plate of Fig. 8.15, with side lengths a and b and isotropic reinforcement of capacity $M_u = M_{ux} = M_{uy}$.

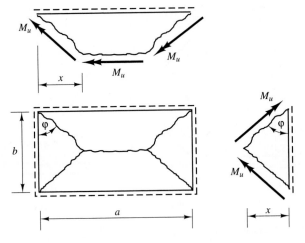

Figure 8.15   Example 8.5: Simply supported rectangular plate.

Solution   This plate will probably fail in a pattern as indicated in Fig. 8.15, where the angle ϕ or length x is yet unknown. Moment equilibrium of a trapezoidal plate segment around the simple support at failure requires

$$M_u a = w_u(a - 2x)\frac{b}{2}\frac{b}{4} + 2w_u\frac{bx}{4}\frac{b}{6} \quad \text{or} \quad M_u = \frac{b^2}{24}\left(3 - \frac{4x}{a}\right)w_u$$

Similarly, the equilibrium condition for a triangular plate segment is

$$M_u b = w_u\frac{xb}{2}\frac{x}{3} \quad \text{or} \quad M_u = w_u\frac{x^2}{6}$$

These two equations can be solved simultaneously for the two unknowns, w_u and x,

$$x = \left(\sqrt{1 + 3\frac{a^2}{b^2}} - 1\right)\frac{b^2}{2a} \quad \text{and} \quad w_u = \frac{6M_u}{x^2}$$

For example, if $a = 2b$, we find

$$x = \left(\sqrt{13} - 1\right)\frac{b}{4} \quad \text{and} \quad w_u = \frac{14.14}{b^2}M_u$$

This compares with $w_u = (24/b^2)M_u$ for a square plate (see Example 8.4) and with $w_u = (8/b^2)M_u$ for a one-way slab (see Example 8.1). Thus, the behavior of a rectangular plate with aspect ratio 2 is slightly closer to that of a one-way slab rather than of a square plate. ■

The previous examples employed exclusively equilibrium conditions in order to relate failure load to flexural capacities. We shall use the next examples to demonstrate that the Principle of Virtual Work can be used to arrive at the same results.

EXAMPLE 8.6

Repeat Example 8.3 using the Principle of Virtual Work.

Solution   After assuming the same pattern of three yield lines, we introduce a virtual displacement of unit value for the center yield line, $\delta = 1$ (Fig. 8.16). The angle changes within the yield lines (or plastic hinges in an equivalent beam) follow as

$$\theta_a = \theta_{b1} = \frac{1}{x} \quad \text{and} \quad \theta_c = \theta_{b2} = \frac{1}{L-x}$$

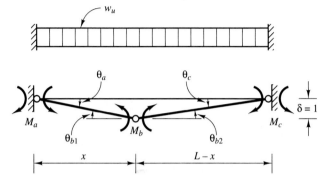

Figure 8.16   Example 8.6: Virtual displacements of slab with fixed ends.

The internal work done by the plastic moments inside the yield lines becomes

$$W_I = \sum_i M_i \theta_i = M_a \cdot \frac{1}{x} + M_b \cdot \left(\frac{1}{x} + \frac{1}{L-x} \right) + M_c \cdot \frac{1}{L-x}$$

while the external work done by the uniform load is given by

$$W_E = w_u x \cdot \frac{1}{2} + w_u (L-x) \cdot \frac{1}{2} = \frac{w_u L}{2}$$

Equating the external and internal work, $W_I = W_E$, we find

$$\frac{w_u L}{2} = \frac{M_a + M_b}{x} + \frac{M_b + M_c}{L-x}$$

Since the plate will locate the position x for the center yield line such that w_u becomes a minimum, we form $dw_u/dx = 0$, that is,

$$-\frac{1}{x^2}(M_a + M_b) + \frac{M_b + M_c}{(L-x)^2} = 0$$

This is a quadratic equation for x and has the same solution as found earlier,

$$x = \frac{L\sqrt{M_a + M_b}}{\sqrt{M_a + M_b} + \sqrt{M_b + M_c}}$$ ∎

EXAMPLE 8.7

Repeat Example 8.5 using the Principle of Virtual Work.

Solution   Again, we start by introducing a unit virtual displacement along the center yield line (Fig. 8.17). When determining the internal work of the four plate segments separately,

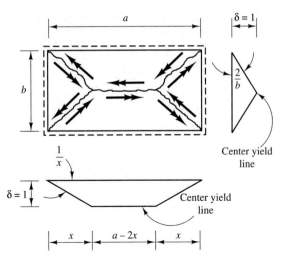

Figure 8.17 Example 8.7: Virtual displacements of rectangular slab.

we can verify that only projections of the angles are required, that is,

$$W_I = 2M_u b \cdot \frac{1}{x} + 2M_u a \cdot \frac{1}{b/2} = \frac{2M_u}{bx}(b^2 + 2ax)$$

$$W_E = 2w_u \frac{bx}{2} \cdot \frac{1}{3} + 2w_u(a - 2x)\frac{b}{2} \cdot \frac{1}{2} + 4w_u \frac{xb}{4} \cdot \frac{1}{3} = \frac{w_u b}{6}(3a - 2x)$$

Equating external and internal work leads to: $w_u = \dfrac{12M_u}{b^2} \dfrac{b^2 + 2ax}{3ax - 2x^2}.$

Differentiating, $dw_u/dx = 0$, and solving for x gives, as before,

$$x = \left(\sqrt{1 + 3\frac{a^2}{b^2}} - 1\right)\frac{b^2}{2a}$$

■

If more than one geometric unknown are needed to uniquely describe the yield line pattern, say, $x_1, x_2, \ldots$, we can differentiate the expression for w_u with respect to each one of them. This leads to as many equations as there are unknowns, $\partial w_u/\partial x_1 = 0$, $\partial w_u/\partial x_2 = 0, \ldots$, which can be solved simultaneously for the unknowns $x_1, x_2, \ldots$.

For geometrically more complex problems, the location of yield lines and identification of geometric parameters is often not straightforward. Therefore, the following rules can serve as helpful guidelines.

1. Plane surfaces always intersect in straight lines, therefore yield lines can be assumed to be approximately straight.
2. Axes of rotation generally follow lines of support. These axes may be real hinges or plastic hinges.
3. Axes of rotation must pass over any columns.
4. A yield line between two slab parts passes through the point of intersection of the axes of rotation of the two parts.

With these rules in mind, we can construct yield line patterns and identify the unknown geometric parameters. Figure 8.18 illustrates a few examples. Readers are encouraged to solve additional problems to practice their skills in applying the above guidelines.

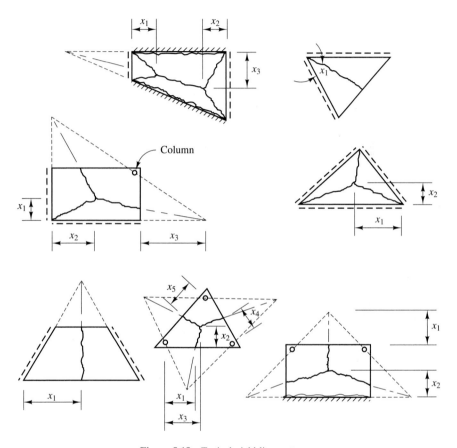

Figure 8.18   Typical yield line patterns.

So far it has been assumed that yield lines enter into the corners between two supported intersecting boundaries. In reality, however, this seldom occurs. Yield lines can be observed to branch out before reaching a corner, forming *corner levers* (Fig. 8.19). Two or more additional unknowns are introduced in this case, making the accurate solution of such problems more time consuming. For a square plate, for example, the more accurate solution leads to the critical load, $w_u = (22/a^2)M_u$, as compared with $w_u = (24/a^2)M_u$, obtained earlier when the corner levers were neglected. An error of such magnitude is not likely to warrant a time-consuming refined analysis, but in plates with acute corners the error may be more significant.

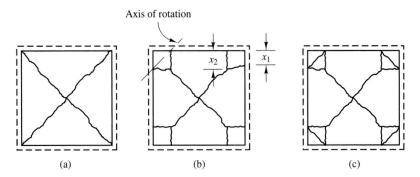

Figure 8.19 Corner levers in square plate: (a) simplified case; (b) corners free to lift; (c) corners held down.

The boundary conditions along a free edge require that both the bending moment and twisting moment are zero, normal to the edge. The same applies to a simply supported edge. This boundary condition demands that any yield line intersect the edge at a right angle. In reality, yield lines will approach the edge at an angle determined by the crack pattern, but at some distance x it will turn into a right angle (Fig. 8.20a). To simplify the analysis, this complication may be ignored, but the violation in statics may not. The conditions of equilibrium can be restored by introducing two fictitious correctional forces P, which equilibrate the unbalanced moment component, $Pe = M_u \cos\alpha$, where the distance e between the two forces can be arbitrarily small (Fig. 8.20b).

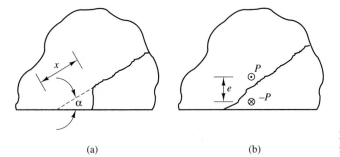

Figure 8.20 Correctional boundary forces for inclined crack.

(a)   (b)

In some cases, more than one apparently permissible crack pattern is possible. However, both the equilibrium method and the method of virtual work provide an upper bound, that is, the correct failure load cannot exceed the value for any of the assumed failure patterns. Thus, each possible mechanism must be investigated separately, and the lowest failure load constitutes the correct answer. The plate of Fig. 8.21, for example, could conceivably fail in either one of the two crack patterns shown. Depending on the aspect ratio, either pattern may yield the critical failure load. Thus, the failure load will have to be computed for both cases.

Yield line theory is a powerful tool that helps us to visualize the formation of crack patterns and failure mechanisms of simple plate structures. For complex plan

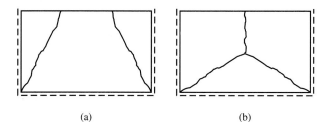

(a)   (b)

Figure 8.21   Failure patterns for rectangular slab.

configurations and boundary condition, solutions require considerable time and effort. Computer programs that trace the response of concrete slabs through their linear and nonlinear range to failure are now available, but at present they are rarely used in design practice because of the skill required for their proper use and verification of results.

8.4 CONCRETE SLABS AND PLATE STRUCTURES

In concrete construction practice, slabs or plates are used for a variety of purposes. *Slabs on grade* are supported directly by the ground, as in the case of highway pavements, airport runways, foundation mats, and building floor slabs at or below ground level. In all of these cases, the sub-base for such slabs usually consists of a well-compacted layer of gravel or crushed stone for uniform support and drainage.

In bridge construction, the roadway slabs are called *bridge decks*. These carry the design loads, primarily the wheel loads, to the supporting elements, such as stringers, floor beams, or girders (Fig. 8.22). In building construction, slabs are usually supported by beams and girders, which by themselves are supported by columns or walls (Fig. 8.23). If supported directly by columns (Fig. 8.23a), the system is called a *flat plate*. It is an economical system if the span lengths are not too large (typically up to 20 ft) and the loads not too heavy. It is very popular for multistory apartment building construction because of the flat ceiling surfaces, reduced construction depth, simplified formwork, and unobstructed lighting. However, the high shear stresses at the column supports can cause problems. To prevent a column from punching through, special shear reinforcement may need to be provided, a problem we will discuss later.

A different way to reduce the high local shear stresses is to thicken the slab around the column with a *drop panel* or a conical column taper known as a *capital*, or both (Fig. 8.23b). Such a system is often referred to as a *flat slab*. It has a markedly higher carrying capacity than a flat plate and is frequently used for buildings with heavy loads, such as warehouses.

A different slab system results if the following logical line of thought is pursued. We recall that the concrete below the neutral axis can be considered cracked, therefore adding a lot of dead weight and little strength. Thus it may be omitted altogether except for a minimum needed to provide sufficient shear strength and to protect the flexural reinforcement against corrosion. The result is a one-way ribbed slab or joist construction (Fig. 8.23c), which is an efficient and popular system in building construction. The ribs are formed using simple U-shaped forms, typically of metal, which can be overlapped to

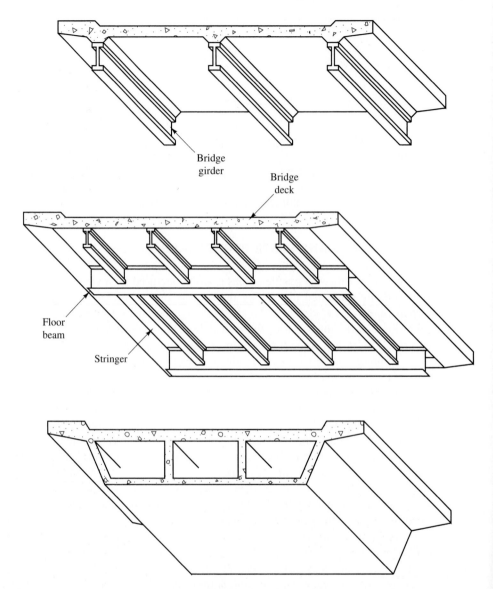

Figure 8.22 Typical bridge decks.

adjust for different lengths (Fig. 8.23c). Near the supports, the joists are usually tapered to increase their shear strength.

By the same logic, a two-way slab becomes a two-way ribbed slab or *waffle slab* (Fig. 8.23d), which is suitable to carry heavy loads. The ribs are formed by positioning mass-produced pans of metal, wood, or fiberglass in a square pattern atop the usual horizontal plywood (Fig. 8.23d). Where additional shear capacity is needed, such as in

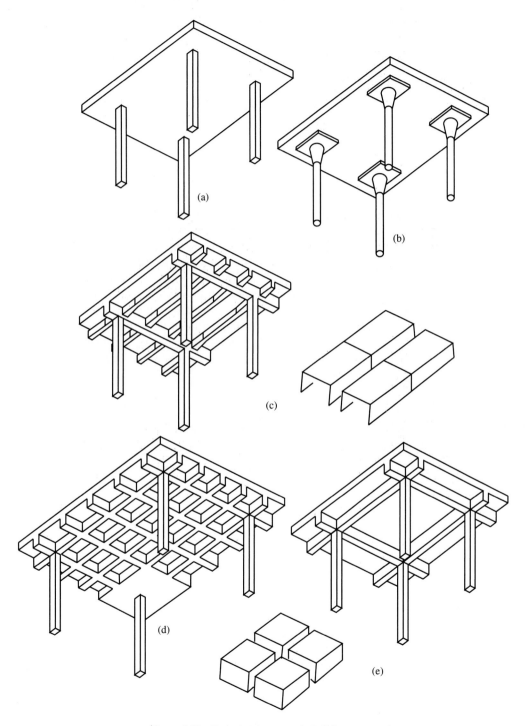

Figure 8.23 Typical slab systems in building construction.

the vicinity of columns, the pans are simply omitted, thereby resulting in a solid drop panel.

Another common type of slab construction is shown in Fig. 8.23e. This is a two-way slab supported by beams, which in turn are supported directly by either columns or walls.

Before proceeding to the commonly used analysis and design methods for two-way slab systems, it is necessary to point out some important load-carrying characteristics of such structures.

A rectangular slab supported by nonyielding members (such as walls) satisfies quite closely the conditions under which it can be analyzed by the classical methods discussed earlier. The boundary conditions can often be assumed to be simply supported or free. Where continuity with supporting walls or neighboring slab panels prevents the slab from rotating, fixed boundary conditions may be appropriate.

If a slab is supported on only two opposite sides, the applied loads are carried to the supports primarily in one direction. Such a *one-way slab* behaves essentially like a wide beam, and its deflected shape is closely approximated by a cylindrical surface. In Chapter 4 we designed them by considering simple strips of unit width.

A slab supported on all four sides can deflect only by bending in two directions. A *two-way slab* thus attempts to carry load in two directions. If the two sides are of different length, the stiffer span (usually the one with the shorter length) attracts a greater share of the total load. If we visualize the slab to be made up of two imaginary orthogonal sets of beams (Fig. 8.24), compatibility requires that each beam of length a deflect the same

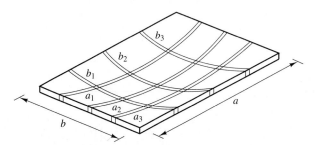

Figure 8.24   Two-way slab.

amount as a beam of span length b wherever they intersect. Designating the shares of the total load carried by the two sets of beams by w_a and $w_b = w_{\text{total}} - w_a$, respectively, compatibility of midspan deflections requires that, for uniformly distributed load,

$$\frac{5}{384}\frac{w_a a^4}{E_a I_a} = \frac{5}{384}\frac{w_b b^4}{E_b I_b}$$

or, if $E_a I_a = E_b I_b$,

$$\frac{w_a}{w_b} = \left(\frac{b}{a}\right)^4$$

For example, for $b = 2a$, we find that $w_a = 16w_b$, that is, the load is carried almost entirely in the short a-direction, so for practical purposes the slab behaves like a one-way slab.

This example illustrates the importance of relative stiffnesses. A stiff member, in this case the beam with the shorter span, attracts a larger share of the load and must be designed accordingly. However, the designer has the means to control relative stiffnesses and therefore load paths. For instance, providing more reinforcement in the long b-direction may result in $E_b I_b > E_a I_a$, so the b-beams will attract load that they otherwise would not, because of their longer spans. In an extreme case, we could stiffen the long span with ribs such that $E_b I_b = 16 E_a I_a$, with the result that even for a 2 to 1 aspect ratio, the load shares would be equal, that is, $w_a = w_b$. The designer can force the slab (just like any other structure) to carry the load in any way desired. However, structural solutions tend to be less economical (and often less esthetically pleasing), the more the forced behavior deviates from the natural behavior. The structure always attempts to choose the shortest load path. Prolonging this load path may sometimes be necessary, but it is likely to be more costly.

The imaginary grid of orthogonal sets of beams illustrates the high intrinsic redundancy of a two-way slab. In case of local overstress or material flaws as severe as a missing reinforcing bar, the slab can easily readjust and carry the load in the other direction. The high redundancy brings numerous benefits and is recognized by the ACI Code in many ways.

To visualize the basic behavior of a slab supported only by columns instead of nonyielding walls, we again imagine it to be made up of two orthogonal sets of beams (Fig. 8.25). These may or may not be stiffened by ribs or beams along the column lines. The most significant fact about such systems is that each of the orthogonal sets of beams must carry the full load, whereas in the case of a system with nonyielding

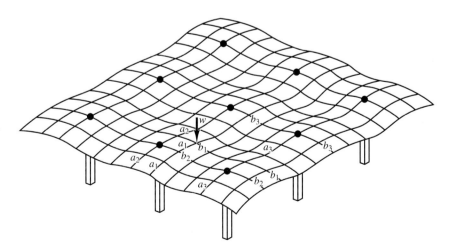

Figure 8.25   Slab supported by columns.

supports, each set of beams carries only a part of the total load, which depends on the relative stiffnesses of the two sets of beams. To demonstrate this important fact, consider Fig. 8.25, with a set of a-beams spanning in the a-direction, and a set of b-beams spanning in the b-direction. Let us apply a load W at the intersection of beams a_1 and b_1, where beam a_1 carries the share αW, and beam b_1 carries the rest, $(1 - \alpha)W$, where the factor α is arbitrary. Beam a_1 carries its load share αW to its support beams b_2 and b_3, so the three b-beams together carry the load $(1 - \alpha)W + \alpha W = W$. Likewise, beam b_1 carries its load share $(1 - \alpha)W$ to its support beams a_2 and a_3, which in turn carry it to the columns, so the three a-beams carry a combined total load of $\alpha W + (1 - \alpha)W = W$. For this reason, the relative stiffnesses of the a- and b-beams do not play the important role they do in slabs, where the continuous supports do not deflect.

The analysis and design of slabs in realistic building construction is complicated by the fact that the slabs cannot be considered separately from the supporting beams, columns, and walls—they all act together as a continuous structure. The columns and walls above as well as below the slab have an effect on slab behavior. Accurate analyses of realistic building structures are next to impossible and numerical solutions are often impractical. To arrive at reasonable yet safe designs, gross oversimplifications are usually inevitable. Before turning to the practical design methods of the ACI Code, we shall consider in some detail two-way slabs on nonyielding supports.

8.5 DESIGN OF TWO-WAY SLABS ON NONYIELDING SUPPORTS

A slab supported on all four sides on nonyielding supports, with spans of comparable length in two directions, will attempt to carry loads in both directions and therefore must be reinforced accordingly. Of course, we could elect to place reinforcement in one direction only, but the bending deformations would lead to unsightly cracking in the other direction if not resisted by some nominal amount of reinforcing steel. As we saw earlier, the ratio of load shares in the two directions, based strictly on a deflection compatibility requirement, is $w_a/w_b = (b/a)^4$. In a uniformly loaded square plate with equal reinforcement in two directions, the maximum moment would be, accordingly, $M = (1/2)(wa^2/8) = .0625wa^2$. This result is based on an oversimplification of plate behavior, because compatibility between the two sets of beams should be satisfied not only for vertical deflections but also rotations. Consider the slab of Fig. 8.24, idealized as an imaginary grid of orthogonal beams. At the supports, beam a_1 is held against twisting, whereas at the intersections with beams b_1, b_2, and b_3, beam a_1's cross section is made to twist. In the imaginary pin-connected grid system of beams, their resistance against such twisting is ignored. In an actual plate, the twisting moments are sufficiently significant to reduce the midspan moment of a square plate by almost 25%, namely, from $.0625wa^2$ to $.048wa^2$. The twisting moments that enforce this compatibility make the slab a much more redundant and stiff continuum than an equivalent grid of orthogonal pin-connected beams. They also add to the high degree of inelastic moment redistribution in case the plate's moment capacity is exceeded locally in either direction. We have already utilized this valuable property of slabs in our discussion of yield line theory.

Figure 8.26 shows the qualitative variations of M_x and M_y moments in a slab of dimensions a and b, where $a > b$. If the load exceeds the slab capacity in the y-direction, in which the moments are larger due to the shorter span, the steel will yield, the neighboring slab strips will pick up the excess load that the overloaded strip cannot carry, and failure will occur when both M_x and M_y reach the ultimate capacity of the slab over a substantial region. For this reason, it is common to base the design moment capacity on the factored moments averaged over a substantial width of the slab (Fig. 8.26). The slab may be divided in each direction into a *middle strip* and two *edge strips*, and each strip is designed for the moment averaged over the entire width of the strip. As long as each slab strip contains a sufficient amount of steel to resist such average moments, we can expect the slab to behave satisfactorily. Tables B.4a through B.4d in Appendix B contain coefficients for a variety of different boundary conditions and span ratios and thus permit a rapid determination of maximum moments and shears. These are computed on the basis of elastic plate theory and were included in the 1963 edition of the ACI Code.

Once the moments are known, it is straightforward to compute the required reinforcement using the procedures outlined in Chapter 4. It is possible to proceed in one of two ways: (a) compute the total moment for an entire strip (e.g., in ft-lb), determine the total amount of steel needed to carry this moment, then select a reasonable bar spacing to evenly distribute the bars within the strip; or (b) compute the average moment for a one-foot-wide strip (e.g., in ft-lb/ft) and determine the amount of steel needed to resist this moment, using, for example, Table B.3.

Note that the moments are larger in the short direction, and therefore it would be logical to let the corresponding steel form the bottom layer. The steel in the longer span will then form the upper layer, with a somewhat smaller internal lever arm d. The choice of which layer is on top barely affects the overall cost, but the different d-values for the two steel layers should be recognized.

EXAMPLE 8.8

A 10-by-15-ft rectangular slab is simply supported along all four sides and carries a live load of 200 psf in addition to its own weight. Design the reinforcement in both directions, using

    a) classical plate theory;
    b) the moment coefficients of Tables B.4a through B.4d;
    c) yield line theory.

Given: $f_c' = 4$ ksi; $f_y = 60$ ksi; $\nu = .15$; slab thickness = 6 in.

Solution

Slab dead weight: $w_D = \dfrac{(150)(6)}{12} = 75$ psf;

Factored load: $w_u = (1.4)(75) + (1.7)(200) = 445$ psf.

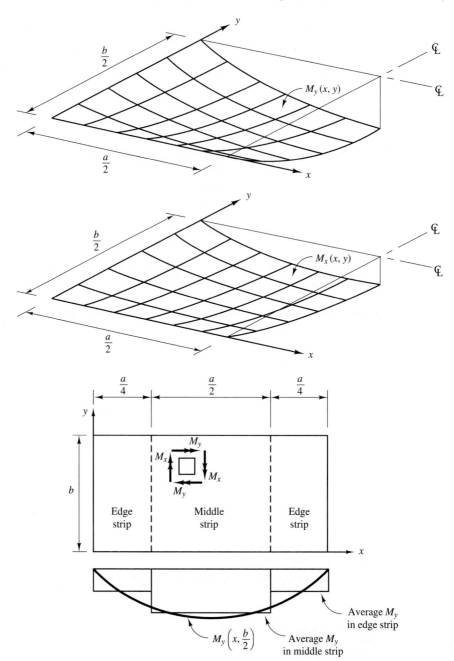

Figure 8.26   Bending moments in two-way slab.

a) *Plate Theory*   We combine Eqs. 8.39 and 8.41 to obtain the following closed-form solution for the deflection of a plate with uniform load p_o.

$$w(x, y) = \frac{16 p_o}{D \pi^6} \sum_{n=1}^{\infty} \sum_{m=1}^{\infty} \frac{1}{nm \left(\frac{n^2}{a^2} + \frac{m^2}{b^2} \right)^2} \sin \frac{n \pi x}{a} \sin \frac{m \pi y}{b} \tag{8.46}$$

The bending moments follow from Eqs. 8.14, 8.15, and 8.16:

$$M_x = \frac{16 p_o}{\pi^4} \sum_{n=1,3,\ldots} \sum_{m=1,3,\ldots} \frac{\left(\frac{n}{a} \right)^2 + v \left(\frac{m}{b} \right)^2}{nm \left(\frac{n^2}{a^2} + \frac{m^2}{b^2} \right)^2} \sin \frac{n \pi x}{a} \sin \frac{m \pi y}{b} \tag{8.47}$$

$$M_y = \frac{16 p_o}{\pi^4} \sum_{n=1,3,\ldots} \sum_{m=1,3,\ldots} \frac{\left(\frac{m}{b} \right)^2 + v \left(\frac{n}{a} \right)^2}{nm \left(\frac{n^2}{a^2} + \frac{m^2}{b^2} \right)^2} \sin \frac{n \pi x}{a} \sin \frac{m \pi y}{b} \tag{8.48}$$

$$M_{xy} = -(1 - v) \frac{16 p_o}{ab \pi^4} \sum_{n=1,3,\ldots} \sum_{m=1,3,\ldots} \frac{1}{\left(\frac{n^2}{a^2} + \frac{m^2}{b^2} \right)^2} \cos \frac{n \pi x}{a} \cos \frac{m \pi y}{b} \tag{8.49}$$

Using only three *harmonics* in each direction (i.e., $n = 1, 3, 5$ and $m = 1, 3, 5$) yields the results shown in Fig. 8.27a, normalized with respect to $p_o a^2$. We can readily verify that the inclusion of additional harmonics changes the results by no more than a few percent.

Average middle strip M_y-moment:

$$M_y^{\text{mid}}/p_o a^2 = [(2)(.0660) + (2)(.0743) + .0773]/5 = .0716$$
$$M_y^{\text{mid}} = (.0716)(445)(10)^2 = 3185 \text{ ft-lb/ft}$$

Average edge strip M_y-moment:

$$M_y^{\text{edge}}/p_o a^2 = [(2)(.0522) + (2)(.0300) + (1)(0)]/5 = .0329$$
$$M_y^{\text{edge}} = (.0329)(445)(10)^2 = 1463 \text{ ft-lb/ft}$$

Average middle strip M_x-moment:

$$M_x^{\text{mid}}/p_o a^2 = [(2)(.0375) + (2)(.0386) + .0396]/5 = .0384$$
$$M_x^{\text{mid}} = (.0384)(445)(10)^2 = 1707 \text{ ft-lb/ft}$$

Average edge strip M_x-moment:

$$M_x^{\text{edge}}/p_o a^2 = [(2)(.0352) + (2)(.0234) + (1)(0)]/5 = .0234$$
$$M_x^{\text{edge}} = (.0234)(445)(10)^2 = 1043 \text{ ft-lb/ft}$$

These values are shown in Fig. 8.27b.

Slab reinforcement:

Short direction, middle strip ($M_u = M_y^{\text{mid}} = 3185$ ft-lb/ft):

$$\text{Let } d = 5, a = \frac{1}{4}, \text{ then } A_s = \frac{(3.185)(12)}{(.9)(60)(5 - .125)} = .145 \text{ in}^2/\text{ft}$$

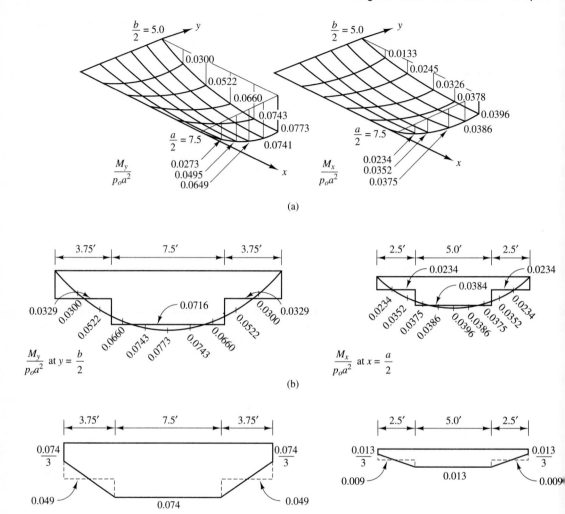

Figure 8.27 Example 8.9: Design of simply supported slab. (a) Moments from plate theory; (b) average moments (plate theory); (c) moments from coefficients.

Short direction, edge strip ($M_u = 1463$ ft-lb/ft):

$$A_s = \frac{(1.463)(12)}{(.9)(60)(5 - .125)} = .067\text{in}^2/\text{ft}$$

Note that the minimum reinforcement in slabs with Grade 60 deformed bars, also referred to as *shrinkage and temperature reinforcement* is $\rho_{min} = .0018$ [Sec. 7.12], that is,

$$A_{s,min} = (.0018)(5)(12) = .108 \text{ in}^2/\text{ft}$$

and maximum bar spacing $= 18$ in.

Thus, use 12 #4 bars, uniformly distributed over the entire slab width of 15 ft, that is, spaced at 15 in. In the middle strip this provides 6 bars with 1.18 in^2, which exceeds the required $(.145)(7.5) = 1.09$ in^2. In the edge strips, this steel is more than enough.

Long direction, middle strip ($M_u = M_x^{\text{mid}} = 1707$ ft-lb/ft):

Now $d = 4.5$, that is, $A_s = \dfrac{(1.707)(12)}{(.9)(60)(4.5 - .125)} = .087$ in^2/ft $< A_{s,\text{min}}$.

Use #3 bars spaced at 12 in, which provide 0.11 in^2/ft. This requires 10 bars spaced evenly over the 10 ft slab width.

Total weight of steel $= (12)(.668)(10) + (10)(.376)(15) = 137$ lb

b) *Moment Coefficients*   The moment coefficient tables distinguish between factored dead and live load, but in the case of a simply supported slab the coefficients are identical (Fig. 8.27c):

$$M_y^{\text{mid}} = (.074)(445)(10)^2 = 3293 \text{ ft-lb/ft}$$

$$M_y^{\text{edge}} = \frac{2}{3}(3293) = 2195 \text{ ft-lb/ft}$$

$$M_x^{\text{mid}} = (.013)(445)(10)^2 = 579 \text{ ft-lb/ft}$$

$$M_x^{\text{edge}} = \frac{2}{3}(579) = 386 \text{ ft-lb/ft}$$

Slab reinforcement:

Short direction ($M_u = M_y^{\text{mid}} = 3293$ ft-lb/ft):

$$A_s = \frac{(3.293)(12)}{(.9)(60)(5 - .125)} = .150 \text{ in}^2/\text{ft. Use 12 #4 bars @ 15 in.}$$

Long direction ($M_u = M_x^{\text{mid}} = 579$ ft-lb/ft):

$$A_s = \frac{(.579)(12)}{(.9)(60)(4.5 - .125)} = .029 \text{ in}^2/\text{ft} \ll A_{s,\text{min}}. \text{ Use 10 #3 bars @ 12 in.}$$

Total amount of steel, as before, 137 lb.

c) *Yield Line Theory*
From Example 8.5:

$$x = \left(\sqrt{1 + 3\frac{a^2}{b^2}} - 1\right)\frac{b^2}{2a} = \left(\sqrt{1 + 3\frac{15^2}{10^2}} - 1\right)\frac{10^2}{(2)(15)} = 5.95 \text{ ft.}$$

Required moment capacity, uniform throughout the slab:

$$M_u = \frac{w_u x^2}{6} = \frac{(445)(5.95)^2}{6} = 2626 \text{ ft-lb/ft.}$$

Slab reinforcement:

Short direction, with $d = 5$ in, $A_s = \dfrac{(2.63)(12)}{(.9)(60)(5 - .125)} = .120$ in^2/ft.

Use 16 #3 bars, spaced at 11 in.

Long direction, with $d = 4.5$ in, $A_s = \dfrac{(2.63)(12)}{(.9)(60)(4.5 - .125)} = .133$ in^2/ft.

Use 12 #3 bars, spaced at 10 in.

Total amount of steel $= (16 \times 10 + 12 \times 15)(.376) = 128$ lb

This is slightly less steel than in the other solutions, where for convenience we used uniform bar spacing throughout, whereas we could have saved a bar or two in the edge strips by using maximum spacing. ∎

This example illustrates an approach common in the design of plates and slabs. Our concern focuses on total or average loads and moments rather than actual maximum values, which are more of theoretical interest. As long as enough steel is provided to cover the total moments, and this steel is reasonably well distributed, the plate behavior is likely to be satisfactory, as loads will find a path from understrength regions to areas with strength reserves. This design philosophy is recurring in many other areas of structural engineering. In particular, it applies to the slab-and-beam systems commonly found in reinforced concrete building construction.

Before proceeding to the ACI design methods, it is instructive to reconsider a square plate, simply supported along all four edges (Fig. 8.28a). For a load applied at point A, symmetry requires that the imaginary beams of unit width, a_1 and b_1, share equally in carrying the load. Extending that argument across the entire plate, the maximum moment due to uniform load would be $(1/2)(wL^2/8) = .0625wL^2$ along the entire section $x = L/2$ (or $y = L/2$) (Fig. 8.28b). This value is much larger than the value given by plate theory, $.048wL^2$, because the contribution of twisting moments is ignored.

The argument of equal load sharing in the two directions obviously breaks down when we consider a load applied at point B. Since beam a_2 can hardly deflect because of its vicinity to the support, the load will have to be carried almost entirely by beam b_1. This conclusion may be generalized by stating that the load always gets carried to the nearest edge, that is, the entire plate can be subdivided into four triangles (Fig. 8.28c), with the load carried entirely in the direction indicated by the arrows. Some imaginary beam a_3 now carries the load depicted in Fig. 8.28d. Plotting the maximum moments in the y-direction for all beams spanning in the x-direction results in the plot of Fig. 8.28e. The maximum value for beam a_1 is $wL^2/8$, because it is loaded with w over its entire length. If the moments of Fig. 8.28e formed the basis for design, the reinforcement would have to vary across the entire plate, which would not be very practical.

The plate subdivision of Fig. 8.28f suggests, as a compromise, a third method of subdividing the plate: in the shaded zones, load is carried equally in both directions, whereas in the remaining areas, the entire load w is carried only in the directions indicated by the arrows; that is, always to the nearest edge. For example, beams a_4 and a_5 would now carry the loads shown in Fig. 8.28g. The maximum values of all a-beam moments are shown in Fig. 8.28h.

The method suggested in Fig. 8.28f is called the *Strip Method* developed by Hillerborg [6]. Note that the results are conservative, and only two strips with different reinforcement need to be considered in each direction. This is very similar to the approach taken by the ACI design methods discussed in the next section.

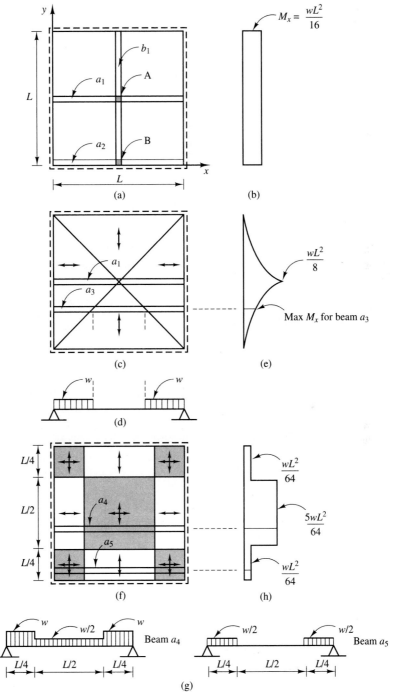

Figure 8.28 Bending moments in plate by strip method [6]: (a) simply supported square plate; (b) max M_x along $x = L/2$; (d) loading for beam a_3; (e) max M_x along $x = L/2$; (g) loading for beams a_4 and a_5; (h) max M_x along $x = L/2$.

EXAMPLE 8.9

The plate shown in Fig. 8.29a is fixed at one edge, simply supported along the other three edges, and supports a unit load of 200 psf. Determine the maximum moments for which the various strips need to be reinforced.

Solution   To subdivide the plate in the y-direction, it is reasonable to assign one-fourth of the total width (5 ft) to each edge strip. In the x-direction, the fixed edge attracts more load, and therefore the edge strip can be a little wider than one-fourth of the total width, say 10 ft (Fig. 8.29b).

Maximum moment in strip b_1:

Reaction $R = (100)(5) = 500$ lb/ft

Moment $M = (500)(10) - (500)(7.5) = 1250$ ft-lb/ft

Maximum moment in strip b_2 (uniformly loaded beam plus load of beam b_1):

Moment $M = \dfrac{(100)(20)^2}{8} + 1250 = 6250$ ft-lb/ft

Maximum moment in strip a_1:

Fixed-end moment for loading shown in Fig. 8.29c [7]:

$$M = -\frac{wc}{8L^2}[4L(a^2 + ab + b^2) - a^3 - a^2b - ab^2 - b^3] + \frac{wc}{2}(a + b)$$

$$= -\frac{(100)(15)}{(8)(30)^2}[(4)(30)(10^2 + 10 \times 25 + 25^2) - 10^3 - (10)^2(25) - (10)(25)^2 - 25^3]$$

$$+ \frac{(100)(15)}{2}(10 + 25) = 7161 \text{ ft-lb/ft}$$

For fixed-end moment due to actual loading, subtract this value from the fixed-end moment for a uniformly loaded beam, that is,

Negative design moment for strip a_1:

$$M = -\frac{(100)(30)^2}{8} + 7161 = -4089 \text{ ft-lb/ft}$$

Pinned support reaction: $R = \dfrac{(100)(10)(5) + (100)(5)(27.5) - 4089}{30} = 488.7$ lb/ft

Moment 5 ft from pinned support:

$$M_5 = (488.7)(5) - \frac{(100)(5)^2}{2} = 1193 \text{ ft-lb/ft}$$

Moment 20 ft from pinned support:

$$M_{20} = (488.7)(20) - (100)(5)(27.5) = 1024 \text{ ft-lb/ft}$$

Positive design moment for strip a_1:

$$M = 1193 \text{ ft-lb/ft}$$

For design moments for strip a_2, add the values for a uniformly loaded beam:

Negative moment $M = -\dfrac{(100)(30)^2}{8} - 4089 = -15,339$ ft-lb/ft

Positive moment $M = \dfrac{(100)(30)^2}{8} + 1193 = 12,443$ ft-lb/ft   ∎

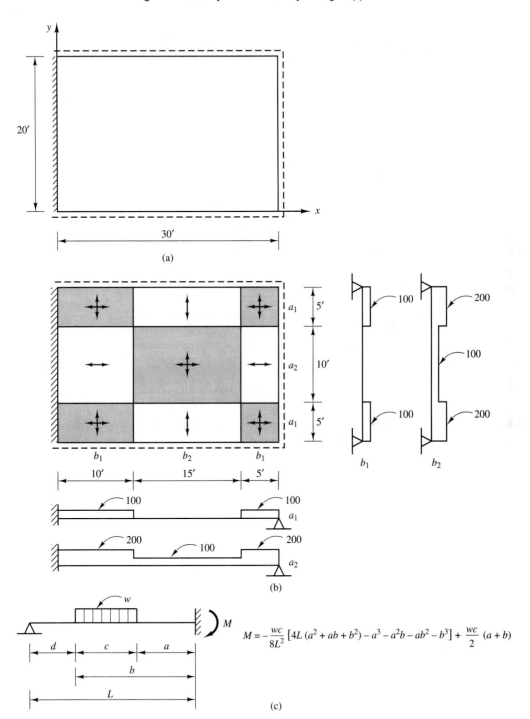

Figure 8.29 Example 8.10: Strip method. (c) Moment in beam for patch load [7].

The method used in the previous example has been referred to as the *Simple Strip Method*. An advanced variant of it applies to plates with point supports. In this case, the load cannot be divided up into two components, each being carried only in one direction. For details, see Hillerborg [6]. It is worth noting that the method is applicable to plates with various geometric shapes and different loadings, whereas the ACI methods described in the next section are limited to rectangular structures with primarily uniform loadings.

8.6 ACI-CODE DESIGN OF TWO-WAY SLAB SYSTEMS

The analysis methods outlined so far in this chapter apply to plates on nonyielding supports. For plates or slabs that are supported on flexible beams and columns instead and that possibly have geometric irregularities dictated by architectural or other considerations, classical plate theory and the moment coefficients derived from it are not readily applicable. Such cases are the rule rather than the exception in design practice. For this reason approximate methods are needed, which are on the safe side without being overly conservative.

Before going into the details of such methods, let us recall the statics of a typical interior span of a continuous beam (Fig. 8.30). The total statical moment, $wL^2/8$, can be divided into a positive (midspan) moment, M_B, and the average of the two negative (support) moments, M_A and M_C,

$$\frac{M_A + M_C}{2} + M_B = \frac{wL^2}{8}$$

In a span completely fixed at both ends, the ratio between support and midspan moment is 2 to 1, that is, the support moment accounts for two-thirds of the total statical moment, and the midspan moment accounts for the remaining one-third.

Let us now consider a rather regular concrete building, which consists of floor slabs supported by beams and columns (Fig. 8.31a). We first isolate a typical structure segment

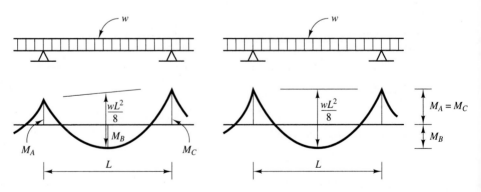

Figure 8.30   Statical moment in continuous beam.

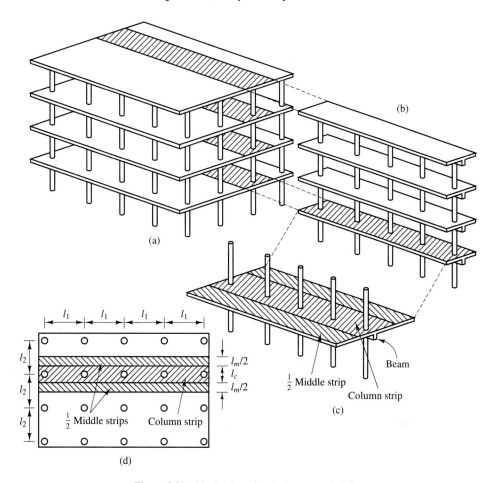

Figure 8.31   Idealization of typical concrete building.

containing half a bay on each side of a column line (Fig. 8.31b), and then a particular story, together with the columns above and below (Fig. 8.31c). The slab segment made up of the two quarter bays immediately adjacent to the column line is called the *column strip*, while the central half bay is referred to as the *middle strip* (Fig. 8.31d). More precisely, the ACI Code defines the width of the column strip as the smaller of $\ell_1/2$ and $\ell_2/2$, that is,

$$\ell_c = \min\left\{\frac{\ell_1}{2}; \frac{\ell_2}{2}\right\} \tag{8.50}$$

so that the width of the middle strip follows as

$$\ell_m = \ell_2 - \ell_c \tag{8.51}$$

ℓ_1 and ℓ_2 are defined in Fig. 8.31d. The structure of Fig. 8.31c is analyzed and designed as one continuous frame. If the effect of the columns is ignored, it becomes a continuous beam. The determination of the total statical moment is straightforward, but the distribution of this moment among the column and middle strips on the one hand, and among the support and midspan moments on the other hand, is not. This depends on the various relative stiffnesses of the structural components involved.

After computing the bending moments in the various structural components and determining the required steel reinforcement for all bays and stories in the building, the structure must be divided into bays in the orthogonal direction, and the entire analysis and design process is repeated. As was pointed out earlier, in contrast to two-way slabs with nonyielding supports in which the two orthogonal spans share the total load, the structure in this case needs to be designed to carry 100% of the load in both directions.

In the ACI Code design methods, a structural segment such as that of Fig. 8.31c is considered as a continuous frame. The stiffnesses of the various slab strips, beams, and columns must be determined for an indeterminate frame analysis, for example, by moment distribution or some computer analysis method. This design method is called the *Equivalent Frame Method*. For structures exhibiting a high degree of regularity, a more simplified method, called the *Direct Design Method*, can be used for gravity load analysis. The ACI Code does not distinguish between flat slab and flat plate structures, as long as all drop panels, column capitals, beams, joists, etc. are properly accounted for. Only slabs with true one-way action are excluded—these are designed like beams, as described earlier.

8.6.1 Direct Design Method

The Direct Design Method is applicable only to regular slab structures, which satisfy the following conditions:

1. There are at least three continuous spans in each direction.
2. Individual panels are rectangular, with ratios of longer to shorter spans not greater than 2.
3. Successive center-to-center span lengths in each direction do not differ by more than one-third the longer span.
4. Columns may be offset no more than 10% of the span (in the direction of the offset) from either axis between centerlines of successive columns.
5. All loads are due only to gravity and uniformly distributed over an entire panel, and live load does not exceed two times dead load.
6. For a panel with beams between supports on all sides, the relative stiffness of beams in two perpendicular directions is not less than 0.2, nor greater than 5.

The Direct Design Method consists of the following basic steps:

Step 1.   Determine the total statical moment for a given panel.
Step 2.   Distribute the total statical moment to the negative (support) and positive (midspan) sections.

Step 3.   Proportion both the negative and positive moments among the column strip, middle strip, and beam, where applicable.

Step 4.   Determine the reinforcement for the column strip, middle strip, and beam to carry the factored moments determined in Step 3.

Step 5.   Repeat Steps 1 through 4 for all other panels in the building.

Step 6.   Repeat Steps 1 through 5 for the orthogonal span direction.

The first three of these steps require some further comment.

Step 1. The total statical moment. This is for a particular panel given by

$$M_o = \frac{(w_u \ell_2) \ell_n^2}{8} \qquad (8.52)[13\text{-}3]$$

where w_u is the factored uniform load, typically in lb/ft^2; ℓ_2 is the center-to-center distance normal to the span direction in which moments are being determined. Where the widths of adjacent panels, ℓ_A and ℓ_B, are different, ℓ_2 is taken as the average of ℓ_A and ℓ_B (Fig. 8.32). ℓ_n is the clear span length under consideration, measured face-to-face of supports (Fig. 8.32). Thus, if c is the column depth, then $\ell_n = \ell_1 - c$, but ℓ_n shall not be less than $0.65\ell_1$. To compute c for nonrectangular columns, the column cross section is to be replaced by a square section having the same area. The total statical moment is thus computed for the shaded area indicated in Fig. 8.32.

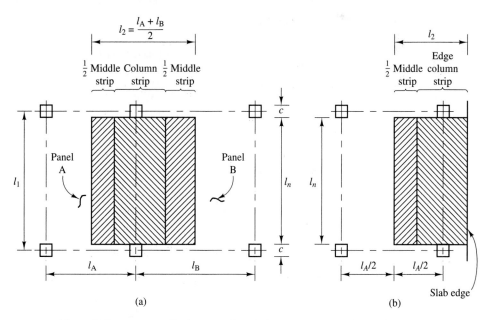

Figure 8.32   Areas contributing to total statical moment: (a) Interior panel; (b) edge panel.

Step 2. Positive and negative moments. When distributing the total statical moment M_o to positive and negative sections, a distinction should be made between a typical interior and an end or exterior span (Fig. 8.33), as was the case with moment coefficients for continuous beams. In an interior span, 65% of M_o are assigned to the negative sections at the support faces and the remaining 35% are taken by the positive section at midspan. In an end span, the distribution of moments depends strongly on the relative stiffness of the end restraint, offered either by a supporting column alone or by a column with spandrel beams, walls, or other structural elements. Table 8.1, reproduced from the ACI Code [Sec. 13.6.3.3], contains the approximate moment coefficients. In the Equivalent Frame Method, which must be used when the Direct Design Method is not applicable, the stiffness of an equivalent column needs to be determined in order to derive the appropriate moment coefficients.

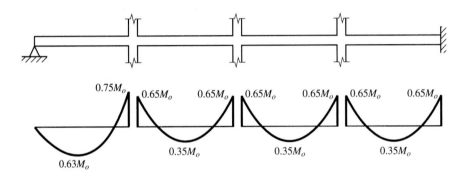

Figure 8.33   Distribution of total statical moment.

Step 3. Column and middle strip moments. When distributing either positive or negative moments between column and middle strips, it is obviously of great consequence whether or not the column strip is stiffened by a beam spanning between the columns, and whether the strip is an interior or edge strip (Fig. 8.34). The influence of the beam is measured by the relative stiffness factor

$$\alpha = \frac{E_{cb} I_b}{E_{cs} I_s} \tag{8.53}$$

where I_b is the moment of inertia of the beam, with effective flange width

$$b_{\text{eff}} = b_w + \min\{2h_w; 8h_f\} \tag{8.54}$$

taken about its centroidal axis, and I_s is the moment of inertia of the entire slab (column and middle strip), that is, $I_s = h_f^3 \ell_2/12$. Note that the definition of beam effective flange width deviates somewhat from that for T-beam design (see Eq. 4.41). E_{cb} and E_{cs} are the elastic moduli, respectively, for beam and slab concrete. If these are identical, as is usually the case, then Eq. 8.53 simplifies to $\alpha = I_b/I_s$.

TABLE 8.1   DISTRIBUTION OF STATICAL MOMENT IN END SPAN

	(1)	(2)	(3)	(4)	(5)
	Exterior edge unrestrained	Slab with beams between all supports	Slab without beams between interior supports		Exterior edge fully restrained
			Without edge beam	With edge beam	
Interior negative factored moment	0.75	0.70	0.70	0.70	0.65
Positive factored moment	0.63	0.57	0.52	0.50	0.35
Exterior negative factored moment	0	0.16	0.26	0.30	0.65

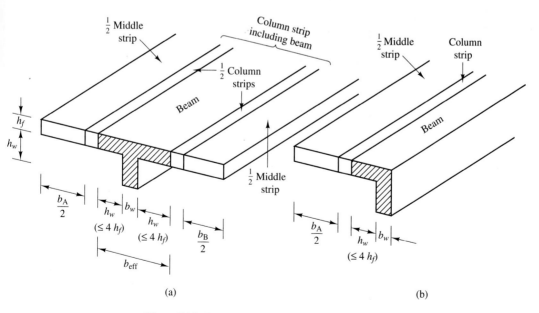

Figure 8.34   Beam and slab strips: (a) interior strip; (b) edge strip.

In addition to α, a second parameter must be computed, which is a measure of the relative torsional stiffness of the edge beam in an end panel. This is defined as

$$\beta_t = \frac{E_{cb}C}{2E_{cs}I_s} \tag{8.55}$$

where C is the torsional constant of the edge beam, approximated by

$$C = \sum \left(1 - 0.63\frac{x}{y}\right)\frac{x^3 y}{3} \tag{8.56}$$

The summation sign extends over all rectangles, into which the section is subdivided (Fig. 8.35); (see also Sec. 5.6).

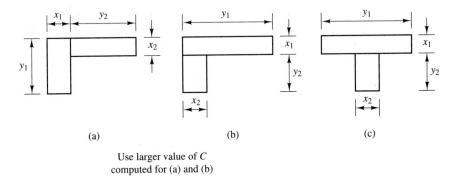

(a)   (b)   (c)

Use larger value of C
computed for (a) and (b)

Figure 8.35 Computation of torsional constant.

With the constants α and β_t known, the moment percentages to be apportioned to the column strip are readily obtained from Table 8.2 (ACI Code Sec. 13.6.4). The balance of 100% is taken by the middle strip. If a beam is part of the column strip, it is to be assigned 85% of the column strip moments given in Table 8.2; the remaining 15% fall to the slab portion of the column strip. With this step, the distribution of moments is complete.

In an edge panel whose end moment is resisted only by columns, the entire slab moment can be assumed to be carried only by the column strip, since the slab is discontinuous. Therefore, the moment that the slab can transfer into the edge columns is limited by the nominal strength of the column strip

$$M_n = A_s f_y \left(d - \frac{a}{2}\right)$$

where A_s is the total negative reinforcement in the column strip.

Before illustrating the Direct Design Method with an example, we should also mention the Code provisions for minimum slab thickness. The reader will recall that, in the case of beams, the Code offers an option to provide a minimum depth to satisfy the deflection control requirements, provided no construction is attached that could be damaged by excessive deflections. In the case of slabs, a similar minimum thickness requirement is offered, bracketed by a lower and upper bound, in lieu of unreliable but nevertheless laborious calculations of deflections. Minimum thicknesses of slabs without interior

TABLE 8.2 MOMENT PERCENTAGE APPORTIONED TO COLUMN STRIP

		ℓ_2/ℓ_1		
		0.5	1.0	2.0
Interior negative moment				
$\alpha_1\ell_2/\ell_1 = 0$		75	75	75
$\alpha_1\ell_2/\ell_1 \geq 1.0$		90	75	45
Exterior negative moment				
$\alpha_1\ell_2/\ell_1 = 0$	$\beta_t = 0$	100	100	100
	$\beta_t \geq 2.5$	75	75	75
$\alpha_1\ell_2/\ell_1 \geq 1.0$	$\beta_t = 0$	100	100	100
	$\beta_t \geq 2.5$	90	75	45
Positive moment				
$\alpha_1\ell_2/\ell_1 = 0$		60	60	60
$\alpha_1\ell_2/\ell_1 \geq 1.0$		90	75	45

beams spanning between the supports shall satisfy the provisions of Table 8.3 (ACI Code Table 9.5c) and shall not be less than 5 in for slabs without drop panels or 4 in for slabs with drop panels. For slabs with or without beams spanning the supports on all four

TABLE 8.3 MINIMUM THICKNESS OF SLABS WITHOUT INTERIOR BEAMS

	Without drop panels Note (2)			With drop panels Note (2)		
	Exterior panels		Interior panels	Exterior panels		Interior panels
Yield stress f_y (psi) Note (1)	Without edge beams	With edge beams Note (3)		Without edge beams	With edge beams Note (3)	
40,000	$\dfrac{\ell_n}{33}$	$\dfrac{\ell_n}{36}$	$\dfrac{\ell_n}{36}$	$\dfrac{\ell_n}{36}$	$\dfrac{\ell_n}{40}$	$\dfrac{\ell_n}{40}$
60,000	$\dfrac{\ell_n}{30}$	$\dfrac{\ell_n}{33}$	$\dfrac{\ell_n}{33}$	$\dfrac{\ell_n}{33}$	$\dfrac{\ell_n}{36}$	$\dfrac{\ell_n}{36}$

(1) For value of reinforcement yield stress between 40,000 and 60,000 psi minimum thickness shall be obtained by linear interpolation.

(2) Drop panel is defined in ACI Code Secs. 13.4.7.1 and 13.4.7.2.

(3) Slabs with beams between columns along exterior edges. The value of α for the edge beam shall not be less than 0.8.

sides, and with span aspect ratios not greater than 2, the minimum slab thickness shall be

$$\frac{F\ell_n}{36+9\beta} \le h_{\min} = \frac{F\ell_n}{36+5\alpha_m\beta - 0.6(1+\beta)} \le \frac{F\ell_n}{36} \qquad (8.57)$$

where

$$F = 0.8 + \frac{f_y}{200,000}$$

(see Code Eqs. 9-11 through 9-13). ℓ_n is again the clear span in the long direction, β is the ratio of clear spans in the long to short direction, and α_m is the average of the α-values (see Eq. 8.53) for all beams on panel edges. If slab thicknesses less than the above limits are used, deflections need to be computed and must remain below the limits of Table 3.2, just as it was required in the case of regular beams. At discontinuous edges, beams should be provided to assure an adequate edge stiffness, or the minimum slab thickness according to Eq. 8.57 should be increased by at least 10%.

EXAMPLE 8.10

An office building has a doubly symmetric plan layout, with three and five spans in the two directions (Fig. 8.36) carrying a live load of 125 psf. Check the adequacy of the 7-in slab thickness and design the reinforcement for a typical exterior panel, using the ACI Code's Direct Design Method. Given: normal-weight concrete with $f'_c = 4$ ksi; $f_y = 60$ ksi.

Solution Preliminary calculations are needed to determine all relative stiffness factors, which are proportional to the I/L ratios because Young's modulus is the same for all members.

Moments of inertia
Edge beam:

Effective flange width, $b_{\text{eff}} = b_w + \min\{h_w; 4h_f\} = 14 + 13 = 27$ in

Centroid, $c_t = \dfrac{(20)(14)(10) + (7)(13)(3.5)}{280 + 91} = 8.4$ in

Moment of inertia, $I = 280\left(\dfrac{20^2}{12} + 1.6^2\right) + 91\left(\dfrac{7^2}{12} + 4.9^2\right) = 12,607$ in^4

Interior beam:

$b_{\text{eff}} = 14 + 26 = 40$ in

$c_t = \dfrac{(20)(14)(10) + (7)(26)(3.5)}{280 + 182} = 7.4$ in

$I = 280\left(\dfrac{20^2}{12} + 2.6^2\right) + 182\left(\dfrac{7^2}{12} + 3.9^2\right) = 14,738$ in^4

20-ft-wide slab strip, $I = \dfrac{(20)(12)(7)^3}{12} = 6,860$ in^4

25-ft-wide slab strip, $I = \dfrac{(25)(12)(7)^3}{12} = 8,575$ in^4

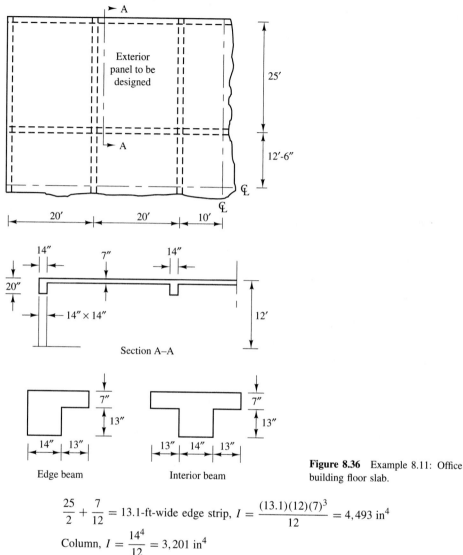

Section A–A

Edge beam   Interior beam

Figure 8.36 Example 8.11: Office
building floor slab.

$$\frac{25}{2} + \frac{7}{12} = 13.1\text{-ft-wide edge strip}, \quad I = \frac{(13.1)(12)(7)^3}{12} = 4,493 \text{ in}^4$$

$$\text{Column, } I = \frac{14^4}{12} = 3,201 \text{ in}^4$$

Relative stiffness factors α

Edge beam, $\alpha = \dfrac{I_b}{I_s} = \dfrac{12,607}{4,493} = 2.81$

20-ft beam, $\alpha = \dfrac{14,738}{8,575} = 1.72$

25-ft beam, $\alpha = \dfrac{14,738}{6,860} = 2.15$

Average α-value, $\alpha_m = \dfrac{2.81 + (2)(2.15) + 1.72}{4} = 2.21$

Check for minimum slab thickness

Span length ratio, $\beta = \dfrac{25 - 14/12}{20 - 14/12} = 1.265$

Clear span, $\ell_n = (25)(12) - 14 = 286$ in

$F = 0.8 + \dfrac{60,000}{200,000} = 1.1$

$h_{min} = \dfrac{(1.1)(286)}{36 + (5)(2.21)(1.265) - (0.6)(1 + 1.265)} = 6.5$ in

Lower bound, $\dfrac{(1.1)(286)}{36 + (9)(1.265)} = 6.6$ in, that is, 7-in slab thickness is OK.

Calculation of moments (refer to Fig. 8.37a)

Step 1. Total statical moment in short direction

Dead weight, $w_d = \dfrac{7}{12}(150) = 87.5$ psf

Factored load, $w_u = (1.4)(87.5) + (1.7)(125) = 335$ psf

Clear span, $\ell_n = 20 - \dfrac{14}{12} = 18.83$ ft

Interior strip moment, $M_o = (.335)(25)(18.83)^2/8 = 371$ ft-kips

Edge strip moment, $M_o = (.335)\left(12.5 + \dfrac{7}{12}\right)(18.83)^2/8 = 194$ ft-kips

Step 2. Positive and negative moments (see Fig. 8.33)

Interior strip negative moment, $M = (.65)(371) = 241$ ft-kips
Interior strip positive moment, $M = (.35)(371) = 130$ ft-kips
Edge strip negative moment, $M = (.65)(194) = 126$ ft-kips
Edge strip positive moment, $M = (.35)(194) = 68$ ft-kips

Step 3. Column and middle strip moments

Span ratio, $\ell_2/\ell_1 = 25/20 = 1.25$
Interior strip, $\alpha\ell_2/\ell_1 = (1.72)(1.25) = 2.15 > 1$
    For $\ell_2/\ell_1 = 1.25$, interpolation in Table 8.2 between 75% and 45% gives
    67.5%, both for negative and positive moments, interior and edge strips.
    Negative beam moment $= (.85)(.675)(241) = 139$ ft-kips
    Negative column strip moment $= (.15)(.675)(241) = 24$ ft-kips
    Negative middle strip moment $= (.325)(241) = 78$ ft-kips
    Positive beam moment $= (.85)(.675)(130) = 75$ ft-kips
    Positive column strip moment $= (.15)(.675)(130) = 13$ ft-kips
    Positive middle strip moment $= (.325)(130) = 42$ ft-kips

Edge strip, $\alpha\ell_2/\ell_1 = (2.81)(1.25) = 3.5 > 1$.

    Thus, the same distribution factor applies as for the interior strip.
    Negative beam moment $= (.85)(.675)(126) = 72$ ft-kips

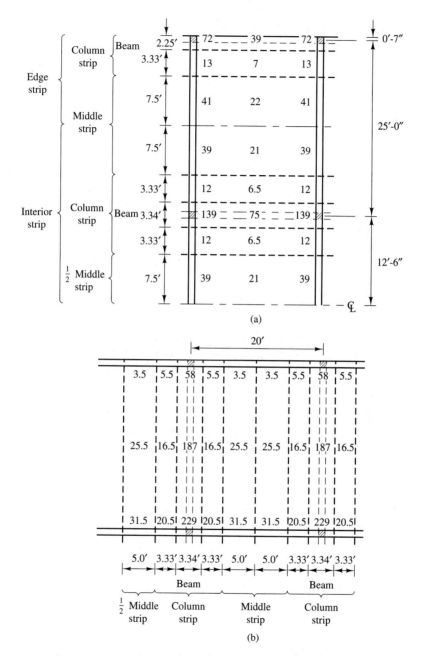

Figure 8.37 Example 8.11: Design moments. (a) Short span; (b) long span.

Negative column strip moment = $(.15)(.675)(126) = 13$ ft-kips
Negative middle strip moment = $(.325)(126) = 41$ ft-kips
Positive beam moment = $(.85)(.675)(68) = 39$ ft-kips

Positive column strip moment = $(.15)(.675)(68) = 7$ ft-kips

Positive middle strip moment = $(.325)(68) = 22$ ft-kips

These design moments are summarized in Fig. 8.37a. Note that column and middle strip moments are equally divided into two half strips.

Step 4. Computation of steel reinforcement

It is convenient to use a tabular format for such computations:

Section	M	d	Trial a	$A_{s,\text{req}}$ $\left[\dfrac{M}{\phi f_y(d-\frac{a}{2})}\right]$	New a $\left[\dfrac{A_s f_y}{.85 f'_c b}\right]$	A_s	$A_{s,\min}$ $\left[\dfrac{200}{f_y}bd \text{ or } .0018bd\right]$	Bars
Edge beam								
	−72	18	1.0	.91	1.15	.92	1.21	2 #6, 2 #4
	39	18	1.0	.50	.33	.49	.67	2 #7
1/2 Column strip								
	−13	6	0.5	.50	.22	.49	.43	3 #4 (@13)
	7	6	0.5	.27	.12	.26	.43	3 #4 (@13)
Middle strip								
	−80	6	0.5	3.09	.30	3.04	1.94	16 #4 (@1
	43	6	0.5	1.66	.16	1.61	1.94	10 #4 (@18
1/2 Column strip								
	−12	6	0.5	.46	.20	.45	.43	3 #4 (@13
	6.5	6	0.5	.25	.11	.24	.43	3 #4 (@13
Interior beam								
	−139	18	1.0	1.76	2.22	1.83	2.40	3 #7, 4 #4
	75	18	1.0	.95	.42	.94	.84	2 #7

Step 5. Design of panel in long direction (refer to Fig. 8.37b):

Total statical moment, $M_o = (.335)(20)\left(25 - \dfrac{14}{12}\right)^2 /8 = 476$ ft-kips

Positive and negative moments (Table 8.1),

Interior negative moment = $(.70)(476) = 333$ ft-kips

Positive moment = $(.57)(476) = 271$ ft-kips

Exterior negative moment = $(.16)(476) = 76$ ft-kips

Column and middle strip moments (Table 8.2)

Torsional constant of edge beam

$$C = \left(1 - .63\frac{14}{20}\right)\frac{(14)^3(20)}{3} + \left(1 - .63\frac{7}{13}\right)\frac{(7)^3(13)}{3} = 11,208 \text{ in}^4 \text{ (controls)}$$

$$C = \left(1 - .63\frac{7}{27}\right)\frac{(7)^3(27)}{3} + \left(1 - .63\frac{13}{14}\right)\frac{(13)^3(14)}{3} = 6,838 \text{ in}^4$$

$$\beta_t = \frac{11,208}{(2)(4493)} = 1.25, \alpha = 2.15$$

$$\ell_2/\ell_1 = \frac{20}{25} = 0.8, \quad \alpha\ell_2/\ell_1 = (2.15)(0.8) = 1.72 > 1$$

To find the interior negative moment for $\ell_2/\ell_1 = 0.8$, interpolation in Table 8.2 between 90% and 75% gives 81%, applicable for both negative and positive moments.

Negative beam moment = $(.85)(.81)(333) = 229$ ft-kips

Negative column strip moment = $(.15)(.81)(333) = 41$ ft-kips

Negative middle strip moment = $(.19)(333) = 63$ ft-kips

Positive beam moment = $(.85)(.81)(271) = 187$ ft-kips

Positive column strip moment = $(.15)(.81)(271) = 33$ ft-kips

Positive middle strip moment = $(.19)(271) = 51$ ft-kips

For exterior negative moments, with $\ell_2/\ell_1 = 0.8$, and $\beta_t = 1.25$, two-way linear interpolation leads to 90.5%.

Negative beam moment = $(.85)(.905)(76) = 58$ ft-kips

Negative column strip = $(.15)(.905)(76) = 11$ ft-kips

Negative middle strip = $(.095)(76) = 7$ ft-kips

These moments are summarized in Fig. 8.37b. Reinforcing bars are determined in the same fashion as was done in the short direction (see Table on page 464). Details are omitted here, but the final reinforcing scheme is summarized in Fig. 8.38. ∎

8.6.2 Equivalent Frame Method

Building structures that are not readily modeled for the Direct Design Method, or that violate some of the limiting assumptions, need to be designed by the Equivalent Frame Method. These two methods are fundamentally similar, except that in the Equivalent Frame Method, bending moments are computed by moment distribution or some other analysis method, whereas in the Direct Design Method, the simple moment coefficients of Table 8.1 are used. The relationship between these two methods is comparable to that

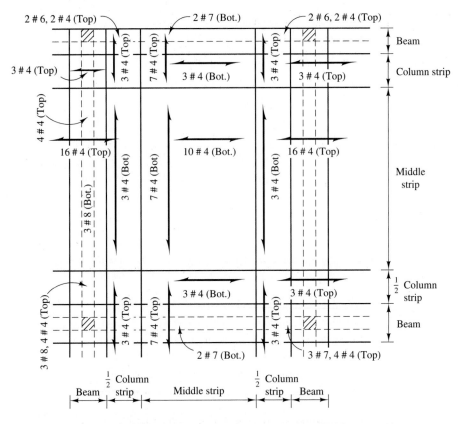

Figure 8.38 Example 8.11: Reinforcing schedule.

between the use of simple moment coefficients for continuous beams (Fig. 6.18) and an actual indeterminate beam analysis. The Equivalent Frame Method applies in particular to structures for which the bending moments are influenced by the torsional rigidity of beams in the orthogonal direction, and by slab sections with moments of inertia that vary at drop panels and column capitals.

The subdivision of a building into segments (Fig. 8.31a and b) follows the same principles, regardless of whether the Direct Design or Equivalent Frame Method is used. In the Equivalent Frame Method, however, a further subdivision into separate frames for each story is not necessary, and a single multistory frame model (Fig. 8.31b), may be analyzed instead of separate frames for each story (Fig. 8.31c).

Whether a building frame such as the one shown in Fig. 8.39 is to be analyzed by moment distribution or some other method, the stiffness properties of all structural members need to be known. For a prismatic member, the stiffness coefficient associated with a rotational degree of freedom is simply $4EI/L$, with a carry-over factor of 0.5, and fixed-end moments, say for uniformly distributed load, of $\pm wL^2/12$. These simple factors obviously do not apply in the case of a frame with drop panels, which introduce

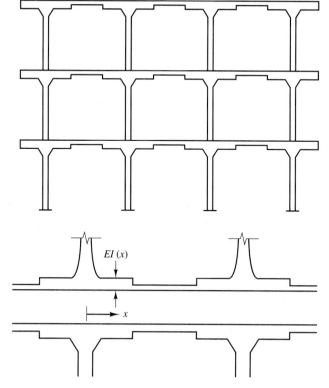

Figure 8.39   Multistory building frame.

Figure 8.40   Variation of slab-beam moment of inertia.

clear discontinuities of the slab moment of inertia, (Fig. 8.40). A second complication is caused by the large stiffness increase of the slab within a joint region. Moreover, as centerline dimensions are used in structural analysis, the joint regions cannot be "counted twice." It is common practice to assign the joint region only to the slab or *slab-beam* part of the frame, which then has a span length equal to the centerline spacing of the columns. For the columns, clear-height dimensions must then be used, together with rigid links that connect the column ends with the points at which the beam and column centerlines intersect (Fig. 8.41).

The moments of inertia of slab-beams outside the joints or column capitals may be based on the gross area of concrete, and variations in moments of inertia along the span (e.g., those caused by drop panels) need to be taken into account. The moment of inertia of a slab-beam within the joint region is approximated, according to the ACI Code [Sec. 13.7.3.3], as

$$I_{jt} = \frac{I_{sb}}{\left(1 - \frac{c_2}{\ell_2}\right)^2} \tag{8.58}$$

where I_{jt} and I_{sb} are the moment of inertia of the slab-beam within and outside the joint region, respectively. c_2 is the column width and ℓ_2 the centerline column spacing, both measured transversely to the direction of the span for which moments are being

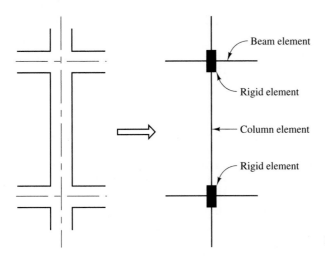

Beam element

Rigid element

Column element

Rigid element

Figure 8.41   Modeling of columns.

determined. Note that in the case of a wall support, with $c_2 = \ell_2$, the joint region is assumed to be infinitely rigid, according to Eq. 8.58.

The increase in slab moment of inertia in the joint region and in the presence of a drop panel increase the stiffness factors, the carry-over factors, and the fixed-end moments. Results for slabs without drop panels are given in Table 8.4. This table is reproduced from [8], where similar tables are also given for slabs with drop panels.

EXAMPLE 8.11

In the building with the floor plan shown in Fig. 8.42a, 14-by-16-in beams are spanning between all columns as shown.

    a) For the slab-beam along line B–B, determine all stiffness coefficients and carry-over factors.

    b) For a computer frame analysis, show in a sketch of the computer model all member lengths and moments of inertia.

Solution

a) *Stiffness coefficients and carry-over factors.*

Beams B1–B2 and B3–B4:

$$\text{Centroid, } c_t = \frac{(18)(12)(7)(3.5) + (14)(16)(15)}{1512 + 224} = 4.98 \text{ in}$$

Moment of inertia,

$$I = (1512)\left(\frac{7^2}{12} + 1.48^2\right) + (224)\left(\frac{16^2}{12} + 10.02^2\right) = 36{,}754 \text{ in}^4$$

In joint regions, with $\dfrac{c_2}{\ell_2} = \dfrac{18}{(12)(16+20)/2} = .083, \quad \dfrac{c_1}{\ell_1} = \dfrac{14}{(12)(18)} = .065,$

find from Table 8.4 by interpolation: $k = 4.1\dfrac{EI}{L}$, carry-over factor = .507.

TABLE 8.4 MOMENT DISTRIBUTION FACTORS FOR SLAB-BEAM ELEMENTS WITHOUT DROP PANELS (FLAT PLATES) [8]

FEM (uniform load w) = $Mw\ell_2\ell_1^2$ K (stiffness) = $kE\ell_2t^3/12\ell_1$
Carryover factor = COF

c_1/ℓ_1		0.00	0.05	0.10	0.15	0.20	0.25
				c_2/ℓ_2			
0.00	M	0.083	0.083	0.083	0.083	0.083	0.083
	k	4.000	4.000	4.000	4.000	4.000	4.000
	COF	0.500	0.500	0.500	0.500	0.500	0.500
0.05	M	0.083	0.084	0.084	0.084	0.085	0.085
	k	4.000	4.047	4.093	4.138	4.181	4.222
	COF	0.500	0.503	0.507	0.510	0.513	0.516
0.10	M	0.083	0.084	0.085	0.085	0.086	0.087
	k	4.000	4.091	4.182	4.272	4.362	4.449
	COF	0.500	0.506	0.513	0.519	0.524	0.530
0.15	M	0.083	0.084	0.085	0.086	0.087	0.088
	k	4.000	4.132	4.267	4.403	4.541	4.680
	COF	0.500	0.509	0.517	0.526	0.534	0.543
0.20	M	0.083	0.085	0.086	0.087	0.088	0.089
	k	4.000	4.170	4.346	4.529	4.717	4.910
	COF	0.500	0.511	0.522	0.532	0.543	0.554
0.25	M	0.083	0.085	0.086	0.087	0.089	0.090
	k	4.000	4.204	4.420	4.648	4.887	5.138
	COF	0.500	0.512	0.525	0.538	0.550	0.563
$x = (1 - c_2/\ell_2)^3$		1.000	0.856	0.729	0.613	0.512	0.421

* c_1 and c_2 are the widths of the column measured parallel to ℓ_1 and ℓ_2.

Note that $I_{jt} = \dfrac{I_{sb}}{(1 - c_2/\ell_2)^2} = \dfrac{36{,}754}{(1 - .083)^2} = 43{,}709 \text{ in}^4.$

Beam B2–B3:

Outside joint region, there is no change, that is,

$k = 4.1\dfrac{EI}{L}$, carry-over factor = .507.

Inside joint region, with $c_2/\ell_2 = .083$, $c_1/\ell_1 = \dfrac{14}{(20)(12)} = .058,$

find from Table 8.4: $k = 4.089\dfrac{EI}{L}$,   carry-over factor = .505.

b) *Input data for computer model.*

Slab-beam moments of inertia were determined in problem part a.

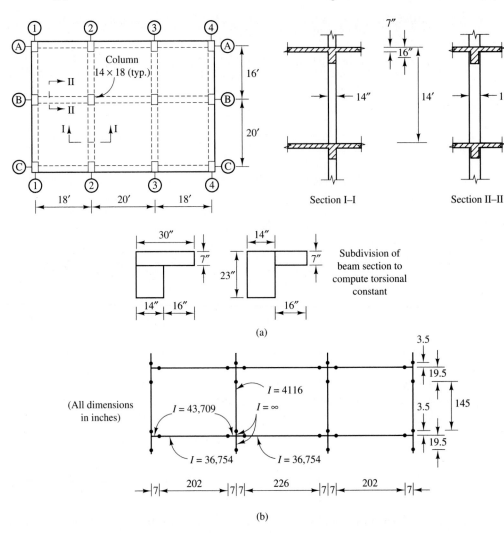

Figure 8.42 Example 8.11: Frame analysis of concrete buildings. (a) Floor plan and section; (b) frame model for computer analysis.

Column moment of inertia, $I = \dfrac{(18)(14)^3}{12} = 4116 \text{ in}^4$.

Within the joint region, the columns are infinitely stiff.

The required input data for the frame model are listed in Fig. 8.42b.   ∎

The frame columns are subjected to bending moments unless the slab-beam moments on the two sides are exactly equal. Exterior columns are the most obvious examples where this is not the case, and as a result, the beam-column joint is expected to rotate. The amount of rotation is partly a function of the relative stiffnesses involved. Consider the slab-column joint of Fig. 8.43. Equilibrium requires that the column moment be equal to the total slab moment. However, the deformational pattern reveals that the slab tends to rotate more where it is further removed from the column. This causes the slab edge to twist, and consequently, if the edge is reinforced with a spandrel beam, the rotational stiffness of this edge beam needs to be accounted for.

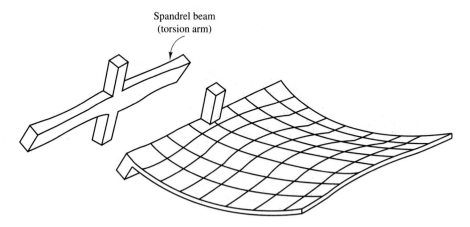

Spandrel beam
(torsion arm)

Figure 8.43   Twisting of torsion arms.

It is common practice to combine the actual external columns and torsion arms into an *equivalent column*. Since these are arranged like springs in series, their respective flexibilities are additive (not their stiffnesses, as in the case of parallel springs), that is,

$$F_{ec} = \sum F_c + F_t$$

or

$$\frac{1}{K_{ec}} = \frac{1}{\sum K_c} + \frac{1}{K_t} \tag{8.59}$$

where K_{ec} is the stiffness of the equivalent column in question, $\sum K_c$ is the sum of the stiffnesses of the column above and below the floor under consideration, and K_t is the stiffness of the attached torsion members.

The stiffness of an actual column must take into account the difference between center-to-center story height ℓ_c and clear story height, ℓ_u. For most practical cases, this effect can be approximated by

$$K_c = 4\frac{E I_c}{\ell_c}\left(\frac{\ell_c}{\ell_u}\right)^{2.5} \tag{8.60}$$

To determine the stiffness of the torsion members, consider a column with attached edge beams extending halfway to the neighbor columns (Fig. 8.44). A twisting moment of unit value applied at the column is equilibrated by a distributed torque in the torsion members. For symmetry reasons, the torque must be zero halfway between columns.

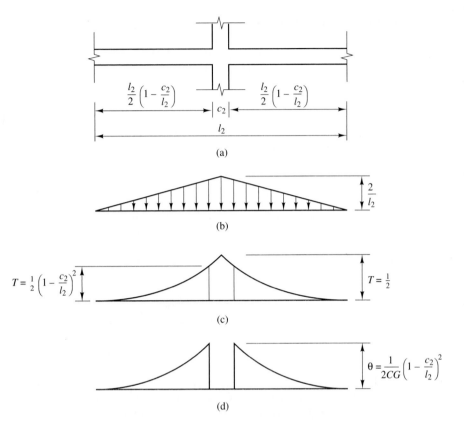

Figure 8.44 Moments in torsion arms: (a) torsion members; (b) distribution of twisting moment; (c) twisting moment diagram; (d) variation of angle of twist.

Because the column is the center of resistance, it can be assumed that the torque per unit length is highest at the column. Thus, it appears reasonable to assume the linear variation of twisting moments shown in Fig. 8.44b. Integration of these distributed moments leads to the torsional moment in the edge beam, as shown in Fig. 8.44c, and the variation of the angle of twist, shown in Fig. 8.44d, which, if integrated, gives

$$\theta_{\text{tot}} = \frac{1}{3} \frac{(1 - c_2/\ell_2)^2}{2GC} \left[\frac{\ell_2}{2} \left(1 - \frac{c_2}{\ell_2} \right) \right]$$

The average twist is approximately one-third of the total angle, and with $G \approx E/2$, given by

$$\theta_{\text{ave}} = \frac{1}{18} \frac{(1 - c_2/\ell_2)^3 \ell_2}{CE}$$

that is, the stiffness of one torsion arm of length $\ell_2/2$ is

$$K_t = \frac{9EC}{\ell_2(1 - c_2/\ell_2)^3}$$

and the stiffness of both arms is

$$K_t = \sum \frac{9E_{cs}C}{\ell_2(1 - c_2/\ell_2)^3} \tag{8.61}$$

where E_{cs} is Young's modulus of the slab concrete, and C is the member's torsional constant, given by Eq. 8.56.

If a beam spans between columns in the direction in which moments are being determined, a much larger share of the slab-beam moment is resisted by the column than otherwise. Equation 8.59 underestimates the column stiffness in this case. Therefore, the torsion arm stiffness K_t should be multiplied by the ratio I_{sb}/I_s, where I_{sb} and I_s are, respectively, the moment of inertia of the entire slab strip with and without the beam.

EXAMPLE 8.12

Determine the stiffness of the equivalent column for column B1 of Example 8.12 (Fig. 8.42a). Given: $f'_c = 4$ ksi.

Solution

Column stiffness, $\sum K_c$

Cross-sectional area, 14 by 18 in

Moment of inertia: $I = \dfrac{(18)(14)^3}{12} = 4116$ in^4

Clear story height: $\ell_u = (14)(12) - 23 = 145$ in

Young's modulus: $E_c = 57,000\sqrt{4000} = 3605$ ksi

Stiffness of column above and below:

$$\sum K_c = 2 \left(\frac{4E_cI}{\ell_c} \right) \left(\frac{\ell_c}{\ell_u} \right)^{2.5} = \frac{(2)(4)(3605)(4116)}{168} \left(\frac{168}{145} \right)^{2.5} = 1.02 \times 10^6 \text{ in-kips}$$

Torsion arm stiffness, K_t

Spandrel beam section (see Fig. 8.42a), 30 × 7 in slab on 14 × 16 in web. Torsional constant,

$$C = \left(1 - .63\frac{7}{30} \right) \frac{(30)(7)^3}{3} + \left(1 - .63\frac{14}{16} \right) \frac{(16)(14)^3}{3} = 9,493 \text{ in}^4$$

$$C = \left(1 - .63\frac{14}{23} \right) \frac{(23)(14)^3}{3} + \left(1 - .63\frac{7}{16} \right) \frac{(16)(7)^3}{3} = 14,295 \text{ in}^4 \text{ (controls)}$$

Stiffness, $K_t = \sum \dfrac{9E_cC}{\ell_2(1 - c_2/\ell_2)^3} = \dfrac{(9)(3605)(14,295)}{(192)(1 - 18/192)^3} + \dfrac{(9)(3605)(14,295)}{(240)(1 - 18/240)^3} =$

5.687×10^6 in-kips

Moment of inertia of slab strip without beam, $\dfrac{16 + 20}{2} \times 12 = 216$ in wide,

$$I_s = \frac{(216)(7)^3}{12} = 6174 \text{ in}^4$$

Slab strip with 14 × 16 in beam (see Example 8.12), $I_{sb} = 36,754$ in^4

Modification of torsional stiffness,

$$K_t \frac{I_{sb}}{I_s} = 5.687 \times 10^6 \times \frac{36,754}{6174} = 33.85 \times 10^6 \text{ in-kips}$$

Equivalent column stiffness, K_{ec}

$$\frac{1}{K_{ec}} = \frac{1}{1.02 \times 10^6} + \frac{1}{33.85 \times 10^6} = 1.01 \times 10^{-6}, \text{ from which,}$$

$$K_{ec} = 990,000 \text{ in-kips}$$

Note that, in this example, the equivalent column stiffness is barely affected by the torsional flexibility of the spandrel beam, since the relatively stiff beam transfers most of the moment directly into the column, bypassing the torsion arms.   ∎

8.7 SHEAR DESIGN AND MOMENT TRANSFER

Building structures in which floor slabs are supported by columns with relatively small cross-sectional dimensions pose some obvious load transfer challenges. We shall first consider the shear design problem and investigate moment transfer thereafter.

Consider the slab of Fig. 8.45a. Vertical equilibrium requires that a column carry the load distributed over its tributary area. The column reaction force must be transmitted by shear stresses, and if these are large, it is not difficult to visualize the column punching through the slab. In fact, this is a significant failure mode that needs to be accounted for in the design. The failure surface has the shape of a truncated pyramid (Fig. 8.45b), with

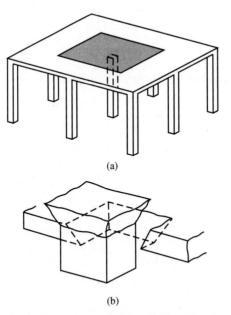

(a)

(b)

Figure 8.45 Column with tributary floor area and punching shear failure; (a) tributary area; (b) punching failure.

the individual surfaces inclined approximately 45°, just like the shear cracks in a beam. If a and b are the column's cross sectional dimensions and d the effective slab thickness, then the critical shear area can be approximated by $2(a + b + 2d)d$ (Fig. 8.46a). In

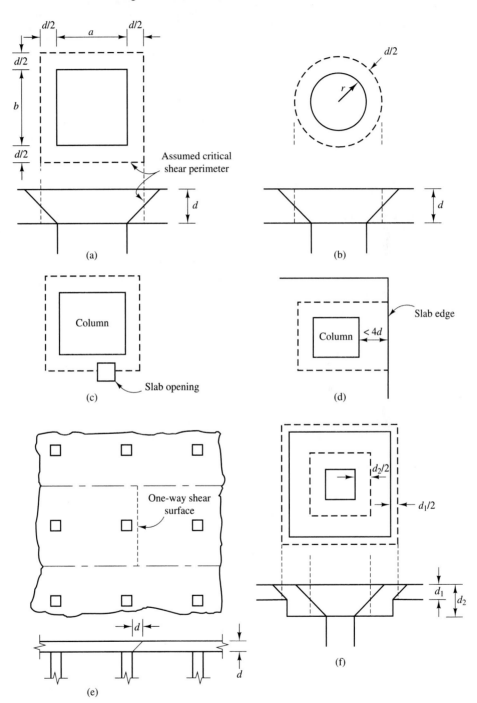

Figure 8.46 Critical shear perimeter.

the case of a circular column with radius r, the failure surface is a truncated cone, with the approximate surface area $2\pi[r + (d/2)]d$ (Fig. 8.46b). For a column placed near a slab edge or with an opening situated within the critical perimeter, the critical area must be modified accordingly (Fig. 8.46c). The critical perimeter or shear plane is always the smallest one possible, and in some cases several different possibilities need to be evaluated. For example, for a column placed not at but near a slab edge, a fourth failure surface may not form if the edge is closer than about $4d$ to the column (Fig. 8.46d). Likewise, the entire slab can fail in one-way shear like a beam, should this turn out to be critical (Fig. 8.46e). In the presence of a drop panel, two different failure perimeters need to be investigated (Fig. 8.46f).

Whereas the nominal shear strength of beams can be conservatively approximated by $2\sqrt{f_c'}$ (see Sec. 5.2), the ACI Code lists three formulas for the concrete's nominal shear capacity in slabs, and the smallest value obtained is to be used for design:

$$V_c = \left(2 + \frac{4}{\beta_c}\right)\sqrt{f_c'}\, b_o d \qquad\qquad (8.62a)[11\text{-}36]$$

where β_c is the ratio of the long to the short side of the column, and b_o is the critical perimeter;

$$V_c = \left(2 + \frac{\alpha_s d}{b_o}\right)\sqrt{f_c'}\, b_o d \qquad\qquad (8.62b)[11\text{-}37]$$

where α_s is 40 for interior columns (with four-sided failure surface), 30 for edge columns (with three sides), and 20 for corner columns (with two-sided failure surface);

$$V_c = 4\sqrt{f_c'}\, b_o d \qquad\qquad (8.62c)[11\text{-}38]$$

Equation 8.62a reflects the shear capacity reduction relative to Eq. 8.62c for columns with side ratios in excess of 2. Similarly, Eq. 8.62b reflects the results of tests according to which V_c decreases for large b_o/d ratios. Note that the punching shear problem is essentially the same, whether a slab is supported by a column (Fig. 8.47a), a concentrated load is applied to a slab (Fig. 8.47b), or a column is supported by a footing (Fig. 8.47c). The latter case shall be considered further in Chapter 9.

EXAMPLE 8.13

An 8-in floor slab, carrying 200-psf live load in addition to its own weight, is supported by circular columns spaced in a regular grid of 20 ft. The 14-in diameter columns are capped with 28-in diameter capitals and reinforced with 5-in drop panels measuring 8 by 8 ft (Fig. 8.48a). Determine the floor's adequacy with regard to shear capacity. Given: $f_c' = 4$ ksi.

Solution

a) *Check one-way shear* (Fig. 8.48b; $d = 7$ in)

Length of shear surface: 20 ft = 240 in

Tributary area: $(20)\left(10 - 4 - \dfrac{7}{12}\right) = 108.3$ ft^2

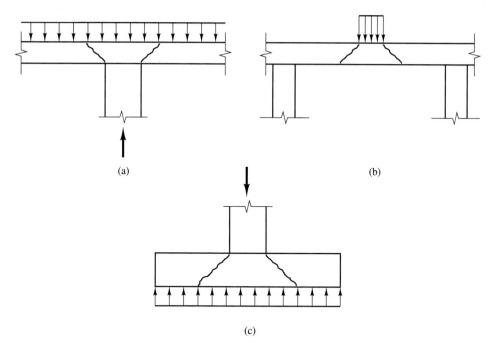

Figure 8.47   Various forms of punching shear failure.

Factored load: $w_u = (1.4)\left(\dfrac{8}{12}\right)(150) + (1.7)(200) = 480$ psf

Factored shear: $V_u = (.48)(108.3) = 52$ kips

Shear capacity: $\phi V_n = (.85)(2)\dfrac{\sqrt{4000}}{1000}(240)(7) = 180.6$ kips > 52 kips <u>OK</u>

b) *Check drop panel perimeter ($d = 7$ in)*

Tributary area: $20^2 - \left(8 + \dfrac{7}{12}\right)^2 = 326.3$ ft^2

Factored shear: $(.48)(326.3) = 156.6$ kips

Critical perimeter: $b_o = (4)\left(8 + \dfrac{7}{12}\right) = 34.3$ ft $= 412$ in

Concrete shear capacity:

Eq. 8.62a, $V_c = \left(2 + \dfrac{4}{1}\right)\sqrt{f'_c}b_od = 6\sqrt{f'_c}b_od$

Eq. 8.62b, $V_c = \left(2 + \dfrac{(40)(7)}{412}\right)\sqrt{f'_c}b_od = 2.68\sqrt{f'_c}b_od$ (controls)

Eq. 8.62c, $V_c = 4\sqrt{f'_c}b_od$

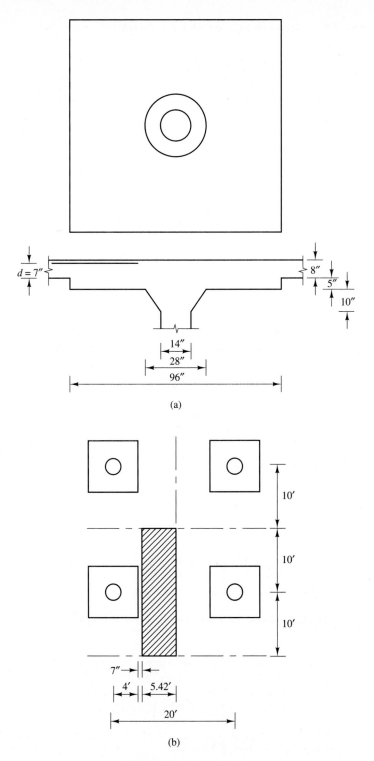

(a)

(b)

Figure 8.48 Example 8.13.

Reduced shear capacity:

$$\phi V_n = (.85)(2.68)\frac{\sqrt{4000}}{1000}(412)(7) = 415 \text{ kips} > 156.6 \text{ kips } \underline{\text{OK}}$$

c) *Check column capital perimeter* ($d = 12$ in)

Tributary area: $20^2 - \dfrac{\pi(14 + 12/2)^2}{144} = 391 \text{ ft}^2$

Factored shear: $(.48)(391) = 188 \text{ kips}$

The drop panel dead weight adds $(1.4)\left(\dfrac{5}{12}\right)(8^2)(.15) = 5.6 \text{ kips}$

Critical perimeter: $b_o = 2\pi\left(14 + \dfrac{12}{2}\right) = 125.7 \text{ in}$

Concrete shear capacity:

$$V_c = (.85)(4)\frac{\sqrt{4000}}{1000}(125.7)(12) = 324 \text{ kips} > 194 \text{ kips } \underline{\text{OK}}$$ ∎

EXAMPLE 8.14

Based on only punching shear strength and the ACI Code formulas for V_c, determine the minimum bridge deck thickness required to safely carry a 16-kip wheel load. Given: $f'_c = 4000$ psi; tire contact area $= 8$ by 20 in.

Solution

Critical shear area: $A_v = b_o d = 2(a + b + 2d)d = 2(28 + 2d)d$

Concrete shear capacity:

Eq. 8.62a, $V_c = \left(2 + \dfrac{4}{2.5}\right)\sqrt{f'_c}b_o d = 3.6\sqrt{f'_c}b_o d$

Eq. 8.62b, $V_c = \left(2 + \dfrac{40d}{56 + 4d}\right)\sqrt{f'_c}b_o d = \left(2 + \dfrac{1}{1.4/d + 0.1}\right)\sqrt{f'_c}b_o d$

Eq. 8.62c, $V_c = 4\sqrt{f'_c}b_o d$ (not controlling)

By equating Eqs. 8.62a and b, we can show that Eq. 8.62b controls only if $d < 2.67$ in, which is not practical, that is, Eq. 8.62a controls.

Factored shear, $V_u = (1.7)(16) = 27.2 \text{ kips}$

Concrete shear strength, $\phi V_c = (.85)(3.6)\dfrac{\sqrt{4000}}{1000}(2)(28 + 2d)d$

From $V_u = \phi V_c$, find $d = 2.2$ in.

For this result, Eq. 8.62b controls, but the difference in shear strength is minimal. Adding 1 in for cover, punching shear requires a minimum bridge deck thickness of 3.2 in. Since bridge decks are typically at least 6 in thick, the probability of a truck wheel load punching through is very small indeed, provided the concrete is sound.   ∎

The nominal shear strength of concrete, $V_c/b_o d$, is independent of the slab thickness d, according to the current ACI Code. This implies that doubling the slab thickness should double the slab's shear capacity. However, this is not the case, because concrete is a brittle material, which cannot distribute the shear stresses at failure evenly over the entire failure surface. If the material were ductile, the most highly stressed points near the column would yield, and any stress increases would be transferred to neighboring material points further up the failure surface until the entire failure surface yields. A brittle material like concrete cracks at the highly stressed column junction long before stress points further removed are stressed to capacity (Fig. 8.49). By this reasoning, doubling the slab thickness would add little capacity. A rational assessment of the concrete's shear capacity must be based on principles of fracture mechanics and crack propagation in brittle materials [9]. We can anticipate that future editions of the ACI Code will account for the *size effect* in an appropriate manner.

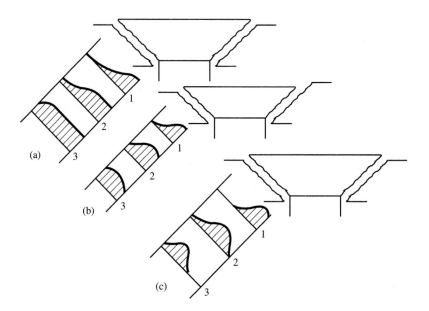

Figure 8.49 Size effect on brittle punching shear failure (stress distributions at three stages of crack propagation) [9]: (a) ductile material; (b) brittle material, small thickness; (c) brittle material, large thickness.

The strength of flat plates without drop panels is often controlled by punching shear. If this capacity is considerably less than that associated with other failure modes, it can be increased by adding shear reinforcement (Fig. 8.50). Standard reinforcing cages with closed stirrups may be appropriate in relatively thick slabs with adequate anchorage of the stirrups (Fig. 8.50a). If shear capacity is very critical, it may be more advantageous to use structural steel shapes as shear heads (Fig. 8.50b). Recently, shear studs have been shown to offer effective solutions [10] (Fig. 8.50c). Note that if shear reinforcement is used, the ACI Code requires that it be designed to carry all shear in excess of a stress of $2\sqrt{f'_c}$, that is, the concrete is credited with no more nominal shear strength than in beams.

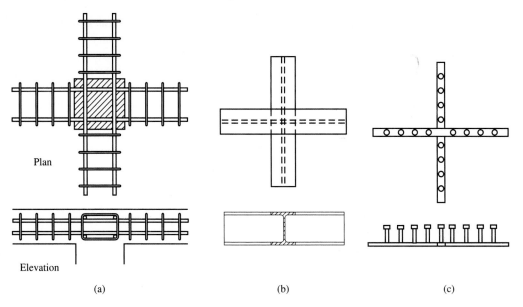

Plan

Elevation

(a)   (b)   (c)

Figure 8.50   Shear reinforcement: (a) reinforcement cages; (b) welded structural steel shear heads; (c) shear studs.

In sum, the following options are available in case of insufficient shear strength:

a) use a thicker slab;

b) add a drop panel and/or a column capital;

c) use a larger column;

d) use shear reinforcement; and

e) use higher strength concrete.

The moment transfer between slab and column poses another design challenge to cope with the severe stress concentration at the slab-column junction (Fig. 8.51). A possible failure mode is the tear-out of a slab segment, not unlike a punching shear failure, except that the slab tears out in opposite directions at two opposing column faces.

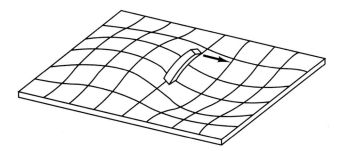

Figure 8.51   The moment transfer design challenge.

Thus, the moment transfer effectively adds to the shear stresses at a section that is already shear critical. For this reason, the moment transfer capacity of such a slab-column connection is rather limited. Together with its high flexibility, this is a strong reason against relying exclusively on flat slab construction in multistory buildings to resist lateral loads. Figure 8.51 also illustrates the difficulty of defining the effective width of the "beam," that is, the part of the column strip that is effectively transferring moment into the column. The Code provides guidance only in the case where a real beam spans between columns (Fig. 8.34).

Any unbalanced moment M_u due to gravity or lateral loads is transferred by a combination of flexure and *eccentric shear* (Fig. 8.52a). According to the ACI Code, the moment fraction transferred by flexure is estimated to be

$$M_f = \gamma_f M_u$$

with

$$\gamma_f = \frac{1}{1 + \frac{2}{3}\sqrt{\frac{b_1}{b_2}}} \qquad (8.63)[13\text{-}1]$$

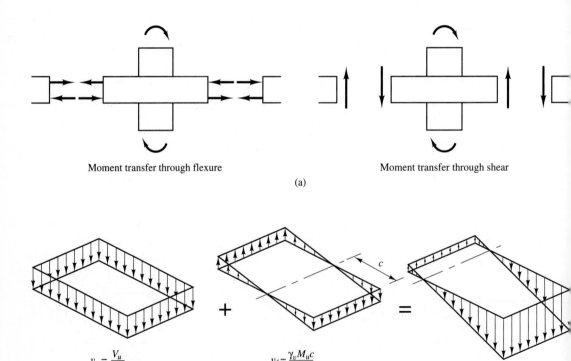

Figure 8.52   Combined shear and moment transfer; (a) moment transfer; (b) superposition of shear stresses.

where b_1 and b_2 are, respectively, the width of the critical section in the direction of the moment being considered, and in the direction perpendicular to it. The part of the moment to be transferred by eccentric shear is then

$$M_v = \gamma_v M_u$$

with

$$\gamma_v = 1 - \frac{1}{1 + \frac{2}{3}\sqrt{\frac{b_1}{b_2}}} \tag{8.64}$$

The resulting shear stress distributions are illustrated in Fig. 8.52b. The maximum shear stress occurs at the point farthest removed from the critical shear area's centroidal axis, that is,

$$v_f = \pm \frac{\gamma_v M_u c}{J}$$

where c is the maximum distance from the centroidal axis of the critical shear area and J is the polar moment of inertia of the shear critical area about its centroidal axis. These stresses need to be added to the stresses due to the punching shear force

$$v_u = \frac{V_u}{b_o d} \pm \frac{\gamma_v M_u c}{J} \tag{8.65}$$

and this combined stress shall not exceed the concrete's shear strength, that is,

$$v_u \le \phi v_c$$

To obtain the polar moment of inertia of a rectangular shear critical perimeter, consider a typical interior column with dimensions a and b. The shear critical surface has side lengths $b_1 = a + d$ and $b_2 = b + d$ (Fig. 8.53a). Because of symmetry, the axis of rotation is centrally located, and the total shear area is $2(b_1 + b_2)d$. The polar

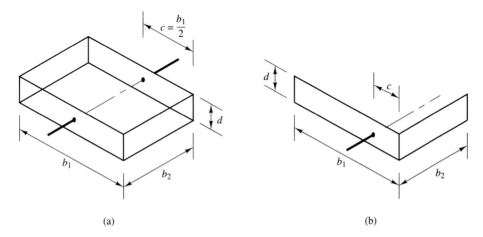

(a)   (b)

Figure 8.53 Geometric properties of shear surfaces; (a) four-sided shear surface; (b) two-sided shear surface.

moment of inertia follows as

$$J = 2\left(\frac{b_1^3 d}{12} + \frac{b_1 d^3}{12}\right) + 2(b_2 d)\left(\frac{b_1}{2}\right)^2$$

For a corner column, with the dimensions of Fig. 8.53b, we locate the centroidal axis as

$$c = \frac{b_1 d(b_1/2)}{b_1 d + b_2 d} = \frac{b_1^2}{2(b_1 + b_2)}$$

and

$$J = \frac{b_1^3 d}{12} + \frac{b_1 d^3}{12} + b_1 d\left(\frac{b_1}{2} - c\right)^2 + b_2 d c^2$$

Similar expressions can be derived for edge columns and columns with irregularly shaped shear perimeter areas. For a circular column, it is common to base the calculations on an equivalent square column with the same cross sectional area.

EXAMPLE 8.15

A column with a 16-in diameter is placed 12 in away from the edge of a 6-in-thick slab (Fig. 8.54). The total factored shear force to be transmitted is 40 kips, and the unbalanced factored moment is 30 ft-kips. Check whether the slab-column connection is adequate to resist the combined shear stresses. Given: $d = 5$ in; $f_c' = 4$ ksi.

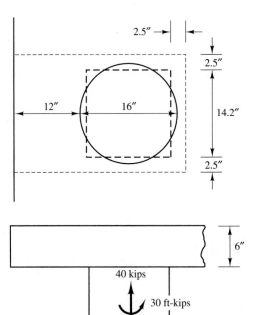

Figure 8.54 Example 8.15.

Solution

Replace round column by equivalent square: $a = r\sqrt{\pi} = 8\sqrt{\pi} = 14.2$ in

Edge distance: $20 - 7.1 = 12.9 < 4d$, that is, shear perimeter has only three sides.

Critical shear area: $A_v = b_o d = [2 \times (20 + 7.1 + 2.5) + 14.2 + 5](5) = 392$ in^2

Fraction of eccentric-shear moment: $\gamma_v = 1 - \dfrac{1}{1 + 2/3\sqrt{29.6/19.2}} = .453$

Centroid of shear area: $c = \dfrac{(2)(29.6 \times 5)(29.6/2)}{(2)(29.6 \times 5) + (19.2)(5)} = 11.2$ in

Moment of inertia: $J = (2)\left[\dfrac{(29.6)^3(5)}{12} + \dfrac{(29.6)(5)^3}{12} + (29.6 \times 5)\left(\dfrac{29.6}{2} - 11.2\right)^2\right] +$

$(19.2 \times 5)(11.2)^2 = 38,107$ in^4

Maximum shear stress:

$$v_u = \frac{V_u}{b_o d} + \frac{\gamma_v M_u c}{J} = \frac{40}{392} + \frac{(.453)(30 \times 12)(11.2)}{38,107} = .102 + .047 = .149 \text{ ksi}$$

Concrete shear strength:

Eq. 8.62a, $\dfrac{V_c}{b_o d} = \left(2 + \dfrac{4}{29.6/19.2}\right)\sqrt{f_c'} = 4.59\sqrt{f_c'};$

Eq. 8.62b, $\dfrac{V_c}{b_o d} = \left(2 + \dfrac{30 \times 5}{78.4}\right)\sqrt{f_c'} = 3.91\sqrt{f_c'}$ (controls);

Eq. 8.62c, $\dfrac{V_c}{b_o d} = 4\sqrt{f_c'}$

$\dfrac{\phi V_c}{b_o d} = (.85)(3.91)\sqrt{4000} = 210$ psi > 149 psi OK   ∎

SUMMARY

1. The bending moments and shears in a plate or slab are defined as

$$M_x = \int_{-t/2}^{t/2} \sigma_x dz \qquad M_y \int_{-t/2}^{t/2} \sigma_y z dz \qquad M_{xy} = M_{yx} = \int_{-t/2}^{t/2} \tau_{xy} z dz$$

$$Q_x = \int_{-t/2}^{t/2} \tau_{xz} dz \qquad Q_y = \int_{-t/2}^{t/2} \tau_{yz} dz$$

Note that they are expressed in their respective units per unit length.

2. The plate stiffness is defined by the expression

$$D = \frac{Et^3}{12(1 - v^2)}$$

and is equivalent to the beam stiffness EI.

3. Where the plate shear is specified as a boundary condition, the Kirchhoff shear should be used, which is defined as

$$V_x = Q_x + \frac{\partial M_{xy}}{\partial y} \quad \text{or} \quad V_y = Q_y + \frac{\partial M_{xy}}{\partial x}$$

4. Yield line theory is to plates what plastic analysis is to frame structures. Both are methods of analysis consistent with ultimate strength design, but neither is widely used in practice.

5. In a plate with isotropic reinforcement, that is, equal reinforcement in the x- and y-directions, the moment capacity is the same in any direction (i.e., $M_u = M_{ux} = M_{uy}$).

6. The following guidelines are helpful for the identification of yield line patterns:
   - plane surfaces always intersect in straight lines, therefore yield lines can be assumed to be approximately straight;
   - axes of rotation generally follow lines of support; these axes may be real hinges or plastic hinges;
   - axes of rotation must pass over any columns;
   - a yield line between two slab parts passes through the point of intersection of the axes of rotation of the two parts.

7. Two-way slab systems supported by beams and columns must be designed to carry 100% of the load in each direction. Slabs supported on nonyielding supports share the load in proportion to the stiffnesses in both directions.

8. The Direct Design Method of the ACI Code is applicable to relatively regular building structures and comparable to the use of moment coefficients for continuous beams. When this method is not applicable, the Equivalent Frame Method may have to be used.

9. The Direct Design Method consists of the following basic steps:

   Step 1.   Determine the total statical moment for a given panel,

   $$M_o = \frac{(w_u \ell_2)\ell_n^2}{8}$$

   where w_u is the factored uniform load, ℓ_2 is the center-to-center distance normal to the span direction in which moments are being determined, ℓ_n is the clear span length under consideration.

   Step 2.   Distribute the total statical moment to the negative (support) and positive (midspan) sections. In an interior span, 65% of M_o are assigned to the negative sections at the support faces and the remaining 35% are taken by the positive section at midspan. In an end span, the moment coefficients of Table 8.1 may be used.

   Step 3.   Proportion both the negative and positive moments among the column strip, middle strip, and beam, where applicable. Use the percentages of Table 8.2, with the relative stiffness factor $\alpha = E_{cb}I_b/E_{cs}I_s$ and the relative torsional stiffness factor $\beta_t = E_{cb}C/2E_{cs}I_s$, where $C = \sum[1 - 0.63(x/y)](x^3 y/3)$ is the torsional constant of the edge beam. If a beam is part of the column strip, it is to be assigned 85% of the column strip moments given in Table 8.2, the remaining 15% fall to the slab portion of the column strip.

   Step 4.   Determine the reinforcement for the column strip, middle strip, and beam to carry the factored moments determined in Step 3.

Step 5. Repeat Steps 1 through 4 for all other panels in the building.

Step 6. Repeat Steps 1 through 5 for the orthogonal span direction.

10. For the Equivalent Frame Method, the moment of inertia of a slab-beam within the joint region can be approximated by

$$I_{jt} = \frac{I_{sb}}{(1 - \frac{c_2}{\ell_2})^2}$$

where I_{jt} and I_{sb} are the moment of inertia of the slab-beam within and outside the joint region, respectively, c_2 is the column width and ℓ_2 the centerline column spacing.

11. Where an external column is stiffened by spandrel beams acting as torsion arms, the stiffenss of the equivalent column is determined as

$$\frac{1}{K_{ec}} = \frac{1}{\sum K_c} + \frac{1}{K_t}$$

where $\sum K_c$ is the sum of the stiffnesses of the column above and below the floor under consideration, with $K_c = 4(EI_c/\ell_c)(\ell_c/\ell_u)^{2.5}$, and K_t is the stiffness of the attached torsion arms

$$K_t = \sum \frac{9E_{cs}C}{\ell_2(1 - c_2/\ell_2)^3}$$

12. The shear strength of slabs provided by the concrete alone is given by

$$V_c = \min \left\{ 2 + \frac{4}{\beta_c}; \ 2 + \frac{\alpha_s d}{b_o}; \ 4 \right\} \times \sqrt{f_c'} b_o d$$

where β_c is the ratio of the long to the short side of the column; b_o is the critical perimeter; α_s is 40 for interior columns, 30 for edge columns, and 20 for corner columns.

13. The following options are available in case of insufficient shear strength:
    a) use a thicker slab;
    b) add a drop panel and/or a column capital;
    c) use a larger column;
    d) use shear reinforcement;
    e) use higher strengh concrete.

14. The portion of any unbalanced moment due to gravity or lateral loads to be transferred by flexure is given by

$$M_f = \frac{1}{1 + \frac{2}{3}\sqrt{\frac{b_1}{b_2}}} M_u$$

where b_1 and b_2 are, respectively, the width of the critical section in the direction of the moment being considered, and in the direction perpendicular to it. The part of the moment to be transferred by eccentric shear is then $M_v = M_u - M_f$.

REFERENCES

1. Timoshenko, S., and Woinowsky-Krieger, S., *Theory of Plates and Shells*, McGraw-Hill Book Co., New York, 1959.
2. Westergaard, H. M., "Computation of Stresses in Bridge Slabs Due to Wheel Loads," Public Roads, vol. 11, no. 1, March 1930.
3. Johansen, K. W., *Yield Line Theory*, Cement and Concrete Association, London, 1962.
4. Hodge, P. G., *Plastic Analysis of Structures*, McGraw-Hill Book Co., New York, 1959.
5. Park, R., and Gamble, W. L., *Reinforced Concrete Slabs*, Wiley-Interscience, New York, 1980.
6. Hillerborg, A., *Strip Method of Design*, Viewpoint Publications, Cement and Concrete Association, Wexham Springs, Slough, England, 1974.
7. Roark, R. J., *Formulas for Stress and Strain*, 4th ed., McGraw-Hill Book Co., New York, 1965.
8. Simmonds, S. H., and Misic, J., "Design Factors for the Equivalent Frame Method," *ACI Journal*, vol. 68, no. 11, Nov. 1971.
9. Bazant, Z. P., ed., *Fracture Mechanics of Concrete*, Elsevier Applied Science, London, 1992.
10. Elgabry, A. E., and Ghali, A., "Design of Stud-Shear Reinforcement for Slabs," *ACI Structural Journal*, vol. 87, no. 3, May-June 1990.

PROBLEMS

For all problems below, unless otherwise specified, assume normal-weight concrete weighing 150 pcf, with $f'_c = 4000$ psi and steel with $f_y = 60,000$ psi. Concrete cover $= 1$ in.

8.1 According to yield line theory, what is the ultimate moment capacity required to support a uniform load of 100 psf on the

    (a) triangular

    (b) hexagonal

plate shown in Fig. P8.1, simply supported all around?

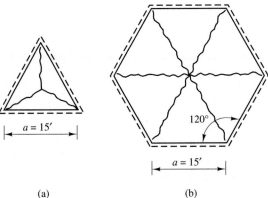

      (a)   (b)   **Figure P8.1**

8.2 Using yield line theory, determine the flexural reinforcement required for the 8-in-thick square plate shown in Fig. P8.2 to carry 100-psf dead weight and 300-psf live load. (Hint: Show that the yield line at the free edge starts at $x_1 = a/2$.)

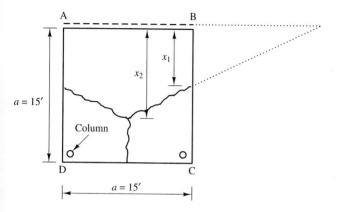

Figure P8.2

8.3 A rectangular slab of 8-in thickness is fully supported by nonyielding supports (walls). The two span lengths are 15 and 20 ft; live load = 500 psf.
  (a) Check the adequacy of the plate thickness, if no deflections are to be computed.
  (b) Design the slab as a one-way slab. In the orthogonal direction, provide Code-specified minimum reinforcement.
  (c) Design the slab as a two-way slab, based on moments computed by elastic plate theory. Include two harmonics in each direction and divide the slab into edge and middle strips to average moments.
  (d) Design the slab as a two-way slab, using moment coefficients.
  (e) Design the slab by yield line theory, assuming full plastic moment redistribution.
  (f) Compute the required steel tonnage for cases (b), (c), (d), and (e). Compare and discuss the results.

8.4 A floor slab is supported by 18-in square columns, spaced 20 and 25 ft on center in the two orthogonal directions, without any beams. Based on a corner panel, determine the required minimum slab thickness.

8.5 Design an 8-in two-way slab stiffened by 12-by-16-in beams along all column lines, as shown in Fig. P8.5, for a 60-psf live load. All columns are 12 by 12 in, and the total floor heights are 15 ft.
  (a) Determine the effective stiffness K_{ec} for the exterior column 2A.
  (b) Check whether the 8-in slab thickness satisfies the ACI Code minimum.
  (c) Using the direct design method, compute positive and negative design moments for the exterior and an interior panel along column line 2.

8.6 A 9-in-thick flat-plate floor system is supported by 18-in square columns. The building has four bays of 20 ft in one direction, four bays of 25 ft in the orthogonal direction, and three stories of 12-ft overall height each. Using the Equivalent Frame Design Method, design a typical interior frame in the direction with the 25-ft spans for a live load of 80 psf.

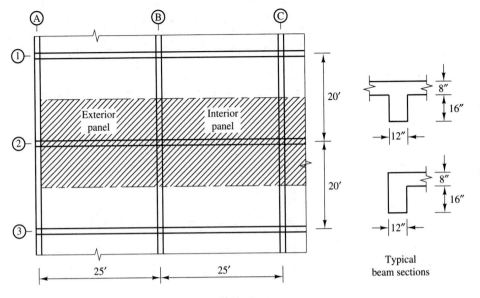

Figure P8.5

8.7 Given that a typical interior column of Problem 8.6 must transfer a shear force of 157 kips and an unbalanced moment of 24 ft-kips, check whether the slab shear strength is sufficient.

8.8 Assuming that an edge column of Problem 8.6 must transfer a shear force of 75 kips and a bending moment of 120 ft-kips, check the adequacy of the slab's shear strength.

8.9 Compute the column strip and middle strip moments for the interior panel of a flat plate building shown in Fig. P8.9. Given: Slab thickness = 6 in; story height = 9 ft; live load = 60 psf.

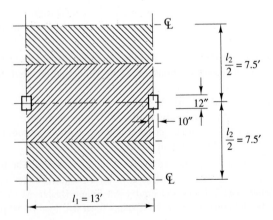

Figure P8.9

8.10 Compute column strip and middle strip moments for the exterior panel of the flat plate building shown in Fig. P8.10. Given: Slab thickness = 8 in; live load = 80 psf; 12-by-16-in edge beam; $\alpha = 0.3$; $\alpha_c = 0.2$.

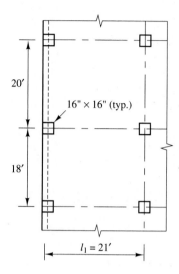

16" × 16" (typ.)

20′

18′

$l_1 = 21'$

Figure P8.10

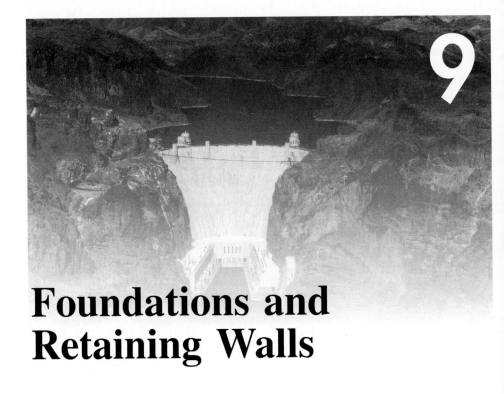

Foundations and Retaining Walls

Preceding chapters covered the basic elements of concrete design—design for flexure, shear, axial loads, and the special considerations due to the bond between concrete and reinforcing bars. Most other conceivable topics basically call for the *application* of these basic design principles to special situations, such as the design of a bridge or a building. In this chapter, we deal with a very special application, namely, the design of *foundations*. The application of basic design principles to this task is special only because of foundations' proverbial importance regarding the integrity of entire structures, since the loads of most civil engineering structures need to be transmitted safely into the ground. This task requires some knowledge of the properties of the underlying soils. Often, the expertise of a geotechnical engineer is needed to design the foundation of a structure.

The design of retaining walls will also be included in this chapter, as this is an application closely related to the design of foundations. For a detailed treatment of the subject, reference is made to the specialized literature on foundation engineering and earth-retaining structures [1, 2, 3].

9.1 GENERAL

Like other materials, soils undergo deformations when compressed. The vertical displacements are called *settlements*. Since many soils are of a viscoplastic nature, settlements generally increase as a function of time. For the structural engineer, it is particularly important to keep relative settlements of different structural components (e.g., different columns) as small as possible, because these can cause unacceptable stresses and cracking in the structure and even jeopardize its safety. It is an important objective of foundation design, therefore, to aim for the same soil pressure under each foundation component.

Unless we have the rare good fortune of building on solid rock, the load-bearing capacities of most soils are much lower than the strength of the foundation concrete. Therefore, the loads need to be spread over larger areas, using *spread footings*. If this is not feasible or practical, for example, if the soil beneath the structure is of very poor load-bearing quality, then it may be necessary to transfer the loads to stronger or firmer soil layers at greater depths, using a deep foundation such as piles or caissons. Herein, we shall restrict ourselves to only spread footings.

Types of foundations.   The choice of a particular type of foundation depends on a number of factors, such as the properties of the soil underlying the structure, the water table, the magnitudes and combinations of loads to be transmitted to the soil, and site constraints such as property lines.

Columns may be supported by single square or rectangular footings (Fig. 9.1a). If the columns are closely spaced or the required footing sizes too large because of poor bearing capacity of the soil, then two or more columns may be supported by a single combined footing (Fig. 9.1b). This solution may be a necessity if a column is placed near a property line and it is necessary to avoid eccentric loading on the footing. Walls are placed on continuous strip footings (Fig. 9.1c). If loads are heavy or soil conditions poor, it may be economical to support the entire structure on a single slab, or foundation *mat* (Fig. 9.1d). In special cases, such a mat can be several feet thick. Instead of using such a great thickness uniformly throughout, it may be more practical to form a *raft* by stiffening the mat with two orthogonal sets of beams. When piles are used to transfer the loads to deeper soil layers, it may be necessary to group several piles together and top them with a *pile cap* (Fig. 9.1e). In the case of a high water table, a foundation mat can be combined with perimeter walls to form a floating box. Such a box, properly waterproofed and designed to safely resist the water pressure, creates valuable basement space.

Bearing pressure distribution.   The exact stress distribution under a footing depends on the soil properties, the footing stiffness, and the loading. Assuming the footing is very stiff and symmetrically loaded, the pressure distribution on a sandy, cohesionless soil has the shape of Fig. 9.2a. Soil particles near the footing edges are likely to be squeezed into lesser stressed regions, thereby lowering the stresses near the edges. In deeper foundations, such soil movement is more restricted, resulting in a more uniform stress distribution. In a clayey soil, the cohesion makes it possible for shear stresses to develop, a process by which the loaded soil induces the neighboring unloaded soil to help support the load. The result is a stress distribution like the one depicted in Fig. 9.2b. For practical purposes, it is both necessary and acceptable to disregard such differences in the soil pressure distributions, because these are not likely to influence the footing design. This is true regardless of the footing's own stiffness, which, if it were taken into account, would further complicate the determination of the correct stress distribution. As a result, we may simply assume it to be uniform, as shown in Fig. 9.2c.

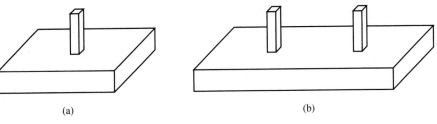

(a)   (b)

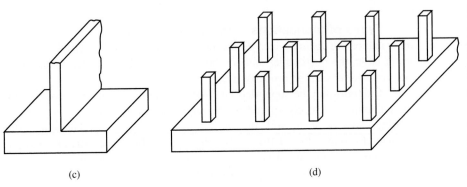

(c)   (d)

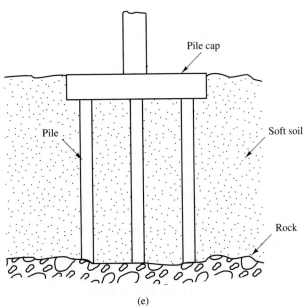

(e)

Figure 9.1 Types of foundations: (a) single footing; (b) combined footing; (c) strip footing; (d) mat foundation; (e) pile foundation.

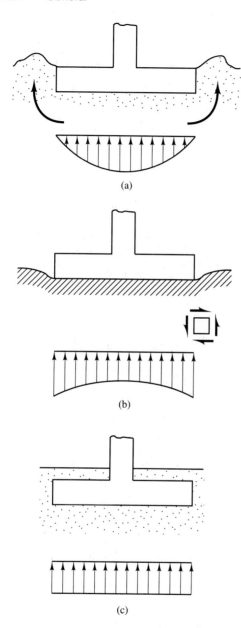

(a)

(b)

(c)

Figure 9.2 Bearing pressure distributions:
(a) sandy soil; (b) clayey soil; (c) assumed
stress distribution.

The situation is different if the footing is not loaded concentrically, or the column
axial load is accompanied by a moment (Fig. 9.3a). The standard column stress formula,

$$f(y) = \frac{P}{A} \pm \frac{My}{I} = \frac{P}{A}\left(1 \pm \frac{Aey}{I}\right) \tag{9.1}$$

applies only as long as no tensile stresses are generated, because the footing cannot

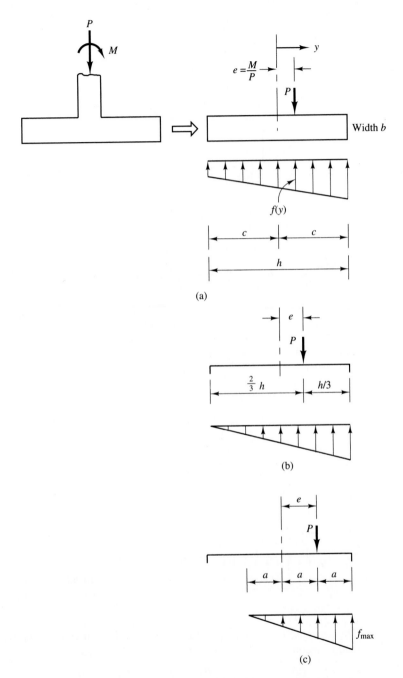

Figure 9.3 Eccentrically loaded footing: (b) $e = k = h/6$; (c) $e > k$.

transfer any tension into the soil. Equation 9.1 indicates that, for $e = I/Ac$, a triangular stress distribution results (Fig. 9.3b). In Sec. 7.2, this characteristic eccentricity was referred to as the *kern* distance k. For a rectangular section of height h it was found to be $k = h/6$. If $e > k$, equilibrium can still be maintained as long as the triangular stress block of depth $3a$, width b, and maximum intensity f_{max} equals the applied load P (Fig. 9.3c), that is,

$$P = \frac{3abf_{max}}{2} \tag{9.2}$$

Also, the centroid of the stress block must line up with load P, and the maximum stress f_{max} cannot exceed the soil bearing capacity. Since the stress-strain relationships of soils are typically highly nonlinear, the stress distribution depicted in Fig. 9.3c deviates further from reality, the higher the eccentricity e and the higher the maximum stress f_{max}.

   Of primary concern is the fact that a footing subject to a high moment tends to tilt unless the soil is very stiff. Therefore, it is usually a design objective to keep the eccentricity as small as possible, that is, to load each footing concentrically. Also, because of the soil flexibility, it is not appropriate to assume columns to be fixed against rotation at the base, that is, to count on perfect rotational restraint, unless the footing is supported on solid rock or very firm soil.

EXAMPLE 9.1

A 4-by-6-ft rectangular footing is loaded by a 150-kip force with an eccentricity of **a)** 6 in; **b)** 18 in (Fig. 9.4). Determine the resulting soil pressure distributions.

Solution

a) $e = \dfrac{1}{2}$ ft $< \dfrac{h}{6} = 1$ ft, that is, Eq. 9.1 applies.

$A = 4 \times 6 = 24$ ft^2, $I = \dfrac{(4)(6)^3}{12} = 72$ ft^4

$f_{max} = \dfrac{150}{24} \pm \dfrac{(150)(0.5)(3)}{72} = 6.25 \pm 3.125 = 9.375; \ 3.125$ kips/ft^2.

b) $e = 1.5$ ft > 1 ft, that is, Eq. 9.2 applies.

$a = \dfrac{h}{2} - e = 1.5$ ft, $f_{max} = \dfrac{P}{1.5ab} = \dfrac{150}{(1.5)(3)(4)} = 16.7$ kips/ft^2. ∎

EXAMPLE 9.2

Two 16-by-16-in building columns are spaced 15 ft apart (Fig. 9.5). The left column carries a 150-kip load, the right column carries a 180-kip load. The left column is so close to the property line that the footing may not extend beyond the column edge. Therefore, is is decided to place both columns onto a common *combined footing*. Determine the required length of the footing if the soil pressure is to be uniform.

Solution   We have available two equations of equilibrium, namely, $\sum P = 0$ and $\sum M = 0$.

$$\sum P = 0: \quad fl = 150 + 180 = 330 \text{ kips} \tag{a}$$

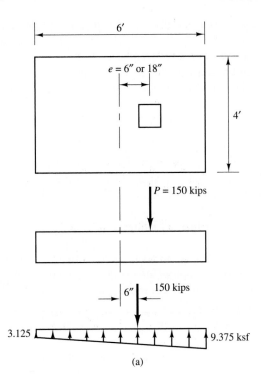

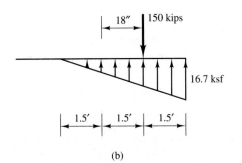

Figure 9.4   Example 9.1: (a) $e = 6''$;
(b) $e = 18''$.

(a)

(b)

$$\sum M = 0: \quad \frac{fl^2}{2} = (150)\left(\frac{8}{12}\right) + (180)\left(15 + \frac{8}{12}\right) = 2920 \text{ ft-kips} \tag{b}$$

Substituting Eq. (a) into Eq. (b), $330(l/2) = 2920$, from which $l = 17.7$ ft, that is, $\underline{l = 17'\text{-}8''}$.   ∎

Before turning to the design details of specific footing types, it is instructive to view one "upside down" (Fig. 9.6). It is obvious that the structural behavior and design criteria are the same, whether the soil pressure and column forces are active or reactive. This means the design considerations dealt with in Chapter 8 apply here as well: the

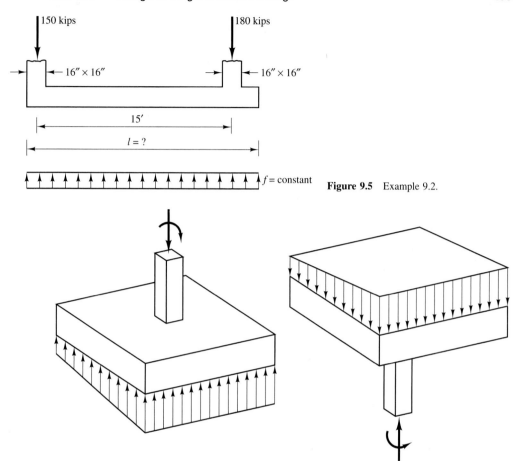

Figure 9.5 Example 9.2.

Figure 9.6 Inverted footing design problem.

footing "slab" is subjected to two-way bending and needs to be reinforced accordingly. Also, the punching shear problems are identical in the two cases. In fact, the applicable section in the ACI Code refers to "slabs and footings" in one phrase.

9.2 DESIGN OF SINGLE-COLUMN FOOTINGS

All fundamental aspects of this problem have been covered at some point in the previous chapters, except that nothing has been said about allowable soil bearing stresses. These are obviously dependent on the type of soil. The values specified in various building codes have safety factors typically between 2.5 and 3. Note that the term "allowable" connotes pressures caused by (unfactored) service loads. This means footing design utilizes both allowable stress and ultimate strength methods: the size of the footing is determined on the basis of unfactored service loads and allowable bearing pressures (increased by 33% if the design loads include the effects due to wind or earthquake); the footing thickness and reinforcement are determined on the basis of ultimate strength design for factored loads.

In determining soil pressures, a distinction needs to be made between gross and net soil pressure. The difference between the two is the weight of the soil (if any) above the base of the footing.

The footing itself must be designed with a sufficient margin of safety against the various possible failure modes (i.e., flexure, shear, and bond). Since the strength of concrete used in footings is usually considerably less than that for columns, the compressive stress that a column exerts on the footing also needs to be checked. The bearing strength of concrete (ACI Code Sec. 10.15.1) is $\phi(0.85 f'_c A_1)$, where A_1 is the loaded area and $\phi = 0.7$. However, due to the confinement of the loaded area provided by the surrounding concrete, the actual strength is much higher. This strength enhancement is taken into account by the factor $\sqrt{A_2/A_1} \leq 2$, where A_2 is the largest area, geometrically similar to the column section, that fits below the loaded area such that a truncated pyramid (frustum) with 1:2 sloping faces is formed (Fig. 9.7). If the contact pressure between column and footing, P_u/A_1, exceeds the strength of the column concrete, dowels will be needed to transfer the load difference, $P_u - \phi(0.85 f'_c A_1)$, into the footing. These dowels can be thought of as extensions of the column reinforcement (Fig. 9.8), with adequate splice length provided.

The use of shear reinforcement in footings is usually uneconomical. Therefore, the thickness is typically controlled by the required shear strength. Both the beam shear and punching shear strengths need to be investigated, as was done for two-way slabs. Strip footings for walls are subjected to one-way bending, while square and rectangular footings under single columns experience two-way bending and need to be reinforced accordingly.

Figure 9.9 identifies the three different kinds of potential shear failure planes. The perimeter surface for punching shear is determined by the truncated pyramid (Fig. 9.9a;

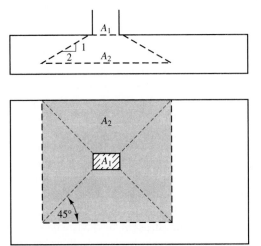

Figure 9.7 Determination of area A_2 for bearing strength enhancement.

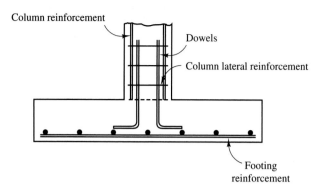

Column reinforcement

Dowels

Column lateral reinforcement

Footing reinforcement

Figure 9.8 Dowels for transferring column loads into footing.

see also Fig. 8.46c), with top area A_1 and bottom area A_3. It can be approximated for a column of cross-sectional dimensions a and b by $b_o d$, where $b_o = 2(a + b + 2d)$. The failure planes for beam shear in the short and long directions are assumed to be located a distance d away from the column faces (Fig. 9.9b), where d is the footing depth, or more precisely, the distance from the compression face to the centroid of the flexural reinforcement.

The nominal shear strength of the concrete, according to the ACI Code, is to be taken as the smallest value obtained by Eqs. 8.62, restated here for convenience:

$$V_c = \left(2 + \frac{4}{\beta_c}\right)\sqrt{f_c'}b_o d \qquad (9.3a)[11\text{-}36]$$

$$V_c = \left(2 + \frac{\alpha_s d}{b_o}\right)\sqrt{f_c'}b_o d \qquad (9.3b)[11\text{-}37]$$

$$V_c = 4\sqrt{f_c'}b_o d \qquad (9.3c)[11\text{-}38]$$

where β_c is the ratio of the long to the short side of the column; and $\alpha_s = 40$ for interior columns, 30 for edge columns, and 20 for corner columns.

The punching shear strength criterion then reads

$$V_u \le \phi V_c \qquad (9.4)$$

where $\phi = 0.85$, and

$$V_u = P_u - p_u A_3 \qquad (9.5)$$

that is, V_u is equal to the factored column load minus the soil pressure acting over the bottom area of the pyramid frustum formed if the column were to punch through the footing (Fig. 9.9a). Equation 9.4 may be used, in conjunction with Eq. 9.3c, to determine the required footing thickness by iteration, if no other guidance is available. For this purpose, we substitute Eq. 9.3c into Eq. 9.4 and rewrite it as

$$d = \frac{V_u}{8\phi\sqrt{f_c'}(a + b + 2d)} \qquad (9.6)$$

We substitute a first guess for d into the right-hand side of Eq. 9.6 and compute a better estimate.

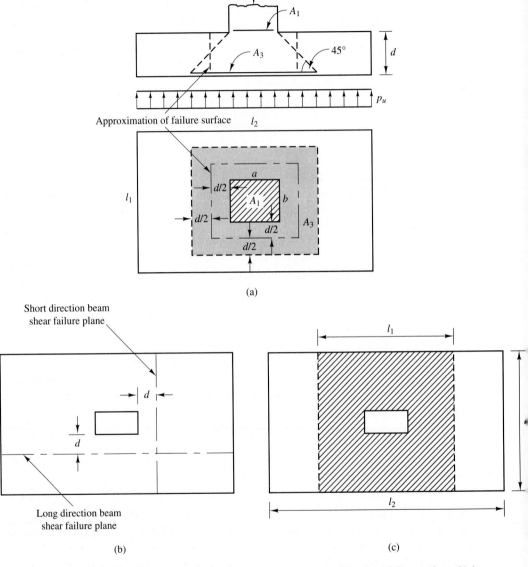

Figure 9.9 Shear failure planes and column strip: (a) punching shear failure surface; (b) beam shear failure planes; (c) column strip.

Flexural reinforcement must be provided to cover the factored bending moments in both directions. In the long direction, the steel is distributed uniformly across the footing width. For the short direction, we recall the design of two-way building slab systems, which were divided into column and middle strips. The column strips, being much stiffer than the middle strips, were assigned the bulk of the flexural reinforcement. Similarly, in footing design, a column strip of width l_1 (Fig. 9.9c) is to receive a portion of the total reinforcement given by $2/(\beta + 1)$, where $\beta = l_2/l_1$, and l_1 and l_2 are the footing dimensions in the short and long directions, respectively.

The design of a footing can now be summarized by the following step-by-step procedure.

Step 1. Determine the required footing area by dividing the unfactored service loads by the allowable soil bearing pressure f_{all}.

$$A_{req} = \frac{P_D + P_L}{f_{all}} \quad \text{or} \quad \frac{P_D + P_L + P_W}{1.33 f_{all}} \tag{9.7}$$

where P_D, P_L, and P_W are the column axial forces due to dead load, live load, and wind, respectively. A square footing is optimal, but if site restrictions apply, rectangular areas are suitable as well.

Step 2. After selecting the footing dimensions, compute p_u, which is the soil pressure due to factored column loads.

Step 3. Estimate the required footing thickness. The ACI Code requires a minimum of $d = 6$ in, or 12 in for footings on piles. If no better estimate is available, determine d iteratively, using Eq. 9.6. Recall that the overall thickness h must allow for a 3-in protection of the steel, that is, $h = d + 3$ in cover $+ \frac{1}{2}$ bar diameter.

Step 4. Check concrete bearing strength. For the footing concrete this implies $P_u \leq \phi(0.85 f_c' A_1)$, with modification factor $\sqrt{A_2/A_1} \leq 2$. For the column concrete, if $P_u > \phi(0.85 f_c' A_1)$, dowels need to transmit the excess load. Such dowels are generally provided for a positive connection between footing and column.

Step 5. Check whether the footing has sufficient strength to prevent a punching shear failure, that is, check whether $P_u - p_u A_3 \leq \phi V_c$, with the shear capacity V_c determined from Eq. 9.3. If the shear strength is insufficient, increase the footing thickness.

Step 6. Check whether the footing has sufficient shear strength to prevent beam-shear failure in both the short and long direction, that is, compare the shear force due to the factored soil pressure p_u with the shear capacity $\phi V_c = 2\phi \sqrt{f_c'} \, l_i d$, where l_i is the footing width under consideration.

Step 7. Design the flexural reinforcement required in the long direction. The critical section is at the column or wall face. If the footing supports a masonry wall, the critical section shall be taken halfway between the middle and edge of the wall. Because a masonry wall is not built monolithically with the footing, a potential failure plane may be located within the wall footprint. The flexural reinforcement is to be distributed uniformly across the footing width. Note that there is no load sharing possible between the two orthogonal directions. Also, according to the ACI Code Commentary, soil-supported slabs are not considered as structural slabs. Therefore, the standard minimum flexural reinforcement limit applies, rather than minimum temperature and shrinkage reinforcement.

Step 8. Determine the flexural reinforcement required in the short direction. Distribute the fraction $2/(\beta + 1)$ within the column strip of width l_1, with $\beta = l_2/l_1$, and distribute the rest uniformly in the areas outside the column strip. Note that d will be one bar diameter less than in the other direction.

Step 9. Check whether the flexural reinforcement has sufficient development length.

EXAMPLE 9.3

Design the footing for a 16-by-20-in column, carrying the service loads $P_D = 250$ kips, $P_L = 175$ kips, and $P_W = 110$ kips (Fig. 9.10a). The base of the footing is to be 5 ft below grade. The soil weighs 125 pcf and has an allowable bearing pressure of 5 ksf. Also given: $f'_c = 5000$ psi (column concrete); 3000 psi (footing concrete with 150 pcf); reinforcing steel with $f_y = 60$ ksi.

Solution

Step 1. Determine required footing area.

Net soil pressure (estimate 2-ft-thick footing):
$$p_n = 5.0 - 2 \times 0.15 \text{ (concrete)} - 3 \times 0.125 \text{ (soil)} = 4.325 \text{ ksf}$$

Required area: $A = \dfrac{250 + 175}{4.325} = 98.3 \text{ ft}^2$ or $\dfrac{250 + 175 + 110}{(1.33)(4.325)} = 93.0 \text{ ft}^2$

Use 10×10 ft footing with $A = 100 \text{ ft}^2$.

Step 2. Soil pressure due to factored loads.

$$P_u = (1.4)(250) + (1.7)(175) = 647.5 \text{ kips (controls)} \quad \text{or}$$
$$= (0.75)[(1.4)(250) + (1.7)(175) + (1.7)(110)] = 625.9 \text{ kips}$$

$$p_u = \frac{647.5}{100} = 6.5 \text{ ksf}$$

Step 3. Establish footing thickness.

Try $h = 24$, $d = 24 - 3 - 1 = 20$ in (average of d in two directions).

Bottom area of pyramid: $A_3 = \dfrac{(16 + 2 \times 20)(20 + 2 \times 20)}{144} = 23.3 \text{ ft}^2$ (Fig. 9.10b)

Punching shear force: $V_u = 647.5 - (6.5)(23.3) = 495.8 \text{ kips}$

Equation 9.6: $d = 495.8/(8)(.85)\dfrac{\sqrt{3000}}{1000}(16 + 20 + 40) = 17.5 \text{ in}$

Try $h = 24$.

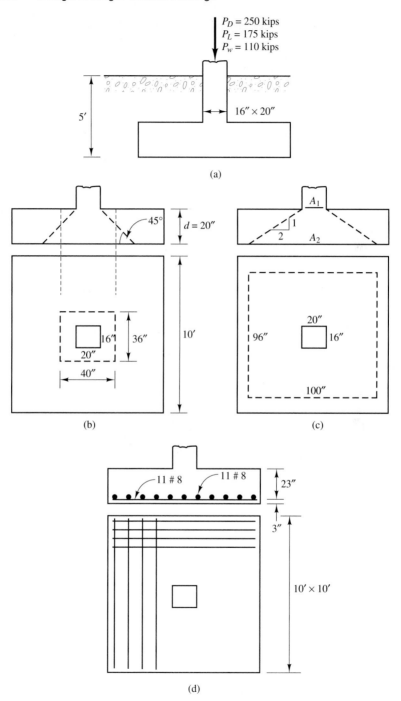

Figure 9.10 Example 9.3: (a) problem statement; (b) punching shear check; (c) bearing strength check; (d) final design.

Step 4.   Check concrete bearing strength.

Column: $\phi(0.85 f_c' A_1) = (0.7)(.85)(5)(16)(20) = 952$ kips > 647 kips OK.

Footing: $(0.7)(.85)(3)(16)(20) = 571$ kips < 647 kips NG, but with factor $\sqrt{A_2/A_1}$: $A_2 = (16+4 \times 20)(20+4 \times 20) = 9600$ in^2 (Fig. 9.10c) $\sqrt{A_2/A_1} = \sqrt{9600/320} = 5.48 > 2$, use 2; $2 \times 571 = 1142$ kips > 647 kips OK.

Although column concrete strength is sufficient to transfer the load, it is common to provide at least four dowel bars for a positive connection between column and footing.

Step 5.   Check punching shear.

$$\beta_c = \frac{20}{16} = 1.25, \ \alpha_s = 40, \ b_o = (2)(20 + 16 + 40) = 152 \text{ in}$$

Equation 9.3a: $V_c = \left(2 + \frac{4}{1.25}\right) \frac{\sqrt{3000}}{1000}(152)(20) = 866$ kips

Equation 9.3b: $V_c = \left(2 + \frac{40 \times 20}{160}\right) \frac{\sqrt{3000}}{1000}(152)(20) = 1166$ kips

Equation 9.3c: $V_c = 4\frac{\sqrt{3000}}{1000}(152)(20) = 666$ kips (controls)

$\phi V_c = (.85)(666) = 566$ kips < 582 kips NG. Use $h = 26$ in, $d = 22$ in (average)

Step 6.   Check beam shear.

Shear strength is equal in both directions, but shear force is slightly larger on failure plane parallel to long column face.

Shear strength: $2\phi\sqrt{f_c'}l_1 d = (2)(.85)\frac{\sqrt{3000}}{1000}(120)(22) = 246$ kips

Shear force: $V_u = (6.5)(10)\left(5 - \frac{8 + 22}{12}\right) = 162.5$ kips < 246 kips OK.

Steps 7 and 8.   Determine flexural reinforcement.

Since the two spans are approximately equal, use the longer $l = 5 - \frac{8}{12} = 4.33$ ft and the smaller $d = 21.5$ in for both (estimating 1-in bar diameter).

Moment: $M_u = p_u l_1 \frac{l^2}{2} = (6.5)(10)(4.33)^2/2 = 609$ ft-kips

Required steel:

Let $a = 1$, $A_s = \dfrac{M_u}{\phi f_y \left(d - \dfrac{a}{2}\right)} = \dfrac{(609)(12)}{(.9)(60)(21.5 - 0.5)} = 6.4$ in^2

Check $a = \dfrac{A_s f_y}{.85 f_c' b} = \dfrac{(6.4)(60)}{(.85)(3)(120)} = 1.26$ in OK

Check minimum steel:

$$\left(\frac{200}{f_y}\right)bd = \left(\frac{200}{60,000}\right)(120)(21.5) = 8.6 \text{ in}^2 > 7.1 \text{ in}^2 \text{ (controls)}$$

Use 11 #8 bars, spaced at 11 in, in both directions, with $A_s = 8.7 \text{ in}^2$.

Note that the weight of the soil above the footing is disregarded in the flexural and shear design, because it creates an additional soil pressure below the footing of equal and opposite magnitude, that is, there is no net effect on the design shears or moments.

Step 9.   Check development length.

Basic development length for #8 bar:

$$l_{db} = \frac{3}{40}\frac{f_y}{\sqrt{f_c'}}d_b = \frac{(3)(60,000)(1.0)}{40\sqrt{3000}} = 82.2 \text{ in}$$

with $\alpha = \beta = \gamma = \lambda = 1.0$ and $\left(\dfrac{c+k_{tr}}{d_b}\right) = \dfrac{3.0+0}{1.0} = 3.0 > 2.5$. Use 2.5.

$$l_d = \frac{82.2}{2.5} = 32.9 \text{ in required} < \frac{120}{2} - 10 = 50 \text{ in available } \underline{\text{OK}}$$

For final design, see Fig. 9.10d.   ■

9.3 DESIGN OF COMBINED FOOTINGS

Soil or site conditions can prohibit the assignment of a separate footing for each column, in which case a *combined* footing may be needed to support two or more columns. The soil pressure under such a footing can be kept uniform by aligning the centroid of the footing with the resultant of the column forces. Rectangular, trapezoidal, and T-shaped footings have enough free design variables to make their selection a simple exercise in geometry (Fig. 9.11). For example, consider the two columns in Fig. 9.11a, with loads P_1 and P_2 and spaced the distance $a+b$ apart. The resultant $R = P_1 + P_2$ acts at a distance $b = P_1(a+b)/(P_1+P_2)$ from the right column. Suppose the distance c is limited by the location of the property line. The length of the footing is then readily determined as $l_2 = 2(b+c)$. If the allowable bearing pressure of the soil, f_{all}, is given, the minimum footing width follows as $l_1 = (P_1+P_2)/(l_2 f_{all})$. This solution is unique because there are two equations of equilibrium available to determine the two design variables l_1 and l_2. The case of Fig. 9.11c has an infinite number of solutions, because there are four design variables for the same two equations of equilibrium (note that $l_5 = a + b$ is a dependent variable). For example, we may select practical dimensions for the right footing area (l_4+l_5 and l_2) and then determine the dimensions for the left portion (l_1 and l_3) to satisfy the equilibrium conditions.

A variation of the combined footing is the *strap* footing (Fig. 9.12), in which a beam (or strap) connects two separate column footings. Such an arrangement may become necessary to stabilize an eccentrically loaded footing, as shown in Fig. 9.12, and to prevent it from tilting.

We shall illustrate the design of combined footings with the following example.

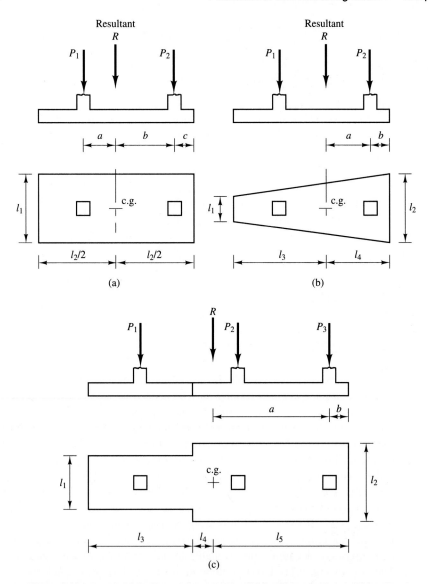

Figure 9.11 Combined footings: (a) rectangular footing for two columns; (b) trape-zoidal footing for two columns; (c) T-shaped footing for three columns.

EXAMPLE 9.4

Design a combined footing for an exterior and interior column, spaced 20 ft apart (Fig. 9.13). The exterior column has cross-sectional dimensions of 24 by 18 in and carries the design loads $P_D = 180$ kips and $P_L = 150$ kips. The interior column with a 24-by-24-in section transmits the design loads $P_D = 300$ kips and $P_L = 250$ kips. The property line is located 12 in from the face of the exterior column. The soil has an allowable bearing pressure of

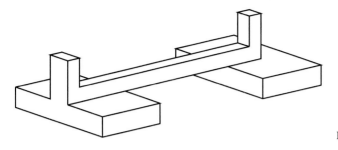

Figure 9.12   Strap footing.

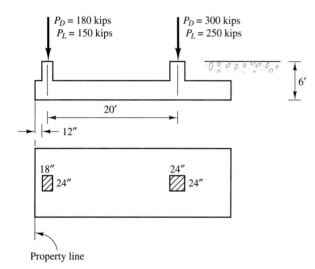

Figure 9.13   Example 9.4: Problem statement.

5 ksf and weighs 130 pcf. The base of the footing is located 6 ft below grade. The footing concrete weighs 150 pcf and is of strength $f'_c = 3000$ psi. Yield strength of reinforcement is $f_y = 60$ ksi.

Solution

Step 1.   Determine footing dimensions.

Net soil pressure (estimate footing thickness = 3 ft):
$p_n = 5.0 - 3 \times 0.15$ (concrete) $- 3 \times 0.13$ (soil) $= 4.16$ ksf

Column service loads: $P_1 = 180 + 150 = 330$ kips; $P_2 = 300 + 250 = 550$ kips; $R = 330 + 550 = 880$ kips

Distance of resultant R from centroid of exterior column:
$$b = \frac{(550)(20)}{880} = 12.5 \text{ ft}; \ c = 12 + 9 = 21 \text{ in} = 1.75 \text{ ft}$$

Footing length: $l_2 = 2(12.5 + 1.75) = 28.5$ ft

Footing width: $l_1 = \dfrac{880}{(4.16)(28.5)} = 7.42$. Say 7.5 ft (Fig. 9.14a).

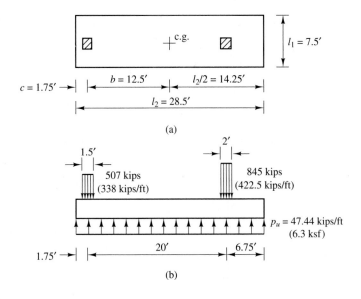

(a)

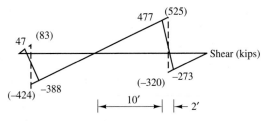

(b)

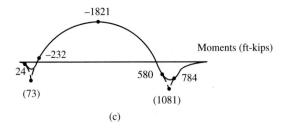

(c)

Figure 9.14 Example 9.4: Combined footing. (a) Footing dimensions; (c) moment and shear diagram.

Step 2. Soil pressure due to factored loads.

$$P_{u1} = (1.4)(180) + (1.7)(150) = 507 \text{ kips}$$

$$P_{u2} = (1.4)(300) + (1.7)(250) = 845 \text{ kips}$$

$$R_u = 507 + 845 = 1352 \text{ kips}$$

$$p_u = \frac{1352}{(7.5)(28.5)} = 6.3 \text{ ksf} \quad \text{or} \quad p_u = \frac{1352}{28.5} = 47.44 \text{ kips/ft (Fig. 9.14b)}.$$

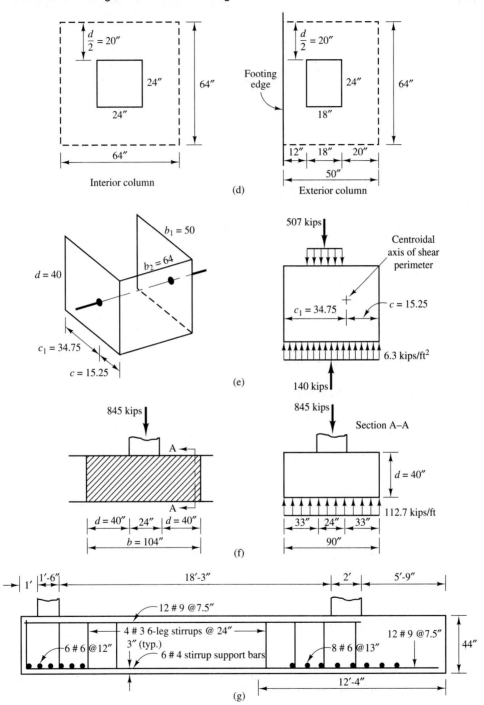

Figure 9.14 (cont.) (d) shear perimeters; (e) shear perimeter of exterior column; (f) transverse column strip; (g) reinforcement details.

Step 3.   Determine shear and moment diagrams.

Having the uniform soil pressure $p_u = 6.3$ ksf (or 47.44 kips/ft) and the column loads, the shear and moment diagrams are readily determined, considering the column loads either as concentrated or (more accurately) distributed (Fig. 9.14c).

Step 4.   Establish footing thickness.

We shall use the beam shear failure criterion.

Let $h = 36$ in, $d = 36 - 4 = 32$ in, then $V_{u,\max} = 477 - \dfrac{32}{12}(47.44) = 350$ kips

$$\phi V_c = \phi 2\sqrt{f_c'}bd = (.85)(2)\frac{\sqrt{3000}}{1000}(90)(32) = 268 \text{ kips} < 350 \text{ kips NG}$$

Try $h = 44$ in, $d = 40$ in, $V_{u,\max} = 477 - \dfrac{40}{12}(47.44) = 319$ kips

$$\phi V_c = (.85)(2)\frac{\sqrt{3000}}{1000}(90)(40) = 335 \text{ kips} > 319 \text{ kips OK.}$$

Use $\underline{h = 44 \text{ in.}}$

Step 5.   Check bearing strength.

Footing, interior column: capacity $= (0.7)(.85)(3)(24)^2 = 1028$ kips > 845 kips OK. This means that the column with higher strength concrete will also be OK.

Exterior column: $(0.7)(.85)(3)(18)(24) = 771$ kips > 507 kips OK

Step 6.   Check punching shear.

Interior column perimeter of shear failure surface (Fig. 9.14d):
$b_o = (4)(24 + 40) = 256$ in

Critical shear force:

$$V_u = P_u - p_u A_3 = 845 - (6.3)\left(\frac{64}{12}\right)^2 = 845 - 179 = 666 \text{ kips};$$

$$\beta_c = \frac{24}{24} = 1; \; \alpha_s = 40$$

Equation 9.3a: $V_c = \left(2 + \dfrac{4}{1}\right)\dfrac{\sqrt{3000}}{1000}(256)(40) = 3365$ kips

Equation 9.3b: $V_c = \left(2 + \dfrac{40 \times 40}{256}\right)\dfrac{\sqrt{3000}}{1000}(256)(40) = 4627$ kips

Equation 9.3c: $V_c = 4\dfrac{\sqrt{3000}}{1000}(256)(40) = 2243$ kips (controls)

$\phi V_c = (.85)(2243) = 1907$ kips > 666 kips OK.

Exterior column (Fig. 9.14d; see also Fig. 8.44d):
$b_o = (2)(12 + 18 + 20) + 64 = 164$ in

$$V_u = 507 - (6.3)\frac{(64)(50)}{144} = 507 - 140 = 367 \text{ kips}$$

$$\phi V_c = (4)\frac{\sqrt{3000}}{1000}(164)(40) = 1437 \text{ kips} > 367 \text{ kips OK.}$$

The safety margin seems adequate, but as Fig. 9.14e shows (and recall Sec. 8.7), the column force produces an eccentric shear, part of which needs to be transferred by flexure, the rest by shear stresses.

Distance of centroidal axis of shear perimeter from footing edge:

$$c_1 = \frac{(2)(50)(25) + (64)(50)}{50 + 50 + 64} = 34.75 \text{ in}; \ c = 50 - 34.75 = 15.25 \text{ in}$$

Unbalanced moment:

$$M_u = (507)(29 - 15.25) - (140)(25 - 15.25) = 5606 \text{ in-kips}$$

Moment of inertia of shear perimeter:

$$J = 2\left[\frac{b_1^3 d}{12} + \frac{b_1 d^3}{12} + b_1 d\left(\frac{b_1}{2} - c\right)^2\right] + b_2 d c^2 =$$

$$2\left[\frac{(50)^3(40)}{12} + \frac{(50)(40)^3}{12} + (50)(40)(25 - 15.25)^2\right] + (64)(40)(15.25)^2 =$$

$$2,342,000 \text{ in}^4$$

Fraction of moment to be transferred by flexure (Eq. 8.62):

$$\gamma_f = \frac{1}{1 + \frac{2}{3}\sqrt{\frac{b_1}{b_2}}} = \frac{1}{1 + \frac{2}{3}\sqrt{\frac{50}{64}}} = 0.63$$

Fraction of moment to be transferred by shear: $\gamma_v = 1 - \gamma_f = 0.37$

Shear stress caused by moment transfer:

$$v_f = \frac{\gamma_v M_u c_1}{J} = \frac{(.37)(5606)(34.75)}{2,342,000} = .031 \text{ ksi}$$

Combined with punching shear stress:

$$v_u = \frac{V_u}{b_o d} + v_f = \frac{367,000}{(164)(40)} + 31 = 87 \text{ psi}; \ \phi v_c = 4\sqrt{3000} = 219 \text{ psi} > 87 \text{ psi}$$

OK

Step 7.   Determine longitudinal flexural reinforcement.

Midspan, top reinforcement (negative moment):

$$\text{Let } a = 2 \text{ in, } A_s = \frac{M_u}{\phi f_y\left(d - \frac{a}{2}\right)} = \frac{(1821)(12)}{(.9)(60)(40 - 1)} = 10.4 \text{ in}^2.$$

$$\text{Check } a = \frac{A_s f_y}{.85 f_c' b} = \frac{(10.4)(60)}{(.85)(3)(90)} = 2.7 \text{ in OK}$$

$$\text{Minimum steel: } \left(\frac{200}{60,000}\right)(40)(90) = 12.0 \text{ in}^2 > 10.4 \text{ in}^2 \text{ (controls).}$$

Use 12 #9 bars (@ 7.5 in).

Interior column, bottom reinforcement (positive moment): As this moment is less than half of the midspan moment, minimum steel also governs, that is, Use 12 #9 bars (@ 7.5 in).

Step 8. Determine transverse reinforcement.

In the transverse direction, it is not reasonable to assume that each one-foot-wide strip carries the same load. As in a two-way slab system, the "column strip" can be expected to carry a much larger portion of the load. In fact, we shall assume that the entire load of the interior column is transferred to the soil over a transverse strip of width $24 + 2d = 24 + (2)(40) = 104$ in $= 8.67$ ft. The uniform load on this strip is then (Fig. 9.14f) $p_u = 845/7.5 = 112.7$ kips/ft.

$$M_u = \frac{(112.7)(2.75)^2}{2} = 426.1 \text{ ft-kips}$$

Let $a = 2$ in, with $d = 39$ in, $A_s = \dfrac{(426.1)(12)}{(.9)(60)(39 - 1)} = 2.5$ in^2

$A_{s,\min} = \dfrac{200}{60,000}(104)(39) = 13.52$. Use instead $\dfrac{4}{3}(2.5) = 3.3$ in^2.
Use 8 #6 bars (@ 13 in).

Under the exterior column, the column strip is only $12 + 18 + 40 = 70$ in wide, that is, 6 #6 bars (@ 12 in) will suffice.

Step 9. Check development lengths.

Basic development length for #9 bar:
$$l_{db} = \frac{3}{40}\frac{f_y}{\sqrt{f_c'}}d_b = \frac{(3)(60,000)(1.128)}{40\sqrt{3000}} = 92.7 \text{ in}$$

Factor for top reinforcement $\alpha = 1.3$

Factor for excess reinforcement $= \dfrac{10.4}{12.0} = .87$;

$\dfrac{c + k_{tr}}{d_b} = \dfrac{3.0 + 0}{1.128} = 2.66 > 2.5$. Use 2.5

$l_d = \dfrac{(92.7)(1.3)(.87)}{2.5} = 41.9$ in. $= 3.5$ ft $\ll 10$ ft available.

Checking flexural bond near the exterior column (Eq. 5.37):
$$\frac{M_n}{V_u} + l_a = \frac{(1821)(12)/0.9}{388} + 30 = 92 \text{ in available}$$

Development length demand $= 41.9$ in < 92 in OK

Bottom reinforcement:
Inflection points are spaced $\sqrt{8M_u/p_u} = \sqrt{(8)(1821)/(47.44)} = 17.5$ ft apart, that is, right inflection point is positioned $10 - (17.5/2) = 1.25$ ft $= 15$ in from face of interior column. Add distance d, $15 + 40 = 55$ in for cut-off point for bottom reinforcement.

Check development length. For excess reinforcement factor, use moment ratio:

$(.87)\dfrac{580}{1821} = .28$; $l_d = \dfrac{(92.7)(.28)}{2.5} = 10.4$ in $\ll 55$ in OK

Transverse reinforcement. Basic development length of #6 bars:
$$l_{db} = \frac{(3)(60,000)(.75)}{40\sqrt{3000}} = 61.6 \text{ in};$$

with $\gamma = 0.8$ and $\dfrac{c + k_{tr}}{d_b} = 2.5$, $l_d = \dfrac{(61.6)(0.8)}{2.5} = 19.7$ in. < 33 in OK

Development length is adequate even without the significant reduction factor for excess reinforcement.

Fig. 9.14g shows the details of the reinforcement, complete with 8 #3 6-leg stirrups to support the top reinforcement and 6 #4 stirrup support bars.

■

9.4 DESIGN OF RETAINING WALLS

When it is necessary to alter the natural topography of a terrain for the construction of a building, highway, or other facility, special structures or retaining walls may be needed to hold the soil in place. For example, a highway routed along a slope requires such retaining walls, whether it is built with only a cut, only a fill, or cut-and-fill (Fig. 9.15a). Also, buildings with basements below grade require walls designed to resist the soil pressure (Fig. 9.15b). Here, we will restrict ourselves to free-standing retaining walls and exclude walls that are parts of structures. An important consideration for the design of such walls is their external stability, that is, adequate safety against horizontal sliding and overturning.

If the wall is not too high, the easiest way of achieving such equilibrium is to rely solely on the wall's own weight. Such walls are termed *gravity walls* (Fig. 9.16a). For walls exceeding about 10 ft in height, *cantilever walls* are likely to be more economical than gravity walls (Fig. 9.16b). These consist of a vertical wall (or stem, or arm), rigidly connected to a footing or base slab. The portion of the slab on the side of the retained soil is called the *heel*, and the remaining portion on the other side, the *toe*. If the wall is placed too close to a property line, there may be no room for a toe. However, the heel is more important, because the weight of the soil above it helps to counterbalance the overturning moment. If the safety against horizontal sliding is deemed insufficient, the base slab may be fitted with a *key*, which increases the resistance to such sliding.

As we shall see in a moment, the maximum bending moment in the wall due to soil pressure increases with the third power of the wall height. As a result, walls higher than about 20 ft must be so thick and heavily reinforced as to be uneconomical. In such cases it will be advantageous to stiffen the wall with *counterforts* on the soil side (Fig. 9.16c), spaced apart no more than the wall height h and no less than $\frac{1}{2}h$. If the triangular wall supports are arranged on the side opposite to the soil, they are referred to as *buttresses*. The horizontal soil pressure is transferred to the ground in two steps. A typical horizontal strip of the retaining wall of, say, a unit width acts like a continuous horizontal beam and transfers the load to the supports provided by the counterforts. The counterforts themselves act as vertical cantilever beams and transfer the load from the horizontal wall strips to the foundation slab through bending. Part of the wall acts as a compression flange of the T-shaped vertical cantilever beam, and much of the triangular web (i.e., the counterfort) is in tension, so that the reinforcement must be concentrated near the sloping tension face (Fig. 9.16c).

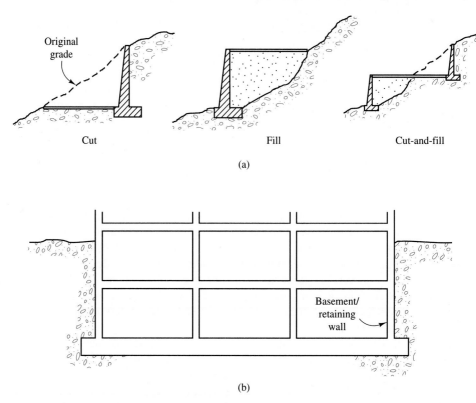

Figure 9.15 Examples of retaining walls: (a) retaining walls for highway on slope; (b) basement walls.

For very deep excavations, especially for temporary purposes during construction, it is common to retain the soil with soldier pile walls, slurry walls, etc., stabilized through soil or rock anchors as shown in Fig. 9.16d. Details can be found in texts on foundation and geotechnical engineering and construction. Here, we shall restrict ourselves to gravity and cantilever walls.

In order to relieve the build-up of ground water pressure behind any kind of retaining wall, it is advisable to use crushed stone or gravel for backfill, in conjunction with a drain pipe and/or weep holes, which are small dewatering pipes penetrating the wall at intervals of 8 to 10 ft. The importance of such a drainage system increases with the wall height but obviously depends on the soil properties and water table.

Earth pressure. The mechanical behavior of soils varies somewhere between those of solids and liquids. The horizontal pressure they exert can be expressed as

$$p_H(y) = k\gamma y \qquad (9.8)$$

where γ is the unit weight of the soil, and y is the depth measured from the soil surface (Fig. 9.17a). k is the *coefficient of earth pressure at rest* and depends on the type of soil. For rocks and hard stone, it can be as small as zero. Of course, no wall is needed to

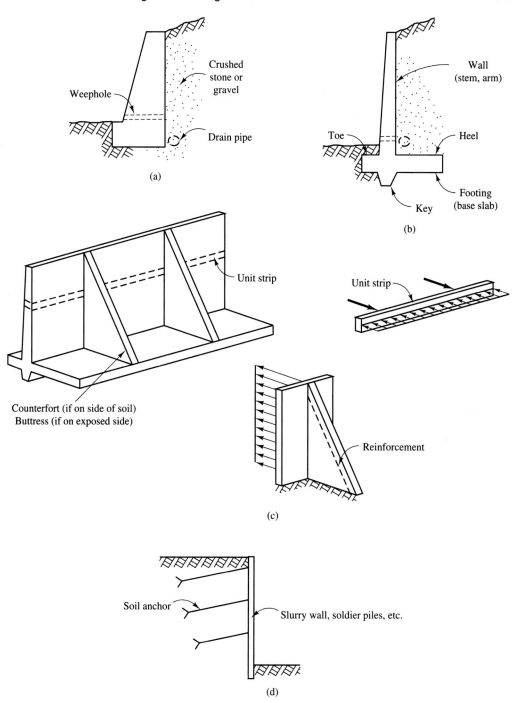

Figure 9.16 Types of retaining walls: (a) gravity wall; (b) cantilever wall; (c) counterfort wall; (d) deep excavation.

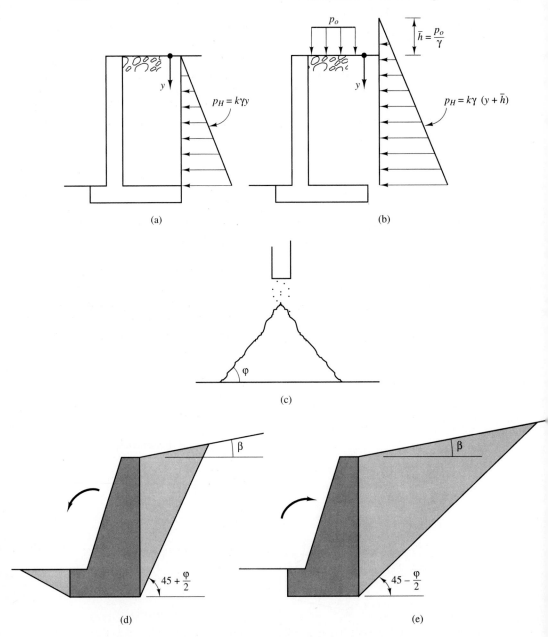

Figure 9.17 Active and passive soil pressure and internal angle of friction: (a) earth pressure; (b) earth pressure with overburden; (c) angle of repose; (d) approximate failure surface for active soil pressure; (e) approximate failure surface for passive soil pressure.

hold back solid rock. For cohesive soils such as clays that are saturated with water, k can be as high as 1.0. In this case, the soil behaves like a true liquid, and the horizontal pressure is a truly hydraulic pressure. For noncohesive soils such as sands or gravels, k may be around 0.4 to 0.5, and higher if compacted. This is one of the reasons why only noncohesive soils should be used for backfill of retaining walls, if possible. If the soil surface is loaded with some overburden p_o, expressed in lb/ft^2 (Fig. 9.17b), it may be converted into an equivalent additional height of soil, $\bar{h} = p_o/\gamma$, and the horizontal pressure at depth y becomes

$$p_H(y) = k\gamma(y + \bar{h}) \tag{9.9}$$

If a sandy soil pours out from a chute or dump truck, it tends to form a pile with surfaces sloping at an angle ϕ with the horizontal. This angle is called the angle of repose and is equal to the angle of internal friction of the noncompacted soil (Fig. 9.17c). Now consider a retaining wall that under the earth pressure tilts slightly or moves to the left (Fig. 9.17d). The failure surface of the soil will form an approximate angle of $45 + (\phi/2)$ with the horizontal. The soil within the shaded wedge exerts *active pressure* on the wall. If for some reason the wall is pushed into the soil (Fig. 9.17e), then the force necessary to achieve this must overcome the internal angle of friction, that is, the expected soil failure plane is as shown, and the *passive pressure* is much higher than the active pressure. As such motion is not really possible without some compaction of the soil, the internal angle of friction will increase also. Retaining walls are generally subjected to active earth pressure. Note, however, that in the case of Fig. 9.17d, a small passive soil wedge forms at the toe of the wall foundation, which helps the wall's resistance against sliding.

According to the theories of Rankine and Coulomb, the coefficients for active and passive earth pressure are, respectively,

$$
\begin{aligned}
k_a &= \cos\beta \frac{\cos\beta - \sqrt{\cos^2\beta - \cos^2\phi}}{\cos\beta + \sqrt{\cos^2\beta - \cos^2\phi}} \\[2mm]
k_p &= \cos\beta \frac{\cos\beta + \sqrt{\cos^2\beta - \cos^2\phi}}{\cos\beta - \sqrt{\cos^2\beta - \cos^2\phi}}
\end{aligned}
\tag{9.10}
$$

where β is the angle that the retained soil surface forms with the horizontal (Fig. 9.17d and e). If that angle is zero, Eqs. 9.10 simplify to

$$k_a = \frac{1 - \sin\phi}{1 + \sin\phi} \qquad k_p = \frac{1 + \sin\phi}{1 - \sin\phi} \tag{9.11}$$

These solutions are valid for noncohesive soils (sand, gravel), but can be modified to be applicable to cohesive soils (clays). In this latter case, it is problematic to define the internal angle of friction ϕ, since it is difficult in laboratory experiments to separate intergranular friction from internal cohesion. In addition, the actual hydraulic pressures due to the water table during steady-state conditions and after rainstorms are difficult to quantify. For these reasons it is common to select conservative values for ϕ or for k in Eq. 9.8. Terzaghi and Peck [1, 4] have prepared charts for estimating the vertical and

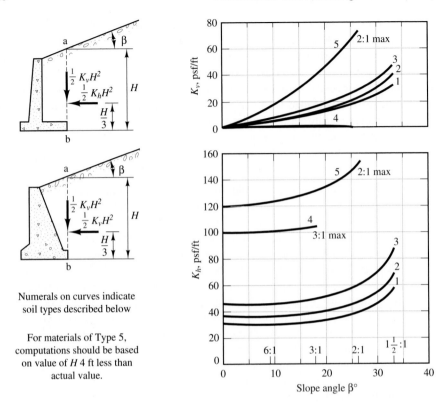

Numerals on curves indicate
soil types described below

For materials of Type 5,
computations should be based
on value of H 4 ft less than
actual value.

Soil type	Description
1	Coarse-grained soil without admixture of fine soil particles, very free-draining (clean sand, gravel, or broken stone)
2	Coarse-grained soil of low permeability due to admixture of particles of silt size
3	Fine silty sand; granular materials with conspicuous clay content; or residual soil with stones
4	Soft or very soft clay; organic silt; or soft silty clay
5	Medium or stiff clay that may be placed in such a way that a negligible amount of water will enter the spaces between the chunks during floods or heavy rains

Figure 9.18   Soil pressures for retaining walls with $h \leq 20$ ft. (Printed with permission of American Railway Engineering Association.)

horizontal soil pressures for different soil categories and soil profiles, which are applicable for walls up to 20 ft in height (Fig. 9.18). Reference is made to the specialized literature for other cases, including those with overburdens [1–4].

External stability.   In addition to having sufficient strength to resist the internal shear forces and bending moments, a retaining wall must be capable of withstanding the tendency of the soil to move or overturn it. While the internal strength is assured by standard ultimate strength design, that is, by making the ultimate strength no less than the effect of the factored loads, the stability checks are performed for actual (i.e., unfactored) loads—just as we sized a footing for actual (not factored) loads.

The equilibrium checks need to include the soil pressure resultants of Fig. 9.18, as well as the weights of the retaining wall, the base slab, and the soil above the base slab (Fig. 9.19). In order to prevent sliding, the horizontal earth pressure resultant P_H should be no greater than the frictional resistance between the base slab and the soil below, which is $\mu(\sum W_i + P_V)$, where $\sum W_i$ is the weight of the wall, base slab, and soil above the base slab; P_V is the vertical component of the soil pressure resultant; and μ is the coefficient of friction, which is typically 0.4 to 0.6 for sandy soils and gravel, and varies from 0.2 to 0.4 for clayey soils. Adopting a commonly used safety factor of 1.5, the check against sliding reads,

$$1.5 P_H \leq \mu \left(\sum W_i + P_V \right) \tag{9.12}$$

If the wall of Fig. 9.19 were to slide, it could do so only by pressing against the soil to the left, that is, by activating a counteracting horizontal soil pressure resultant Q_H. Unless the availability of this soil resistance is guaranteed, it is advisable to be conservative and to discount it in the equilibrium check. If it is difficult to satisfy Eq. 9.12, a shear key may have to be provided, which forces the soil to fail along line b–b instead of line a–a (Fig. 9.19), that is, a larger reactionary force Q_H is activated. Also, the friction

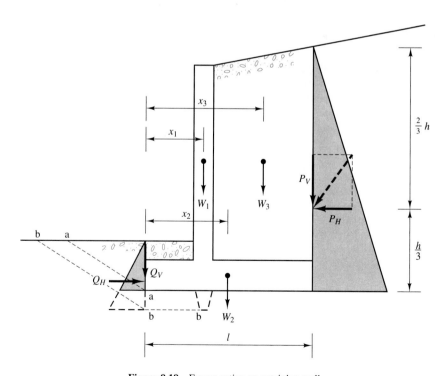

Figure 9.19 Forces acting on retaining wall.

coefficient along the horizontal portion of the failure plane b–b is larger than between base slab and soil, namely, $\mu = \tan \phi$.

The safety against overturning is checked readily by comparing the overturning moment $P_H \times (h/3)$ with the restoring moment provided by the W_i forces and P_V, the vertical component of the earth pressure resultant. Each force is to be multiplied with its respective lever arm x_i, measured from the tip of the toe, which is the point about which the wall would rotate if it were to overturn. Again, a safety factor of 1.5 is commonly used in this stability check, that is,

$$1.5 P_H \frac{h}{3} \leq \sum W_i x_i + P_V l \tag{9.13}$$

It is also necessary to assure that the bearing capacity of the soil under the base slab is not exceeded. We can replace all vertical forces acting on the foundation by a single force $P = \sum W_i + P_V$, acting at the centroid and a moment M, or the same force P applied at a distance $e = M/P$ from the centroid. The determination of the soil pressure distribution is then reduced to the problem of Sec. 9.1 (Fig. 9.3). If the equivalent load P acts within the kern area of the foundation, that is, the middle third, then it causes a trapezoidal soil pressure distribution, which is readily determined using Eq. 9.1. If the force resultant P falls outside the kern area (Fig. 9.3c), then the soil pressure distribution is triangular, with its centroid located below the load resultant and the maximum soil pressure determined from Eq. 9.2. Note that in this case the soil is stressed very unevenly and likely to result in uneven settlement and rotation, especially if the soil is highly compressible.

EXAMPLE 9.5

Design a 10-ft-high gravity wall. The surface of the retained soil has a 20° slope. The backfill consists of a sandy soil weighing 120 pcf with an internal angle of friction of 30° (type 2, Fig. 9.18) and a coefficient of friction with concrete $\mu = 0.5$. The soil has an allowable bearing pressure of 7000 psf and is sufficiently compressible so that the force resultant shall fall within the foundation's kern area. The footing base shall be 4 ft below the lower soil grade level. Use normal-weight concrete @ 150 pcf.

Solution   A preliminary design leads to a wall with the dimensions shown in Fig. 9.20a. For soil type 2 and $\beta = 20°$, Fig. 9.18 gives the soil pressures

$k_h = 40$ psf/ft, $k_v = 12$ psf/ft.

With maximum soil depth $H = 16.5$ ft, soil forces are

$$P_H = \frac{1}{2} k_h H^2 = (0.5)(40)(16.5)^2 = 5445 \text{ lb/ft}$$

$$P_V = \frac{1}{2} k_v H^2 = (0.5)(12)(16.5)^2 = 1633 \text{ lb/ft}$$

Here we shall consider only a one-foot-deep slice of the wall. The above values then have pounds as units. The weight components are

$W_1 = (12)(1)(150) = 1800 \text{ lb}$
$W_2 = (12)(6)(150)/2 = 5400 \text{ lb}$

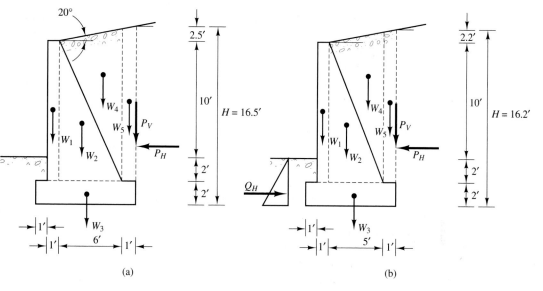

Figure 9.20   Example 9.5: Gravity wall. (a) Preliminary design; (b) final design.

$W_3 = (9)(2)(150) = 2700$ lb

$W_4 = (14.5)(6)(120)/2 = 5220$ lb

$W_5 = (14.5)(1)(120) = 1740$ lb

$\sum W_i + P_V = 16,860 + 1633 = 18,493$ lb

Check against sliding.

$$\mu \left(\sum W_i + P_V \right) = (0.5)(18,493) = 9246 \text{ lb}$$

Safety factor against sliding: S.F. $= \dfrac{9246}{5445} = 1.7 > 1.5$ <u>OK</u>

S.F. is sufficient even without counting on passive soil pressure at left of toe.

Check against overturning.

Overturning moment: $M_{OT} = P_H \dfrac{H}{3} = (5445)\dfrac{16.5}{3} = 29,947$ ft-lb

Restoring moment:

$M_R = (1800)(1.5) + (5400)(4) + (2700)(4.5) + (5220)(6) + (1740)(8.5) + (1633)(9)$
$= 97,257$ ft-lb

Safety factor against overturning $= \dfrac{97,257}{29,947} = 3.2 > 1.5$ <u>OK</u>

Check soil pressure below base slab.

Distance of vertical load resultant from tip of toe:

$\dfrac{M_R - M_{OT}}{\sum W_i + P_V} = \dfrac{97,257 - 29,947}{18,493} = 3.64$ ft

Eccentricity: $e = 4.5 - 3.64 = 0.86$ ft

Kern distance: $k = \dfrac{l}{6} = \dfrac{9}{6} = 1.5$ ft > 0.86 ft $\underline{OK}$

Maximum soil pressure:

$$f_{max} = \frac{P}{A}\left(1 + \frac{e}{k}\right) = \frac{18,493}{9}\left(1 + \frac{0.86}{1.5}\right) = 3233 \text{ psf} < 7000 \text{ psf } \underline{OK}$$

The high safety factors indicate that a smaller wall may suffice. We shall try a wall with the dimensions indicated in Fig. 9.20b. Reanalysis gives:

$$P_H = (0.5)(40)(16.2)^2 = 5249 \text{ lb/ft}$$

$$P_V = (0.5)(12)(16.2)^2 = 1575 \text{ lb/ft}$$

$$W_1 = (12)(1)(150) = 1800 \text{ lb}$$

$$W_2 = (12)(5)(150)/2 = 4500 \text{ lb}$$

$$W_3 = (8)(2)(150) = 2400 \text{ lb}$$

$$W_4 = (14.2)(5)(120)/2 = 4260 \text{ lb}$$

$$W_5 = (14.2)(1)(120) = 1704 \text{ lb}$$

$$\sum W_i + P_V = 16,239 \text{ lb}, \quad \mu\left(\sum W_i + P_V\right) = (0.5)(16,239) = 8120 \text{ lb}$$

Safety factor against sliding: S.F. $= \dfrac{8120}{5249} = 1.55 > 1.5$ $\underline{OK}$

Note that if we were to count on the passive soil pressure, with $\beta = 20°$ and $\phi = 30°$ from Eq. 9.10b, we would get $k_p = 2.13$. Then (see Fig. 9.20b)

$$Q_H = \frac{1}{2}(120)(4)^2(2.13) = 2045 \text{ lb}$$

Safety factor $= \dfrac{8120 + 2045}{5249} = 1.9 > 1.5$ $\underline{OK}$

Overturning moment: $M_{OT} = P_H\dfrac{H}{3} = (5249)\dfrac{16.2}{3} = 28,345$ ft-lb

Restoring moment: $M_R = (1800)(1.5) + (4500)(3.67) + (2400)(4) + (4260)(5.33) + (1704)(7.5) + (1575)(8) = 76,900$ ft-lb

Safety factor against overturning: $\dfrac{76,900}{28,345} = 2.7 > 1.5$ $\underline{OK}$

Eccentricity of vertical load resultant: $4 - \dfrac{76,900 - 28,345}{16,239} = 1.0$ ft $< \dfrac{8}{6}$ $\underline{OK}$

Maximum soil pressure: $f_{max} = \dfrac{16,239}{8}\left(1 + \dfrac{1.0}{1.33}\right) = 3556$ psf < 7000 psf $\underline{OK}$

If it can be assured that the passive soil pressure to the left of the wall is reliable, the wall size can be reduced even further. ∎

EXAMPLE 9.6

Design a cantilever wall to retain a 20-ft-deep cut. The upper soil surface is level but subject to a 300 psf overburden (live load). Assume the same soil properties as in Example 9.5 and again, the pressure resultant shall fall within the middle third of the footing, and the footing base shall be 4 ft below the lower soil grade level. Use 150 pcf concrete with $f'_c = 4000$ psi and reinforcing steel with $f_y = 60$ ksi.

Solution

   a) *Preliminary Design*

From Fig. 9.18, find $k_h = 37$ psf, $k_v = 0$.

Overburden is equivalent to $\bar{h} = 300/120 = 2.5$ ft

Horizontal pressure resultant: $P_H = \dfrac{1}{2}(37)(24 + 2.5)^2 = 12{,}992$ lb

For preliminary stability check purposes, assume wall height-to-thickness ratio of 10, that is, a 2-ft wall thickness, and also a 2-ft-thick footing of 12-ft width (Fig. 9.21a).

Vertical forces:

$$W_1 = (2)(22)(150) = 6600 \text{ lb}$$

$$W_2 = (2)(12)(150) = 3600 \text{ lb}$$

$$W_3 = (7)(22)(120) = 18{,}480 \text{ lb}$$

$$W_4 = (300)(7) = 2100 \text{ lb}$$

$$\sum W_i = 30{,}780 \text{ lb}, \quad \mu \sum W_i = (0.5)(30{,}780) = 15{,}390 \text{ lb}$$

Safety factor against sliding: $\dfrac{15{,}390}{12{,}992} = 1.18 < 1.5$ NG

Counting on passive soil pressure to the left of toe, with Eq. 9.11,

$$k_p = \frac{1 + \sin 30}{1 - \sin 30} = 3.0, \quad Q_H = \frac{1}{2}(120)(4)^2(3) = 2880 \text{ lb}$$

Safety factor: $\dfrac{15{,}390 + 2880}{12{,}992} = 1.41 > 1.5$ still NG

Check without overburden:

$$P_H = \frac{1}{2}(37)(24)^2 = 10{,}656 \text{ lb}$$

$$\sum W_i = 6600 + 3600 + 18{,}480 = 28{,}680 \text{ lb}, \quad \mu \sum W_i = 14{,}340 \text{ lb}$$

Safety factor against sliding: $\dfrac{14{,}340 + 2880}{10{,}656} = 1.61 > 1.5$ OK

Overturning moment: $(12{,}992)\left(\dfrac{26.5}{3}\right) = 114{,}760$ ft-lb

Restoring moment: $(6600)(4) + (3600)(6) + (18{,}480 + 2100)(8.5) = 222{,}930$ ft-lb

Safety factor against overturning: $\dfrac{222{,}930}{114{,}760} = 1.94 > 1.5$ OK

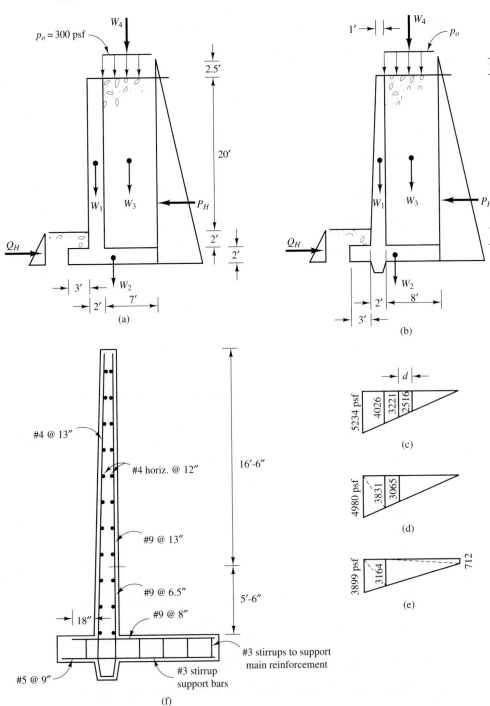

Figure 9.21 Example 9.6: Cantilever wall. (a) Preliminary design; (b) final dimensions; (c) assumed triangular soil pressure; (d) upward soil pressure (corrected); (e) soil pressure without overburden; (f) reinforcing details.

Position of load resultant: $\dfrac{222{,}930 - 114{,}760}{30{,}780} = 3.51$ ft

Eccentricity: $e = 6 - 3.51 = 2.49$ ft

Kern distance: $k = \dfrac{l}{6} = \dfrac{12}{6} = 2.0$ ft < 2.49 NG

Increase footing width to 13 ft to pick up more soil weight, and add a shear key of 1.5-ft depth for increased resistance against sliding (Fig. 9.21b).

To size wall thickness, select $\rho = 0.01$ and use Eq. 4.14 to solve for wall thickness d.

Factored horizontal pressure resultant: $P_u = \dfrac{1}{2}(37)(22 + 2.5)^2(1.7) = 18{,}878$ lb

Factored moment at base of wall: $M_u = (18{,}878)\dfrac{24.5}{3} = 154{,}169$ ft-lb

$bd^2 = \dfrac{M_u}{\phi \rho f_y[1 - \rho f_y/1.7 f_c']} = \dfrac{(154)(12)}{(0.9)(.01)(60)[1 - (.01)(60)/(1.7)(4)]} = 3753 \ \text{in}^3$

$d = \sqrt{\dfrac{3753}{12}} = 17.7$ in. Add 3 in for cover. $h = 24$ in is OK.

Taper thickness down to 12 in at top of wall.

Check shear at distance $d = 20.5$ in $= 1.71$ ft above footing.

$V_u = \dfrac{1}{2}(37)(24.5 - 1.71)^2(1.7) = 16{,}335$ lb

$V_c = 2\phi\sqrt{f_c'}bd = (2)(0.85)\sqrt{4000}(12)(20.5) = 26{,}449$ lb $> 16{,}335$ lb OK

Size footing thickness for shear.

Total downward load: $\sum W_i = 30{,}780 + (2)(150) + (22)(120) + 300 = 34{,}020$ lb

Assume triangular soil pressure distribution below footing, with

$f_{\max} = \dfrac{(34{,}020)(2)}{13} = 5234$ psf (Fig. 9.21c)

Critical section is at $d = 21$ in $= 1.75$ ft from wall or 6.25 ft from right edge.

Soil pressure at critical section: $\dfrac{6.25}{13}(5234) = 2516$ psf

Downward load: $(22)(6.25)(120) + (300)(6.25) + (2)(6.25)(150) = 20{,}250$ lb

Net shear (factored conservatively):

$V_u = (1.7)\left[20{,}250 - \dfrac{2516}{2}(6.25)\right] = 21{,}059$ lb

$V_c = 2\phi\sqrt{f_c'}bd = (2)(0.85)\sqrt{4000}(12)(21) = 27{,}094$ lb $> 21{,}059$ lb OK

Check footing thickness for bending.

Total downward load: $(20{,}250)\dfrac{8}{6.25} = 25{,}920$ lb

Soil pressure at wall face: $(5234)\dfrac{8}{13} = 3221$ psf

$$M_u = (1.7)\left[(25{,}920)(4) - \frac{(3221)(4)}{2}\frac{4}{3}\right] = 161{,}654 \text{ ft-lb}$$

With $\rho = 0.01$, $bd^2 = \dfrac{(162)(12)}{(0.9)(.01)(60)[1 - (0.01)(60)/(1.7)(4)]} = 3948 \text{ in}^3$

$$d = \sqrt{\frac{3948}{12}} = 18.1 \text{ in } \underline{OK}$$

We shall adopt the dimensions of Fig. 9.21b.

b) *External Stability*

Load Case: Dead Load + Soil + Overburden

$$P_H = \frac{1}{2}(37)(26.5)^2 = 12{,}992 \text{ lb}$$

$$W_1 = \frac{1+2}{2}(22)(150) = 4950 \text{ lb}$$

$$W_2 = (2)(13)(150) = 3900 \text{ lb}$$

$$W_3 = (8)(22)(120) = 21{,}120 \text{ lb}$$

$$W_4 = (8)(300) = 2400 \text{ lb}$$

$$\sum W_i = 32{,}370 \text{ lb}, \quad \mu \sum W_i = 16{,}185 \text{ lb}$$

Overturning moment unchanged: 114,760 ft-lb

Restoring moment: $(4950)(4) + (3900)(6.5) + (21{,}120 + 2400)(9) = 256{,}830 \text{ ft-lb}$

Safety factor against overturning: $\dfrac{256{,}830}{114{,}760} = 2.2 > 1.5 \text{ } \underline{OK}$

Position of vertical load resultant: $\dfrac{256{,}830 - 114{,}760}{32{,}370} = 4.4 \text{ ft}$

Eccentricity: $e = 6.5 - 4.4 = 2.1 \text{ ft}$

Kern distance: $k = \dfrac{13}{6} = 2.17 \text{ ft} > e \text{ } \underline{OK}$

Maximum soil pressure: $\dfrac{P}{A}\left(1 + \dfrac{e}{k}\right) = \dfrac{32{,}370}{13}(1 + 1) = 4980 \text{ psf} < 7000 \text{ psf } \underline{OK}$

(See Fig. 9.21d)

Friction force on toe: $\dfrac{4980 + 3831}{2}(3)(\tan 30) = 7631 \text{ lb}$

Friction force on heel and key: $\dfrac{3831}{2}(10)(0.5) = 9577 \text{ lb}$

Passive earth pressure, disregarding 2 ft of soil above toe as less reliable:

$$\frac{1}{2}(120)(3.5)^2(3) = 2205 \text{ lb}$$

Safety factor against sliding: $\dfrac{7631 + 9577 + 2205}{12{,}992} = 1.49 \approx 1.5 \text{ } \underline{OK}$

Load Case: Dead Load + Soil

$$P_H = \frac{1}{2}(37)(24)^2 = 10{,}656 \text{ lb}$$

$$\sum W_i = 32{,}370 - (8)(300) = 29{,}970 \text{ lb}$$

Overturning moment: $(10{,}656)\dfrac{24}{3} = 85{,}248 \text{ ft-lb}$

Restoring moment: $256{,}830 - (2400)(9) = 235{,}230 \text{ ft-lb}$

Safety factor against overturning: $\dfrac{235{,}230}{85{,}248} = 2.7 > 1.5 \text{ OK}$

Position of vertical load resultant: $\dfrac{235{,}230 - 85{,}248}{29{,}970} = 5.0 \text{ ft}$

Eccentricity: $e = 6.5 - 5.0 = 1.5 \text{ ft}$

Maximum soil pressure: $\dfrac{29{,}970}{13}\left(1 + \dfrac{1.5}{2.17}\right) = 3899 \text{ psf} < 7000 \text{ psf OK}$

Minimum soil pressure: $\dfrac{29{,}970}{13}\left(1 - \dfrac{1.5}{2.17}\right) = 712 \text{ psf (see Fig. 9.21e)}$

Friction force on toe: $\dfrac{3899 + 3164}{2}(3)(\tan 30) = 6117 \text{ lb}$

Friction force on heel and key: $\dfrac{3164 + 712}{2}(10)(0.5) = 9690 \text{ lb}$

Passive earth pressure, same as before: 2205 lb

Safety factor against sliding: $\dfrac{6117 + 9690 + 2205}{10{,}656} = 1.69 > 1.5 \text{ OK.}$

c) Wall Design

Maximum moment unchanged: $M_u = 154{,}169 \text{ ft-lb}$

With $d = 24 - 3 \text{ in (cover)} - 0.5 \text{ in (half bar diameter)} = 20.5 \text{ in and } a = 1 \text{ in.}$

Required steel: $A_s = \dfrac{M_u}{\phi f_y(d - a/2)} = \dfrac{(154)(12)}{(0.9)(60)(20.5 - 0.5)} = 1.71 \text{ in}^2/\text{ft}$

Check: $a = \dfrac{A_s f_y}{.85 f'_c b} = \dfrac{(1.71)(60)}{(.85)(4)(12)} = 2.52 \text{ in};$

$A_s = \dfrac{(154)(12)}{(0.9)(60)(20.5 - 1.26)} = 1.78 \text{ in}^2/\text{ft}$

Use #9 bars @ $6\dfrac{1}{2}$ in with 1.85 in²/ft.

The bending moment decreases with the cube of depth, therefore we shall discontinue every other bar where they are no longer needed.

Capacity of wall with #9 bars @ 13 in, or $A_s = .925 \text{ in}^2/\text{ft}$ and $d - (a/2) = 17 \text{ in (estimate)}$:

$$M_u = (0.9)(60)(0.925)\left(\dfrac{17}{12}\right) = 70.8 \text{ ft-kips}$$

Soil pressure as function of depth y: $p_H(y) = (37)[2.5 + y]$

Horizontal soil force: $P_H(y) = (37)\left[2.5y + \dfrac{y^2}{2}\right]$

Factored moment: $M_u(y) = (1.7)(37)\left[\dfrac{2.5y^2}{2} + \dfrac{y^3}{6}\right] = 70{,}800$ ft-lb, from which,

by iteration, $y = 16.7$ ft

Wall thickness at $y = 16.7$: $12 + \dfrac{16.7}{22}(12) = 21.1$ in, that is, $d = 21.1 - 3.5 = 17.6$ in

Check $a = \dfrac{A_s f_y}{.85 f'_c b} = \dfrac{(.925)(60)}{(.85)(4)(12)} = 1.36$ in,

$d - \dfrac{a}{2} = 17.6 - \dfrac{1.36}{2} = 16.9$ in ≈ 17 in $\underline{\text{OK}}$

Determine bar cut off point.

Basic development length of #9 bars:

$\dfrac{3}{40}\dfrac{f_y}{\sqrt{f'_c}}d_b = \dfrac{(3)(60{,}000)(1.128)}{(40)\sqrt{4000}} = 80.3$ in

with factor $\dfrac{c + k_{tr}}{d_b} = \dfrac{6.5/2 + 0}{1.128} = 2.9 > 2.5$

Development length $= \dfrac{80.3}{2.5} = 32.1$ in

Cut off point at $22 - 16.7 = 5.3$ ft $= 63.6$ in > 32.1 in above base provides ample development length.

<u>Cut off every other #9 bar 5'-6" above top of base</u>

Check minimum reinforcement (although not applicable to walls):

$\rho = \dfrac{.925}{(12)(18.8)} = .0041 > \dfrac{200}{60{,}000} = .0033$ $\underline{\text{OK}}$

Check shear strength: Shear force at critical section is unchanged, $V_u = 16{,}335$ lb, shear strength is slightly less than $V_c = 26{,}449$ lb, because of taper. But the change is much less than the excess capacity, that is, shear strength is $\underline{\text{OK}}$.

For horizontal steel, provide minimum to cover temperature and shrinkage reinforcement:

With average $d = \dfrac{9 + 21}{2} = 15$ in, $(.0012)(12)(15) = .216$ in^2

<u>Use #4 bars @ 12 in</u> on both faces of wall. Also use #4 vertical bars @ 13 in on exposed wall face, see Fig. 9.21f.

d) Toe Design

The toe slab is subjected to the soil pressure of Fig. 9.21d (to be factored with 1.7) and the slab's own weight (to be factored with 0.9 because it counteracts the load). We ignore the weight of the soil above the toe.

Maximum moment:

$M_u = 1.7\left[\dfrac{(4980)(3)}{2}(2) + \dfrac{(3831)(3)}{2}(1)\right] - (0.9)(2)(150)\dfrac{3^2}{2} = 33{,}952$ ft-lb

Required steel, with $d = 20.5$ in, $a = 1$ in:

$A_s = \dfrac{(33.95)(12)}{(.9)(60)(20.5 - 0.5)} = .377$ in^2

Check $a = \dfrac{(.377)(60)}{(.85)(4)(12)} = .55$ in OK

<u>Use #5 bars @ 9 in</u> with 0.41 in^2/ft.

Check shear strength at critical section, $d = 20.5$ in from face of wall:

$$V_u \approx \frac{36 - 20.5}{12}(4980)(1.7) = 10,900 \text{ lb}$$

Shear strength is similar to the wall's, that is, $V_c \approx 26,449$ lb $> 10,900$ lb OK

Development length of #5 bars: $\dfrac{(3)(60,000)(0.8)}{(40)\sqrt{4000}(2.5)} = 22.8$ in

We may safely cut off bars at the back side of wall, see Fig. 9.21f.

e) Heel Design

Downward load: $(2)(150) + (22)(120) + 300 = 3240$ lb/ft

Upward soil pressure is shown in Fig. 9.21d.

Net bending moment: $M_u = (1.7)\left[\dfrac{(3240)(8)^2}{2} - \dfrac{(3065)(8)^2}{6}\right] = 120,677$ ft-lb

Required steel, with $d = 20.5$ in, $a = 1$ in: $A_s = \dfrac{(120.68)(12)}{(.9)(60)(20.5 - 0.5)} = 1.34 \text{ in}^2$

Check $a = \dfrac{(1.34)(60)}{(.85)(4)(12)} = 1.97$ in, revise $A_s = \dfrac{(120.68)(12)}{(.9)(60)(20.5 - .98)} = 1.37 \text{ in}^2$

Use #9 bars @ 8 in with 1.50 in^2/ft.

Check shear strength at face with wall (since reaction does not produce compression):

Downward load: $(3240)(8) = 25,920$ lb

Upward soil force: $\left(\dfrac{3065}{2}\right)(8) = 12,260$ lb

Net factored shear: $V_u = (1.7)(25,920 - 12,260) = 23,222$ lb $< 26,449$ lb OK

Development length for #9 bars (see wall design) was 32.1 in; with factor 1.3 for top reinforcement, get 42 in; cut off bars 18 in to left of wall face (see Fig. 9.21f). ∎

SUMMARY

1. If the load resultant falls within the kern area of a foundation, the bearing pressure is determined by the standard column stress formula

$$f(y) = \frac{P}{A} \pm \frac{My}{I} = \frac{P}{A}\left(1 \pm \frac{Aey}{I}\right)$$

If the eccentricity of the load resultant falls outside the kern area, the maximum bearing pressure follows as

$$f_{max} = \frac{2P}{3ab}$$

where a is the distance from load P to the edge with maximum soil pressure.

2. The design of single-column footings can be summarized by the following step-by-step procedure.

Step 1. Determine the required footing area

$$A_{req} = \frac{P_D + P_L}{f_{all}} \quad \text{or} \quad \frac{P_D + P_L + P_W}{1.33 f_{all}}$$

where P_D, P_L, and P_W are the column axial forces due to unfactored dead load, live load, and wind, respectively. f_{all} is the allowable soil bearing pressure.

Step 2. After selecting the footing dimensions, compute p_u, which is the soil pressure due to factored column loads.

Step 3. Estimate the required footing depth. If no better estimate is available, determine d iteratively:

$$d = \frac{V_u}{8\phi\sqrt{f_c'}(a + b + 2d)}$$

Step 4. Check concrete bearing strength. For the footing concrete, this implies $P_u \leq \phi(0.85 f_c' A_1)$, with modification factor $\sqrt{A_2/A_1} \leq 2$. For the column concrete, if $P_u > \phi(0.85 f_c' A_1)$, dowels need to transmit the excess load.

Step 5. Check whether the footing has sufficient strength to prevent a punching shear failure, that is, check whether $P_u - p_u A_3 \leq \phi V_c$, with the shear capacity V_c determined from

$$V_c = \min\left\{2 + \frac{4}{\beta_c}; 2 + \frac{\alpha_s d}{b_0}; 4\right\}\sqrt{f_c'} b_0 d$$

where β_c is the ratio of long to short side of the column; and $\alpha_s = 40$ for interior columns, 30 for edge columns, and 20 for corner columns.

Step 6. Check whether the footing has sufficient shear strength to prevent beam-shear failure in both the short and long direction, that is, compare the shear force due to the factored soil pressure p_u with the shear capacity $\phi V_c = 2\phi\sqrt{f_c'} l_i d$, where l_i is the footing width under consideration.

Step 7. Design the flexural reinforcement required in the long direction. The flexural reinforcement is to be distributed uniformly across the footing width. Note that there is no load sharing possible between the two orthogonal directions.

Step 8. Determine the flexural reinforcement required in the short direction. Distribute the fraction $2/(\beta + 1)$ within the column strip of width l_i, with $\beta = l_2/l_1$, and distribute the rest uniformly in the areas outside the column strip. Note that d will be one bar diameter less than in the other direction.

Step 9. Check whether the flexural reinforcement has sufficient development length.

3. External stability of retaining walls is assured by satisfying the check against sliding

$$1.5 P_H \leq \mu\left(\sum W_i + P_V\right)$$

and against overturning

$$1.5 P_H \frac{h}{3} \leq \sum W_i x_i + P_V l$$

4. The design of a gravity retaining wall may be achieved with the following steps:

Step 1.   Preliminary design to establish wall dimensions.
Step 2.   Check against sliding and overturning.
Step 3.   Check maximum bearing pressure on soil.
Step 4.   Revise wall dimensions if necessary.

REFERENCES

1. Terzaghi, K., and Peck, R. B., *Soil Mechanics in Engineering Practice*, 2d ed., John Wiley and Sons, Inc., New York, 1967.

2. Peck, R. B., Hanson, W. E., and Thornton, T. H., *Foundation Engineering*, 2d ed., John Wiley and Sons, Inc., New York, 1974.

3. Das, B. M., *Theoretical Foundation Engineering*, Elsevier Science Publishers, Amsterdam, 1987.

4. American Railway Engineering Association, "Retaining Walls and Abutments," Chap. 8, Part 5, *1994 Manual for Railway Engineering*, American Railway Engineering Association, Washington, D.C., 1994.

PROBLEMS

9.1 Design a square footing for a 20-by-20-in column, reinforced with 6 #11 bars and carrying the service loads $P_D = 300$ kips and $P_L = 400$ kips. The base of the footing shall be 6 ft below grade. Given: Allowable soil pressure = 8 ksf; soil weight = 130 pcf; $f'_c = 4000$ psi (column concrete); 3000 psi (footing concrete @ 150 pcf); reinforcing steel with $f_y = 60$ ksi.

9.2 A rectangular footing of dimensions 8 by 10 ft and 30-in thickness is reinforced with 10 #8 bars in both directions (Fig P9.2). The base is 5 ft below grade. The soil weighs 120 pcf and has a bearing capacity of 9 ksf. The footing has been designed to support a rectangular column of 16-by-20-in section, reinforced with 8 #11 bars and carrying the service loads $P_D = 250$ kips, and $P_L = 300$ kips. The owner wishes to add a few floors to the building. Assuming the added live and dead loads are equal, determine the additional loads that the footing can safely support. Given: $f'_c = 3000$ psi (footing); 4000 psi (column); $f_y = 60$ ksi.

9.3 A single rectangular footing is intended to support two columns with the dead and live loads indicated in Figure P9.3. The two columns are spaced 15 ft apart, and the footing may not extend beyond 3 ft from the centerline of the right column. Given: allowable soil pressure = 8 ksf; footing base 6 ft below grade; soil weight 125 pcf; concrete weight 150 pcf; footing thickness (estimated) 2 ft. Determine the required footing dimensions.

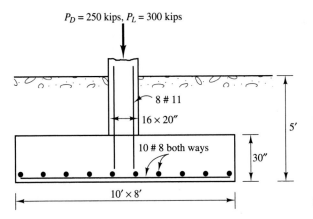

$P_D = 250$ kips, $P_L = 300$ kips

8 # 11

16 × 20″

10 # 8 both ways

30″

5′

10′ × 8′

Figure P9.2

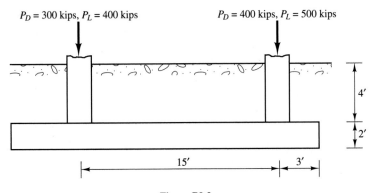

$P_D = 300$ kips, $P_L = 400$ kips $P_D = 400$ kips, $P_L = 500$ kips

4′

2′

15′   3′

Figure P9.3

9.4 Based on the punching shear requirement, determine the footing thickness required to support a factored column load of $P_u = 2310$ kips. Given: soil pressure due to factored load 11 ksf; $f'_c = 4000$ psi; column size 22 by 22 in.

9.5 Design a footing to support a 10-in-thick concrete wall carrying a dead load of 12 kips/ft and a live load of 8 kips/ft. The footing base shall be 4 ft below grade. Given: soil weight 130 pcf; allowable soil bearing pressure 6 ksf; $f'_c = 3000$ psi; $f_y = 60$ ksi.

9.6 Design a combined footing to support three columns with the combined service loads shown in Fig. P9.6. The allowable soil bearing pressure is 10 ksf.

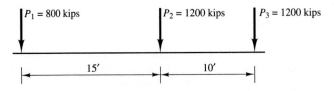

$P_1 = 800$ kips $P_2 = 1200$ kips $P_3 = 1200$ kips

15′   10′

Figure P9.6

9.7 Design an 11-ft 6-in high gravity wall. The retained soil surface is horizontal but subject to a live load surcharge of 350 psf. The backfill consists of a sand-and-gravel mixture with some silty fines (type 2, Fig. 9.18), that has the following properties: internal angle of friction, $\phi = 30°$ (assuming adequate drainage is provided); weight = 120 pcf; base coefficient of friction = 0.5; allowable bearing pressure 8 ksf; the force resultant shall strike within kern area. Concrete weight 150 pcf.

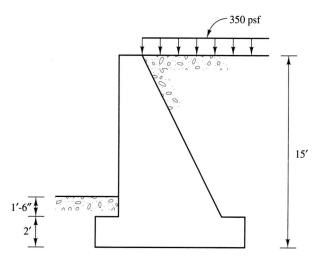

Figure P9.7

The Concrete ABC*

Abrasion resistance. Ability of a surface to resist being worn away by rubbing and friction.

Abutment. In bridges, the end structure (usually of concrete) that supports the beams, girders, and deck of the bridge, and sometimes retains the earthen embankment, or supports the end of the approach pavement slab.

Accelerator. A chemical substance which, when added to concrete, mortar, or grout, increases the rate of hydration of the hydraulic cement, shortens the time of setting, or increases the rate of hardening.

Additive. (See Admixture.)

Admixture. A chemical or mineral material other than water, aggregate, and hydraulic cement, added to the concrete immediately before or during mixing for certain desired effects.

Aggregate. Granular material, such as sand, gravel, crushed stone, and iron blast-furnace slag, used with a cementing medium to form a concrete or mortar; relatively inexpensive and generally available.

Aggregate interlock. The mechanism by which concrete transfers shear forces across a rough crack.

*Many of the definitions listed here are taken from Report 116R-90, "Cement and Concrete Terminology," American Concrete Institute, Detroit, MI, 1990.

Air entrainment. The addition of some chemical to the concrete or mortar mix, which forms minute air bubbles during mixing (generally smaller than 1 mm) to increase workability and frost resistance.

Alkali-aggregate reaction. Chemical reaction between alkalies (sodium and potassium) present in portland cement and certain constituents of some aggregates; this reaction may cause concrete expansion and thereby impair its durability.

Alkali-silica reaction (ASR). Chemical reaction between alkalies present in portland cement and silica present in some aggregates; this reaction may cause concrete expansion and thereby reduce its durability.

Allowable stress. Maximum permissible stress used in design of members of a structure and based on a factor of safety against yielding or failure of any type.

Allowable stress design (ASD). Design principle according to which stresses resulting from service or working loads are not allowed to exceed specified allowable values.

Alternate design method. (See Allowable stress design.)

Anchorage. Device used to anchor wires, cables, or tendons in prestressed concrete; for nonprestressed reinforcement, minimum embedment length (see Development length) or end hooks are used.

Architect-engineer (A/E). The architectural or engineering firm that issues project drawings and specifications or administers the work under contract specifications and drawings, or both.

Architectural concrete. Concrete that will be permanently exposed to view and therefore requires special care in selection of the concrete materials, forming, placing, and finishing to obtain the desired architectural appearance.

Autoclave. A pressure vessel in which an environment of steam at high pressure may be produced to cure concrete products or test hydraulic cement.

Bag (of cement). A quantity of portland cement: 94 lb in the United States, and 50 kg in most other countries.

Balanced load. Combination of axial force and bending moment that causes simultaneous crushing of concrete and yielding of tension steel.

Balanced reinforcement. An amount and distribution of flexural reinforcement such that the tensile reinforcement reaches its specified yield strength simultaneously with the concrete in compression reaching its assumed ultimate strain of 0.003.

Bar schedule. A list of the reinforcement, showing the shape, number, size, and dimensions of every different element required for a structure.

Batch. Quantity of concrete or mortar mixed at one time.

Beam. A structural member subjected primarily to flexure; depth-to-span ratio is limited to 2/5 for continuous spans, or 4/5 for simple spans, otherwise the member is to be treated as a deep beam (see also Girder, Joist, Spandrel beam, Stringer).

Beam-column. A structural member that is subjected simultaneously to bending and substantial axial forces.

Bent. Two-dimensional frame that is self-supporting within its own plane, having at least two legs and usually placed at right angles to the length of the structures that it supports.

Blast-furnace slag. The nonmetallic (waste) product that is developed in a molten condition in a blast furnace for iron production.

Bleeding. The self-induced flow of mixing water within, or its emergence from newly placed concrete; caused by the settlement of the solid materials within the mass.

Bond. Adhesion and grip of concrete or mortar to reinforcement or to other surfaces against which it is placed; to enhance bond strength, ribs or other deformations are added to reinforcing bars.

Buckling. Failure by lateral or torsional instability of a structural member at a stress below the yield or ultimate strength.

Cable-stayed bridge. A bridge suspended from cables that are anchored in the bridge deck.

Caisson pile. A cast-in-place pile made by driving a tube, excavating it, and filling the cavity with concrete.

Camber. A deflection that is intentionally built into a structural element or form to improve appearance or to offset the deflection of the element under the effects of loads, shrinkage, and creep.

Cantilever construction. In bridge construction, balanced segmental construction of bridge superstructure, cantilevering simultaneously to both sides of a pier, thereby eliminating the need for falsework (usually prestressed).

Cast-in-place concrete. Concrete poured in its final or permanent location; also called in-situ concrete; opposite of precast concrete.

Cathodic protection. A scheme to protect reinforcing bars against corrosion by providing an electric potential that neutralizes the naturally occurring electrical currents associated with corrosion; used, for example, in bridge decks subjected to deicing salt in winter.

Cellular concrete. A lightweight concrete with large air content, produced by adding gas-forming chemicals or foaming agents to the mix.

Cement. (See Hydraulic cement.)

Cement gel. The colloidal material that results from hydration of cement paste.

Cement paste. Cement plus water.

Clinker. A partially fused product of a kiln that is ground to make cement.

Coarse aggregate. Gravel or aggregate retained on the U.S. Standard No. 4 sieve (4.75 mm).

Cold joint. A joint or discontinuity resulting from a delay in placement of sufficient time to preclude a union of the material in two consecutive lifts.

Column. A member used to support primarily axial compression loads with a height of at least three times its least lateral dimension; the capacity of short columns is controlled by strength; the capacity of long columns is limited by buckling.

Column capital. An enlargement of a column below a slab to increase the slab's shear strength.

Column strip. The portion of a flat slab over a row of columns consisting of the two adjacent quarter panels on each side of the column centerline.

Compaction. The process whereby the volume of freshly placed concrete is reduced, usually by vibration, centrifugation, tamping, or some combination of these, to reduce the air voids other than entrained air; also called consolidation.

Composition construction. A type of construction using members made of different materials (e.g., concrete and structural steel), or combining members made of cast-in-place concrete and precast concrete such that the combined components act together as a single member; strictly speaking, reinforced concrete is also composite construction.

Compression member. A member subjected primarily to longitudinal compression; often synonymous with "column."

Compressive strength. Strength, typically measured on a standard 6×12 inch cylinder of concrete in an axial compression test, 28 days after casting.

Concrete. A composite material that consists essentially of a binding medium within which are embedded particles or fragments of aggregate; in portland cement concrete, the binder is a mixture of portland cement and water.

Condensed silica fume. (See Silica fume.)

Confined concrete. Concrete enclosed by closely spaced transverse reinforcement, which restrains the concrete expansion in directions perpendicular to the applied stresses.

Construction joint. The surface where two successive placements of concrete meet, across which it may be desirable to achieve bond, and through which reinforcement may be continuous.

Continuous beam or slab. A beam or slab that extends as a unit over three or more supports in a given direction and is provided with the necessary reinforcement to develop the negative moments over the interior supports; a redundant structure that requires a statically indeterminant analysis (opposite of simply supported beam or slab).

Contraction joint. Formed, sawed, or tooled groove in a concrete structure to create a weakened plane and regulate the location of cracking resulting from dimensional change of different parts of the structure.

Corbel. A very short cantilever projecting from the face of a column, beam, or wall and serving as a beam support.

Core. A cylidrical sample of hardened concrete or rock obtained by means of a core drill, taken to test the in-situ properties of the concrete or rock; also the molded open space in a hollow concrete masonry unit.

Corrosion. Disintegration or deterioration of steel reinforcement by electrolysis or chemical attack.

Cover. In reinforced concrete, the shortest distance between the surface of the reinforcement and the outer surface of the concrete; minimum values are specified to protect the reinforcement against corrosion and to assure sufficient bond strength.

Cracks. Result of stresses exceeding concrete's tensile strength capacity; cracks are ubiquitous in reinforced concrete and needed to develop the strength of the reinforcement, but a design goal is to keep their widths small (hairline cracks).

Cracked section. A section designed or analyzed on the assumption that concrete has no resistance to tensile stress.

Cracking load. The load that causes tensile stress in a member to be equal to the tensile strength of the concrete.

Creep. Time-dependent long-term deformation under sustained load; can be a multiple of the instantaneous elastic deformation.

Cryogenic concrete. Concrete for extremely low temperatures used, for example, for liquid natural gas (LNG) storage tanks.

C-S-H gel. The hydration product of portland cement, vaguely described as $3CaO \cdot 2SiO_2 \cdot 8H_2O$.

Culvert. A drain or waterway of pipe or masonry crossing under a road or embankment; in effect a very short bridge.

Curing. Maintenance of satisfactory moisture content and temperature of concrete during its early age so that desired properties can develop.

Dead load. Constant load due to a structure's own weight and permanent attachments or accessories.

Deflection. A variation in position or shape of a structure due to the effects of load or volume change; influenced by many uncontrollable environmental effects and therefore difficult to predict.

Deformed bar. Reinforcing bar with a manufactured pattern of surface ridges intended to prevent slip when the bar is embedded in concrete.

Design strength. Ultimate load-bearing capacity of a member multiplied by a strength reduction factor.

Development length. The length of embedded reinforcement required to develop the design strength of the reinforcement; a function of bond strength.

Diagonal crack. An inclined crack caused by diagonal tension, usually at about 45 degrees to the neutral axis of a concrete member.

Diagonal tension. The principal tensile stress resulting from the combination of normal and shear stresses acting upon a structural element.

Dowel. A pin, commonly a plain steel bar, which extends into two adjoining portions of concrete, so as to connect the portions and transfer shear loads; also a reinforcing bar intended to transmit tension, compression, or shear through a construction joint.

Drift. Horizontal deflection of a building or individual building story under lateral load.

Drop-in beam. Precast element, simply supported on adjacent cantilever elements.

Drop panel. The portion of a flat slab in the area surrounding a column, column capital, or bracket which is thickened in order to reduce the intensity of stresses.

Dry packing. Placing of zero or near-zero slump concrete, mortar, or grout by ramming into a confined space.

Ductility. Capability of a material or structural member to undergo large inelastic deformations without distress; opposite of brittleness; very important material property, especially for earthquake-resistant design; steel is naturally ductile, concrete is brittle but can be made ductile if well confined.

Durability. The ability of concrete to maintain its qualities over long time spans while exposed to weather, freeze-thaw cycles, chemical attack, abrasion, and other service load conditions.

Effective depth. Depth of a beam or slab section measured from the compression face to the centroid of the tensile reinforcement.

Effective flange width. Width of slab adjoining a beam stem assumed to function as the flange of a T-section.

Effective prestress. The stress remaining in the prestressing steel or in the concrete due to prestressing after all losses have occurred.

Effective span. The lesser of the distance between centers of supports and the clear distance between supports plus the effective depth of the beam or slab.

Elastic modulus. (See Modulus of elasticity.)

End block. In prestressed concrete, an enlarged end section of a member designed to reduce anchorage stresses to allowable values and to provide space needed for anchoring post-tensioning tendons.

Ettringite. A mineral, high-sulfate calcium sulfoaluminate; occurs in nature or by sulfate attack on mortar and concrete or is the result of alkali-silica reaction; causes volume expansion and cracking and thereby reduces durability.

Euler load. Buckling load of a column pin-connected at both ends and subjected to only an axial force ($= \pi^2 EI/L^2$).

Expansion joint. A separation between adjoining parts of a structure, provided to allow small relative movements such as those caused by temperature changes.

Expansive cement. A cement that when mixed with water forms a paste which, after setting, tends to increase in volume to a significantly greater degree than portland cement paste; used to compensate for volume decrease due to shrinkage or to induce tensile stress in reinforcement.

Falsework. The temporary structure erected to support work in the process of construction; composed of shoring or vertical posting, lateral bracing, and sometimes also includes the formwork.

Fatigue. The weakening of a material caused by repeated load applications.

Ferrocement. A composite structural material consisting of thin sections of cement mortar reinforced by a number of very closely spaced layers of steel wire mesh.

Fiber-reinforced concrete. Concrete containing dispersed, randomly oriented fibers made of steel, polypropylene, or other materials, typically in addition to regular reinforcement.

Fine aggregate. Aggregate passing the U.S. standard No. 4 sieve (4.75 mm) and predominantly retained on the No. 200 sieve (75 μm). (See Sand.)

Fineness modulus. Measure of fineness of aggregate; the sum of the cumulative fractions retained on standard sieves, that is, inversely proportional to fineness.

Fines. Aggregate particles that pass through the No. 200 sieve (75 μm size or smaller.)

Finishing. Leveling, smoothing, compacting, and otherwise treating surfaces of freshly or recently placed concrete to produce desired appearance and service.

Flat plate. A flat slab without column capitals or drop panels (see also Flat slab).

Flat slab. A concrete slab reinforced in two or more directions, generally without beams or girders to transfer the loads to supporting members, but with drop panels or column capitals or both (see also Flat plate.)

Fly ash. The finely divided residue resulting from the combustion of ground or powdered coal and which is transported from the firebox through the boiler by flue gases; used as admixture in concrete.

Flying form. Large prefabricated units of formwork and supporting falsework for an entire building floor or part of a floor, which can be lifted by crane in a single operation for reuse on the next floor or structure segment to be cast.

Folded plate. A framing assembly composed of sloping slabs in a hipped or gabled arrangement; or a prismatic shell with open polygonal section.

Footing. The structural element that transmits loads directly to the soil.

Form. A temporary structure or mold that contains the concrete while it is setting and gaining sufficient strength to be self-supporting (see also Formwork).

Form coating. A liquid applied to interior formwork surfaces to promote easy release from the concrete, which preserves the form material or retards the set of the near-surface matrix for preparation of exposed-aggregate finishes.

Form oil. Oil applied to interior formwork surfaces to promote easy release from the concrete when forms are removed.

Form pressure. Lateral pressure acting on vertical or inclined formed surfaces, resulting from the fluid-like behavior of the unhardened concrete confined by the forms.

Formwork. Total system of support for freshly placed concrete, including the mold or sheathing that contacts the concrete as well as all supporting members, hardware, and necessary bracing (sometimes called shuttering in the U.K.).

Gel. (See Cement gel.)

Girder. A large beam that serves as a main structural member.

Grade beam. A reinforced concrete beam, usually at ground level, which forms a foundation for the walls of a superstructure.

Grading. The distribution of sand and gravel particles among various sizes; usually expressed in terms of cumulative percentages passing through or retained on each of a series of sieves of decreasing mesh sizes.

Gravel. (See Coarse aggregate.)

Grout. A mixture of cementitious material and water, with or without fine aggregate, proportioned to produce a pourable consistency.

Gunnite. (See Shotcrete.)

Half-cell potential. Electrical potential that can be measured to assess the susceptibility of concrete reinforcement to corrosion.

Haunch. The deepened portion of a beam that increases in depth toward a support.

Heat of hydration. Heat evolved by chemical reaction with water, such as that evolved during the setting and hardening of portland cement; important in massive concrete structures.

High-early strength cement. Cement producing strength in mortar or concrete earlier than regular cement (also: Type III cement).

Hinge joint. A joint that permits rotation with no appreciable moments developed in the adjacent members.

Honeycomb. Voids left in concrete due to failure of the mortar to effectively fill the spaces between coarse aggregate particles and reinforcing bars.

Hook. A bend in the end of a reinforcing bar; needed if sufficient development length is not available.

Hoop. A one-piece closed reinforcing tie or continuously wound tie that encloses the longitudinal reinforcement.

Hydration. The chemical reaction between hydraulic cement and water.

Hydraulic cement. Cement that needs water to undergo chemical reaction to develop strength.

Hyperbolic paraboloid. Saddle-shaped surface popular in shell roof construction.

Incremental launching. Method of bridge construction; building the bridge at one end, segment by segment, and pushing the partially completed structure toward the other end after each new segment has been completed.

In-situ concrete. (See Cast-in-place concrete.)

Interaction diagram. Failure curve for a member subjected to both axial force and bending moment, indicating the moment capacity for a given axial load and vice versa; used to develop design charts for reinforced concrete compression members.

Joist. A comparatively narrow beam, used in closely spaced arrangements to support floor or roof slabs.

Kern area. The area within a section in which a compressive force can be applied without causing tensile stresses anywhere in the section.

Laitance. A layer of weak and nondurable material containing cement and fines from aggregates, brought by bleeding water to the top of overwet concrete if the concrete is improperly finished by overworking.

Lap splice. A connection of reinforcing steel made by overlapping the ends of the bars.

Lift. The concrete placed between two consecutive horizontal construction joints, usually consisting of several layers or courses.

Lift slab. A method of concrete construction in which floor and roof slabs are cast on the ground or at ground level and hoisted into position by jacking.

Lightweight concrete. Concrete of substantially lower unit weight than that made using normal-weight gravel or crushed stone aggregate.

Limit analysis. (See Plastic analysis.)

Limit design. A method of proportioning structural members based on satisfying certain strength and serviceability limit states.

Lintel. A horizontal supporting member above an opening such as a window or a door.

Load and resistance factor design (LRFD). (See Ultimate strength design.)

Load factor. A factor by which a service load is multiplied to determine the factored load used in ultimate strength design.

Mass concrete. Any volume of concrete with dimensions large enough to require that measures be taken to cope with the generation of heat from hydration of the cement and attendant volume change to minimize cracking.

Mat foundation. A continuous slab resting directly on soil, supporting several rows of columns and/or walls.

Matrix. In the case of mortar, the cement paste in which the fine aggregate particles are embedded; in the case of concrete, the mortar in which the coarse aggregate particles are embedded.

Maturity. Measure of strength development of concrete, combining effects of hydration time and curing temperature, typically expressed in units of degree-hours.

Middle strip. The portion of a flat slab that occupies the middle half of the span between columns (see also Column strip).

Modular ratio. The ratio of the modulus of elasticity of steel to the modulus of elasticity of concrete, usually denoted by the symbol n.

Modulus of elasticity. The ratio of normal stress to corresponding strain for tensile of compressive stresses below the proportional limit of the material; for steel, $E_s = 29,000$ ksi, for concrete it is highly variable with stress level and strength f'_c; for normal-weight concrete and low stresses, a common approximation is $E_c = 57,000\sqrt{f'_c}$.

Modulus of rupture. The tensile strength of concrete as measured in a flexural test of a small prismatic specimen of plain concrete.

Mortar. A mixture of cement paste and fine aggregate; in fresh concrete, the material occupying the interstices among particles of coarse aggregate.

Neutral axis. A line in a section of a flexural member along which the flexural stress is zero.

Nominal strength. The strength of a structural member based on its assumed material properties and sectional dimensions, before application of any strength reduction factor.

Ottowa sand. Sand from open-pit deposits near Ottowa, Illinois, processed to consist almost entirely of naturally rounded grains of nearly pure quartz; used in mortar for testing of hydraulic cement.

Overlay. A layer of concrete or mortar, seldom thinner than one inch, placed on and usually bonded onto the worn or cracked surface of a concrete slab to restore or improve the function of the previous surface.

Over-reinforced beam. A beam with more than balanced reinforcement, such that the concrete crushes in compression before the reinforcement yields in tension.

Parapet. That part of a wall that extends above the roof level; a low wall along the top of a dam.

Partial prestressing. Prestressing to a stress level such that, under design loads, tensile stresses exist in the precompressed tensile zone of a prestressed member.

Pedestal. An upright compression member whose height does not exceed three times its average least lateral dimension, such as a short pier used as the base for a column.

Petrography. The branch of the science of rocks that describes and classifies rocks, mainly by chemical and microscopical laboratory methods.

Pier. Isolated foundation member of plain or reinforced concrete.

Pilaster. Column built within a wall, usually projecting beyond the wall.

Pile. A slender structural element driven into the ground for the purpose of supporting a load or of compacting the soil.

Pile cap. A relatively thick slab placed on top of a group of piles used to transmit loads from a structure through the pile group into the soil.

Plain concrete. Concrete without reinforcement.

Plastic analysis. A method of structural analysis to determine the intensity of a specified load distribution at which the structure forms a collapse mechanism.

Plastic hinge. Region of a flexural member where the ultimate moment capacity can be developed and maintained with corresponding significant inelastic rotation, as main tensile steel is stressed beyond the yield point.

Poisson's ratio. The ratio of transverse strain to axial strain due to uniformly distributed axial stress; for concrete usually between 0.15 and 0.2.

Polymer concrete. Concrete in which an organic polymer replaces hydraulic cement as the binder.

Polymer-modified concrete. Concrete to which a polymer (large-molecule chemical) has been added to improve its strength or imperviousness or both.

Ponding. The increasing deflections of flat roofs under rainwater or snow melt, which may lead to failure.

Portland cement. A hydraulic cement produced by pulverizing clinker consisting essentially of hydraulic calcium silicates, and usually containing one or more of the forms of calcium sulfate.

Post-tensioning. A method of prestressing reinforced concrete in which the tendons are tensioned after the concrete has hardened (opposite of pretensioning).

Pozzolan. A siliceous material that in itself possesses little or no cementitious value but will, in finely divided form and in the presence of moisture, chemically react with calcium hydroxide to form a material with cementitious properties.

Precast concrete. Concrete cast elsewhere than in its final position, usually in factories or factory-like shop sites near the final site (opposite of: cast-in-place concrete).

Preplaced-aggregate concrete. Concrete produced by placing coarse aggregate in a form and later injecting a portland cement-sand grout, usually with admixtures, to fill the voids.

Prestressed concrete. Concrete in which internal stresses of such magnitude and distribution are introduced that the tensile stresses resulting from the service loads are counteracted to a desired degree; in reinforced concrete the prestress is commonly introduced by tensioning embedded tendons.

Prestressing steel. High strength steel used to prestress concrete, commonly seven-wire strands, single wires, bars, rods, or groups of wires or strands.

Pretensioning. A method of prestressing reinforced concrete in which the tendons are tensioned before the concrete has hardened (opposite of: post-tensioning).

Prismatic beam. A beam with constant cross-sectional dimensions and material properties along its axis.

Pumped concrete. Concrete that is transported through hose or pipe by means of a pump.

Quality assurance. Actions taken by an owner or his representative to provide assurance that what is being done is in accordance with the applicable standards of good practice.

Quality control. Actions taken by a producer or contractor to provide control over what is being done or provided so that the applicable standards of good practice are followed.

Radius of gyration. Square root of the ratio of moment of inertia over cross-sectional area; a measure of a section's capacity to resist bending for a given cross-sectional area.

Raft foundation. A continuous two-way grid of beams that support columns at the points of beam intersection.

Ready-mixed concrete. Concrete manufactured for delivery to a purchaser in a plastic and unhardened state; usually delivered by truck.

Rebar. (Short for reinforcing bar; see Reinforcement.)

Redundancy. Statical indeterminancy; that is, structural components are arranged in parallel such as to permit alternative load paths in case of local distress or failure.

Refractory concrete. Concrete suitable for use at high temperatures (generally about 315 to 1315°C), in which the binding agent is a hydraulic cement.

Reinforced concrete. Concrete containing adequate reinforcement (prestressed or not) and designed on the assumption that the two materials act together in resisting forces.

Reinforced masonry. Unit masonry in which reinforcement is embedded in such a manner that the two materials act together in resisting forces.

Reinforcement. Bars, wires, strands, and other slender members that are embedded in concrete in such a manner that the reinforcement and the concrete act together in resisting forces.

Relaxation. Decrease in stress as a function of time due to creep of a material under constant strain.

Retarding admixture. An admixture that decreases the rate of hydration of hydraulic cement and lengthens the time of setting.

Roller-compacted concrete (RCC). Zero-slump concrete for massive structures (such as dams) or slabs on grade, to be compacted by heavy rollers.

Sack. (See Bag.)

Safety factor. The ratio of a load producing an undesirable state (such as collapse) and an expected or service load.

Sand. (See Fine aggregate.)

Secondary moment. In statically indeterminate structure, a bending moment caused by effects other than direct loads, such as deformations due to temperature, creep, and shrinkage, foundation settlements, and joint rigidity in trusses.

Segmental construction. Construction method in which precast structure segments are joined, typically by prestressing, to act as a monolithic unit under service load.

Service loads. Loads on a structure with high probability of occurrence; such as dead weight supported by a member or the live loads specified in building codes and bridge specifications.

Set. The condition reached by a cement paste, mortar, or concrete when it has lost its plasticity to an arbitrary degree.

Shear friction. The transfer of shear across a given plane, such as an interface between concretes cast at different times and between precast concrete elements.

Shear key. A recess or groove in a joint between successive lifts or placements of concrete, which is filled with concrete of the adjacent lift, giving shear strength to the joint.

Shear span. The distance from a support of a simply supported beam to the nearest concentrated load.

Shear wall. (See Structural wall.)

Shell. A structural member or roof whose middle surface is curved in one or two directions and whose thickness is generally small compared with the other two dimensions (thin shell).

Shore. A temporary support for formwork and fresh concrete or for recently built structures that have not developed full design strength, or for damaged structures.

Shotcrete. Mortar or concrete pneumatically projected at high velocity onto a surface (also called Gunnite).

Shrinkage. Volume decrease caused by drying and chemical changes; a function of time and temperature, but not stress due to external load.

Silica fume. Very fine noncrystalline silica produced in electric arc furnaces as a byproduct of the production of metallic silicon and various silicon alloys (also known as condensed silica fume); used as a mineral admixture in concrete.

Skin reinforcement. Longitudinal reinforcement distributed uniformly along the two faces of the tension zone of a deep flexural member.

Slab. A flat, horizontal (or nearly so) molded layer of plain or reinforced concrete, usually of uniform thickness, either on the ground or supported by beams, columns, walls, or other frame work (see also Flat plate and Flat slab).

Slenderness ratio. Effective unsupported length of a column divided by the radius of gyration of its cross section.

Slipform. A form that is advanced as concrete is placed; may move in a generally horizontal direction to lay concrete evenly for highway paving or on slopes and inverts of canals, tunnels, and siphons; or vertically to form walls, bins, towers, or silos.

Slump. A measure of consistency of freshly mixed concrete equal to the subsidence of the molded specimen immediately after removal of the slump cone, expressed in inches.

Slurry. Very liquid mixture of water and finely divided insoluble material such as portland cement; used, for example, in slurry-wall construction, that is, walls sunk into the ground.

Slurry-infiltrated fiber-reinforced concrete (SIFCON). Concrete with a large percentage of fiber reinforcement (generally more than 10%), produced by first placing the fiber reinforcement, then infiltrating the fiber network with a cement-based slurry.

Soffit. The underside of a part or member of a structure, such as a beam, stairway, or arch.

Spandrel beam. A beam in the perimeter of a building, spanning between columns and usually supporting floors or roof.

Spiral. A reinforcing bar continuously bent into the form of a helix; used in spirally reinforced compression members for effective concrete confinement.

Splice. Connection of one reinforcing bar to another by lapping, welding, mechanical couplers, or other means.

Split cylinder test. Test for tensile strength of concrete in which a standard cylinder is loaded to failure in diametral compression applied along the entire length (also called Brazilian test).

Standard cylinder. Cylindric specimen of 12-inch height and 6-inch diameter, used to determine standard compressive strength and splitting tensile strength of concrete.

Steam curing. Curing of concrete or mortar in water vapor at atmospheric or higher pressures and at temperatures between about 100 and 420°F.

Stiffness. The force necessary to produce a unit displacement or rotation.

Stiffness coefficient. The coefficient k_{ij} of stiffness matrix $\mathbf{K}$ for a multi-degree of freedom structure is the force needed to hold the ith degree of freedom in place, if the jth degree of freedom undergoes a unit of displacement, while all others are locked in place.

Stirrup. A reinforcement used to resist shear and diagonal tension stresses in a structural member; typically a steel bar bent into a U or rectangular shape and installed perpenducular to or at an angle to the longitudinal reinforcement, and properly anchored; the term "stirrup" is usually applied to lateral reinforcement in flexural members and the term "tie" to lateral reinforcement in compression members (see Tie).

Strand. A prestressing tendon composed of a number of wires twisted about a center wire or core.

Strength design. (See Ultimate strength design.)

Strength reduction factor. Capacity reduction factor (typically designated as ϕ) by which the nominal strength of a member is to be multiplied to obtain the design strength; specified by the ACI Code for different types of members.

Stringer. A secondary flexural member that is parallel to the longitudinal axis of a bridge or other structure (see also Beam).

Structural concrete. Concrete used to carry load or to form an integral part of a structure (opposite of, for example, insulating concrete).

Structural wall. A wall portion of a structural frame intended to resist gravity loads as well as lateral loads due to earthquake, wind, blast, etc., acting in the plane of the wall (formerly called shear wall).

Superplasticizer. (See Water-reducing admixture.)

T-beam. A beam composed of a stem and a flange in the form of a "T," with the flange usually provided by a slab.

Tendon. A wire, cable, bar, rod, or strand (usually steel) used to apply prestress to a structure or structural member.

Tension stiffening effect. The added stiffness of a single reinforcing bar due to the surrounding uncracked concrete between bond cracks.

Tie. Reinforcing bar bent into a loop to enclose the longitudinal steel in columns; tensile bar to hold a form in place while resisting the lateral pressure of unhardened concrete.

Tilt-up construction. A method of construction in which concrete elements are cast horizontally at the job site and tilted into place after removal of molds.

Tolerance. The permitted variation from a given dimension, quantity, or alignment.

Transformed section. A hypothetical section of one material arranged so as to have the same elastic properties as a section of two different materials.

Traveller. A structure usually mounted on tracks, which permit it to move from one location to another to facilitate the construction of an arch, bridge, tower, or building; can be enclosed for year-round construction.

Tremie concrete. Underwater concrete placed by means of a pipe or tube, with a hopper for filling at its upper end.

Ultimate strength design (USD). Design principle such that the actual (ultimate) strength of a member or structure, multiplied by a strength reduction factor, is no less than the effects of all service load combinations, multiplied by respective overload factors.

Unbonded tendon. A tendon that is not bonded to the concrete.

Under-reinforced beam. A beam with less than balanced reinforcement such that the reinforcement yields before the concrete crushes in compression.

Vacuum concrete. Concrete from which excess water and entrapped air are extracted by a vacuum process before hardening occurs.

Viaduct. A bridge, usually of several spans.

Vibration. Employment of mechanical devices to consolidate freshly poured concrete in the form or mold.

Waffle slab. A floor slab built integrally with two orthogonal sets of closely spaced ribs or joists.

Water-cement ratio. Ratio by weight of water to cement in a mixture; inversely proportional to concrete strength.

Water-reducing admixture. An admixture capable of lowering the mix viscosity, thereby allowing a reduction of water (and increase in strength) without lowering the workability (also called superplasticizer).

Welded wire fabric. A series of longitudinal and transverse wires arranged substantially at right angles to each other and welded together at all points of intersection; used for slab and wall reinforcement.

Whitney stress block. A rectangular area of uniform stress intensity $0.85 f_c'$, whose area and centroid are similar to that of the actual stress distribution in a flexural member near failure.

Wire mesh. (See Welded wire fabric.)

Workability. General property of freshly mixed concrete that defines the ease with which it can be placed into forms without honeycombs; closely related to slump.

Working stress design. (See Allowable stress design.)

Yield-line theory. Method of structural analysis of plate structures at the verge of collapse under factored loads.

Yield point. The point during increasing stress at which the stress-strain curve of a material experiences a sharp drop in slope.

Young's modulus. (see Modulus of elasticity.)

B

Design Aids

TABLE B.1a ASTM STANDARD REINFORCING BARS

Bar size	Nominal diameter (in)	Nominal area (sq in)	Nominal diameter weight (lb per ft)
#3	0.375	0.11	0.376
4	0.500	0.20	0.668
5	0.625	0.31	1.043
6	0.750	0.44	1.502
7	0.875	0.60	2.044
8	1.000	0.79	2.670
9	1.128	1.00	3.400
10	1.270	1.27	4.303
11	1.410	1.56	5.313
14	1.693	2.25	7.650
18	2.257	4.00	13.600

TABLE B.1b AREAS OF GROUPS OF STANDARD BARS (in^2)

Bar No.	Number of Bars											
	1	2	3	4	5	6	7	8	9	10	11	12
4	0.20	0.39	0.59	0.78	0.98	1.18	1.37	1.57	1.77	1.96	2.16	2.36
5	0.31	0.61	0.92	1.23	1.53	1.84	2.15	2.45	2.76	3.07	3.37	3.68
6	0.44	0.88	1.32	1.77	2.21	2.65	3.09	3.53	3.98	4.42	4.86	5.30
7	0.60	1.20	1.80	2.41	3.01	3.61	4.21	4.81	5.41	6.01	6.61	7.22
8	0.79	1.57	2.36	3.14	3.93	4.71	5.50	6.28	7.07	7.85	8.64	9.43
9	1.00	2.00	3.00	4.00	5.00	6.00	7.00	8.00	9.00	10.00	11.00	12.00
10	1.27	2.53	3.79	5.06	6.33	7.59	8.86	10.12	11.39	12.66	13.92	15.19
11	1.56	3.12	4.68	6.25	7.81	9.37	10.94	12.50	14.06	15.62	17.19	18.75
14	2.25	4.50	6.75	9.00	11.25	13.50	15.75	18.00	20.25	22.50	24.75	27.00
18	4.00	8.00	12.00	16.00	20.00	24.00	28.00	32.00	36.00	40.00	44.00	48.00

TABLE B.2   ASTM STANDARD WIRE REINFORCEMENT

W&D Size		Nominal diameter (in)	Nominal area (sq in)	Nominal weight (lb per ft)	Area (sq in per ft of width) for various spacings						
					Center-to-center spacing (in)						
Plain	Deformed				2	3	4	6	8	10	12
W31	D31	0.628	0.310	1.054	1.86	1.24	0.93	0.62	0.465	0.372	0.31
W30	D30	0.618	0.300	1.020	1.80	1.20	0.90	0.60	0.45	0.366	0.30
W28	D28	0.597	0.280	0.952	1.68	1.12	0.84	0.56	0.42	0.336	0.28
W26	D26	0.575	0.260	0.934	1.56	1.04	0.78	0.52	0.39	0.312	0.26
W24	D24	0.553	0.240	0.816	1.44	0.96	0.72	0.48	0.36	0.288	0.24
W22	D22	0.529	0.220	0.748	1.32	0.88	0.66	0.44	0.33	0.264	0.22
W20	D20	0.504	0.200	0.680	1.20	0.80	0.60	0.40	0.30	0.24	0.20
W18	D18	0.478	0.180	0.612	1.08	0.72	0.54	0.36	0.27	0.216	0.18
W16	D16	0.451	0.160	0.544	0.96	0.64	0.48	0.32	0.24	0.192	0.16
W14	D14	0.422	0.140	0.476	0.84	0.56	0.42	0.28	0.21	0.168	0.14
W12	D12	0.390	0.120	0.408	0.72	0.48	0.36	0.24	0.18	0.144	0.12
W11	D11	0.374	0.110	0.374	0.66	0.44	0.33	0.22	0.165	0.132	0.11
W10.5		0.366	0.105	0.357	0.63	0.42	0.315	0.21	0.157	0.126	0.105
W10	D10	0.356	0.100	0.340	0.60	0.40	0.30	0.20	0.15	0.12	0.10
W9.5		0.348	0.095	0.323	0.57	0.38	0.285	0.19	0.142	0.114	0.095
W9	D9	0.338	0.090	0.306	0.54	0.36	0.27	0.18	0.135	0.108	0.09
W8.5		0.329	0.085	0.289	0.51	0.34	0.255	0.17	0.127	0.102	0.085
W8	D8	0.319	0.080	0.272	0.48	0.32	0.24	0.16	0.12	0.096	0.08
W7.5		0.309	0.075	0.255	0.45	0.30	0.225	0.15	0.112	0.09	0.075
W7	D7	0.298	0.070	0.238	0.42	0.28	0.21	0.14	0.105	0.084	0.07
W6.5		0.288	0.065	0.221	0.39	0.26	0.195	0.13	0.097	0.078	0.065
W6	D6	0.276	0.060	0.204	0.36	0.24	0.18	0.12	0.09	0.072	0.06
W5.5		0.264	0.055	0.187	0.33	0.22	0.165	0.11	0.082	0.066	0.055
W5	D5	0.252	0.050	0.170	0.30	0.20	0.15	0.10	0.075	0.06	0.05
W4.5		0.240	0.045	0.153	0.27	0.18	0.135	0.09	0.067	0.054	0.045
W4	D4	0.225	0.040	0.136	0.24	0.16	0.12	0.08	0.06	0.048	0.04
W3.5		0.211	0.035	0.119	0.21	0.14	0.105	0.07	0.052	0.042	0.035
W3		0.195	0.030	0.102	0.18	0.12	0.09	0.06	0.045	0.036	0.03
W2.9		0.192	0.029	0.098	0.174	0.116	0.087	0.058	0.043	0.035	0.029
W2.5		0.178	0.025	0.085	0.15	0.10	0.075	0.05	0.037	0.03	0.025
W2		0.159	0.020	0.068	0.12	0.08	0.06	0.04	0.03	0.024	0.02
W1.4		0.135	0.014	0.049	0.084	0.056	0.042	0.028	0.021	0.017	0.014

TABLE B.3   REINFORCING AREAS IN SLABS (in^2/ft)

Spacing (in)	\ Bar No. 3	4	5	6	7	8	9	10	11
3	0.44	0.78	1.23	1.77	2.40	3.14	4.00	5.06	6.25
$3\frac{1}{2}$	0.38	0.67	1.05	1.51	2.06	2.69	3.43	4.34	5.36
4	0.33	0.59	0.92	1.32	1.80	2.36	3.00	3.80	4.68
$4\frac{1}{2}$	0.29	0.52	0.82	1.18	1.60	2.09	2.67	3.37	4.17
5	0.26	0.47	0.74	1.06	1.44	1.88	2.40	3.04	3.75
$5\frac{1}{2}$	0.24	0.43	0.67	0.96	0.31	0.71	2.18	2.76	3.41
6	0.22	0.39	0.61	0.88	1.20	0.57	2.00	2.53	3.12
$6\frac{1}{2}$	0.20	0.36	0.57	0.82	1.11	1.45	1.85	2.34	2.89
7	0.19	0.34	0.53	0.76	1.03	1.35	1.71	2.17	2.68
$7\frac{1}{2}$	0.18	0.31	0.49	0.71	0.96	1.26	1.60	2.02	2.50
8	0.17	0.29	0.46	0.66	0.90	1.18	1.50	1.89	2.34
9	0.15	0.26	0.41	0.59	0.80	1.05	1.33	1.69	2.08
10	0.13	0.24	0.37	0.53	0.72	0.94	1.20	1.52	1.87
12	0.11	0.20	0.31	0.44	0.60	0.78	1.00	1.27	1.56

TABLE B.4a   COEFFICIENTS FOR NEGATIVE MOMENTS IN SLABS[†]

$$M_{a,\text{neg}} = C_{a,\text{neg}}\,w\,l_a^2$$
$$M_{b,\text{neg}} = C_{b,\text{neg}}\,w\,l_b^2$$

where w = total uniform dead plus live load

Ratio $m = l_a/l_b$		Case 1	Case 2	Case 3	Case 4	Case 5	Case 6	Case 7	Case 8	Case 9
1.00	$C_{a,\text{neg}}$	0.045			0.050	0.075	0.071		0.033	0.061
	$C_{b,\text{neg}}$	0.045		0.076	0.050			0.071	0.061	0.033
0.95	$C_{a,\text{neg}}$	0.050			0.055	0.079	0.075		0.038	0.065
	$C_{b,\text{neg}}$	0.041		0.072	0.045			0.067	0.056	0.029
0.90	$C_{a,\text{neg}}$	0.055			0.060	0.080	0.079		0.043	0.068
	$C_{b,\text{neg}}$	0.037		0.070	0.040			0.062	0.052	0.025
0.85	$C_{a,\text{neg}}$	0.060			0.066	0.082	0.083		0.049	0.072
	$C_{b,\text{neg}}$	0.031		0.065	0.034			0.057	0.046	0.021
0.80	$C_{a,\text{neg}}$	0.065			0.071	0.083	0.086		0.055	0.075
	$C_{b,\text{neg}}$	0.027		0.061	0.029			0.051	0.041	0.017
0.75	$C_{a,\text{neg}}$	0.069			0.076	0.085	0.088		0.061	0.078
	$C_{b,\text{neg}}$	0.022		0.056	0.024			0.044	0.036	0.014
0.70	$C_{a,\text{neg}}$	0.074			0.081	0.086	0.091		0.068	0.081
	$C_{b,\text{neg}}$	0.017		0.050	0.019			0.038	0.029	0.011
0.65	$C_{a,\text{neg}}$	0.077			0.085	0.087	0.093		0.074	0.083
	$C_{b,\text{neg}}$	0.014		0.043	0.015			0.031	0.024	0.008
0.60	$C_{a,\text{neg}}$	0.081			0.089	0.088	0.095		0.080	0.085
	$C_{b,\text{neg}}$	0.010		0.035	0.011			0.024	0.018	0.006
0.55	$C_{a,\text{neg}}$	0.084			0.092	0.089	0.096		0.085	0.086
	$C_{b,\text{neg}}$	0.007		0.028	0.008			0.019	0.014	0.005
0.50	$C_{a,\text{neg}}$	0.086			0.094	0.090	0.097		0.089	0.088
	$C_{b,\text{neg}}$	0.006		0.022	0.006			0.014	0.010	0.003

[†]A crosshatched edge indicates that the slab continues across, or is fixed at, the support; an unmarked edge indicates a support at which torsional resistance is negligible.

Authorized reprint from ACI 318-63, *Building Code Requirements for Reinforced Concrete*, Appendix A, Section A2003.

TABLE B.4b COEFFICIENTS FOR DEAD LOAD POSITIVE MOMENTS IN SLABS[†]

$$M_{a,\text{pos},dl} = C_{a,dl}\, wl_a^2$$
$$M_{b,\text{pos},dl} = C_{b,dl}\, wl_b^2 \qquad \text{where } w = \text{total uniform, dead load}$$

Ratio $m = l_a/l_b$		Case 1	Case 2	Case 3	Case 4	Case 5	Case 6	Case 7	Case 8	Case 9
1.00	$C_{a,dl}$	0.036	0.018	0.018	0.027	0.027	0.033	0.027	0.020	0.023
	$C_{b,dl}$	0.036	0.018	0.027	0.027	0.018	0.027	0.033	0.023	0.020
0.95	$C_{a,dl}$	0.040	0.020	0.021	0.030	0.028	0.036	0.031	0.022	0.024
	$C_{b,dl}$	0.033	0.016	0.025	0.024	0.015	0.024	0.031	0.021	0.017
0.90	$C_{a,dl}$	0.045	0.022	0.025	0.033	0.029	0.039	0.035	0.025	0.026
	$C_{b,dl}$	0.029	0.014	0.024	0.022	0.013	0.021	0.028	0.019	0.015
0.85	$C_{a,dl}$	0.050	0.024	0.029	0.036	0.031	0.042	0.040	0.029	0.028
	$C_{b,dl}$	0.026	0.012	0.022	0.019	0.011	0.017	0.025	0.017	0.013
0.80	$C_{a,dl}$	0.056	0.026	0.034	0.039	0.032	0.045	0.045	0.032	0.029
	$C_{b,dl}$	0.023	0.011	0.020	0.016	0.009	0.015	0.022	0.015	0.001
0.75	$C_{a,dl}$	0.061	0.028	0.040	0.043	0.033	0.048	0.051	0.036	0.031
	$C_{b,dl}$	0.019	0.009	0.018	0.013	0.007	0.012	0.020	0.013	0.007
0.70	$C_{a,dl}$	0.068	0.030	0.046	0.046	0.035	0.051	0.058	0.040	0.033
	$C_{b,dl}$	0.016	0.007	0.016	0.011	0.005	0.009	0.017	0.011	0.006
0.65	$C_{a,dl}$	0.074	0.032	0.054	0.050	0.036	0.054	0.065	0.044	0.034
	$C_{b,dl}$	0.013	0.006	0.014	0.009	0.004	0.007	0.014	0.009	0.005
0.60	$C_{a,dl}$	0.081	0.034	0.062	0.053	0.037	0.056	0.073	0.048	0.036
	$C_{b,dl}$	0.010	0.004	0.011	0.007	0.003	0.006	0.012	0.007	0.004
0.55	$C_{a,dl}$	0.088	0.035	0.071	0.056	0.038	0.058	0.081	0.052	0.037
	$C_{b,dl}$	0.008	0.003	0.009	0.005	0.002	0.004	0.009	0.005	0.003
0.50	$C_{a,dl}$	0.095	0.037	0.080	0.059	0,039	0,061	0.089	0.056	0.038
	$C_{b,dl}$	0.006	0.002	0.007	0.004	0.001	0.003	0.007	0.004	0.002

[†]A crosshatched edge indicates that the slab continues across, or is fixed at, the support; an unmarked edge indicates a support at which torsional resistance is negligible.

Authorized reprint from ACI 318-63, *Building Code Requirements for Reinforced Concrete*, Appendix A, Section A2003.

TABLE B.4c   COEFFICIENTS FOR LIVE LOAD POSITIVE MOMENTS IN SLABS[†]

$$M_{a,\text{pos},ll} = C_{a,ll},\, wl_a^2$$
$$M_{b,\text{pos},ll} = C_{b,ll}\, wl_b^2$$
   where w = total uniform, dead load

Ratio $m = l_a/l_b$		Case 1	Case 2	Case 3	Case 4	Case 5	Case 6	Case 7	Case 8	Case 9
1.00	$C_{a,ll}$	0.036	0.027	0.027	0.032	0.032	0.035	0.032	0.028	0.030
	$C_{b,ll}$	0.036	0.027	0.032	0.032	0.027	0.032	0.035	0.030	0.028
0.95	$C_{a,ll}$	0.040	0.030	0.031	0.035	0.034	0.038	0.036	0.031	0.032
	$C_{b,ll}$	0.033	0.025	0.029	0.029	0.024	0.029	0.032	0.027	0.025
0.90	$C_{a,ll}$	0.045	0.034	0.035	0.039	0.037	0.042	0.040	0.035	0.036
	$C_{b,ll}$	0.029	0.022	0.027	0.026	0.021	0.025	0.029	0.024	0.022
0.85	$C_{a,ll}$	0.050	0.037	0.040	0.043	0.041	0.046	0.045	0.040	0.039
	$C_{b,ll}$	0.026	0.019	0.024	0.023	0.019	0.022	0.026	0.022	0.020
0.80	$C_{a,ll}$	0.056	0.041	0.045	0.048	0.044	0.051	0.051	0.044	0.042
	$C_{b,ll}$	0.023	0.017	0.022	0.020	0.016	0.019	0.023	0.019	0.017
0.75	$C_{a,ll}$	0.061	0.045	0.051	0.052	0.047	0.055	0.056	0.049	0.046
	$C_{b,ll}$	0.019	0.014	0.019	0.016	0.013	0.016	0.020	0.016	0.013
0.70	$C_{a,ll}$	0.068	0.049	0.057	0.057	0.051	0.060	0.063	0.054	0.050
	$C_{b,ll}$	0.016	0.012	0.016	0.014	0.011	0.013	0.017	0.014	0.011
0.65	$C_{a,ll}$	0.074	0.053	0.064	0.062	0.055	0.064	0.070	0.059	0.054
	$C_{b,ll}$	0.013	0.010	0.014	0.011	0.009	0.010	0.014	0.011	0.009
0.60	$C_{a,ll}$	0.081	0.058	0.071	0.067	0.059	0.068	0.077	0.065	0.059
	$C_{b,ll}$	0.010	0.007	0.011	0.009	0.007	0.008	0.011	0.009	0.007
0.55	$C_{a,ll}$	0.088	0.062	0.080	0.072	0.063	0.073	0.085	0.070	0.063
	$C_{b,ll}$	0.008	0.006	0.009	0.007	0.005	0.006	0.009	0.007	0.006
0.50	$C_{a,ll}$	0.095	0.066	0.088	0.077	0.067	0.078	0.092	0.076	0.067
	$C_{b,ll}$	0.006	0.004	0.007	0.005	0.004	0.005	0.007	0.005	0.004

[†]A crosshatched edge indicates that the slab continues across, or is fixed at, the support; an unmarked edge indicates a support at which torsional resistance is negligible.

Authorized reprint from ACI 318-63, *Building Code Requirements for Reinforced Concrete*, Appendix A, Section A2003.

TABLE B.4d RATIO OF LOAD W IN l_a- AND l_b-DIRECTIONS FOR SHEAR IN SLAB AND LOAD ON SUPPORTS[†]

Ratio $m = l_a/l_b$		Case 1	Case 2	Case 3	Case 4	Case 5	Case 6	Case 7	Case 8	Case 9
1.00	W_a	0.50	0.50	0.17	0.50	0.83	0.71	0.29	0.33	0.67
	W_b	0.50	0.50	0.83	0.50	0.17	0.29	0.71	0.67	0.33
0.95	W_a	0.55	0.55	0.20	0.55	0.86	0.75	0.33	0.38	0.71
	W_b	0.45	0.45	0.80	0.45	0.14	0.25	0.67	0.62	0.29
0.90	W_a	0.60	0.60	0.23	0.60	0.88	0.79	0.38	0.43	0.75
	W_b	0.40	0.40	0.77	0.40	0.12	0.21	0.62	0.57	0.25
0.85	W_a	0.66	0.66	0.28	0.66	0.90	0.83	0.43	0.49	0.79
	W_b	0.34	0.34	0.72	0.34	0.10	0.17	0.57	0.51	0.21
0.80	W_a	0.71	0.71	0.33	0.71	0.92	0.86	0.49	0.55	0.83
	W_b	0.29	0.29	0.67	0.29	0.08	0.14	0.51	0.45	0.17
0.75	W_a	0.76	0.76	0.39	0.76	0.94	0.88	0.56	0.61	0.86
	W_b	0.24	0.24	0.61	0.24	0.06	0.12	0.44	0.39	0.14
0.70	W_a	0.81	0.81	0.45	0.81	0.95	0.91	0.62	0.68	0.89
	W_b	0.19	0.19	0.55	0.19	0.05	0.09	0.38	0.32	0.11
0.65	W_a	0.85	0.85	0.53	0.85	0.96	0.93	0.69	0.74	0.92
	W_b	0.15	0.15	0.47	0.15	0.04	0.07	0.31	0.26	0.08
0.60	W_a	0.89	0.89	0.61	0.89	0.97	0.95	0.76	0.80	0.94
	W_b	0.11	0.11	0.39	0.11	0.03	0.05	0.24	0.20	0.06
0.55	W_a	0.92	0.92	0.69	0.92	0.98	0.96	0.81	0.85	0.95
	W_b	0.08	0.08	0.31	0.08	0.02	0.04	0.19	0.15	0.05
0.50	W_a	0.94	0.94	0.76	0.94	0.99	0.97	0.86	0.89	0.97
	W_b	0.06	0.06	0.24	0.06	0.01	0.03	0.14	0.11	0.03

[†]A crosshatched edge indicates that the slab continues across, or is fixed at, the support; an unmarked edge indicates a support at which torsional resistance is negligible.

Authorized reprint from ACI 318-63, *Building Code Requirements for Reinforced Concrete*, Appendix A, Section A2003.

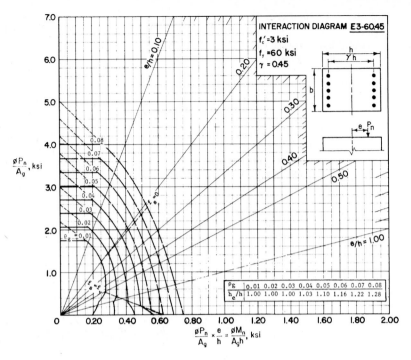

Figure B.1 Column strength interaction diagram E3-60.45. (Authorized reprint from ACI Design Handbook, Vol. 2—Columns.)

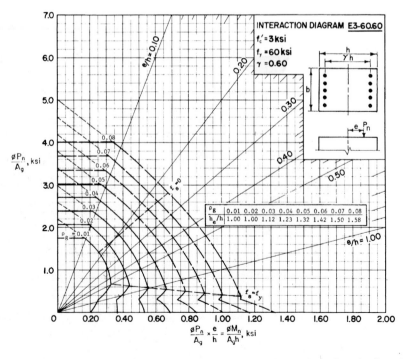

Figure B.2 Column strength interaction diagram E3-60.60. (Authorized reprint from ACI Design Handbook, Vol. 2—Columns.)

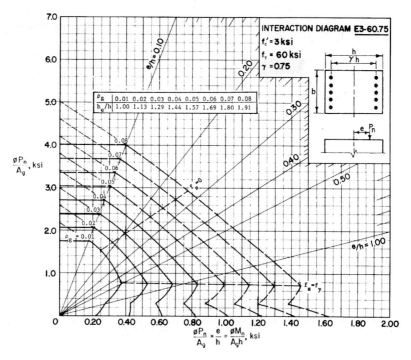

Figure B.3 Column strength interaction diagram E3-60.75. (Authorized reprint from ACI Design Handbook, Vol. 2—Columns.)

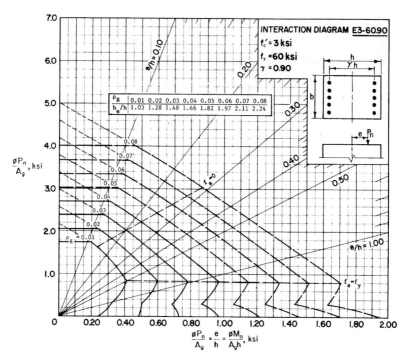

Figure B.4 Column strength interaction diagram E3-60.90. (Authorized reprint from ACI Design Handbook, Vol. 2—Columns.)

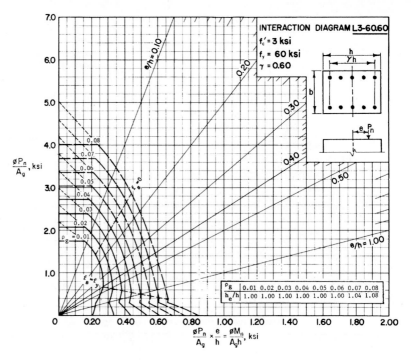

Figure B.5 Column strength interaction diagram L3-60.60. (Authorized reprint from ACI Design Handbook, Vol. 2—Columns.)

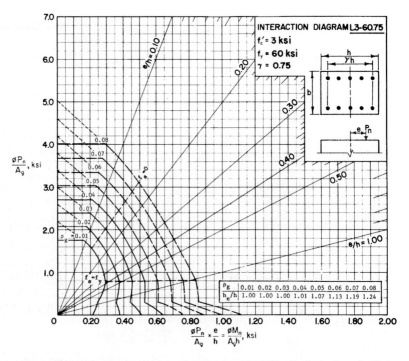

Figure B.6 Column strength interaction diagram L3-60.75. (Authorized reprint from ACI Design Handbook, Vol. 2—Columns.)

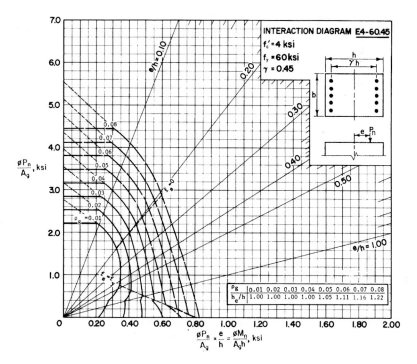

Figure B.7 Column strength interaction diagram E4-60.45. (Authorized reprint from ACI Design Handbook, Vol. 2—Columns.)

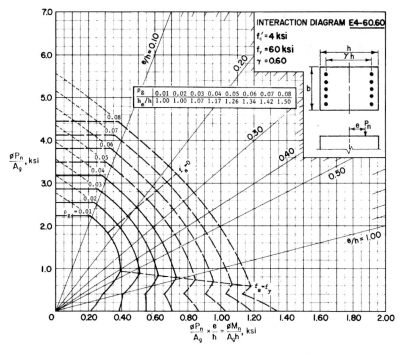

Figure B.8 Column strength interaction diagram E4-60.60. (Authorized reprint from ACI Design Handbook, Vol. 2—Columns.)

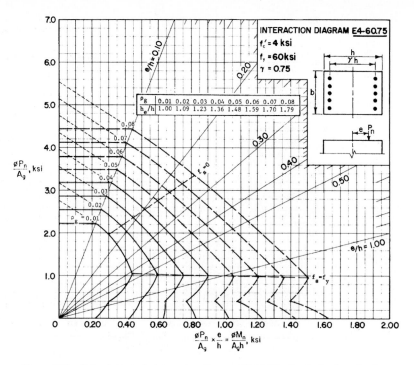

Figure B.9 Column strength interaction diagram E4-60.75. (Authorized reprint from ACI Design Handbook, Vol. 2—Columns.)

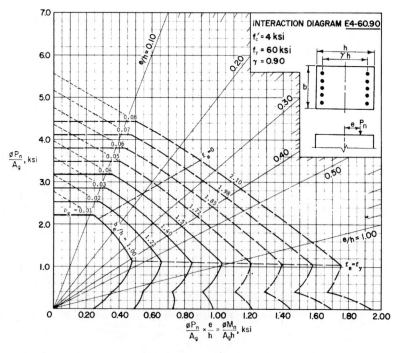

Figure B.10 Column strength interaction diagram E4-60.90. (Authorized reprint from ACI Design Handbook, Vol. 2—Columns.)

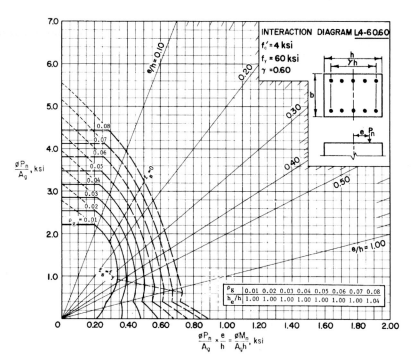

Figure B.11 Column strength interaction diagram L4-60.60. (Authorized reprint from ACI Design Handbook, Vol. 2—Columns.)

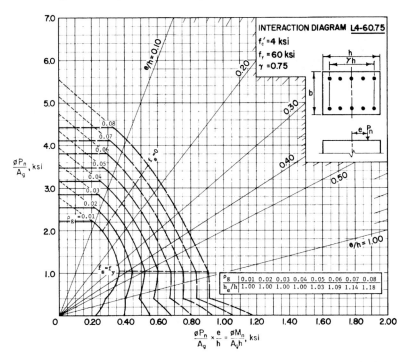

Figure B.12 Column strength interaction diagram L4-60.75. (Authorized reprint from ACI Design Handbook, Vol. 2—Columns.)

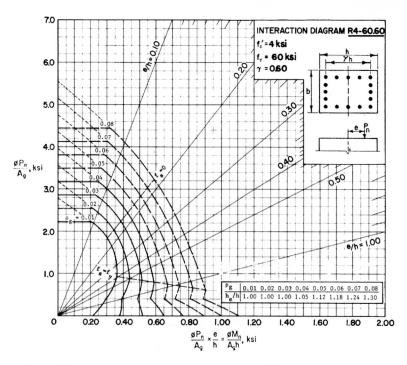

Figure B.13 Column strength interaction diagram R4-60.60. (Authorized reprint from ACI Design Handbook, Vol. 2—Columns.)

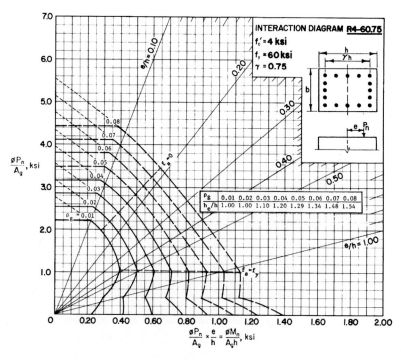

Figure B.14 Column strength interaction diagram R4-60.75. (Authorized reprint from ACI Design Handbook, Vol. 2—Columns.)

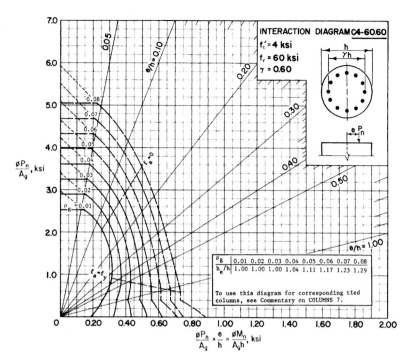

Figure B.15 Column strength interaction diagram C4-60.60. (Authorized reprint from ACI Design Handbook, Vol. 2—Columns.)

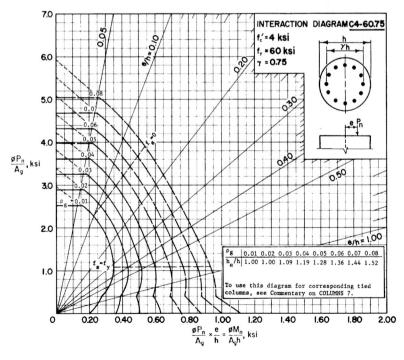

Figure B.16 Column strength interaction diagram C4-60.75. (Authorized reprint from ACI Design Handbook, Vol. 2—Columns.)

Index